Stuart Rice

43

Springer Series in Chemical Physics

Edited by Manuel Cardona

Springer Series in Chemical Physics

Editors: Vitalii I. Goldanskii Fritz P. Schäfer J. Peter Toennies

Volume 40 **High-Resolution Spectroscopy of Transient Molecules**
By E. Hirota

Volume 41 **High Resolution Spectral Atlas of Nitrogen Dioxide 559–597 nm**
By K. Uehara and H. Sasada

Volume 42 **Antennas and Reaction Centers of Photosynthetic Bacteria**
Structure, Interactions, and Dynamics
Editor: M. E. Michel-Beyerle

Volume 43 **The Atom-Atom Potential Method. Applications to Organic Molecular Solids**
By A. J. Pertsin and A. I. Kitaigorodsky

Volume 44 **Secondary Ion Mass Spectrometry SIMS V**
Editors: A. Benninghoven, R. J. Colton, D. S. Simons, and H. W. Werner

Volume 45 **Thermotropic Liquid Crystals, Fundamentals**
By G. Vertogen and W. H. de Jeu

Volume 46 **Ultrafast Phenomena V**
Editors: G. R. Fleming and A. E. Siegman

Volumes 1–39 are listed on the back inside cover

A. J. Pertsin   A. I. Kitaigorodsky

The Atom-Atom Potential Method

Applications to Organic Molecular Solids

With 77 Figures

Springer-Verlag Berlin Heidelberg New York
London Paris Tokyo

Dr. Alexander J. Pertsin
Professor Alexander I. Kitaigorodsky †

Institute of Elemento-Organic Compounds, USSR Academy of Sciences
GSP-1, V-334, Vavilov 28, SU-117813 Moscow, USSR

Guest Editor

Professor Dr. Manuel Cardona

Max-Planck-Institut für Festkörperforschung, Heisenbergstrasse 1
D-7000 Stuttgart 80, Fed. Rep. of Germany

Series Editors

Professor Dr. Fritz Peter Schäfer

Max-Planck-Institut für
Biophysikalische Chemie
D-3400 Göttingen-Nikolausberg, FRG

Professor Vitalii I. Goldanskii

Institute of Chemical Physics
Academy of Sciences, Kosygin Street 3
Moscow V-334, USSR

Professor Dr. J. Peter Toennies

Max-Planck-Institut für Strömungsforschung
Böttingerstraße 6–8
D-3400 Göttingen, FRG

ISBN 3-540-16246-1 Springer-Verlag Berlin Heidelberg New York
ISBN 0-387-16246-1 Springer-Verlag New York Berlin Heidelberg

Library of Congress Cataloging-in-Publication Data. Pertsin, A.J. (Alexander J.), 1948-. The atom-atom potential method. Applications to organic molecular solids. (Springer series in chemical physics ; v. 43) 1. Solid state physics. 2. Solid state chemistry. 3. Lattice theory. 4. Chemistry, Physical organic. I. Kitaĭgorodskiĭ, A.I. (Aleksandr Isaakovich), 1914-. II. Title. III. Series. QC176.P44 1986 530.4'1 86-1840

Offsetprinting: Druckhaus Beltz, 6944 Hemsbach/Bergstr.
Bookbinding: J. Schäffer GmbH & CO. KG., 6718 Grünstadt
2153/3150-543210

Preface

The history of physics furnishes many examples of how a simple semiempirical method, essentially based on intuitive considerations, may prove to be much more successful than a rigorous theoretical approach. A pertinent example is the method of atom-atom potentials, which treats the intermolecular interactions between polyatomic molecules in terms of pairwise interactions between their constituent atoms. Despite a few conceptual shortcomings, the method provides a fairly reliable practical means of handling, on a microscopic level, a wide range of problems that arise in the solid-state physics and chemistry of organic compounds.

This monograph is an attempt to generalize the experience gained in the past twenty years in interpreting the static and dynamic properties of organic molecular solids in terms of atom-atom potentials. It embraces nearly all aspects of the application of the method, including an evaluation of cohesive energies, equilibrium crystal structures, phonon spectra, thermodynamic functions, and crystal defects. Many related topics such as the effect of the crystal field on molecular conformation, the determination of crystal structures from raw diffraction data, and the problem of polymorphic transitions are also discussed.

We believe that this book will be of use to researchers in solid-state physics, chemistry, crystallography, physical chemistry, and polymer chemistry.

It also gives us an opportunity to acknowledge our indebtedness to those who sent us published as well as unpublished information and suggestions, including A.T. Amos, E.L. Bokhenkov, H. Bonadeo, R.K. Boyd, C.P. Brock, R. Busing, P. Claverie, D.P. Craig, E. Giglio, C. Huiszoon, R. Mason, R.M. Metzger, G.S. Pawley, H.A. Scheraga, E.F. Sheka, M. Simonetta, J.M. Thomas, A. Warshel, V. Zubkov, and P. Zugenmaier.

Moscow, July 1986 *A.J. Pertsin*

Contents

1. Introduction

For a long time interest in the organic solid state was chiefly stimulated by results from crystal-structure determinations about the molecular structure of organic compounds. The fact that molecules form crystals was frequently given only cursory consideration - merely as an obstacle to be overcome - since one had to determine the molecular arrangement. The gaseous or liquid state would have been used for an analysis if this were possible.

Of course, it was well realized that many properties of an organic solid - such as the mechanical and thermal properties - not only depend on the molecular structure but also on the crystal structure. There was, however, no practical need to measure such properties and to interpret them in terms of the crystal structure.

The situation has now completely changed due to the discovery of many interesting phenomena governed by the collective properties of organic molecules and their arrangement in crystals. For example, the specific properties of organic semiconductors and quasi-one-dimensional metals are intimately linked to the peculiarities of their crystal structures. Other examples are organic solid-state reactions which have been shown to be strongly dependent on the crystal structures of the reactants.

The active interest in the organic solid state also stems from the extensive use of organic polymers in various branches of technology and engineering. In these fields, too, one is interested in the organic solid itself, and not in its constituent molecules.

The above circumstances have generated a need for a microscopic model of the organic solid state, which would enable one to predict the structure and properties of an organic solid from the structure and properties of its constituent molecules.

A first step in this direction was made by *Kitaigorodsky* whose work from 1944–1949 has been summarized and generalized in [1.1]. The analysis undertaken by *Kitaigorodsky* [1.1]was based on the finding that the shortest intermolecular contact distance between specified kinds of atoms varies only slightly in different crystals. This allowed *Kitaigorodsky* to introduce the idea of an *intermolecular atomic radius* and to treat it as a characteristic atomic constant. Numerical values for the intermolecular radii have been determined for the most important atomic species, including H, C, N, O, Cl, Br and I.

Based on a set of intermolecular radii, a geometrical model of an organic crystal was constructed. In this model each molecule is represented as a collection of hard spheres centered about individual atoms. Since the intermolecular radius of an atom is always greater than its covalent radius, the spheres are intersecting within the molecule. The external surface that is produced by the intersecting spheres defines the shape and volume of the molecule.

Analyses of organic crystal structures in terms of hard molecules with a definite shape and volume have shown that the molecular arrangement in a crystal is always a kind of *packing*, in the sense that all of the molecules are in contact, none being suspended in empty space and none interpenetrating. Moreover, it is clearly seen that the bumps in one molecule fit the hollows in another, so that the empty space between the molecules has the lowest possible volume.

To assess the density of molecular packing, a *packing coefficient* k was introduced. It is defined as the ratio of the molecular volume to the unit-cell volume per molecule. For the overwhelming majority of organic crystals, the packing coefficient lies between 0.65 and 0.75, with 0.7 being the most typical. That is, the efficiency of molecular packing in organic crystals approaches that of hard spheres ($k = 0.74$). Only a few crystals with k between 0.6 and 0.65 and no crystals with k less than 0.6 are known. All this evidence indicates that organic molecules tend to form closely packed systems upon crystallization.

Many important conclusions can be drawn from the principle of close packing. For example, it is easy to show that the molecules in a crystal cannot be juxtaposed by a mirror plane. This is in full agreement with experiment. Of the many thousands of organic crystal structures known today, not one exhibits such a juxtaposition.

The close-packing principle easily explains why almost half of the organic crystals crystallize in the space group $P2_1/a$ [1.2], why asymmetric molecules prefer to crystallize in the space groups $P2_1$ and $P2_12_12_1$ and why highly symmetrical molecules usually lose their symmetry elements (except for the inversion center) in the crystal.

The existence of certain relations between crystal symmetry and closest packing may be of great assistance in choosing between alternative crystal symmetries. Moreover, if the crystal symmetry and the dimensions of the unit cell are experimentally available and if the molecular geometry has roughly been estimated from stereochemical considerations, the close-packing principle can be used to determine the molecular arrangement in the unit cell. This is frequently done by crystallographers to construct a rough preliminary model of a crystal, which serves as a starting point for subsequent crystal-structure refinement with conventional methods.

A fruitful field for application of the geometrical approach is the field of mixed crystals. In many cases it has been demonstrated that the close-packing principle is capable of explaining and even predicting the mutual solid-state solubility of organic substances [1.3,4]. Good solubility results when the replacement of a host molecule by a guest molecule does not markedly worsen the molecular packing. This can usually be observed when the shortest intermolecular contacts between a guest molecule and its surrounding host molecules differ from the normal ones by no more than 10–15%.

Packing considerations can prove to be useful even for such complicated mixed systems as inclusion compounds. This has been illustrated in a recent monograph [1.5]in which the formation of inclusion compounds was rationalized in terms of closest packing. It was shown that an inclusion compound can only be formed if the packing density of the resulting crystal is greater than that of the mixture of the host and guest.

Despite the success of the close-packing principle in interpreting the structure of organic crystals, it is clear that the geometrical model represents only a first approximation. Of course it is not capable of predicting molecular packings a priori or even of distinguishing between competing packings with similar densities. It cannot explain the small changes in the intermolecular radii in different crystals and, what is most important, it cannot be used to describe the properties of crystals other than structure and density.

When translated into the language of intermolecular interactions, the geometrical model treats the intermolecular potential as a sum of hard-sphere atom-atom potentials:

$$\varepsilon^{AB} = \sum_{a,b} \varphi_{ab}(r_{ab}), \quad \text{where} \tag{1.1}$$

$$\varphi_{ab} = \begin{cases} \infty, & r_{ab} < r_a + r_b, \\ 0, & r_{ab} \geq r_a + r_b. \end{cases} \tag{1.2}$$

In (1.1,2) ε^{AB} represents the interaction energy of the molecules A and B; r_a and r_b are the intermolecular atomic radii of the atoms a and b; and the summation over a and b is carried out over all atoms in the molecules A and B, respectively.

An obvious way to improve the model is to replace the hard-sphere potential φ_{ab} by a more realistic, soft potential, e.g., in the Lennard-Jones or the Buckingham form. The resulting approximation to the intermolecular potential is exactly what we call *the atom-atom potential approximation.* The major advantage of this approximation over the geometrical model is the possibility of expressing the intermolecular potential in terms of energies by appropriately calibrating the atom-atom potential parameters.

The idea to partition the intermolecular interaction energy into pairwise atom-atom interactions was put forward by *Kitaigorodsky* [1.6]as early as 1951. However, it was not until the 1960's that high-speed computers made it possible to use this idea in practical calculations.

The atom-atom potential method has since found widespread use in the various fields of the physics and chemistry of organic solids, liquids, gases and biologically important systems. In this monograph, however, we shall mainly be concerned with organic molecular crystals and will not attempt to cover all of the fields of application. Thus, we shall not dwell on the applications concerned with fluids, solutions and dense gases (including the adsorption of gases on organic solids). For details on such applications the reader is referred to [1.7–12].

Ionic organic crystals will not be covered, too. This subject has recently been discussed by *Metzger* [1.13]– essentially in terms of atom-atom interactions.

In addition, the reader will not find a comprehensive treatment of the conformational-energy problem, for which the advantage of atom-atom potentials has been realized. The interested reader may refer to the classical book by *Eliel* et al. [1.14]or to the more recent work by *Momany* [1.15], and others [1.16].

Finally, we have not found it necessary to include applications concerned with excited electronic states in organic molecular crystals. *Warshel, Craig* et al. [1.17–20]have used atom-atom potentials in a general force-field treatment applicable to both the ground and excited states. Successful calculations of this kind depend primarily on the non-atom-atom part of the force field, so that it is hardly reasonable to regard such calculations as an application of the atom-atom potential method.

The book is arranged as follows. In Chap. 2 we have tried to outline the problems involved in a rigorous quantum-mechanical treatment of the intermolecular forces between organic molecules. Emphasis has been placed on the range of intermolecular distances important for molecular crystals, and on the possibility of partitioning the interaction energy into contributions from small molecular fragments (bonds or atoms).

A formal definition of the atom-atom potential method and ways used to calibrate the parameters of the potential are discussed in Chap. 3. A large number of parameter sets is presented, covering the most important atomic species encountered in organic solids.

Chapters 4, 5 and 6 deal with the applications of the atom-atom potential method to problems encountered in a treatment of the statics, dynamics and thermodynamics of lattices, respectively. In each chapter the theoretical aspects of the problem are introduced first, and thereafter applications are discussed.

Finally, Chap. 7 is concerned with the various kinds of defects in organic crystals. The possibilities and limitations of the atom-atom potential method in predicting the structure and the energetics of defective regions are discussed, mainly with respect to topochemical solid-state reactions.

The units used throughout this book are as follows:

Energy:

$$1\,\mathrm{kcal/mol} = \begin{cases} 6.948\times10^{-21}\,\mathrm{J} \\ 4.184\,\mathrm{kJ/mol} \end{cases}$$

$$1\,\mathrm{a.u.\ (Hartree)} = m_e e^4/\hbar^2 = \begin{cases} 4.360\times10^{-18}\,\mathrm{J} \\ 2625.5\,\mathrm{kJ/mol} \\ 627.7\,\mathrm{kcal/mol} \\ 27.21\,\mathrm{eV} \end{cases}$$

Entropy:

$$1\,\mathrm{e.u.} = \begin{cases} 4.184\,\mathrm{J/(mol\,K)} \\ 1\,\mathrm{cal/(mol\,K)} \end{cases}$$

Pressure:

$$1\,\mathrm{kbar} = \begin{cases} 10^8\,\mathrm{Pa} \\ 986.9\,\mathrm{atm} \end{cases}$$

Length:

$$1\,\mathrm{Å} = 10^{-10}\mathrm{m}$$

$$1\,\mathrm{a.u.\ (Bohr)} = \hbar^2/(m_e e^2) = \begin{cases} 5.292\times10^{-11}\mathrm{m} \\ 0.5292\,\mathrm{Å} \end{cases}$$

Wavenumber:

$$1\,\mathrm{cm}^{-1} = (2.998\times10^{10})^{-1}\mathrm{s}^{-1} = 0.29979\,\mathrm{THz}$$

Charge:

$$1\,\mathrm{a.u.} = \mathrm{e} = \begin{cases} 1.602\times10^{-19}\mathrm{C} \\ 4.803\times10^{-10}\mathrm{esu} \end{cases}$$

Dipole moment:

$$1\,\mathrm{debye\ (D)} = 3.336\times10^{-30}\mathrm{C\,m}$$

$$1\,\mathrm{a.u.} = \begin{cases} 8.479\times10^{-30}\mathrm{C\,m} \\ 2.542\times10^{-18}\mathrm{esu\,cm} \\ 2.542\,\mathrm{D} \end{cases}$$

Polarizability:

$$1\,\mathrm{a.u.} = \begin{cases} 14.83\times10^{-30}\mathrm{m}^3 \\ 14.83\,\mathrm{Å}^3. \end{cases}$$

2. Non-Empirical Calculations of Intermolecular Forces Between Organic Molecules

In the last twenty years considerable progress has been made in the non-empirical quantum-mechanical calculations of intermolecular interaction energies. For the most part, this progress was due to the appearance of high-speed computers which permitted the manipulation of very complicated wave functions. The most reliable theoretical results were actually obtained using formalisms suggested long ago.

Of course, there has also been progress in the theory of intermolecular forces, especially in perturbation theories capable of dealing with exchange interactions. However, the practical applications of these theories are still limited to simple molecular systems because the expressions for the interaction energy are very difficult to compute, even for second-order perturbation.

Despite their limited possibilities, non-empirical calculations are now becoming most reliable sources of information about intermolecular forces. It is of particular importance that such calculations, at least the best of them, can already serve as criteria in assessing the quality of various models and can be used to refine these models both quantitatively and conceptually.

In this chapter we shall briefly discuss the basic non-empirical methods used in intermolecular force calculations and recent theoretical results directly related to organic molecular solids. The reader shall not find a complete account of the theory of intermolecular forces, but only the elementary ideas and formulas needed to understand the material. For a more detailed discussion we recommend to study [2.1–5].

As far as organic solids are concerned, our interest should obviously lie in the region of intermediate intermolecular distances, where a potential minimum occurs. Indeed, the intermediate-range interactions make the dominant contribution to the lattice energy and govern all of the static and dynamic properties of an organic crystal. Unfortunately, the region of intermediate distances is just that region which causes the greatest difficulties. At large intermolecular distances one can neglect any overlap between the molecular wave functions and the intermolecular antisymmetry of the total wave function. Considering the interaction Hamiltonian to be a small perturbation, the problem can be successfully solved by applying the ordinary Rayleigh-Schrödinger perturbation theory or the Ritz variation method [2.1]. At small distances the overlap cannot, of course, be neglected. Instead, one can treat the position of each electron as independent of the positions of the others or, in other words, one can ignore the correlation effects. Estimates of the correlation energy usually yield values of the order of 1 eV per doubly-occupied orbital [2.1]. Such values can be neglected at small separations. Hence, one can express the wave function of the system in terms of one-electron functions and resort to the powertful Hartree-Fock formalism widely used in quantum chemistry. At small distances the interacting molecules may be treated as a single supermolecule, and the interaction energy may be taken to be the difference between the total energy of the supermolecule and the sum of the energies of the individual molecules.

The main difficulty involved in the treatment of the intermediate-range interactions consists in simultaneously taking into account the overlap and correlation effects. To surmount this difficulty, two general approaches are now in use. The first is based on the concept of a supermolecule and utilizes one-electron wave functions. However, some special procedures are used to take into account the correlation effects. The supermolecule methods are divided into two categorites according to the manner in which the total wave function is constructed from the one-electron atomic functions. These are the well-known molecular orbital (MO) and Heitler-London or valence bond (VB) methods. The basic principles of these methods and their application to organic dimers and crystals are discussed in Sect. 2.1.

The second approach used for intermediate distances is based on perturbation theory, but unlike the standard Rayleigh-Schrödinger treatment it deals with wave functions having the correct symmetry. The various approaches of this type are usually described as exchange perturbation theories. These theories are briefly discussed in Sect. 2.2.1. In the subsequent three sections we shall consider the major contributions to the interaction potential, as they appear in exchange perturbation theories. Most of the emphasis has been laid on the possibility of expressing the interaction potential in terms of interactions between small molecular subunits (atoms, bonds, lone pairs). Our interest in this point is quite natural because it is in this way that one may find a theoretical justification of the atom-atom potential method.

2.1 The Supermolecule Method

2.1.1 Molecular Orbital Theory

The treatment of interacting molecules within the framework of the supermolecule approach is no different from the treatment of an isolated molecule. The total energy of the system is obtained by solving the Schrödinger equation,

$$H\psi = E\psi, \tag{2.1}$$

where H and ψ are the Hamiltonian and the wave function of the system, respectively. In the Born-Oppenheimer approximation ψ describes the electronic states at a fixed position of the nuclei,

$$H = \sum_{a} -1/2\nabla_a^2 - \sum_{\alpha,a} Z_\alpha/r_{\alpha a} + \sum_{a,b} 1/r_{ab} + \sum_{\alpha,\beta} Z_\alpha Z_\beta/r_{\alpha\beta}, \tag{2.2}$$

where a, b refer to the electrons, and α, β refer to the nuclei; the Z's are the nuclear charges, and the r's are the distances between particles defined by the subscripts (all of the quantities are expressed in atomic units).

Unfortunately, (2.1) cannot be solved exactly, except in the simplest of cases. The main difficulty is caused by the presence of the interelectronic distances r_{ab} in the wave function, so that the motions of the electrons relative to one another have to be taken into account. To our knowledge, the only attempt to include r_{ab} in a wave function explicitly has been made for the interaction of two hydrogen atoms [2.6].

Usually, the wave function of the system is constructed from the one-electron wave functions η_i using the Hartree-Fock (HF) representation, i.e.,

$$\psi = A\prod_i \eta_i \equiv \sum_\lambda (-1)^\lambda P_\lambda \prod_i \eta_i, \tag{2.3}$$

where A represents the antisymmetrizing operator and P_λ is the permutation operator which interchanges electrons in the function it acts upon (λ is the number of permutations). The antisymmetrizer must be applied to the product of η_i since physically allowed wave functions must satisfy the Pauli principle.

In MO theory, as applied to closed-shell molecules, the total wave function is constructed from delocalized molecular orbitals, each occupied by two electrons with opposite spin. In this case (2.3) may be represented as a Slater determinant

$$\begin{aligned}\psi &= [(2n)!]^{-1/2}\begin{vmatrix} \eta_1(1) & \eta_1(2) & \cdots & \eta_1(2n) \\ \bar{\eta}_1(1) & \bar{\eta}_1(2) & \cdots & \bar{\eta}_1(2n) \\ \cdots & \cdots & \cdots & \cdots \\ \bar{\eta}_n(1) & \bar{\eta}_n(2) & \cdots & \bar{\eta}_n(2n) \end{vmatrix} \\ &\equiv [(2n)!]^{-1/2}\,|\,\eta_1(1) \quad \bar{\eta}_1(2) \quad \cdots \quad \bar{\eta}_n(2n)\,|\,, \end{aligned} \tag{2.4}$$

where η_i and $\bar{\eta}_i$ differ from one another only by their spin components ($\eta_i = \varphi_i\alpha$, $\bar{\eta}_i = \varphi_i\beta$); n is the total number of one-electron orbitals; and the factor $[(2n)!]^{-1/2}$ is introduced if ψ is to be a normalized function. Note that the wave function given in (2.4) is an eigenfunction of the total spin operator S^2 and describes a singlet ground state with $S = 0$.

Equations (2.3,4) only define the allowable form of ψ, while the wave function itself is to be found from other considerations. Usually, ψ is determined using the variational principle according to which any trial function

ψ should yield an energy E greater than or equal to the energy E_{exact} corresponding to the exact wave function ψ_{exact}, i.e.,

$$E = \langle\psi|H|\psi\rangle/\langle\psi|\psi\rangle \geq E_{\text{exact}}. \tag{2.5}$$

Applying the variational principle to the wave function given in (2.4) yields a set of differential equations for φ_i, whose solution yields the best antisymmetrized product function. These differential equations, usually referred to as the Hartree-Fock equations, have the form

$$h\varphi_i = e_i\varphi_i \quad (i = 1,\ldots,n), \tag{2.6}$$

where h is the Fock operator and e_i is the orbital energy. The action of h upon φ_i is described by [2.1]

$$h\varphi_i(1) = [-1/2\nabla_1^2 - \sum_\alpha Z_\alpha/r_{1\alpha} + \sum_{j=1}^{n} 2\langle\varphi_j(2)|1/r_{12}|\varphi_j(2)\rangle]\varphi_i(1) - \sum_{j=1}^{n} \langle\varphi_j(2)|1/r_{12}|\varphi_i(2)\rangle\varphi_j(1). \tag{2.7}$$

For orthonormal φ_i's the electronic energy of the system is equal to

$$E = 2\sum_{k=1}^{n} I_{kk} + \sum_{k,l=1}^{n} (2J_{kl} - K_{kl}), \quad \text{where} \tag{2.8}$$

$$I_{kk} = \langle\varphi_k(1)| - 1/2\nabla_1^2 - \sum_\alpha Z_\alpha/r_{1\alpha}|\varphi_k(1)\rangle, \tag{2.9}$$

$$J_{kl} = \langle\varphi_k(1)\varphi_l(2)|1/r_{12}|\varphi_k(1)\varphi_l(2)\rangle, \tag{2.10}$$

$$K_{kl} = \langle\varphi_k(1)\varphi_l(2)|1/r_{12}|\varphi_k(2)\varphi_l(1)\rangle. \tag{2.11}$$

In MO theory each one-electron orbital φ_i is usually a linear combination of "atomic" functions χ_μ centered about the atomic nuclei (the LCAO-MO method),

$$\varphi_i = \sum_{\mu=1}^{m} c_{\mu i}\chi_\mu, \tag{2.12}$$

where m is the number of functions in the atomic basis set. For the orbitals φ_i represented by (2.12) the HF differential equations (2.6) are reduced to a set of algebraic equations for the unknown coefficients $c_{\mu i}$ (Roothaan's equations), i.e.,

$$\sum_{\nu=1}^{m} (h_{\mu\nu} - S_{\mu\nu}e_i)c_{\nu i} = 0 \begin{pmatrix} \mu = 1, & \ldots, & m \\ i = 1, & \ldots, & n \end{pmatrix}, \quad \text{where} \tag{2.13}$$

$$h_{\mu\nu} = I_{\mu\nu} + G_{\mu\nu}, \tag{2.14}$$

$$I_{\mu\nu} = \langle \chi_\mu(1)| - \nabla^2/2 - \sum_\alpha Z_\alpha / r_{1\alpha} | \chi_\nu(1) \rangle, \tag{2.15}$$

$$G_{\mu\nu} = \sum_{\lambda\sigma} P_{\lambda\sigma}[(\mu\nu|\lambda\sigma) - (\mu\sigma|\nu\lambda)/2], \tag{2.16}$$

$$(\mu\nu|\lambda\sigma) = \langle \chi_\mu(1)\chi_\lambda(2)|1/r_{12}|\chi_\nu(1)\chi_\sigma(2)\rangle, \tag{2.17}$$

$$S_{\mu\nu} = \langle \chi_\mu(1)|\chi_\nu(1)\rangle, \tag{2.18}$$

$$P_{\lambda\sigma} = 2\sum_i^n c_{\lambda i} c_{\sigma i}. \tag{2.19}$$

In the above equations $I_{\mu\nu}$ is the one-electron Hamiltonian which includes the kinetic energy of an electron and its potential energy in the field of the nuclei; $G_{\mu\nu}$ is the matrix element of the potential due to other electrons; $S_{\mu\nu}$ is the overlap matrix and $P_{\lambda\sigma}$ is the population matrix. The summation in (2.19) is only carried out over occupied orbitals. The orbital energies e_i are the roots of the secular equation,

$$|h_{\mu\nu} - eS_{\mu\nu}| = 0. \tag{2.20}$$

The total electronic energy is easily obtained by substituting (2.12) for φ_k into (2.8–11). The result is

$$E = \frac{1}{2}\sum_{\mu\nu} P_{\mu\nu}(I_{\mu\nu} + h_{\mu\nu}). \tag{2.21}$$

Finally, the total energy of the system is obtained by taking into account the energy of repulsion between the nuclei,

$$E_{\text{tot}} = E + \sum_{\alpha<\beta} Z_\alpha Z_\beta / r_{\alpha\beta}. \tag{2.22}$$

If only the valence electrons are included in E, Z_α and Z_β represent the core charges (Z_α^c, Z_β^c) on the atoms α and β, respectively.

The Roothaan equations (2.13) are nonlinear in $c_{\mu i}$ since the matrix elements ($h_{\mu\nu}$) themselves depend on the $c_{\mu i}$. This nonlinear problem is usually solved iteratively. In the first step one selects a set of basis functions χ_μ and assumes a set of coefficients $c_{\mu i}$ from which the matrix $h_{\mu\nu}$ is then constructed. This matrix is used in (2.13) to solve for a new set of coefficients. These in turn define a new matrix $h_{\mu\nu}$ which determines a new set of $c_{\mu i}$. This iterative process is repeated until the input set of $c_{\mu i}$ agrees with the output set to within some specified tolerance or the energy of the system becomes sufficiently close to that of the previous iteration. The orbitals φ_i that are obtained in this manner are referred to as self-consistent orbitals and the method is known as the self-consistent field (SCF) method.

It is to be noted that, in addition to n occupied orbitals, (2.13) yields $m-n$ unoccupied orbitals. The latter correspond to higher orbital energies and are called virtual orbitals.

If the basis set $\{\chi_\mu\}$ were complete, the SCF orbitals that are obtained from the Roothaan equations (2.13) would be identical to the exact HF orbitals derived directly from the HF differential equations (2.6). This is usually not the case, however, so that all SCF-LCAO-MO calculations suffer from errors produced by the incompleteness of the basis set.

Because of the tremendous computational difficulties involved in the manipulation of large basis sets, the calculations are most frequently performed with *minimal* basis sets which only contain atomic orbitals for the shells occupied or partly occupied in the atomic ground state (a $1s$ function for hydrogen; $1s$, $2s$ and $2p$ functions for the first-row elements, etc.). Larger basis sets, employing more than one basis function per atomic orbital, are usually called "extended". The choice of a basis set is of crucial importance in the calculations of intermolecular forces. This point will be illustrated somewhat later.

Various analytical expressions may, in principle, be adopted for the atomic functions χ_μ, but, in practice, almost all of the calculations have been carried out using either Slater (STO, Slater-type) orbitals,

$$\chi_{(nlm)} = Cr^{n-1}\exp(-\varsigma r)Y_l^m(\theta,\varphi), \tag{2.23}$$

or Gaussian orbitals (G or GTO),

$$\chi_{(nlm)} = Cr^{n-1}\exp(-\varsigma r^2)Y_l^m(\theta,\varphi), \tag{2.24}$$

where C is a normalization factor; n, l, and m are integers (quantum numbers); $Y_l^m(\theta,\varphi)$ is the spherical harmonic, and ς is the exponent of the orbital.

Slater functions are physically more realistic and generally provide rather short basis sets. In practice, one-exponent (one STO per atomic orbital) and two-exponent "double zeta" basis sets are used. Unfortunately, the use of Slater-type orbitals is restricted to diatomic molecules due to the computational difficulties involved in the evaluation of multicenter integrals such as those given in (2.17). From the computational point of view, Gaussians are much more convenient. However, Gaussians reproduce the behavior of atomic orbitals much more poorly than do Slater-type functions and, as a rule, a large number of Gaussians are required to construct a reliable wave function. This, in turn, increases the number of SCF equations.

Hence, although the required integrals can be computed quite efficiently, the iterative solution of (2.13) may prove to be too time-consuming.

To surmount this difficulty, some of the basis Gaussians are frequently grouped together (or "contracted") and then treated as a single function. A contracted basis set $\{\chi_\nu\}$ is defined by

$$\chi_\nu = \sum_\mu a_{\nu\mu}\chi_\mu, \tag{2.25}$$

where $\{\chi_\mu\}$ denotes the original (primitive) basis set. For example, the primitive basis set $(9s5p/4s)$ consisting of nine s-type and five p-type Gaussians on first-row atoms and four s-type Gaussians on hydrogen atoms can be contracted to the basis set $[3s2p/2s]$ consisting of three s-type and two p-type contracted Gaussians on first-row atoms and two s-type contracted Gaussians on hydrogen atoms [2.7]. Primitive basis sets are usually written in parentheses while contracted sets are given in brackets. The two above bases may be also represented by (9, 5/4) and [3, 2/2], respectively.

There are several standard contraction schemes used in the study of intermolecular interactions, of which the schemes STO-nG and n-31G are the most popular. In the STO-nG basis sets each STO is simulated by a linear combination of n Gaussians, whose coefficients and exponents are determined by a least-squares fitting procedure. In the n-31G basis sets, each inner shell basis function is a sum of n Gaussians and each valence shell (hydrogen $1s$, first-row atoms $2s$, $2p$) is described by two functions which are sums of three Gaussians and of one Gaussian, respectively. The n-31G basis sets generally yield results of double-zeta quality.

To improve the long-range behavior of molecular orbitals, which is of particular importance in dealing with intermolecular interactions, polarized basis sets which include basis functions describing excited-state atomic orbitals are frequently used. For example, the 6-31G^* and 6-31G^{**} basis sets both include the d orbitals of carbon. The latter also contains the p orbitals of hydrogen.

2.1.2 MO Treatment of the Interaction Energy

We now assume that the HF equations (2.6) or the Roothaan equations (2.13) have been solved for the energy E of the supermolecule (or its SCF-LCAO-MO estimate). In order to determine the intermolecular energy ε, one should calculate the difference

$$\varepsilon = E - E^A - E^B, \tag{2.26}$$

where E^A and E^B denote the energies of the isolated molecules. A serious shortcoming of the supermolecule approach lies in the fact that the interaction potential ε in (2.26) appears as a comparatively small difference between very large terms. For example, the total energy of two helium atoms is five orders of magnitude larger than the depth of the potential [2.1]. This implies that all of the energies in (2.26) must be known to at least 7 or 8 significant figures or that the errors in the energies must cancel.

The most significant error appearing in the calculations of the HF interaction energy from (2.26) results from the fact that the calculations of E, E^A and E^B are actually made using different basis sets, namely $\{AB\}$, $\{A\}$ and $\{B\}$, respectively. In dealing with the supermolecule and determining E, the basis set $\{B\}$ improves the energy of A, and vice versa. Thus, the total energy E is the sum of the improved energies of A and B, and the interaction potential ε. If one substitutes into (2.26) the non-improved energies E^A and E^B, i.e., the energies calculated using only their own basis sets, a non-negligible error is introduced into ε [usually referred to as a basis-set superposition error (BSSE).]As a rule, the smaller the basis set used, the larger is the error. For sufficiently large basis sets the energies of the isolated molecules are close to the corresponding HF limits. Hence, no further improvement in these energies is observed in the calculation of the energy of the entire system.

In order to reduce the BSSE, all three energies in (2.26) may be calculated using the same basis set $\{AB\}$. Such an approach, known as the counterpoise method, yielded good results for two H_2 molecules [2.8]. The results for some hydrogen-bonded systems were less satisfactory [2.9a].

Sadlej [2.9b]has shown that there exists a "second-order" BSSE that affects the multipole moments and the polarizability of the monomers, and that cannot be handled with the counterpoise method. Fortunately, this contribution to BSSE seems to be an order of magnitude smaller than the whole error and can be neglected in most cases.

A further shortcoming of the supermolecule approach in its customary SCF-LCAO-MO form is the result of the general shortcoming of the single-determinant HF method – the neglect of the correlation between electrons of opposite spin, particularly between electrons in the same orbital [2.1]. Hence such an important contribution to the intermolecular potential as the dispersion energy is not considered [2.1,10]. There are several possible ways of taking into account the dispersion energy in the framework of the supermolecule approach. In practice, only two of them have found an application.

In the first the wave function is represented by two or more Slater determinants,

$$\psi = \sum_i c_i \psi_i . \qquad (2.27)$$

Equation (2.27) is a formal attempt to expand the actual wave function ψ, a function of all the interelectron distances r_{ab}, in terms of a sum of antisymmetrized products of one-electron functions. The component wave functions ψ_i in (2.27) are referred to as "configurations". The mixing of these configurations is called "configuration interaction" (CI). Besides the original (ground-state) configuration, (2.27) also contains excited configurations which are constructed from the original configuration by promoting electrons from occupied to virtual orbitals. The mixing coefficients c_i are usually obtained using the variational principle, by minimizing the total energy given in (2.5).

Although the CI method, as applied to MO theory, is frequently claimed to yield the correct potential, this is, strictly speaking, not so. Numerous calculations on simple atomic systems have shown that the intermolecular energy calculated with the wave function given in (2.27) converges very slowly to the correct potential and that it usually takes a tremendous number of configurations to accurately reproduce the dispersion energy. For example, we refer the reader to the study of a water dimer by *Diercksen* et al. [2.11], where the wave function involved as many as 56268 configurations. (The correlation correction to the dimerization energy was found to be ~1 kcal/mol which accounts for approximately 16% of the total dimerization energy.)

It is to be noted that the absence of a dispersion component in the SCF-LCAO-MO interaction energy is frequently compensated by a BSSE which leads to an artificial stabilization of the supermolecule. This fact has been directly verified in [2.9a,12,13].

The slow convergence of the CI expansion within the framework of the MO approach is due to the fact that the MO description does not allow one to separate intra- and inter-correlation effects [2.1]and to add only those configurations which are responsible for inter-correlation effects. It is essential that intra-correlation energies are substantially larger than inter-correlation energies. For example, in the water dimer the former are estimated to be about 5×10^3kcal/mol while the latter are only ~1 kcal/mol [2.11]. As a result, most of the added MO configurations correct the intra-correlation energies first and, hence, the convergence of the intermolecular energy to the correct potential is very slow.

Despite the limited possibilities of the CI method, the use of multi-determinant wave functions leads to a considerable improvement in the intermolecular potential. This has been convincingly demonstrated by *Clementi* et al. [2.14,15] in their calculations of the water dimer and the properties of liquid water.

Another way of taking into account the dispersion energy is trivial. It consists of an independent evaluation of this energy component using a perturbation technique and the subsequent addition of the result to the SCF interaction energy.

In contrast to the perturbation methods in which the interaction potential is obtained as a sum of components, each having a clear physical nature, the supermolecule approach yields a net value for the interaction energy, so that a partitioning of the energy into different contributions necessitates independent calculations. To illustrate this point, we shall now briefly discuss the partitioning scheme suggested by *Morokuma* [2.16]. According to this scheme, the interaction energy can be written as a sum of electrostatic, exchange, induction and charge-transfer terms,

$$\varepsilon = \varepsilon_{\text{elst}} + \varepsilon_{\text{exch}} + \varepsilon_{\text{ind}} + \varepsilon_{\text{CT}}. \tag{2.28}$$

Let ψ_0^A and ψ_0^B be the HF wave functions of the isolated molecules, and ψ_1^A and ψ_1^B the wave functions individually optimized in the electrostatic field of the other molecule. The energy components in (2.28) may then be obtained from the following set of equations,

$$\varepsilon_{\text{elst}} = \langle \psi_0^A \psi_0^B | H | \psi_0^A \psi_0^B \rangle - E^A - E^B, \tag{2.29}$$

$$\varepsilon_{\text{elst}} + \varepsilon_{\text{exch}} = \langle A(\psi_1^A \psi_1^B) | H | A(\psi_1^A \psi_1^B) \rangle - E^A - E^B, \tag{2.30}$$

$$\varepsilon_{\text{elst}} + \varepsilon_{\text{ind}} = \langle \psi_1^A \psi_1^B | H | \psi_1^A \psi_1^B \rangle - E^A - E^B, \tag{2.31}$$

$$\varepsilon_{\text{elst}} + \varepsilon_{\text{exch}} + \varepsilon_{\text{ind}} + \varepsilon_{\text{CT}} = E - E^A - E^B. \tag{2.32}$$

An obvious shortcoming of this scheme lies in the definition of ε_{CT}. As can be seen from (2.32), ε_{CT} is the difference between the total interaction energy and the sum of the remaining three terms. Hence, it absorbs, aside from the actual charge-transfer energy, all of the coupling terms between the various energy components. A more sophisticated scheme has been introduced by *Kitaura* and *Morokuma* [2.17], in which (2.28) also involves an exchange polarization energy and a special coupling term.

In general, the possibility of decomposing the total interaction potential into terms having a clear physical meaning is of great significance. On the one hand, such information may help us to better understand the nature of the binding forces in organic systems. On the other hand, a knowledge of the relative size of the individual energy components and their behavior with respect to distance and orientation may help to develop sound model potentials. Unfortunately, the energy partitioning within the framework of the supermolecule approach is not very reliable because the results of this partitioning exhibit a pronounced dependence on the size of the basis set used [2.18].

2.1.3 Application to Organic Systems

The supermolecule approach has been used extensively in studies of the interaction between organic molecules. As a rule, this approach is applied to systems in which the dispersion energy intuitively seems to be of minor importance. Thus, a large number of studies have been devoted to hydrogen-bonded systems. These are, first of all, the studies of small hydrogen-bonded dimers such as $(H_2O)_2$ and $(HF)_2$. A discussion of these systems and an exhaustive review of the most recent work may be found in [2.18].

The problem of hydrogen bonding is also the central one in investigations of alcohols [2.19,20], carboxylic acids [2.21]and peptides [2.20]. Much effort has been devoted to the study of the interaction of biologically active molecules with water and ions [2.22–27]. A number of studies also deals with electron donor-acceptor complexes [2.28–31], heterocyclic molecules [2.32–34]and hydrocarbons [2.35,36]. This list of applications of the supermolecule method is far from complete and can be extended.

Unfortunately, only a few of the calculations presently available may be considered as a source of reliable quantitative information since most of them were not performed with the proper corrections for correlation effects nor were the basis sets sufficiently large. These calculations involve small hydrogen-bonded dimers exclusively [2.11,14]and have little bearing to the subject of this book.

Inspection of the recent literature shows that the construction of wave functions close to the HF limit requires very large basis sets, unless some special methods are used to enhance the flexibility of the basis set. Thus, even the use of extended standard bases with polarization, such as 6-31G^{**}, may appreciably distort the size of the interaction potential. For example, the 6-31G^{**} basis set yields an interaction energy equal to −5.6 kcal/mol for the linear configuration of the water dimer [2.37a], whereas the exact HF value is −3.9 kcal/mol [2.11,14].

Calculations using medium-sized bases, such as the 4-31G, considerably overestimate the absolute value of the interaction energy. For example, a value of −7.7 kcal/mol is obtained for the linear water dimer [2.5]. It is thought that the 4-31G basis set exaggerates the polarity of the monomers and hence the absolute value of the electrostatic contribution to the potential.

Calculations using minimal bases, such as the STO-3G, are the most feasible but the least reliable. Sometimes the STO-3G basis set yields quite satisfactory results for the total interaction energy (including $\varepsilon_{\mathrm{disp}}$). An energy of −5.1 kcal/mol [2.5], which is close to the experimental value, is obtained for the linear water dimer. However, it seems that the good results obtained with the STO-3G basis set are due to a mutual compensation of errors in the individual components of the potential.

Minimal basis sets are the most interesting from a practical point of view because only these sets are applicable to large systems [2.37b]. In this respect, a recent paper by *Kolos* [2.38], in which an effective method for improving the flexibility of the basis set has been introduced, is of great significance. Proceeding from a standard minimal basis (7, 3/3) contracted to [2, 1/1], both the contraction coefficients and the Gaussian exponents were varied. The quality of the basis set was assessed not only by the total energy but also by the charge distribution of the system. To reduce BSSE, the counterpoise method was applied. In spite of the small size of the basis set, the interaction energies that were computed for several dimers proved to be very close to the corresponding HF values. Although the basis set suggested in [2.38]was tested on a limited number of systems, it seems that minimal bases optimized specifically for dealing with intermolecular interactions will help in the near future to reduce the tremendous computational work involved in the treatment of large molecules.

The possibility of obtaining good results with properly optimized minimal bases was also demonstrated in another paper by *Kolos* et al. [2.35] concerned with the interaction of two methane molecules. The calculations were carried out using a fairly extended basis set, (11, 7, 1/6, 1) contracted to [4, 3, 1/3,1], and two minimal bases of the type (7, 3/3) contracted to [2, 1/1]. One of the latter bases, denoted as "old", is a standard basis set [2, 1/1]taken from [2.39]. The second, denoted as "new" set differs from the old one in the contraction coefficients of the 1*s* orbital and the exponents of the 2*s* and 2*p* valence orbitals of carbon. The calculations with the minimal basis sets were carried out using both the conventional and the counterpoise methods to remove BSSE. Several test calculations were also made with the STO-4*G* basis set.

The interaction energy was determined as a function of the distance between the carbon atoms, R_{CC}, for six orientations of the molecules. These orientations are coded *A* to *F* and are shown in Fig. 2.1.

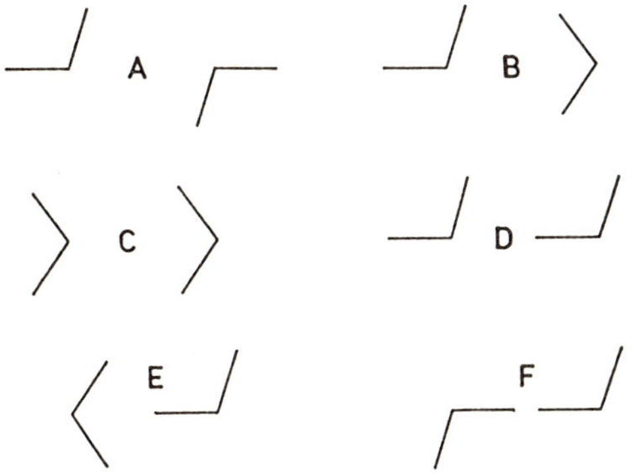

Fig. 2.1. The orientations of pairs of methane molecules. Only C–H bonds in plane are indicated

An important observation made in [2.35]was that even the extended basis set was still not flexible enough to yield a negligible BSSE. Thus, at $R_{CC} = 6$ a.u. this error amounted to 0.54 kcal/mol for the configuration *A*. For the configurations *A*, *B*, *D* and *E*, BSSE led to the appearance of spurious potential minima at large separations.

Interesting results were obtained with the "new" minimal-basis set. In this case BSSE was very small and the interaction energies were, at small and intermediate distances, close to those obtained with the extended basis set. At large distances the minimal-basis calculations were even more reliable since they did not yield spurious minima for the repulsive configurations. (A slightly attractive energy was only obtained for the configuration *D* because of a somewhat overestimated octupole-octupole attraction.) The STO-4G basis set yielded interaction energies very close to those obtained with the "new" minimal-basis set, although much more computer time was required.

Kolos et al. [2.35]noted that properly optimized minimal bases should yield good results in the case of nonpolar molecules where the induction energy, poorly described by small bases, contributes only slightly to the interaction energy.

To correct the resulting SCF-LCAO-MO potential for the dispersion energy, two different methods were examined. In the first, the dispersion energy was calculated as a sum of bond-bond interactions using a London-type equation and experimentally obtained bond polarizabilities. However, the resulting potential curves were not consistent with the experimentally determined isotropic potential. Only at $R_{\mathrm{CC}}>9$ a.u. did the agreement become satisfactory. *Kolos* et al. [2.35]attributed this failure to the use of a London-type equation which underestimated the dispersion energy because of the neglect of higher-order terms in the multipole expansion of $\varepsilon_{\mathrm{disp}}$. In the second method the dispersion energy was computed as

$$\varepsilon_{\mathrm{disp}} = \sum_{n=1}^{3} C_{2(n+2)} R_{\mathrm{CC}}^{-2(n+2)}, \qquad (2.33)$$

with the experimental constants C_m from [2.40]. In this case the orientation-dependent part of $\varepsilon_{\mathrm{disp}}$ was obviously lost so that the results only became reliable at large separations ($R_{\mathrm{CC}}>7$ a.u. [2.35]). The second method gave a potentival curve that is much more consistent with experiment.

It is interesting to see which energy component is most important in stabilizing the CH_4 dimer. The decomposition of the interaction potential into electrostatic, exchange and dispersion terms is shown in Table 2.1. The total potential is the sum of the dispersion energy calculated in the bond-bond approximation and the SCF-LCAO-MO interaction energy calculated using the "new" minimal-basis set and the counterpoise method. The electrostatic part of $\varepsilon_{\mathrm{SCF}}$ was estimated using the point charges resulting from the "new" basis set, while the exchange energy was taken to be the difference $\varepsilon_{\mathrm{SCF}} - \varepsilon_{\mathrm{elst}}$. All of the quantities (Table 2.1) were calculated for $R_{\mathrm{CC}} = 8$ a.u., a distance close to the nearest-neighbor C...C distance in solid methane (7.9 a.u.). As can be seen from Table 2.1, the attractive configurations A to E are mainly stabilized due to $\varepsilon_{\mathrm{disp}}$. Nevertheless, the electrostatic contribution is rather important in all of the configurations. This is especially true for configuration D which is characterized by a strong octupole-octupole attraction, whose magnitude is approximately a third of the total interaction energy. Configuration F is repulsive at $R_{\mathrm{CC}} = 8$ a.u., which is obviously the result of the repulsion between the nearest hydrogen atoms.

Smit et al. [2.21]have been concerned with the dimer and crystal of formic acid. This work represents one of very few attempts to determine

Table 2.1. Separation of the interaction energy of two methane molecules that are 8 a.u. apart into electrostatic, exchange and dispersion energies [kcal/mol][2.35]

Configuration	A	B	C	D	E	F
ε_{elst}	0.05	0.02	-0.04	-0.09	-0.04	0.16
ε_{exch}	0.04	0.09	0.13	0.20	0.30	0.83
ε_{disp}	-0.38	-0.40	-0.42	-0.44	-0.47	-0.52
ε	-0.29	-0.29	-0.33	-0.33	-0.21	0.47

the lattice energy of an organic molecular crystal directly, with no recourse to an analytical potential previously fitted to the potential surface of the dimer. The advantages of such a calculation are quite obvious. Indeed, a direct evaluation of the lattice energy is free from errors associated with the fitting of analytical potentials to the potential surface of the dimer. What is more, the direct use of an ab initio method in calculating the lattice energy allows one to analyze the role of various energy components in stabilizing the crystal structure. Usually this information is lost when one uses analytical potentials.

The paper [2.21]is also of particular interest because the reader can find valuable information concerning the accuracy of evaluating different energy components as well as the total interaction energy. The estimates of the errors seem to be representative of calculations involving truncated basis sets and allow the reader to assess the reliability of such calculations.

The lattice energy of formic acid was calculated using the pair approximation. The interaction between two molecules was written as a sum of the SCF interaction energy (corrected for BSSE) and the dispersion energy:

$$\varepsilon = \varepsilon_{SCF} + \varepsilon_{disp}. \tag{2.34}$$

The former was partitioned into first-order and second-order contributions,

$$\varepsilon_{SCF} = \varepsilon_1 + \varepsilon_2, \tag{2.35}$$

where ε_1 takes into account the exchange and electrostatic energies and ε_2 the induction and charge-transfer energies. The calculation of ε_{elst} and ε_{exch} was carried out exactly as described above for the Morokuma scheme in (2.29–32), while the partitioning of ε_2 into ε_{ind} and ε_{CT} was made using the second-order perturbation equation for ε_{ind}. The dispersion energy in (2.34) was also estimated using the second-order perturbation technique.

The lattice energy was obtained by summing up the interaction energies between a reference molecule A and the surrounding molecules B. A summation limit R_{SCF}, within which the interaction energies were determined using (2.34) was chosen. The interaction energies for intermolecular distances greater than R_{SCF} were computed using the multipole expansion of ε_{elst}, ε_{ind}, and ε_{disp} (ε_{exch} and ε_{CT} were neglected). The polar multipole expansion of ε_{elst} was truncated at the sixth moment [2.41]. Only the R^{-6} terms were taken into account in the multipole expansions of ε_{ind} and ε_{disp}. Thus, at large separations ($R > R_{SCF}$) the intermolecular energy can be represented as

$$\varepsilon = \varepsilon_{mult} + \varepsilon_{ind\text{-}6} + \varepsilon_{disp\text{-}6}, \tag{2.36}$$

where ε_{mult}, $\varepsilon_{ind\text{-}6}$ and $\varepsilon_{disp\text{-}6}$ denote the above-specified approximations to ε_{elst}, ε_{ind} and ε_{disp}, respectively.

The SCF calculations of the lattice energy were performed using a (6, 3/3) basis set contracted to [2, 1/1](minimal, from now on Basis I) and contracted to [3, 2/2](split valence, Basis II). In the discussion of accuracy *Smit* et al. [2.21]sometimes referred to (9, 5/4) contracted to [4, 2/2](double zeta, Basis III) and to (9, 5, 1/4, 1) contracted to [4, 2, 1/2, 1](polarized double zeta, Basis IV).

In order to assess the quality of the basis sets and to find the optimum value for R_{SCF}, the energy of the cyclic configuration of the formic acid dimer was calculated for various intermolecular separations. The geometry of the individual molecules was not optimized but was taken from gas-phase electron diffraction data. The results are listed in Table 2.2 and shown in Fig. 2.2 [2.21].

It can be seen from Fig. 2.2 that the most important contribution to the interaction energy at intermediate and large separations is ε_{elst}. The

Table 2.2. Components of the interaction energy [kcal/mol]of the cyclic dimer of formic acid [2.21]

R_{00} [Å]	2.5		2.7			3.0	
Basis set	I	II	I	II	III	I	II
ε_{elst}	-46.47	-50.20	-30.15	-32.07	-36.21	-17.22	-17.74
ε_{mult}	-35.95	-34.03	-25.69	-24.64	-30.74	-16.36	-15.90
ε_{exch}	49.84	61.90	21.45	28.63	34.16	5.63	8.66
ε_{ind}	- 6.34	- 8.66	- 3.99	- 4.71	- 6.58	- 2.00	- 2.33
$\varepsilon_{ind\text{-}6}$	- 0.20	- 0.26	- 0.13	- 0.17	- 0.22	- 0.08	- 0.10
ε_{CT}	-21.09	-21.74	-11.04	-13.39	-10.30	- 3.66	- 6.54
ε_{disp}	- 3.13	- 6.00	- 1.98	- 3.88	- 4.21	- 1.04	- 2.08
$\varepsilon_{disp\text{-}6}$	- 2.30	- 4.61	- 1.52	- 3.06	- 3.25	- 0.86	- 1.74
ε	-27.19	-24.70	-25.71	-25.42	-23.14	-18.29	-20.03
$\varepsilon_{mult,corr}$	-24.0	-23.1	-18.1	-17.3	-19.3	-12.6	-11.7
$\varepsilon_{disp,corr}$	-12.0	-10.7	- 7.6	- 6.9	- 6.7	- 4.0	- 3.7
ε_{corr}	-24.1	-18.5	-23.8	-21.1	-14.2	-17.5	-17.5

contribution of ε_{elst} to the stabilizing (attractive) energy is about 60% at $R_{OO} = 2.7$ Å (the experimental equilibrium distance for the dimer and it becomes dominant with increasing R_{OO}. The magnitude of ε_{elst} is markedly sensitive to changes in the basis set (Table 2.2). In order to determine the

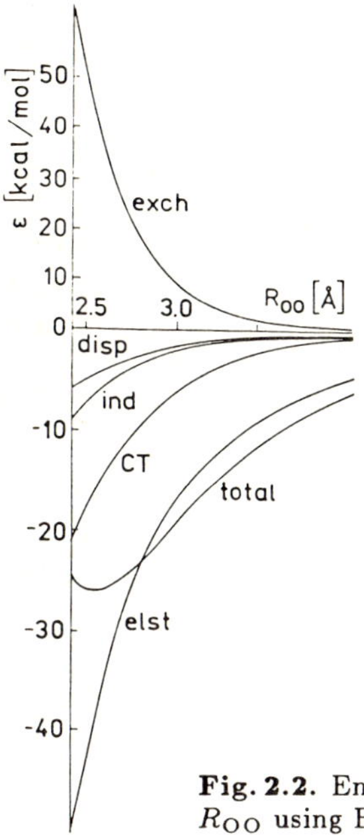

Fig. 2.2. Energy components of the formic acid dimer as a function of R_{OO} using Basis II [2.21]

3.4		3.9		4.4	4.9	5.4	5.7	R_{OO} [Å]
I	II	I	II	I	I	I	I	Basis set
- 9.40	- 9.47	-4.94	-5.04	-2.78	-1.65	-1.04	-0.80	ε_{elst}
- 9.43	- 9.23	-5.00	-5.10	-2.81	-1.67	-1.04	-0.80	ε_{mult}
0.81	1.62	0.06	0.18					ε_{exch}
- 0.83	- 1.00	-0.31	-0.37	-0.13	-0.06	-0.03	-0.02	ε_{ind}
- 0.04	- 0.05	-0.02	-0.02	-0.01	-0.01	0.00	0.00	ε_{ind-6}
- 0.58	- 2.39	0.04	-0.51					ε_{CT}
- 0.49	- 0.97	-0.22	-0.42	-0.11	-0.06	-0.03	-0.02	ε_{disp}
- 0.44	- 0.88	-0.20	-0.41	-0.10	-0.06	-0.03	-0.02	ε_{disp-6}
-10.49	-12.21	-5.37	-6.16					ε
- 7.9	- 7.3	-4.6	-4.2					$\varepsilon_{mult,corr}$
- 1.9	- 1.7	-0.9	-0.8					$\varepsilon_{disp,corr}$
-10.4	-11.0	-5.7	-5.6					ε_{corr}

correct value of ε_{elst}, the latter was expressed as a sum of ε_{mult} and a penetration energy ε_{pen}, a result of the mutual penetration of the charge distributions of the monomers. Inspection of Table 2.2 shows that at intermediate distances the uncertainty in ε_{elst} is mainly a result of the uncertainty in ε_{mult} which, in turn, is determined by the inaccuracies in the calculation of the molecular multipole moments. In order to improve the molecular charge distribution, the basis set was extended to include polarization functions (Set IV). The resulting value for ε_{mult} is -28.76 kcal/mol at $R_{OO} = 2.7$ Å. A further extension of the basis set, obtained by adding a set of diffuse polarization functions, reduced ε_{mult} by only 0.5 kcal/mol. Based on these results, the HF limit for ε_{mult} was estimated to be about -28 kcal/mol. However, even this estimate can not be regarded to be satisfactory because the calculated dipole and quadrupole moments do not agree at all with the experimental values. The final estimate for ε_{mult}, obtained by replacing the calculated dipole and quadrupole moments by their experimental values ($\varepsilon_{mult,corr}$ in Table 2.2), is -19 ± 4 kcal/mol, – a value which differs considerably from the above-mentioned HF limit.

The changes in the penetration energy ε_{pen} with respect to the basis set used are much smaller than those observed with ε_{mult}. At $R_{OO} = 2.7$ Å ε_{pen} is equal to -4.5, -7.4, and -5.5 kcal/mol for Bases I, II and III, respectively. The best estimate of ε_{pen} is -6 ± 2 kcal/mol. ε_{pen} decreases rapidly with increasing R_{OO} so that at $R_{OO}>3.9$ Å the electrostatic interactions are well described by ε_{mult} alone.

The largest uncertainty was observed for ε_{exch}. At the equilibrium distance the magnitude of ε_{exch} varies from 21.45 to 34.16 kcal/mol, depending on the basis set used. Unfortunately a reliable estimate of ε_{exch} at the HF limit could not be made so that the mere mean of the above two values (-28 ± 6 kcal/mol) had to be taken.

The induction and charge-transfer energies are rather insensitive to changes in the basis set (Table 2.2). An even lesser sensitivity was observed for ε_2 due to the mutual compensation of errors in ε_{ind} and ε_{CT}. The HF limit for ε_2 was estimated to be -17 ± 3 kcal/mol at $R_{OO} = 2.7$ Å.

As far as the dispersion energy is concerned, – it nearly doubled on changing from the minimal to the extended basis set. The best estimate for ε_{disp} has been made by relating ε_{disp} to the square of the dipole polarizability and scaling the calculated values of ε_{disp} by a factor of $(\alpha_{exp}/\alpha_{calc})^2$. The corrected vlaues of ε_{disp} are given in Table 2.2. At $R_{OO} = 2.7$ Å the best estimate adopted for ε_{disp} is -7 ± 2 kcal/mol.

Assuming the above estimates for the individual energy components and that the uncertainties are independent of each other, the total interaction energy was estimated to be -21 ± 8 kcal/mol. This value necessitated the introduction of a further correction in order to account for the molecular distortions that take place upon dimerization. According to [2.42], the energy needed to distort the geometry of the free molecule to the geometry it adopts in the dimer is about 7 kcal/mol. The addition of this quantity to the estimate for ε yields the final dimerization energy of 14 kcal/mol with an uncertainty of 8 kcal/mol. Although a value of 14 kcal/mol is in good agreement with experiment (16±1.5 kcal/mol), the extremely large uncertainty suggests that this agreement is quite likely to be fortuitous.

It is to be noted that the above estimate for ε refers to the distance of 2.7 Å, which is the experimentally observed but not calculated equilibrium distance for the dimer. Strictly speaking, ε may only be used to estimate the dimerization energy if the calculated equilibrium distance is close to 2.7 Å. By plotting ε as a function of R_{OO} it was found [2.21]that the minimum of ε occurred at 2.56 and 2.61 Å for Bases I and II, respectively. The corresponding (uncorrected) well depths were 27.3 and 25.8 kcal/mol which gave 20.3 and 18.8 kcal/mol, respectively, for the dimerization energy. This is in poor agreement with the experiment. The use of ε_{mult} and ε_{disp} corrected, as described above by substituting the experimental values for the multipole moments and polarizability), yielded equilibrium distances of 2.59 and 2.68 Å and well depths of 24.8 and 21.1 kcal/mol for Bases I and II, respectively. In this case the dimerization energies are equal to 17.8 and 14.1 kcal/mol, which is in good agreement with the experiment.

From an analysis of Table 2.2 it can be seen that the approximate potential given in (2.36) may be successfully employed instead of the one given in (2.34) at distances beyond 4 Å. At these distances ε_{exch} and ε_{CT} are negligible, while ε_{elst} and ε_{disp} can be replaced by ε_{mult} and $\varepsilon_{disp\text{-}6}$, respectively. $\varepsilon_{ind\text{-}6}$ fails to give an adequate representation of ε_{ind}. So, it is only the smallness of ε_{ind} itself that justifies the replacement of ε_{ind} by $\varepsilon_{ind\text{-}6}$. Of course, the applicability of (2.36) to the interactions beyond 4 Å was only tested for a single configuration. Hence, it is likely that the potential given in (2.36) is not as good for all of the other configurations encountered in the crystal. Considering these circumstances, *Smit* et al. [2.21]adopted a somewhat larger value for R_{SCF} (6 Å) in their calculations of the lattice energy.

We now turn to a discussion of the lattice-energy calculation. Examination of the crystal structure of formic acid has shown that there are 14 molecular pairs within the 6 Å limit, of which only 7 are symmetrically nonequivalent. Thus, to compute the lattice energy, only 7 configurations had to be treated using the SCF approach, while the contribution from the other configurations might be estimated using the multipole approximation. The results of the calculations are summarized in Table 2.3 [2.21]. To identify the molecules surrounding the reference molecule in the crystal four integer numbers are used. The first number labels the symmetry operation by which the coordinates of the given molecule are obtained from those of the reference molecule (the sequence of the operations is the same as in [2.43]for the space group *Pna*). The remaining three numbers represent lattice translations. The reference molecule is labeled 1000.

About 80% of the lattice energy is accounted for by the hydrogen-bonded pair (1000, 3000) (Table 2.3). The remaining six pairs within the 6 Å limit account for approximately 10% of the lattice energy. For these pairs, the distribution of the interaction energy over the individual energy components is very different from that of the pair (1000, 3000). The dispersion interactions are much more important, while the charge-transfer energy becomes negligible. It is interesting that for some of the pairs the electrostatic interactions prove to be destabilizing, which results in positive values of the interaction energies.

The molecules outside of the 6 Å sphere account for approximately 10% of the lattice energy. The dispersion energy is even more important for these molecules. It makes up about 45% of their total contribution to the lattice energy.

As can be seen from Table 2.3, the corrections to ε_{elst} and ε_{disp} have a minor effect on the lattice energy. *Smit* et al. [2.21]attributed this fact to a mutual compensation of the corrections on the summation of the interaction energies over the lattice.

Inasmuch as the geometry of the molecules in the crystal differs from the geometry in the free state, the energy values in Table 2.3 require further correction to account for the molecular distortion energy. According to [2.44], this energy is about 2 kcal/mol. The final estimates of the lattice energy are thus 11 and 13 kcal/mol for Bases I and II, respectively. *Smit* et al. [2.21]adopted the same relative uncertainty for the lattice energies as they did for the dimer (40%). This yielded an uncertainty of 5 kcal/mol.

Table 2.3. Components of the lattice energy [kcal/mol]of the formic-acid crystal. The reference molecule for each pair is molecule 1000 [2.21]

Pair	3000		1010		1001		2011		2001	
Basis set	I	II	I	II	I	II	I	II	I	II
ε_{elst}	-18.21	-19.61	1.35	1.47	0.64	0.74	-1.19	-0.95	-1.62	-1.35
ε_{mult}	-15.56	-15.05	1.33	1.60	0.64	0.74	-1.14	-0.83	-1.60	-1.30
ε_{exch}	13.92	17.51	0.55	0.90	0.00	0.01	0.34	0.56	0.18	0.28
ε_{ind}	- 1.89	- 2.50	-0.06	-0.12	-0.02	-0.03	-0.05	-0.06	-0.03	-0.04
ε_{ind-6}	- 0.07	- 0.12	-0.03	-0.06	0.00	-0.01	0.00	-0.04	-0.02	-0.03
ε_{CT}	- 4.91	- 5.94	0.07	0.00	0.00	-0.01	0.02	-0.11	0.06	0.06
ε_{disp}	- 1.18	- 2.42	-0.28	-0.71	-0.04	-0.10	-0.16	-0.39	-0.15	-0.36
ε_{disp-6}	- 0.53	- 1.05	-0.28	-0.70	-0.03	-0.07	-0.05	-0.14	-0.12	-0.21
ε	-12.27	-12.96	1.63	1.54	0.58	0.61	-1.04	-0.96	-1.56	-1.41
$\varepsilon_{mult,corr}$		-11.5		0.9		0.5		-0.2		-0.5
$\varepsilon_{disp,corr}$		- 4.3		-1.3		-0.2		-0.7		-0.7
ε_{corr}		-11.3		0.3		0.3		-0.6		-0.9

Thus, both final estimates of the lattice energy prove to be in acceptable agreement with experiment (14.8±2 kcal/mol).

An important question that always arises in calculations of the lattice energy is the applicability of the pair approximation. *Smit* et al. [2.21]noted that the calculations of the lattice energy as a sum of energies between isolated pairs of molecules involve two kinds of errors. The first kind arises from the neglect of the crystal symmetry. Clearly, insofar as all of the molecules in the formic acid crystal are symmetrically equivalent, the charge distribution should be the same in each. However, this is not so if one deals with an isolated molecular pair and does not impose any symmetry constraints upon the wave function of the system. In this case a rearrangement and a transfer of charge may occur during the SCF process, which would be impossible if the proper symmetry of the wave function were maintained. This, in turn, results in an overestimation of ε_2. The effect is obviously the largest for the hydrogen-bonded pair. An analysis of the Mulliken populations [2.21]for this pair showed that a charge transfer occurred from the acceptor molecule 1000 to the donor molecule 3000 to yield a net molecular charge of 0.04 and 0.05 e using the Bases I and II, respectively. The largest difference in charge on crystallographically equivalent atoms was 0.04 and 0.07 e with the Bases I and II, respectively. The total effect of this spurious charge transfer on ε_2 was estimated to be less than 1 kcal/mol.

The second source of error, inherent to the pair approximation, is the neglect of many-body interactions. *Smit* et al. [2.21]estimated the energy of three-body interactions using the semiempirical CNDO/2 method. It was argued that the CNDO/2 method should overestimate the three-body interaction energy, thus yielding an upper limit. As expected, the largest three-body energy was found for the trimer (1000, 3000, 1011) which is part of a

3001		4100		Sum over the 7 pairs		Lattice sum		Pair
I	II	I	II	I	II	I	II	Basis set
-1.48	-2.24	1.57	1.21	-18.95	-20.73	-19.61	-21.25	ε_{elst}
-1.41	-2.08	1.57	1.22	-16.17	-15.70	-16.83	-16.22	ε_{mult}
0.61	1.08	0.08	0.09	15.68	20.43	15.68	20.43	ε_{exch}
-0.16	-0.26	-0.08	-0.11	- 2.29	- 3.12	- 2.31	- 3.16	ε_{ind}
-0.02	-0.05	-0.01	-0.02	- 0.15	- 0.33	- 0.17	- 0.37	ε_{ind-6}
0.15	0.06	0.06	0.16	- 4.55	- 5.78	- 4.55	- 5.78	ε_{CT}
-0.30	-0.74	-0.16	-0.29	- 2.27	- 5.01	- 2.42	- 5.46	ε_{disp}
-0.24	-0.68	-0.11	-0.17	- 1.38	- 3.02	- 1.53	- 3.47	ε_{disp-6}
-1.18	-2.10	1.47	1.06	-12.38	-14.21	-13.21	-15.22	ε
	-1.2		0.3		-11.7		-12.1	$\varepsilon_{mult,corr}$
	-1.3		-0.5		- 9.0		- 9.8	$\varepsilon_{disp,corr}$
	-1.8		-0.1		-14.1		-15.4	ε_{corr}

hydrogen-bonded chain in the crystal. The total three-body energy of the crystal was estimated to be about −0.25 kcal/mol. More rigorous ab initio calculations for water trimers were made by *Clementi* et al. [2.45]and yielded somewhat higher values for the three-body energy (~1 kcal/mol).

Thus, the systematic errors in the lattice energy due to the pair approximation were very likely to be within 1–2 kcal/mol. Hence, they were not as important as the errors involved in the calculation of the pair interaction energies.

Smit et al. [2.21]illustrated how the SCF-LCAO-MO approach can be used to study intermolecular forces in organic crystals. It is clearly seen that some of the energy terms are extremely sensitive to changes in the basis set. Although the total potential and the total lattice energy do not exhibit marked sensitivity, there is no guarantee that the mutual compensation of errors in the individual energy terms will always occur. The reported estimates of the error [2.21](8 kcal/mol for the dimerization energy, 5 kcal/mol for the lattice energy and 1–2 kcal/mol due to the pair approximation) speak for themselves; it is hardly possible to hope that the method employed will be useful for large systems.

It seems that the widespread use of the SCF-LCAO-MO approach in the studies of intermolecular forces mainly stems from the fact that the SCF-LCAO-MO computer programs are comparatively simple and easily adaptable to the calculations of interaction energies. In our opinion, the methods to be discussed below have a number of conceptual advantages, although the corresponding computer programs are more complicated and do not possess much universality.

2.1.4 Valence Bond Method

We now turn to an alternative approach to the calculation of the interaction energy, which still remains within the framework of the supermolecule method. This approach is usually referred to as the Heitler-London (HL) or valence bond (VB) method.

In the simplest HL theory the wave function of the system is taken to be the antisymmetrized product of the ground-state functions of the isolated molecules, ψ_0^A and ψ_0^B,

$$\psi_{\mathrm{HL}} = A\psi_0^A\psi_0^B . \tag{2.37}$$

Comparing with the MO approximation we notice that ψ_{HL} is constructed from functions localized on the individual molecules, not from delocalized "supermolecular" orbitals. In practice, the functions ψ_0^A and ψ_0^B are usually obtained from SCF-LCAO-MO calculations on the isolated molecules. The total HL interaction energy is given by the relation [2.1]

$$\begin{aligned}\varepsilon_{\mathrm{HL}} = &\langle\psi_{\mathrm{HL}}|H|\psi_{\mathrm{HL}}\rangle/\langle\psi_{\mathrm{HL}}|\psi_{\mathrm{HL}}\rangle \\ &- \langle\psi_0^A|H^A|\psi_0^A\rangle - \langle\psi_0^B|H^B|\psi_0^B\rangle,\end{aligned} \tag{2.38}$$

where H^A and H^B denote the Hamiltonians for the molecules A and B. If ψ_0^A and ψ_0^B are the eigenfunctions of H^A and H^B, (2.38) may be rewritten as

$$\varepsilon_{\mathrm{HL}} = \langle\psi_{\mathrm{HL}}|H|\psi_{\mathrm{HL}}\rangle / \langle\psi_{\mathrm{HL}}|\psi_{\mathrm{HL}}\rangle - E_0^A - E_0^B, \tag{2.39}$$

where E_0^A and E_0^B are the ground-state energies of molecules A and B.

Inasmuch as ψ_{HL} does not involve any excited molecular states, it will obviously fail to describe polarization effects responsible for the attractive long-range van der Waals interaction. Thus, for the triplet state of two hydrogen atoms the HL wave function,

$$\psi_{\mathrm{HL}} = A[1s_A(1)\alpha(1)1s_B(2)\alpha(2)], \tag{2.40}$$

yields positive interaction energies at all separations [2.1]. Clearly, to take into account polarization effects, the excited molecular states, ψ_i^A and ψ_i^B $(i \neq 0)$, must be also included in the total wave function. This can be done in complete analogy to the CI method by writing the total wave function in the following form

$$\psi = \sum_{\kappa} c_\kappa A \psi_\kappa^A \psi_\kappa^B, \tag{2.41}$$

where κ denotes a particular set of molecular wave functions chosen from a set of basis functions $\{\psi_0^A, \psi_1^A, \ldots; \psi_0^B, \psi_1^B, \ldots\}$. The individual configurations in (2.41), each labeled by a unique value of κ, are usually referred to as VB structures. The coefficients c_κ are found using the variational principle, see (2.5). The resulting equations for c_κ are

$$\sum_{\kappa} H_{\kappa\lambda} c_\kappa = E \sum_{\kappa} S_{\kappa\lambda} c_\kappa, \quad \text{where} \tag{2.42}$$

$$H_{\kappa\lambda} = \langle A\psi_\kappa^A \psi_\kappa^B | H | A\psi_\lambda^A \psi_\lambda^B \rangle, \tag{2.43}$$

$$S_{\kappa\lambda} = \langle A\psi_\kappa^A \psi_\kappa^B | A\psi_\lambda^A \psi_\lambda^B \rangle. \tag{2.44}$$

In practice, two sets of one-electron orbitals, for example, SCF molecular orbitals, are available. The problem is the construction of a total wave function that corresponds to a desired value of the total spin and is antisymmetric. This problem is not at all trivial and a number of methods have been suggested to resolve it. Most of the methods are based on group theory and involve a rather complicated mathematical technique. We shall not discuss these methods here and refer the reader to [2.46,47]. The only point we would like to note here is that it is mainly the difficulties involved in the construction and manipulation of many-electron eigenfunctions of the total spin operator, which prohibit the widespread use of the VB method in the studies of intermolecular forces.

Another problem involved in the use of VB wave function (2.41) lies in the non-orthogonality of the molecular wave functions ψ^A and ψ^B. This non-orthogonality leads to a considerable increase in the number of non-zero matrix elements in (2.43,44) and a corresponding increase in the computational work needed to solve the secular equations in (2.42). To overcome this difficulty, various orthogonalization procedures have been suggested to transform the initial (non-orthogonal) orbitals into orthogonal ones. Details of these procedures may be found in [2.48–51].

In a sense, the VB method may be considered as a compromise of the supermolecule method and perturbation theory. On the one hand, the interaction energy is not computed explicitly but is considered to be the difference between the total energy of the system and the sum of the energies of the individual molecules. In this respect the VB method may be indeed regarded as a version of the supermolecule method. On the other hand, however, VB calculations use the unperturbed wave functions of the isolated molecules. Hence, the VB method has something in common with perturbation methods as well. Moreover, it can be shown [2.52]that the VB interaction potential can be decomposed into first- and second-order terms which approach the standard first- and second-order perturbation energies as the intermolecular distance goes to infinity.

Despite the mathematical difficulties involved, the VB method seems to be a very promising means of studying intermolecular forces. The fact that the VB method is actually based upon a localized orbital description (with respect to the interacting molecules), allows an effective separation to be made between intra- and inter-molecular effects [2.1]. As a result, the method enables one to reproduce the van der Waals attraction using a comparatively small number of VB structures, provided that the choice of these structures is judicious.

To illustrate this point, we refer the reader to the paper by *Wormer* et al. [2.53]. It was shown that a considerable fraction of the van der Waals energy between two helium atoms (~35% of the potential well) could be obtained using only two VB structures, i.e., [$1s^A$, $1s^A$, $1s^B$, $1s^B$] and [$1s^A$, $2p_z^A$, $1s^B$, $2p_z^B$]. Addition of two more VB structures, [$1s^A$, $2p_x^A$, $1s^B$, $2p_x^B$] and [$1s^A$, $2p_y^A$, $1s^B$, $2p_y^B$], gave as much as 66% of the potential well. These results are quite demonstrative and provide weighty arguments in favor of the VB method. For a comparison, we may quote the SCF-LCAO-MO study by *Phillipson* [2.54], in which a sixty-four-term CI wave function failed to predict the correct depth of the van der Waals well.

The only application of the VB method to organic molecules has been made with the ethylene dimer [2.52].

The calculations were performed using a rather short AO basis set, (6, 3/3), contracted to [3, 2/2](split valence). This basis set was first used in the usual SCF procedure to obtain the occupied and virtual MO's of the isolated monomer. These were then used to construct the ground- and excited-state molecular wave functions $\psi_0^A, \psi_i^A, \psi_0^B, \psi_j^B$. Only singly excited states, ψ_i^A and ψ_j^B, were included in the calculations. The VB structures were generated by applying the Young operator Y to the orbital products $\psi_i^A \psi_j^B$ (Young's operator Y is essentially equivalent to an antisymmetrizer and spin-adapting operator [2.55,56]). In discussing the role of the various energy terms in the stabilization of the structures of the dimer, the total VB energy E_{VB}, was partitioned into the first- and second-order terms, namely

$$E_{\mathrm{VB},1} = \langle Y\psi_0^A\psi_0^B|H|Y\psi_0^A\psi_0^B\rangle, \tag{2.45}$$

$$E_{\mathrm{VB},2} = E_{\mathrm{VB}} - E_{\mathrm{VB},1}. \tag{2.46}$$

The first-order energy is defined as the expectation value of the Hamiltonian over the ground-state of the dimer structure. The energy lowering due to the inclusion of excited VB structures is defined as the second-order energy. Among the excited structures, it was also possible to distinguish between singly-excited structures, $Y(\psi_i^A\psi_0^B)$ and $Y(\psi_0^A\psi_j^B)$, responsible for the induction energy, and doubly-excited structures $Y(\psi_i^A\psi_j^B)$ responsible for the dispersion energy.

The first- and second-order contributions to the interaction potential were defined as

$$\varepsilon_1 = E_{\mathrm{VB},1} - E_0^A - E_0^B, \tag{2.47}$$

$$\varepsilon_2 = E_{\mathrm{VB},2}. \tag{2.48}$$

By comparing (2.45,47) with (2.39) it can be seen that ε_1 coincides with the HL interaction energy. As the intermolecular distance tends to infinity, ε_1 and ε_2 go asymptotically over into the first- and second-order interaction energies, as given by the standard Rayleigh-Schrödinger perturbation theory.

The total potential energy and its first-order term are illustrated in Fig. 2.3 as a function of the intermolecular distance for two configurations of the dimer [2.52]. An interesting result is that the perpendicular configuration, not the parallel one, is preferred at all separations. Such a behavior of the potential cannot, in principle, be reproduced by a sum of isotropic atom-atom potentials of the Buckingham or Lennard-Jones type.

One can also conclude from Fig. 2.3 that ε_1 is comparable with or even more important than ε_2 at intermediate and large separations even though the interacting molecules possess no permanent dipoles. Using the calculated values for the multipole moments (up to hexadecupole), it was possible to estimate the electrostatic multipole part of ε_1. It turns out that ε_1 is only sufficiently well represented by $\varepsilon_{\mathrm{mult}}$ for $R > 12$ a.u., i.e., for separations much greater than the equilibrium distance for the perpendicular configuration (9.4 a.u.). It is also important to note that the leading (quadrupole-

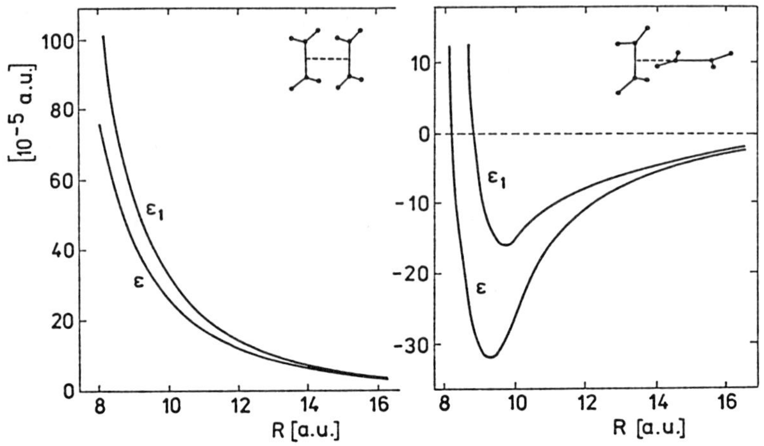

Fig. 2.3. VB interaction energy and its first-order component for two configurations of the ethylene dimer [2.52]

quadrupole) term of ε_{mult} is not sufficient to describe the behavior of ε_{mult} at large separations. Up to $R = 16$ a.u. the higher-order contributions, especially the quadrupole-hexadecupole one, are still important.

The second-order interaction energy ε_2 was decomposed into induction and dispersion components; the latter were additionally decomposed into components corresponding to a multipole expansion of ε_{disp}. It is interesting to note that ε_{ind} only gave 2–3% of ε_2 at the equilibrium distance and decreased rapidly with increasing separation. As with the electrostatic energy, the leading term of the multipole expansion of ε_{disp} (dipole-dipole dispersion) cannot adequately reproduce the behavior of ε_{disp}. Thus, at the equilibrium distance the dipole-quadrupole and the quadrupole-quadrupole terms account for approximately 25 and 10% of ε_{disp}, respectively. The latter two terms decrease with increasing intermolecular separation but remain important up to $R = 16$ a.u. (the largest separation examined).

It is to be noted that the potential curves obtained in [2.52](Fig. 2.3) agree rather poorly with the viscosity data on the molecular scattering diameter and the well depth ($\sigma = 7.7$–8.0 a.u., $d = 72.9$–65.0×10^{-5} a.u. [2.57,58]). For the perpendicular configuration of the dimer the calculated values of σ and d are 8.3 and 33.5×10^{-5} a.u., respectively i.e., the former is larger while the latter is much smaller than the observed values. If we assume the perpendicular configuration to be the most stable, any rotational averaging, as it occurs in viscosity measurements, can only result in even larger σ and smaller d values. It is true that the experimental values for σ and d were obtained in a rather approximate way by fitting an isotropic Lennard-Jones potential to the viscosity data. Nevertheless, the calculated depth of the well seems to be substantially underestimated.

In more recent studies by *Wasiutinski* et al. [2.59], *Mulder* et al. [2.60] and *Wormer* et al. [2.61] it was concluded that the AO basis set employed in [2.52] was too small. Separate calculations of the first- and second-order interaction energies using bases of polarized and doube-zeta quality have shown that the values reported for ε_1 as well as the absolute values of ε_2 were substantially underestimated [2.52]. In the intermediate range ($R = 8$–10 a.u.) the earlier results for ε_1 differed from the more recent ones by a factor of 1.5 on the average. More significant differences were observed for the long-range values of ε_2. At $R = 15$ a.u., for example, the earlier calculations [2.52]yielded $\varepsilon_2 \approx -0.006$ kcal/mol for both configurations, while the extended basis calculations gave -0.02 and -0.03 kcal/mol for the parallel and perpendicular configurations, respectively [2.60,61].

Thus, it is seen that the VB method is not free from the same drawback as inherent to the SCF-LCAO-MO method – the dependence of the interaction energy and its components on changes in the basis set. In [2.61]it was concluded that an adequate description of the van der Waals forces within the framework of the VB method necessitates the inclusion of polarization functions in the basis set and the use of excited monomer orbitals specially optimized for a calculation of the long-range interaction. For large molecules, however, such an optimization still remains impracticable and one has to be content with various approximations [2.61]to assess the quality of a basis set.

2.2 Perturbation Methods and Simplified Equations for the Interaction Energy

2.2.1 Exchange Perturbation Theories

Let us represent the total Hamiltonian of two interacting molecules, A and B, as

$$H = H^A + H^B + V, \tag{2.49}$$

where H^A and H^B are the Hamiltonians of A and B, and V is the interaction Hamiltonian defined by

$$V = \sum_{\alpha\beta} Z_\alpha Z_\beta / r_{\alpha\beta} - \sum_{\alpha b} Z_\alpha / r_{\alpha b} - \sum_{\beta a} Z_\beta / r_{a\beta} + \sum_{ab} 1/r_{ab}. \tag{2.50}$$

The subscripts a, α and b, β refer to the electrons and nuclei in the molecules A and B, respectively. Clearly, if ψ_i^A and ψ_j^B are eigenfunctions of H^A and H^B, then the products $\psi_j^A \psi_j^B$ will be eigenfunctions of $H_0 \equiv H^A + H^B$. In the standard Rayleigh-Schrödinger (RS) perturbation theory the perturbed wavefunction is expanded in terms of the products $\psi_i^A \psi_j^B$. H_0 is considered to be an unperturbed Hamiltonian and V a small perturbation.

The interaction energy through second order in V is given by

$$\varepsilon_{\mathrm{RS}} = \langle\psi_0^A\psi_0^B|V|\psi_0^A\psi_0^B\rangle + \sum_{i>0}\frac{|\langle\psi_0^A\psi_0^B|V|\psi_i^A\psi_0^B\rangle|^2}{E_i^A - E_0^A} + \sum_{j>0}\frac{|\langle\psi_0^A\psi_0^B|V|\psi_0^A\psi_j^B\rangle|^2}{E_j^B - E_0^B} + \sum_{i>0}\sum_{j>0}\frac{|\langle\psi_0^A\psi_0^B|V|\psi_i^A\psi_j^B\rangle|^2}{E_i^A + E_j^B - E_0^A - E_0^B}. \tag{2.51}$$

The first term in (2.51), appearing in first-order perturbation, is nothing but the electrostatic energy

$$\varepsilon_{\mathrm{RS},1} = \varepsilon_{\mathrm{elst}}. \tag{2.52}$$

The second-order terms represent the polarization energy which, in turn, can be divided into induction and dispersion energies

$$\varepsilon_{\mathrm{RS},2} = \varepsilon_{\mathrm{ind}} + \varepsilon_{\mathrm{disp}}. \tag{2.53}$$

At intermediate separations, where the overlap between ψ_i^A and ψ_j^B cannot be neglected, (2.51) becomes inadequate. The origin of this inadequacy is usually attributed to the fact that the simple products $\psi_i^A\psi_j^B$ are not antisymmetric with respect to intermolecular permutations of the electrons and, hence, do not provide a good basis for the expansion of the exact wave function. A more convincing argument against the RS treatment has been put forward by *Claverie* [2.62]. He has shown that the perturbed wave function expanded in terms of the products $\psi_i^A\psi_j^B$ converges towards a wave function which does not represent the physical ground state of AB, but some mathematical ground state with an incorrect symmetry. The success of RS theory at large separations results from the fact that the difference between the physical and mathematical ground states rapidly decreases with increasing distance.

At first sight, the difficulties encountered in the perturbation treatment of intermediate-range interactions could easily be surmounted by expanding the perturbed wave function in terms of the antisymmetric products $A\psi_i^B\psi_j^B$. This is not, however, convenient because the zero-order wave function $A\psi_0^A\psi_0^B$ would not be an eigenfunction of H_0 which we would like to use as a zero-order Hamiltonian. Besides, the functions $A\psi_i^A\psi_j^B$ are nonorthogonal and, what is especially important, are linearly dependent. Since an expansion over a set of linearly dependent functions is not unique, a number of different ways to construct perturbation theory based on the functions $A\psi_i^A\psi_j^B$ should be possible.

In the last twenty years, a large variety of perturbation theories applicable to intermediate-range interactions have been suggested. All of these theories, usually referred to as exchange perturbation theories, may be divided into "symmetric" and "non-symmetric" theories [2.63].

In a symmetric theory the total Hamiltonian is arbitrarily divided into an unperturbed Hamiltonian H_s ($H_s \neq H_0$) and a perturbation V. The only requirement is that the zero-order wave function $A\psi_0^A\psi_0^B$ be an eigenfunction of H_s (i.e., the zero-order Hamiltonian and the wave function have the correct symmetry properties from the very beginning). Thereafter, the ordinary RS perturbation theory is applied and certain additional requirements are imposed upon H_s to define its final form. Actually, the various symmetric theories suggested so far differ from one another by the choice of H_s [2.62,64,65].

A non-symmetric approach in which the zero-order Hamiltonian does not exhibit the correct symmetry and is usually taken as $H_0 = H^A + H^B$, is used more often. It is possible to discuss most of the non-symmetric theories within the framework of a single perturbation formalism. This has been done by *Chipman* et al. [2.66]and, more recently, by *Jeziorski* and *Kolos* [2.67]. Let us write down the exact wave function for a pair of interacting molecules as [2.66]

$$\psi = {}^{\nu}\!A\xi, \tag{2.54}$$

where ${}^{\nu}\!A$ is a projection operator which imposes the proper symmetry upon the primitive function ξ (ν labels the particular irreducible representation to which ψ must belong). The equation which ξ must satisfy is

$$(H - E)\psi = (H - E)\,{}^{\nu}\!A\xi = {}^{\nu}\!A(H - E)\xi = 0, \tag{2.55}$$

where we have made use of the fact that ${}^{\nu}\!A$ commutes with H. It can easily be shown that (2.55) does not define ξ uniquely and can be rewritten as

$$(H - E)\xi = (1 - {}^{\nu}\!A)\kappa, \tag{2.56}$$

where κ is an arbitrary function.

By applying perturbation theory to the wave function in (2.54) one obtains [2.66],

$${}^{\nu}\!AH_0\xi_0 = E_0\,{}^{\nu}\!A\xi_0, \tag{2.57}$$

$${}^{\nu}\!A(H_0 - E_0)\xi_1 = {}^{\nu}\!A(\varepsilon_1 - V)\xi_0, \tag{2.58}$$

$${}^{\nu}\!A(H_0 - E_0)\xi_k = {}^{\nu}\!A(\varepsilon_1 - V)\xi_{k-1} + \sum_{i=2}^{k} {}^{\nu}\!A\varepsilon_i\xi_{k-i}, \quad k \geq 2. \tag{2.59}$$

Amos [2.68]has shown that the various non-symmetric treatments can be derived by suppressing ${}^{\nu}\!A$ in various positions in (2.57–59). The latter procedure implicitly puts further constraints upon the function ξ, which ensures its unique definition. Thus, the familiar MS-MA theory suggested by *Murrell* and *Shaw* [2.69], and *Musher* and *Amos* [2.70]can be derived by removing ${}^{\nu}\!A$ from the left-hand side of (2.58) and everywhere in (2.59). This is equivalent to writing the following additional equation for ξ [2.68],

$$(1 - {}^{\nu}\!A)(H - E)(\xi - \xi_0) = 0. \tag{2.60}$$

Another general formalism allowing one to derive many of the non-symmetric theories has been suggested by *Jeziorsky* and *Kolos* [2.67]. In it the Schrödinger equation,

$$(H_0 + V)\psi = (E_0 + \varepsilon)\psi, \tag{2.61}$$

is replaced by the equivalent equations,

$$\varepsilon = \langle\phi_0|V|\psi\rangle, \tag{2.62}$$

$$\psi = \phi_0 + R_0(\varepsilon - V)\psi, \tag{2.63}$$

where ϕ_0 is the ground-state eigenfunction of H_0, subject to the normalization condition,

$$\langle\phi_0|\psi\rangle = 1, \tag{2.64}$$

and R_0 is the reduced ground-state resolvent of H_0,

$$R_0 = \sum_{k\neq 0} \frac{|\phi_k\rangle\langle\phi_k|}{E_0 - E_k}. \tag{2.65}$$

(For the sake of brevity, we have used the simplified notations $\phi_0 = \psi_0^A\psi_0^B$, $\phi_k = \psi_i^A\psi_j^B$ and $E_k = E_i^A + E_j^B$.) Equations (2.62,63) can be solved by an iteration procedure of the general form

$$\varepsilon_k = \langle\phi_0|VG|\psi_{k-1}\rangle, \tag{2.66}$$

$$\psi_k = \phi_0 + R_0(\varepsilon_k - V)F\psi_{k-1}, \tag{2.67}$$

in which the operators G and F are intended to accelerate the convergence of the iterative process by forcing the proper symmetry on the wave function at each step. Actually, the various perturbation treatments are different versions of (2.66,67). They differ from one another in the choice of the operators G and F and the function ψ_0 used to initiate the iteration. (For example, the MS-MA theory is obtained by setting $F = G = 1$ and $\psi_0 = A\phi_0$.)

For subsequent discussion, there is no need to present the explicit expressions for the interaction energy, which appear in the various perturbation theories. Most of these theories (both the symmetric and non-symmetric ones) yield very similar results, at least as far as the leading contributions to the interaction energy are concerned. Thus, in many of the theories [2.62,67,69,71]the first-order perturbation energy ε_1 is given by

$$\varepsilon_1 = \frac{\langle\phi_0|VA|\phi_0\rangle}{\langle\phi_0|A|\phi_0\rangle}. \tag{2.68}$$

It can readily be shown that ε_1 is equal to the Heitler-London interaction energy given in (2.39). Indeed,

$$\begin{aligned}\varepsilon_1 &= \frac{\langle\phi_0|VA|\phi_0\rangle}{\langle\phi_0|A|\phi_0\rangle}\\ &= \frac{\langle\phi_0|(H-H_0)A|\phi_0\rangle}{\langle\phi_0|A|\phi_0\rangle}\\ &= \frac{\langle A\phi_0|H|A\phi_0\rangle}{\langle A\phi_0|A\phi_0\rangle} - E_0.\end{aligned} \tag{2.69}$$

The expressions for the second-order energy ε_2, are quite different in the various perturbation theories. It is, however, essential that in most of the theories ε_2 becomes equal to the RS polarization energy $\varepsilon_{\mathrm{RS},2}$ as the intermolecular distance approaches infinity. This property permits the introduction of the "second-order exchange energy" [2.62],

$$\varepsilon_{\mathrm{exch},2} = \varepsilon_2 - \varepsilon_{\mathrm{RS},2}, \tag{2.70}$$

which vanishes as $R \to \infty$ and increases exponentially as the interacting molecules approach one another. The calculation of ε_2 using equations found in exchange perturbation theories is a very complicated problem. Fortunately, in the region of the van der Waals minimum $\varepsilon_{\mathrm{exch},2}$ is still comparatively small. This has been verified in direct calculations on the H...H [2.72]and He...He [2.73]systems. The smallness of $\varepsilon_{\mathrm{exch},2}$ at intermediate distances has also been verified by *Jeziorsky* and *van Hemmert* [2.10]for the water dimer.

Thus, it seems that a sufficiently good approximation to the interaction energy in the intermediate range can be obtained by setting [2.62,74]

$$\varepsilon \approx \varepsilon_{\mathrm{HL}} + \varepsilon_{\mathrm{RS},2}. \tag{2.71}$$

In this approximation the interaction energy can also be represented by

$$\varepsilon \approx \varepsilon_{\mathrm{exch}} + \varepsilon_{\mathrm{RS}}, \tag{2.72}$$

where $\varepsilon_{\mathrm{exch}}$ is the first-order exchange energy, and $\varepsilon_{\mathrm{RS}}$ is the RS interaction energy (through second order in V) see (2.51). In order to prove this and derive an explicit expression for ε_{exch}, let us represent the total antisymmetrizer A as

$$A = A_{int} A^A A^B, \tag{2.73}$$

where A_{int} only involves the intermolecular permutations and A^A and A^B are the permutations within A and B, respectively [2.62]. Considering that

$$A^A \psi_0^A = N_A! \psi_0^A, \tag{2.74}$$

$$A^B \psi_0^B = N_B! \psi_0^B, \tag{2.75}$$

(N_A and N_B represent the number of electrons in A and B), one can rewrite (2.68)

$$\begin{aligned} \varepsilon_1 &= \varepsilon_{HL} \\ &= \frac{\langle \phi_0 | V A_{int} A^A A^B | \phi_0 \rangle}{\langle \phi_0 | A_{int} A^A A^B | \phi_o \rangle} \\ &= \frac{\langle \phi_0 | V A_{int} | \phi_0 \rangle}{\langle \phi_0 | A_{int} A^A A^B | \phi_0 \rangle} . \end{aligned} \tag{2.76}$$

It is now convenient to represent A_{int} as

$$A_{int} = \sum_{\lambda} (-1)^\lambda P_\lambda = P_0 + \sum_{\lambda \geq 1} (-1)^\lambda P_\lambda = 1 - A'_{int}, \tag{2.77}$$

where A'_{int} denotes the exchange part of A_{int}, i.e.,

$$A'_{int} = P_1 - P_2 + P_3 - \dots \ . \tag{2.78}$$

With these definitions, (2.76) can be rewritten as

$$\begin{aligned} \varepsilon_{HL} &= \frac{\langle \phi_0 | V | \phi_0 \rangle - \langle \phi_0 | V A'_{int} | \phi_0 \rangle}{1 - \langle \phi_0 | A'_{int} | \phi_0 \rangle} \\ &= \varepsilon_{RS,1} + \frac{\langle \phi_0 | V | \phi_0 \rangle \langle \phi_0 | A'_{int} | \phi_0 \rangle - \langle \phi_0 | V A'_{int} | \phi_0 \rangle}{1 - \langle \phi_0 | A'_{int} | \phi_0 \rangle} . \end{aligned} \tag{2.79}$$

The second term in the last line is equal to ε_{exch} in (2.72).

Finally, the intermolecular potential in (2.71,72) can be divided into a sum of four terms, each having a distinct physical meaning;

$$\varepsilon = \varepsilon_{exch} + \varepsilon_{elst} + \varepsilon_{ind} + \varepsilon_{disp}, \tag{2.80}$$

where the last three terms have been defined in (2.51).

We now turn to a discussion of the various contributions to the interaction energy. In this discussion we shall mainly be interested in the possibility of deriving simplified equations for ε, which can be applied to large organic molecules at separations important for their cystals. For the most part, the discussion will follow the papers by *Claverie* [2.62]and *Amos* and *Crispin* [2.74], which seem to be the most valuable contributions to this field.

2.2.2 Electrostatic Energy

It is well known that at intermolecular distances much larger than the dimensions of the interacting molecules, the classical electrostatic interaction energy can be represented by a multipole expansion based on an expansion of r^{-1} as an inverse power series in R, see (2.50). The multipole expansion of the interaction potential V has the form

$$V = \sum_{l_A,l_B=0}^{\infty} C^{\text{elst}}_{l_A+l_B+1} R^{-(l_A+l_B+1)}, \tag{2.81}$$

where the $C^{\text{elst}}_{l_A+l_B+1}$ only depend on the relative orientation of the interacting molecules. Expressed in terms of Cartesian tensors [2.75],

$$C^{\text{elst}}_{l_A+l_B+1} = M^{[l_A]}_{(t)} T^{[l_A+l_B]}_{(t);(u)} M^{[l_B]}_{(u)}, \tag{2.82}$$

where $(t) = t_1 t_2 \ldots t_{l_A}$ and $(u) = u_1 u_2 \ldots u_{l_B}$. The summation over repeated indices is implicit. The components of the multipole moment of order l are defined by

$$M^{[l]}_{(t)} = \sum_i q_i r_{t_1}(i) r_{t_2}(i) \ldots r_{t_l}(i), \tag{2.83}$$

where the summation is carried out over all particles in the molecule with charges q_i; and $r_{t_k}(i)$ is the t_kth coordinate of particle i ($t_k = 1,2$ or 3, $r_{t_k} = x, y$ or z). The three lowest-order multipoles are given by

$$M^{[0]} = \sum_i q_i = q$$

(the total molecular charge), (2.84)

$$M^{[1]}_t = \sum_i q_i r_t(i) = \mu_t$$

(the components of the dipole moment), (2.85)

$$M^{[2]}_{tu} = \sum_i q_i r_t(i) r_u(i) = Q_{tu}$$

(the components of the quadrupole moment). (2.86)

The tensor $T^{[l_A+l_B]}$, whose components appear in (2.82) and which is usually referred to as the interaction tensor, takes into accout the relative orientation of the charge distributions. Its components are defined by

$$T^{[l_A+l_B]}_{(t);(u)} = R^{(l_A+l_B+1)} \frac{(-1)^{l_A}}{l_A! l_B!} (\nabla_{t_1} \ldots \nabla_{t_{l_A}} \nabla_{u_1} \ldots \nabla_{u_{l_B}})(1/R). \quad (2.87)$$

It is worthwhile to write down the explicit expressions for the components of a few lowest-order interaction tensors:

$$T^{[0]} = 1, \quad (2.88)$$

$$T^{[1]}_t = R_t/R, \quad (2.89)$$

$$T^{[2]}_{tu} = (\delta_{tu} - 3R_t R_u/R^2). \quad (2.90)$$

Using (2.90,85) we can now write an expression for the electrostatic dipole-dipole term,

$$\begin{aligned} \varepsilon_{\text{dip-dip}} &= \sum_{tu} R^{-3} \mu^A_t T^{[2]}_{tu} \mu^B_u \\ &= \sum_{tu} \mu^A_t (\delta_{tu} - 3R_t R_u/R^2) \mu^B_u \\ &= R^{-3}[(\boldsymbol{\mu}^A \boldsymbol{\mu}^B) - 3(\boldsymbol{\mu}^A \boldsymbol{R})(\boldsymbol{\mu}^B \boldsymbol{R}) R^{-2}]. \end{aligned} \quad (2.91)$$

It coincides with the usual expression for the interaction of two dipoles.

Instead of using the Cartesian coordinates $r_t(i)$, one can specify the positions of the charges q_i using the spherical polar coordinates r_i, θ_i and φ_i. In this case the multipole moments are defined in a more compact form [2.41],

$$Q^m_l = \sum_j q_j r^l_j P^{|m|}_l(\cos\theta_j) \exp(im\varphi_j), \quad m = -l, \ldots, l, \quad (2.92)$$

in which the P^m_l's are the associated Legendre polynomials. The coefficients C^{elst} in (2.81) are now given by

$$C^{\text{elst}}_{l_A l_B} = \sum_{m=-\lambda}^{\lambda} Q^{m^*}_{l_A} Q^m_{l_B} \frac{(-1)^{l_B+m}(l_A + l_B)!}{(l_A + |m|)!(l_B + |m|)!}, \quad (2.93)$$

where λ is the smaller of l_A and l_B. It is essential that the multipole moments in (2.93) are computed in the local molecular coordinate systems $\{X, Y, Z\}$ whose Z axes coincide and whose X and Y axes are parallel. The desired moments Q^m_l can be obtained from the moments $\tilde{Q}^m_l$ calculated in an arbitrary coordinate system $\{\tilde{X}, \tilde{Y}, \tilde{Z}\}$ by the transformation [2.41],

$$Q_l^m = \sum_{m'=-l}^{l} \mathrm{i}^{|m'|-m'-|m|+m} \left(\frac{(l+|m|)!(l-|m'|)!}{(l-|m|)!(l+|m'|)!} \right)^{1/2}$$
$$\times D^l(S)_{m'm} \tilde{Q}_l^{m'}, \qquad (2.94)$$

where S is the rotation matrix that transforms $\{X, Y, Z\}$ to $\{\tilde{X}, \tilde{Y}, \tilde{Z}\}$. The coefficients $D^l(S)_{m'm}$ are the representation coefficients of the three-dimensional rotation group for the rotation S.

There are a number of other ways in which the multipole expansion for V can be written [2.1,58,76].

The quantum-mechanical analogues of the above equations are readily obtained by substituting the expanded interaction operator V into the expression for the electrostatic energy

$$\varepsilon_{\text{elst}} = \langle \phi_0 | V | \phi_0 \rangle. \qquad (2.95)$$

The resulting equations have exactly the same form as (2.81,82,93), except that the multipole moments are replaced by their expectation values $\langle \psi_0 | M_{(t)}^{[l]} | \psi_0 \rangle$ or $\langle \psi_0 | Q_l^m | \psi_0 \rangle$.

The usefulness of the multipole expansion is quite obvious. While the exact matrix element given in (2.95) has to be recalculated for each new configuration of the interacting molecules, the multipole moments have to be computed only once and then they are used as fixed quantities in pure algebraic calculations.

Generally, the estimates of $\varepsilon_{\text{elst}}$ using (2.81,82,93) suffer from two kinds of errors. The first kind is produced by imperfections in the molecular wavefunctions ψ_0^A and ψ_0^B used to calculate the multipole moments. These errors can readily be observed by comparing the multipole moments computed using different basis sets. For example, presented in Table 2.4 are the dipole moment and the components of the quadrupole moment of formic acid calculated by *Smit* et al. [2.41]using the four Bases (I to IV) specified in Sect. 2.1. Also given, for a comparison, are the corresponding observed quantities. The angle α specifies the orientation of the dipole moment. (Since we are only

Table 2.4. Components of the dipole and quadrupole moment of formic acid [2.41]

	Dipole moment [D,deg]		Cartesian components of quadrupole [a.u.]			
Basis	p	α	Q_{xx}	Q_{yy}	Q_{zz}	Q_{xy}
I	1.08	8	-6.2	5.1	1.1	-3.9
II	1.29	14	-5.8	4.7	1.1	-3.8
III	1.48	21	-6.3	5.3	1.0	-4.0
IV	1.63	21	-5.3	4.8	0.5	-3.5
Observed	1.41	11	-3.9	3.9	0.1	

interested in variations in α, we do not give the exact definition for α.) It can be seen that the calculated quantities are extremely sensitive to changes in the basis set and are all in rather poor agreement with the experimental values. Considering that $\varepsilon_{\text{elst}}$ is a quadratic function of the multipole moments, it should exhibit an even worse agreement.

The experience gained with small systems indicates that a reliable calculation of multipole moments calls for bases that are especially flexible in the outer molecular regions [2.59,60]. The desired flexibility can be achieved, for instance, by varying all of the nonlinear parameters (AO exponents) of the basis set including those of the atomic polarization functions. To assess the quality of a basis set, various criteria are now in use. One such criterion is the ratio STM/CM, where STM is the sum over the transition moments

$$\text{STM} = \sum_{k>0} \langle\psi_0|Q_l^m|\psi_k\rangle\langle\psi_k|Q_{l'}^{m'}|\psi_0\rangle, \tag{2.96}$$

and CM is the closure moment

$$\text{CM} = \langle\psi_0|Q_l^m Q_{l'}^{m'}|\psi_0\rangle - \langle\psi_0|Q_l^m|\psi_0\rangle\langle\psi_0|Q_{l'}^{m'}|\psi_0\rangle. \tag{2.97}$$

If the basis set $\{\psi_k\}$ were complete, the STM would obviously be equal to the CM. This follows from the closure relation

$$\sum_{k>0} |\psi_k\rangle\langle\psi_k| = \sum_{k\geq 0} |\psi_i\rangle\langle\psi_i| - |\psi_0\rangle\langle\psi_0| = 1 - |\psi_0\rangle\langle\psi_0|. \tag{2.98}$$

In practice, STM/CM is always less than unity and increases as the basis set becomes more complete.

The second source of error is the non-zero overlap between the charge distributions of the interacting molecules. This overlap always takes place because the molecular wave functions have exponential tails. The penetration of the charge distributions usually becomes important at distances where the overlap integral reaches values of the order of 10^{-2} [2.21]. For large molecules the multipole expansion becomes inadequate at even larger separations, i.e., as soon as the intermolecular distance becomes comparable to the dimensions of the interacting molecules. Let us suppose that $\boldsymbol{r}_A$ and $\boldsymbol{r}_B$ are the position vectors of a charge in molecules A and B, respectively (both vectors are defined in local molecular coordinate systems). As has already been mentioned, the multipole expansion of V is based on an expansion of r^{-1} as an inverse power series in R. The condition that guarantees that the series will converge is [2.74]:

$$|\boldsymbol{r}_A + \boldsymbol{r}_B| < R. \tag{2.99}$$

The existence of this restrictive condition suggests that the multipole expansion may diverge even in the ideal case, i.e., when the charge distributions do not overlap.

The above considerations show that the term "penetration energy", frequently used for the difference

$$\varepsilon_{\text{pen}} = \varepsilon_{\text{elst}} - \varepsilon_{\text{mult}}, \tag{2.100}$$

is not quite correct. Indeed, at intermediate distances ε_{pen} may also involve errors produced by the non-fulfillment of the condition given in (2.99). Besides, ε_{pen} always contains errors that result from a truncation of the multipole expansion.

The most reliable data presently available show that the multipole approximation is very inaccurate in the region of the potential well and that it is even meaningless for large molecules. To verify this, the reader may analyze Table 2.2 in which $\varepsilon_{\text{elst}}$ and $\varepsilon_{\text{mult}}$ are compared for the formic acid dimer [2.21]. Another example is provided by the unpublished data of *Wormer* in [2.77]. For the perpendicular configuration of the ethylene dimer at $R = 8$ a.u. *Wormer* reported that $\varepsilon_{\text{mult}} = -26.11 \times 10^{-5}$a.u., whereas $\varepsilon_{\text{elst}} = -98.32 \times 10^{-5}$a.u. The difference between the expanded and unexpanded energies decreased at $R = 10$ a.u. but could still not be considered negligible (~10% of the exact value).

It is important that even at distances where the penetration effects are negligible and the condition given in (2.99) is fulfilled the leading term of the multipole expansion is frequently insufficient to accurately reproduce $\varepsilon_{\text{elst}}$. Two examples illustrating this point are given in Fig. 2.4 and Table 2.5.

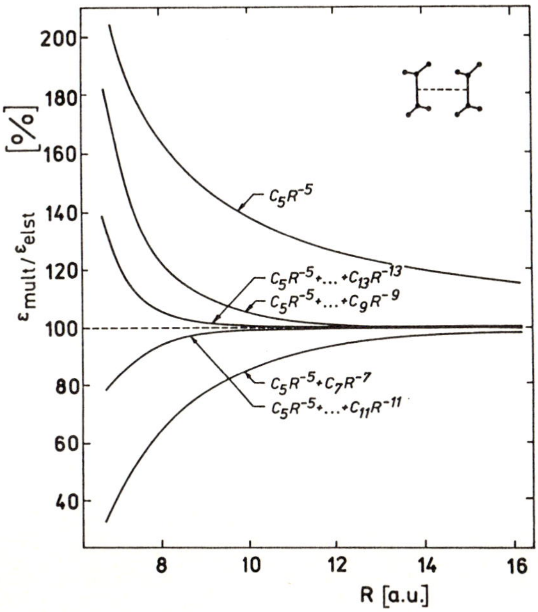

Fig. 2.4. The ratio $\varepsilon_{\text{mult}}/\varepsilon_{\text{elst}}$ as a function of the intermolecular distance for various lengths of the multipole expansion in the ethylene dimer [2.78]

Table 2.5. The energy contributions [kcal/mol]per multipole order for several summation spheres in the formic-acid crystal [2.41]

Summation radius [Å]	Multipole order						Multipole sum
	1	2	3	4	5	6	
0 <r ≤3.64	-0.13	-12.94	-4.27	-2.85	0.83	-1.05	-20.41
3.64 <r ≤6	0.10	0.55	-0.36	-0.15	0.00	-0.04	0.10
6 <r ≤10	-0.27	- 0.15	-0.20	-0.06	0.02	0.01	- 0.54
10 <r ≤15	-0.01	- 0.06	0.01	0.00	0.00	0.00	0.05
15 <r ≤20	0.05	- 0.05	0.00	0.00	0.00	0.00	0.00
20 <r ≤25	0.02	0.00	0.00	0.00	0.00	0.00	0.02
25 <r ≤35	-0.02	0.00	0.00	0.00	0.00	0.00	- 0.02
35 <r <∞	-0.38	0.00	0.00	0.00	0.00	0.00	- 0.38
Total							-21.17

Figure 2.4 shows the dependence of the ratio $\varepsilon_{\text{mult}}/\varepsilon_{\text{elst}}$ on the distance for the parallel configuration of the ethylene dimer [2.78]. The various curves in Fig. 2.4 correspond to multipole expansions of various lengths. It is clearly seen that the leading term of the expansion does not provide a reliable estimate of $\varepsilon_{\text{elst}}$ even at large distances.

Table 2.5 lists the contributions to the lattice energy of the formic acid crystal made by the various multipoles (up to 2^6-pole) [2.41]. The values in the first row ($R<3.64$ Å) are meaningless because of the strong penetration effects (Table 2.2). As seen from Table 2.5, the higher multipoles are very important at intermediate distances. This is especially true of molecules lying within a spherical layer for which 3.64 Å$<R\leq$6 Å. In this case, as much as 40% of the total electrostatic energy is accounted for by such "exotic" multipole moments as 2^6-poles. The contribution from higher multipoles continues to be important in the following layers. Thus, the contribution of hexadecupoles is almost half of that of quadrupoles.

For further examples illustrating the importance of higher multipole terms at intermediate and large distances the reader is referred to [2.79], investigating azabenzene molecules. The only result we would like to quote here is that the maximum electrostatic attraction for pyrimidine was observed at a configuration which had almost nothing in common with the one corresponding to a maximum dipole-dipole attraction. This result shows that a consideration of the leading multipole term alone can not only give rise to quantitative but even qualitative erroneous conclusions.

Thus, the evaluation of $\varepsilon_{\text{elst}}$ within the framework of the multipole approximation is a rather hazardous procedure. In following this approximation, one must take extreme care that the basis set is flexible enough to reproduce the molecular multipole moments and that the multipole series is carried to convergence. At present, only a few calculations have been made in which these requirements are satisfied. These have been carried out by *Wormer* et al. [2.60,61]for the ethylene dimer, by *Mulder* et al. [2.79,77]for the dimers of azabenzenes and the crystals of ehtylene and pyrazine, and by *Smit* et al. [2.41]for the formic acid crystal.

Table 2.6. Contributions to the long-range interaction energy [%]for several azabenzenes at $R = 15$ a.u. (the energies correspond to configurations of the dimers in which the total long-range attraction is a maximum) [2.79]

	Ben-zene	Pyri-dine	Pyri-dazine	Pyri-midine	Pyra-zine	Tri-azine	Tetra-zine
ε_{elst}	35	73	84	74	70	66	76
ε_{disp}	65	23	11	24	27	33	21
ε_{ind}	1	4	5	2	3	1	3

In all of these papers, except the last one, the authors also estimated the second-order long-range energy $\varepsilon_{RS,2}$ (using the multipole expansion). Details of the calculation of $\varepsilon_{RS,2}$ will be discussed in the following subsection. For the moment we should like to focus the reader's attention to the qualitative result that in all of the cases ε_{elst} proved to be a very important, if not dominant contribution to the long-range interaction energy. Table 2.6 lists the first- and second-order contributions to the long-range interaction energy for seven azabenzenes including benzene itself. The above values are taken from [2.79]and refer to the configuration in which the total long-range attraction is the largest. It is seen that in all of the cases, except for benzene, ε_{elst} makes a dominant contribution. Most striking are the results for the last three compounds whose molecules possess no permanent dipole moment but nevertheless exhibit a strong electrostatic attraction.

Despite the obvious importance of electrostatic interactions in organic dimers, it is not clear a priori that these interactions are also as important in crystals. The fact is that the electrostatic interactions are extremely anisotropic [2.41,78,79]and are either attractive or repulsive depending on the orientation of the interacting molecules. (Unlike the dispersion energy, the electrostatic interaction energy averaged over all orientations is zero for neutral molecules.) Since each molecule in a crystal sees its neighbors in a great variety of orientations, the electrostatic interactions may partially cancel. Such is the situation in the ethylene crystal in which the electrostatic energy, as estimated by *Mulder* and *Huiszoon* [2.77], only accounts for 3% of the long-range part of the lattice energy. In the ethylene dimer ε_{elst} is of the same order of magnitude as ε_{disp} [2.52].

Of course, the results of *Mulder* and *Huiszoon* [2.77]are not very reliable because the multipole expansion that was used is hardly applicable to the interactions of the reference molecule with its nearest neighbors in the first coordination shell. In this respect, it is of interest to see whether a cancellation of the electrostatic interactions occurs only for the interactions beyond the first coordination shell, where the multipole approximation is valid. An answer to this question is readily obtained by excluding from the first- and second-order components of the lattice energy the corresponding contributions of the nearest twelve molecules. The results are -0.16×10^{-3}a.u. for the electrostatic energy and -1.4×10^{-3}a.u. for the dispersion energy. In other words, the long-range electrostatic contribution to the lattice energy of ethylene still remains an order of magnitude smaller than the long-range dispersion contribution.

A similar calculation for the pyrazine crystal [2.77] yields -0.22×10^{-3} a.u. for the long-range electrostatic contribution and -3.79×10^{-3}a.u. for the long-range dispersion contribution. That is, the electrostatic interactions only account for 6% of the long-range part of the lattice energy. This is a striking result in view of the fact that $\varepsilon_{\text{elst}}$ is a dominant contribution to the long-range interaction energy in the pyrazine dimer (Table 2.6).

A partial cancellation of the long-range electrostatic interactions even occurs in the formic acid crystal. This can easily be seen by comparing the data listed in Tables 2.2,3. Indeed, in the formic acid dimer the long-range electrostatic energy exceeds the long-range second-order energy by an order of magnitude (Table 2.2). Turning to the crystal (Table 2.3), we observe that the long-range part of the lattice energy (i.e., the sum of the electrostatic interactions of the reference molecule with all molecules outside of the first coordination sphere) becomes almost equal to the long-range part of the second-order contribution to the lattice energy. Thus, using the basis set II yields -0.52 kcal/mol for the electrostatic energy and -0.49 kcal/mol for the second-order energy.

It seems that the partial cancellation (or, in other words, the averaging) of electrostatic interactions is a phenomenon common to all organic molecular crystals. This phenomenon should obviously be more pronounced for distant coordination shells.

We now turn to the problem of calculating the electrostatic interaction energy for separations near the van der Waals minimum. One of the difficulties precluding the use of the multipole approximation is that the intermolecular distance is comparable to the dimensions of the interacting molecules. An obvious way of overcoming this difficulty is to divide each molecule into a number of small units of localized charge. For subsequent discussions it is convenient to introduce the charge-density operator

$$\varrho(r) = \varrho_{\text{nuc}}(r) + \varrho_{\text{el}}(r), \quad \text{where} \tag{2.101}$$

$$\varrho_{\text{nuc}}(r) = \sum_{\alpha} Z_\alpha \delta(r - r_\alpha), \tag{2.102}$$

$$\varrho_{\text{el}}(r) = \sum_{a} \delta(r - r_a). \tag{2.103}$$

The total charge density $f(r)$ at the point r will be given by the expectation value of $\varrho(r)$ over the ground-state wave function ψ_0 (for the sake of simplicity, we omit the superscript labeling the molecule);

$$f(r) = f_{\text{nuc}}(r) + f_{\text{el}}(r)$$

$$
\begin{aligned}
&= \langle\psi_0|\sum_\alpha Z_\alpha \delta(r - r_\alpha)|\psi_0\rangle - \langle\psi_0|\sum_{a=1}^{N}\delta(r - r_a)|\psi_0\rangle \\
&= \sum_\alpha Z_\alpha \delta(r - r_\alpha) - N\langle\psi_0|\delta(r - r_1)|\psi_0\rangle \\
&= \sum_\alpha Z_\alpha \delta(r - r_\alpha) \\
&- N\int \psi_0^*(r, r_2, \ldots, r_N)\psi_0(r, r_2, \ldots, r_N)dr_2 \ldots dr_n.
\end{aligned} \tag{2.104}
$$

With these definitions, the interaction operator may be written as

$$
V = \int\int r_{AB}^{-1} \varrho^A(r_A)\varrho^B(r_B)dr_A dr_B, \tag{2.105}
$$

and the electrostatic energy as

$$
\varepsilon_{\text{elst}} = \int\int r_{AB}^{-1} f^A(r_A) f^B(r_B)dr_A dr_B. \tag{2.106}
$$

Let us consider the case in which the molecular wave function is represented by a Slater determinant. The electronic part of the charge density is equal to

$$
f_{\text{el}}(r) = \sum_a^N |\varphi_a(r)|^2. \tag{2.107}
$$

If we are dealing with doubly occupied orbitals, the above summation over the electrons can be replaced by a summation over the occupied orbitals φ_i $(i = 1, \ldots, n)$

$$
f_{\text{el}}(r) = 2\sum_i^n |\varphi_i(r)|^2. \tag{2.108}
$$

An important property of the charge density in (2.108), as well as of the Slater determinant in (2.4), is that it is invariant to a unitary transformation $\underline{\underline{U}}$ of the orbitals among themselves [2.80]. Using this property, we may attempt to go from the canonical (delocalized) orbitals φ_i to a set of new orbitals, namely

$$
\varphi_k' = \sum_{i=1}^n U_{ki}\varphi_i, \tag{2.109}
$$

that are localized in more or less compact regions of space, e.g., around atoms or along bonds.

There are a number of criteria that can be used to measure localization. Thus, *Boys* [2.81]recommended choosing the φ_i''s so as to minimize the quantity

$$\sum_{i=1}^{n} \langle \varphi_i' | (r - R_i)^2 | \varphi_i' \rangle, \tag{2.110}$$

where $R_i = \langle \varphi_i' | r | \varphi_i' \rangle$ is the centroid of φ_i'. *Edmiston* and *Ruedenberg* [2.82]used the "self-energy" criterion

$$\sum_{i=1}^{n} \langle \varphi_i' \varphi_i' | r_{12}^{-1} | \varphi_i' \varphi_i' \rangle, \tag{2.111}$$

in which the φ_i' have been chosen so as to maximize (2.111). Another alternative is to use φ_i''s such that the charge density $|\varphi_i'(r)|^2$ has a maximum overlap with itself and a minimum overlap with the charge densities of the other orbitals. In other words, the φ_i''s are chosen to maximize

$$\sum_{i=1}^{n} \langle [\varphi_i'(r)]^2 | [\varphi_i'(r)]^2 \rangle, \tag{2.112}$$

or to minimize,

$$\sum_{i \neq j}^{n} \langle \varphi_i'(r) \varphi_j'(r) | \varphi_i'(r) \varphi_j'(r) \rangle. \tag{2.113}$$

There are also other alternative localization procedures. Most of them have been discussed in [2.80].

The diversity of localization criteria may cause some trouble. But, in fact, the localized orbitals that can be derived using different criteria prove to be similar for the majority of molecules. Generally, the localized orbitals break up into three types of orbitals associated with bonds, inner shells and lone pairs, respectively.

Dropping the prime in φ_i', we now assume that the φ_i's in (2.108) are already localized orbitals. To derive an expression for $\varepsilon_{\text{elst}}$ in terms of localized orbitals, let us assign the nuclear charges Z_α to these localized orbitals [2.74]. In principle such an assignment can be made arbitrarily, except that the contribution from Z_α to a given orbital i, $Z_{\alpha i}$, must satisfy

$$Z_\alpha = \sum_i Z_{\alpha i}. \tag{2.114}$$

We can now define the total charge density on the orbital i,

$$f_i(r) = \sum_\alpha Z_{\alpha i} \delta(r - r_\alpha) - 2|\varphi_i(r)|^2, \tag{2.115}$$

so that the total molecular charge density is given by

$$f(r) = \sum_{i=1}^{n} f_i(r). \tag{2.116}$$

It is convenient to choose the $Z_{\alpha i}$'s in the following manner [2.74]:

$$Z_{\alpha i} = \begin{cases} 2 \text{ if } \varphi_i \text{ is a lone pair or inner-shell orbital localized} \\ \quad \text{on atom } \alpha, \\ 1 \text{ if } \varphi_i \text{ is a bond orbital associated with atom } \alpha, \\ 0 \text{ otherwise.} \end{cases} \tag{2.117}$$

With these choices, the net charge of the φ_i is zero,

$$\int f_i(r)dr = 0. \tag{2.118}$$

The electrostatic interaction energy between two molecules can be now written as a sum of pairwise interactions between localized charge distributions

$$\varepsilon_{\text{elst}} = \sum_{i=1}^{n_A} \sum_{j=1}^{n_B} \varepsilon_{\text{elst},ij}, \quad \text{where} \tag{2.119}$$

$$\varepsilon_{\text{elst},ij} = \int \int r_{AB}^{-1} f_i^A(r_A) f_j^B(r_B) dr_A dr_B. \tag{2.120}$$

In computing the sum in (2.119) it is reasonable to neglect all terms which involve inner-shell charge densities. Indeed, the inner-shell charge distributions are very compact, almost spherically symmetrical, and practically compensated for by the respective nuclear contributions. Thus, the electrostatic interaction energy reduces to a sum of pairwise interactions between the bonds and the lone pairs in the molecules A and B.

The bond and lone-pair charge distributions occupy comparatively little space. At intermediate intermolecular distances they can be approximated by their lowest multipoles. In the latter case, instead of expanding about a single (molecular) centre, we obtain a "many-center" multipole expansion. The leading terms of this expansion are dipole-dipole ones because the net charge of the localized charge distributions can always be chosen to be zero, see (2.118). A thorough examination of the many-center expansion has been made by *Bonaccorsi* et al. [2.83]. It has been found that this expansion always works much better than the usual single-center one. However, even with the many-center expansion the leading (dipole-dipole) term alone does not yield satisfactory results.

The main disadvantage of the many-center expansion lies in the fact that the parameters of this expansion (i.e., the bond and lone-pair multipole moments) are not experimentally obtainable quantities, they have to be computed [2.74]. As with molecular multipole moments, the bond and lone-pair moments are markedly sensitive to the particular wave function used to compute them. This point has well been illustrated by *Amos* and *Crispin*

[2.74] for the water molecule. For example, the values for the dipole moment of the OH-bond range from −0.406 to +0.552 a.u., depending on the wave function used.

Another way of decomposing the electrostatic interaction energy into pairwise interactions between localized charge distributions is based on a LCAO expansion of the molecular wave function.

If the molecular orbitals are written in the LCAO form given in (2.12), the electronic charge density will be given by

$$f_{\mathrm{el}}(r) = \sum_{\mu,\nu=1}^{m} P_{\mu\nu} \chi_{\mu}^{*}(r) \chi_{\nu}(r), \tag{2.121}$$

where the population matrix $P_{\mu\nu}$ is defined as in (2.19),

$$P_{\mu\nu} = 2 \sum_{i=1}^{n} c_{i\mu}^{*} c_{i\nu}. \tag{2.122}$$

Again, for the sake of simplicity, the superscript, labeling the molecule A or B, is omitted.

Equation (2.121) represents localized charge distributions only in the simplest case in which all χ_{μ} are floating spherical Gaussians, i.e.,

$$\chi_{\mu}(r) = C_{\mu} \exp\left(-\varsigma_{\mu} |r - R_{\mu}|^2\right). \tag{2.123}$$

This can easily be seen by noting that the product of two such Gaussians is itself a spherical Gaussian,

$$\chi_{\mu}(r) \chi_{\nu}(r) = C_{\mu\nu} \exp\left(-\varsigma_{\mu\nu} |r - R_{\mu\nu}|^2\right), \tag{2.124}$$

with the center

$$R_{\mu\nu} = (\varsigma_{\mu} R_{\mu} + \varsigma_{\nu} R_{\nu}) / (\varsigma_{\mu} - \varsigma_{\nu}), \tag{2.125}$$

and exponent

$$\varsigma_{\mu\nu} = \varsigma_{\mu} + \varsigma_{\nu}. \tag{2.126}$$

By substituting (2.124) into (2.121) it can be shown that the electronic part of the charge density is a sum of $m(m+1)/2$ localized charge distributions.

For atomic orbitals of a more complicated form, such as the Slater-type orbitals in (2.23) or the usual Gaussians given in (2.24), only the diagonal elements in (2.121) will represent localized charge distributions. The off-diagonal elements will not be localized but will represent rather diffuse clouds of charge.

There are a number of more or less accurate ways to reallocate the off-diagonal elements among the diagonal ones (for a complete analysis see [2.62]). The most widely used represents the electron density in terms of

Mulliken's gross populations, namely

$$G_{\mu\mu} = \sum_{\nu} P_{\mu\nu} S_{\mu\nu}, \tag{2.127}$$

where the $S_{\mu\nu}$ are overlap integrals, see (2.18). With the above definition for $G_{\mu\mu}$, the electron density is approximated by

$$f_{\text{el}} \approx \sum_{\mu} G_{\mu\mu} |\chi_{\mu}(r)|^2, \tag{2.128}$$

which neglects the off-diagonal one-center distributions $\chi_{\mu}^{*}\chi_{\mu'}$ (μ and μ' are associated with the same atom) and assumes that the two-center distributions can be represented by

$$\chi_{\mu}^{*}\chi_{\nu} \approx 1/2 S_{\mu\nu}(|\chi_{\mu}|^2 + |\chi_{\nu}|^2) \tag{2.129}$$

(μ and ν are associated with different atoms).

A further simplification of this charge-distribution model is usually made by assuming that the atomic charge densities are delta functions

$$|\chi_{\mu}|^2 = \delta(r - r_{\alpha}). \tag{2.130}$$

In this case, the total charge density becomes

$$f(r) = \sum_{\alpha} q_{\alpha} \delta(r - r_{\alpha}), \quad \text{where} \tag{2.131}$$

$$q_{\alpha} = Z_{\alpha} - \sum_{\mu}^{(\alpha)} G_{\mu\mu} \tag{2.132}$$

is the net charge on atom α (the summation is carried out over all orbitals χ_{α} associated with this atom).

The use of (2.131) for $f(r)$ reduces the electrostatic interaction energy to a particularly simple form,

$$\varepsilon_{\text{elst}} = \sum_{\alpha} \sum_{\beta} q_{\alpha} q_{\beta} / r_{\alpha\beta}. \tag{2.133}$$

Actually, it is this simplicity that explains the extensive use of the point-charge model in calculations of the electrostatic interactions between large organic molecules. However, this model suffers from a few shortcomings. Thus, the approximation in (2.130) obviously implies that all multipoles of $|\chi_{\mu}|^2$, except the monopole, are neglected. For orbitals with a non-zero azimuthal quantum number this may lead to serious errors [2.62]. Furthermore, the approximation in (2.129) implies that the total charge of the distribution $\chi_{\mu}^{*}\chi_{\nu}$ is divided into two equal charges $S_{\mu\nu}/2$ placed at the atoms bearing the μth and νth orbitals. In doing so, the dipole moment of the distribution $\chi_{\mu}^{*}\chi_{\nu}$ is neglected. This dipole moment is very important because the above distribution is not a localized one [2.62].

To provide the reader with an idea of the accuracy of the point-charge model, we quote the point-charge estimates of the electrostatic part of the lattice energy for the formic acid crystal [2.41], namely −13.70 and −14.29 kcal/mol for Bases I and II, respectively. A comparison with the corresponding "exact" values presented in Table 2.3 (−19.61 kcal/mol and −21.25 kcal/mol, respectively) shows that the point-charge model based on gross Mulliken populations is very inaccurate. The quality of this model can also be judged by comparing the atomic point charges in (2.132) with those derived empirically by fitting (2.133) to the exact electrostatic energy. A comparison of this kind is given in Table 2.7 for the formaldehyde molecule [2.84]. It is seen that the atomic charges derived from Mulliken populations differ markedly from those which provide the best representation of the exact ab initio electrostatic energy.

Table 2.7. Atomic point charges [a.u.]in the formaldehyde molecule [2.84]

Atom	Charges a	b
C	0.155	0.542
O	-0.447	-0.485
H	0.146	-0.029

[a] Derived from gross Mulliken populations;
[b] Fitted by the method of least squares to the ab initio electrostatic interaction energy

A more satisfactory point-charge model has been suggested by *Hall* [2.85]. It is based on the use of floating spherical Gaussians given in (2.123) and can be derived by writing (2.124) as

$$\chi_\mu(r)\chi_\nu(r) = S_{\mu\nu}\delta(r - R_{\mu\nu}). \qquad (2.134)$$

In this model the initial charge distribution is represented by a set of m charges of magnitude $P_{\mu\mu}$ at the Gaussian centers and $m(m-1)/2$ charges of magnitude $2P_{\mu\nu}S_{\mu\nu}$ at the points $R_{\mu\nu}$ off the centers. The *Hall* model has been shown to yield very accurate results for the system LiH–H^+ and is believed to be as successful for other systems [2.74]. It is, however, hardly convenient to apply this model to large molecules in which the total number of force centers will be too large.

We conclude this section by noting that the failure of particular point-charge models to describe $\varepsilon_{\rm elst}$ does not mean that the atom-atom point-charge representation itself is bad. In Chap.3 it will be shown that this representation is flexible enough to fit $\varepsilon_{\rm elst}$, provided that the atomic point charges are treated empirically, as adjustable parameters.

2.2.3 Polarization Energy

Analogous to the electrostatic energy, we start the discussion for the case of large intermolecular separations. Again, it is convenient to resort to the multipole expansion for the interaction potential. Substituting (2.81), with $C^{\text{elst}}_{l_A+l_B+1}$ from (2.82), into the Rayleigh-Schrödinger expression for the polarization energy, yields [2.77,79]

$$\varepsilon_{\text{RS},2} \equiv \varepsilon_{\text{pol}} = \sum_{l_A,l'_A;l_B,l'_B} C^{\text{pol}}_{l_A+l'_A+l_B+l'_B+2} R^{-(l_A+l'_A+l_B+l'_B+2)} \;, \quad (2.135)$$

where

$$C^{\text{pol}}_{l_A+l'_A+l_B+l'_B+2} = -\; T^{[l_A+l_B]}_{(t);(u)} T^{[l'_A+l'_B]}_{(t');(u')} \times \sum_{i,j} \langle\psi_0^A|M^{[l_A]}_{(t)}|\psi_i^A\rangle\langle\psi_i^A|M^{[l'_A]}_{(t')}|\psi_0^A\rangle\langle\psi_0^B|M^{[l'_B]}_{(u)}|\psi_j^B\rangle \times\langle\psi_j^B|M^{[l'_B]}_{(u')}|\psi_0^B\rangle/(E_i^A + E_j^B - E_0^A - E_0^B). \quad (2.136)$$

Equivalent equations can be derived by using expressions of the multipole moments in terms of spherical harmonics [2.60,61].

The terms with $l_A \neq l'_A$ and $l_B \neq l'_B$ are usually referred to as cross terms. They have no isotropic components and vanish when averaged over all orientations. In this respect the cross terms are similar to the electrostatic interactions and it is expected that they will mainly cancel in crystals.

The coefficients in (2.136) that relate to the induction energy are readily obtained by collecting the terms with $i = 0$, $j \neq 0$ and $i \neq 0$, $j = 0$. The result is

$$C^{\text{ind}}_{l_A+l'_A+l_B+l'_B+2} = -\frac{1}{2}[\langle\psi_0^A|M^{[l_A]}_{(t)}|\psi_0^A\rangle\langle\psi_0^A|M^{[l'_A]}_{(t')}|\psi_0^A\rangle \times T^{[l_A+l_B]}_{(t);(u)} T^{[l'_A+l'_B]}_{(t');(u')} \alpha^{l_B l'_B}_{(u)(u')} + \alpha^{l_A l'_A}_{(t)(t')} T^{[l_A+l_B]}_{(t);(u)} T^{[l'_A+l'_B]}_{(t');(u')} \times\langle\psi_0^B|M^{[l_B]}_{(u)}|\psi_0^B\rangle\langle\psi_0^B|M^{[l'_B]}_{(u')}|\psi_0^B\rangle] \quad , \quad (2.137)$$

where the $\alpha^{l_A l'_A}_{(t)(t')}$'s are the components of the polarizability tensors defined by

$$\alpha^{l_A l'_A}_{(t)(t')} = 2\sum_{i\neq 0} \frac{\langle\psi_0^A|M^{[l_A]}_{(t')}|\psi_i^A\rangle\langle\psi_i^A|M^{[l'_A]}_{(t')}|\psi_0^A\rangle}{E_i^A - E_0^A}. \quad (2.138)$$

The definition for $\alpha^{l_B l'_B}_{(u)(u')}$ is similar. Recall that in (2.136,137) we have used the tensor notation when the repeated indices $(t) = t_1 t_2 \ldots t_{l_A}$, $(u) = u_1 u_2 \ldots u_{l_B}$ denote a sum.

It is worthwile to write down an explicit expression for the leading term in ε_{ind} :

$$\varepsilon_{\text{ind},6} = -\frac{1}{2} T^{[2]}_{tu} T^{[2]}_{vw} (\mu_t^A \mu_v^A \alpha^B_{uw} + \mu_u^B \mu_w^B \alpha^A_{tv}) R^{-6}, \quad (2.139)$$

where the μ_t's are the components of the molecular dipole moments and the α_{uw}'s are the components of the dipole polarizability tensors ($\alpha_{uw} \equiv \alpha_{uw}^{11}$). The components of the dipole interaction tensor $T_{tu}^{[2]}$ are given in (2.90).

The situation with the dispersion energy is much more complicated. Due to the appearance of both E_i^A and E_j^B in the denominator of (2.136), the terms with $i, j \neq 0$ are not separable into products of terms that relate to one molecule only. Hence, it is not as simple to express the dispersion energy in terms of the properties of the isolated molecules as it is for the induction energy. The first way to overcome the above difficulty is to replace the differences between the excited-state and ground-state energies by certain mean exictation energies (Unsöld's approximation). This can also be done for the polarizability tensors in (2.138):

$$\begin{aligned}\alpha_{(t)(t')}^{l_A l'_A} &= 2 \sum_{i \neq 0} \frac{\langle \psi_0^A | M_{(t)}^{[l_A]} | \psi_i^A \rangle \langle \psi_i^A | M_{(t')}^{[l'_A]} | \psi_0^A \rangle}{E_i^A - E_0^A} \\ &= \frac{2}{\Delta_{(t)(t')}^{l_A l'_A}} \sum_{i \neq 0} \langle \psi_0^A | M_{(t)}^{[l_A]} | \psi_i^A \rangle \langle \psi_i^A | M_{(t')}^{[l'_A]} | \psi_0^A \rangle . \end{aligned} \tag{2.140}$$

A similar expression can be obtained for $\alpha_{(u)(u')}^{l_B l'_B}$. The last line in (2.140) can be considered as a definition for the mean excitation energies $\Delta_{(t)(t')}^{l_A l'_A}$. (Note that, in general, the latter are different for different polarizability tensors and their components.)

In the mean-excitation-energy (Unsöld) approximation the coefficients for the dispersion energy become [2.77,79]

$$\begin{aligned} C_{l_A+l'_A+l_B+l'_B+2}^{\mathrm{disp}} &= -\frac{1}{4} \alpha_{(t)(t')}^{l_A l'_A} \Delta_{(t)(t')}^{l_A l'_A} \frac{T_{(t);(u)}^{[l_A+l_B]} T_{(t');(u')}^{[l'_A+l'_B]}}{\Delta_{(t)(t')}^{l_A l'_A} + \Delta_{(u)(u')}^{l_B l'_B}} \\ &\times \Delta_{(u)(u')}^{l_B l'_B} \alpha_{(u)(u')}^{l_B l'_B} . \end{aligned} \tag{2.141}$$

After averaging over all orientations, (2.141) reduces to

$$C_{2l_A+2l_B+2}^{\mathrm{disp}} = -\frac{1}{4} \frac{(2l_A + 2l_B)!}{(2l_A)!(2l_B)!} \frac{\Delta^{l_A} \Delta^{l_B}}{\Delta^{l_A} + \Delta^{l_B}} \alpha^{l_A} \alpha^{l_B} , \tag{2.142}$$

where the Δ's and α's denote the averaged mean excitation energies and polarizabilities. Explicit expressions for the latter quantities can be found in [2.78].

Equation (2.142) represents a generalization of the known London formula for C_6^{disp} ($l_A = l_B = 1$). Usually, the mean excitation energies are replaced by the first ionization potentials to give

$$\varepsilon_{\text{disp},6} = -\frac{3}{2}\frac{I^A I^B}{I^A + I^B}\alpha^A \alpha^B R^{-6}. \tag{2.143}$$

Another way of dealing with the dispersion energy, given by the multipole expansion in (2.135,136), is based on the identity [2.62]

$$\frac{1}{a+b} = \frac{2}{\pi}\int_0^\infty \frac{ab}{(a^2+\omega^2)(b^2+\omega^2)} d\omega, \quad (a,b>0). \tag{2.144}$$

Applying this identity to the factor $[(E_i^A - E_0^A) + (E_j^B - E_0^B)]^{-1}$ in (2.136) reduces the dispersion energy coefficients to

$$C^{\text{disp}}_{l_A+l'_A+l_B+l'_B+2} = -\frac{1}{2\pi} T^{[l_A+l_B]}_{(t);(u)} T^{[l'_A+l'_B]}_{(t');(u')} \times \int_0^\infty \alpha^{l_A l'_A}_{(t)(t')}(\mathrm{i}\omega)\alpha^{l_B l'_B}_{(u)(u')}(\mathrm{i}\omega) d\omega, \tag{2.145}$$

where $\alpha^{l_A l'_A}_{(t)(t')}(\omega)$ and $\alpha^{l_B l'_B}_{(u)(u')}(\omega)$ are the frequency-dependent polarizabilities defined by

$$\alpha^{l_A l'_A}_{(t)(t')}(\omega) = 2\sum_{i\neq 0} \frac{\langle\psi_0^A|M^{[l_A]}_{(t)}|\psi_i^A\rangle\langle\psi_i^A|M^{[l'_A]}_{(t')}|\psi_0^A\rangle(E_i^A - E_0^A)}{(E_i^A - E_0^A) - \omega^2}. \tag{2.146}$$

A similar definition for $\alpha^{l_B l'_B}_{(u)(u')}(\omega)$ can also be given. The leading term of the dispersion energy is

$$\varepsilon_{\text{disp},6} = -\frac{1}{2\pi} R^{-6} T^{[2]}_{tu} T^{[2]}_{uw} \int_0^\infty \alpha^A_{tv}(\mathrm{i}\omega)\alpha^B_{uw}(\mathrm{i}\omega) d\omega, \tag{2.147}$$

where $\alpha^A_{tv}(\mathrm{i}\omega)$ and $\alpha^B_{uw}(\mathrm{i}\omega)$ are the frequency-dependent dipole polarizabilities at the imaginary frequency $\mathrm{i}\omega$.

After averaging over all orientations of the interacting molecules, (2.147) becomes

$$\varepsilon_{\text{disp},6} = -\frac{3}{\pi} R^{-6} \int_0^\infty \alpha^A(\mathrm{i}\omega)\alpha^B(\mathrm{i}\omega) d\omega, \quad \text{where} \tag{2.148}$$

$$\alpha^A(\mathrm{i}\omega) = \frac{1}{3}\sum_t \alpha^A_{tt}(\mathrm{i}\omega), \tag{2.149}$$

and similarly for $\alpha^B(\mathrm{i}\omega)$.

The above expression enables one to find the isotropic dispersion energy if the theoretical or experimental frequency-dependent polarizabilities are known. A very simple equation for $\varepsilon_{\text{disp},6}$ is obtained if one calculates $\alpha(\mathrm{i}\omega)$ using the Frost model [2.86,87]. In this model the wave function of a molecule consists of a Slater determinant of n doubly occupied floating spherical Gaussians. The exponents ς_μ and the Gaussian centers R_μ are found using the variational principle. For the Frost-model wave function, the frequency-dependent polarizability is given by [2.86,87]

$$\alpha^A(\omega) = \sum_{\mu=1}^{n_A} \frac{2}{(\varsigma_\mu^A)^2 - \omega^2}. \tag{2.150}$$

Substituting (2.150) and a similar expression for $\alpha^B(\omega)$ into (2.148) yields

$$\varepsilon_{\text{disp},6} = -6R^{-6} \sum_{\mu=1}^{n_A} \sum_{\nu=1}^{n_B} \frac{1}{\varsigma_\mu^A \varsigma_\nu^B (\varsigma_\mu^A + \varsigma_\nu^B)}. \tag{2.151}$$

The isotropic C_6^{disp} coefficients computed by *Amos* and *Yoffe* [2.87]for $H_2 - H_2$, $H_2 - CH_4$, $CH_4 - CH_4$, and $H_2O - H_2O$ are in good agreement with the best experimental results available. Thus, the calculation for methane yielded 142 a.u., while the best experimental result is 149 a.u [2.88]. For ethylene, the Frost-model value of 321 a.u. is in good agreement with the quite reliable estimate (341.3 a.u.) reported by *Mulder* and *Huiszoon* [2.77].

An expression of the dispersion energy in terms of frequency-dependent polarizabilities is a useful starting point for the derivation of various approximate equations for $\varepsilon_{\text{disp}}$. We shall now briefly discuss these approximations along the lines of *Amos* and *Crispin* [2.74]. First, let us express the frequency-dependent polarizabilities in terms of oscillator strengths f_s,

$$\alpha^A(\mathrm{i}\omega) = \sum_s f_s^A / [(\omega_s^A)^2 + \omega^2], \tag{2.152}$$

where the summation is carried out over both the discrete and continuous states of the molecule; and ω_s denote the transition frequencies (in atomic units $\omega_s^A = E_s^A - E_0^A$). For many molecules the oscillator strengths that make a significant contribution to the sum in (2.152) lie within a rather narrow frequency range. As a result, (2.152) may be well approximated by a single composite term, namely

$$\alpha^A(\mathrm{i}\omega) \approx F^A / [(\omega^A)^2 + \omega^2], \tag{2.153}$$

$$\varepsilon_{\text{disp}} \approx -\frac{3}{2} R^{-6} \frac{F^A F^B}{\omega^A \omega^B (\omega^A + \omega^B)}. \tag{2.154}$$

Equations (2.153,154) are frequently used to relate the dispersion energy to the refractive index. To this end the experimentally obtained refractive indices n are fitted to the expression,

$$n - 1 \approx C/(\omega_0^2 - \omega^2), \tag{2.155}$$

where C and ω_0 are empirical parameters. Then, the Clausius-Mossotti equation, which at low frequencies assumes the following form,

$$n - 1 = 2\pi\alpha^A(\omega)d^A, \tag{2.156}$$

can be used (d^A is the number density of the solid). By equating (2.156) to (2.155) one can determine both F^A and ω^A and estimate the dispersion energy using (2.154).

The dispersion energy can also be related to the static molecular polarizability defined by (2.153) by setting $\omega = 0$:

$$\alpha^A = F^A/(\omega^A)^2. \tag{2.157}$$

Replacing F^A and F^B in (2.154) by $\alpha^A(\omega^A)^2$ and $\alpha^B(\omega^B)^2$, respectively, one obtains

$$\varepsilon_{\text{disp}} \approx -\frac{3}{2}R^{-6}\alpha^A\alpha^B\frac{\omega^A\omega^B}{\omega^A + \omega^B}. \tag{2.158}$$

The resulting equation can now be reduced to either the London equation (2.143), by replacing the average transition frequencies by the ionization potentials, or to the Slater-Kirkwood equation, by noting that for large ω,

$$\alpha^A(\mathrm{i}\omega) \approx N_A/\omega^2. \tag{2.159}$$

Comparing (2.159) with (2.153), it can be seen that $F^A \approx N_A$ for large ω's. Hence, from (2.157), $\omega^A \approx (\alpha^A/N_A)^{1/2}$. Substituting into (2.158) yields

$$\varepsilon_{\text{disp}} \approx -\frac{3}{2}\alpha^A\alpha^B\frac{R^{-6}}{(\alpha^A/N_A)^{1/2} + (\alpha^B/N_B)^{1/2}}. \tag{2.160}$$

Equations (2.143,160) are used extensively in various semiempirical schemes to calcalute the dispersion energy between organic molecules. It is frequently assumed that (2.143 or 160) is valid for each pair of atoms in the interacting molecules. The atomic polarizabilities are derived from atomic refractive indices [2.89], while the atomic ionization potentials are assumed to be equal to the valence-state ionization potentials, as used in the extended Hückel method. A few representative semiempirical schemes of this kind will be discussed in the following chapter.

In organic molecular crystals the orientationally averaged expressions for $\varepsilon_{\text{disp}}$ obviously become useless in view of the marked anisotropy of the dispersion interactions. In this case a more rigorous approach is required.

Quite reliable data on the dispersion interaction of organic molecules have been reported by *Mulder* et al. [2.79]for azabenzene molecules. The calculations were carried out using the Unsöld approximation. The basis set (9, 5/4) contracted to [4, 2/2](double zeta) was used. To correct the results for a possible incompleteness of the basis et, *Mulder* et al. replaced the sum over the transition moments (STM) in (2.140) by the corresponding closure moments (CM) in (2.97). In general, such a replacement is justified by the fact that both the closure moments and the mean excitation energies are rather insensitive to the quality of the basis set [2.60]. Hence, the use of CM instead of STM allows one to obtain reliable results even with comparatively small bases. The success of this procedure is illustrated by the excellent agreement between the calculated and the observed dipole polarizabilities of the azabenzenes [2.79]. Thus, for benzene the experimental values are $\alpha_{xx} = \alpha_{yy} = 79.2$, $\alpha_{zz} = 44.1$ a.u., while the calculated values are $\alpha_{xx} = \alpha_{yy} = 79.9$, $\alpha_{zz} = 44.8$ a.u.

The dispersion energies calculated by *Mulder* et al. [2.79]showed a pronounced dependence on orientation. Thus, at $R = 15$ a.u. the dispersion interaction energy of two benzene molecules varies from -18×10^{-5} to -53×10^{-5}a.u., depending on the molecular orientation. The smallest variations were observed with tetrazine – from -12×10^{-5} to -35×10^{-5}a.u.

It is worth noting that the mean excitation energies computed by *Mulder* et al. [2.79]were substantially larger than the corresponding first ionization potentials. The ratio Δ/I was about 2.3 on the average. This indicates that the London equation (2.143) would yield substantially smaller values of the dispersion energy; – a well-known fact which has a theoretical explanation [2.74].

Analogous to the electrostatic energy, the multipole expansion for the polarization energy becomes inadequate at intermediate and short separations. This is well illustrated by the data on the ethylene dimer [2.78]. Figure 2.5 illustrates the dependence of the ratio of the multipole-expanded and unexpanded polarization energies on the distance for different lengths of the multipole series. One can see that the multipole series diverges at short distances. It is also clearly seen that even at large distances the leading term of the expansion alone is unsufficient to reproduce the unexpanded value. Thus, up to a distance of 20 a.u. the C_8R^{-8} term is more than 10% of the C_6R^{-6} term.

The convergence of the multipole expansion should obviously be worse for larger molecules. For example, in the pyrazine dimer the distance at which the C_8R^{-8} polarization term becomes less than 10% of the C_6R^{-6} term is as large as 30 a.u. [2.77].

All comments made above concerning the multipole expansion are also true for crystals. Thus, *Mulder* and *Huiszoon* [2.77]reported that the C_6R^{-6} term contributes only half of the dispersion part of the lattice energy of the ethylene crystal. The same also applies to the pyrazine crystal, at least with respect to the interactions beyond the summation radius of 7.21 a.u. [2.77]. Considering that the estimates of the interaction energy of the reference

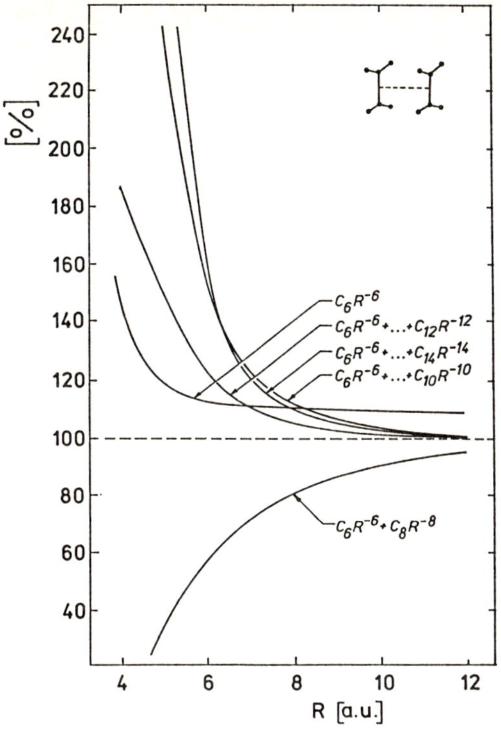

Fig. 2.5. The ratio of the multipole-expanded to non-expanded polarization energy as a function of the intermolecular distance for various lengths of the multipole expansion in the ethylene dimer [2.78]

molecule with its nearest neighbors may be highly inaccurate, it is again reasonable to neglect the nearest-neighbor interactions and to see whether the higher-order multipole terms are important if only for the long-range contribution. A simple manipulation of the data reported in [2.77]shows that an estimate of the long-range part of the lattice energy using only the C_6R^{-6} term is still in a $\approx$ 30% error for both the ethylene and the pyrazine crystals.

Another important result obtained by *Mulder* and *Huiszoon* in their study of ethylene and pyrazine crystals [2.77]is that the cross dispersion terms ($l_A \neq l'_A, l_B \neq l'_B$) cancel when summed over the crystal lattice. Thus, in the pyrazine crystal the C_8R^{-8} cross terms only yield 9% of the total C_8R^{-8} contribution, while in the ethylene crystal the C_8R^{-8} cross-term contribution to the lattice energy is practically zero. In the ethylene and pyrazine dimers the C_8R^{-8} cross terms are of the same order of magnitude as the C_8R^{-8} quadratic terms [2.77]. This means that the relatively small cross-term dispersion contribution to the lattice energy is indeed a result of the cancellation of the contributions from various molecular pairs.

To complete the discussion of the long-range polarization interactions, we should again like to focus the reader's attention on Table 2.6 which lists the various contributions to the long-range interaction energy for seven azabenzene dimers [2.79]. It can be seen that the induction interactions contribute very little to the total long-range energy. The largest contribution (5%) is observed for pyridazine which possesses the largest dipole moment. The induction interactions also seem to be of minor importance in crystals. Thus, in the ethylene and pyrazine crystals the induction contributions are 1.4 and 5.2% of the corresponding dispersion contributions, respectively.

We shall now discuss some simplified equations for the polarization energy that are applicable to large molecules at separations where the multipole expansion becomes inadequate. The derivation of these equations is based on the same idea as used in the preceding subsection for the electrostatic energy, i.e., to represent the interaction energy as a sum of contributions from small molecular units (bonds, lone pairs, etc.). There are several ways of obtaining such a representation [2.62]. The most elegant methods seem to be the ones suggested by *Claverie* [2.62]and *Amos* and *Crispin* [2.90].

As a starting point for the derivation of the simplified equations for ε_{pol}, *Claverie* used the standard Rayleigh-Schrödinger equation (2.51). The molecular ground-state wave functions ψ_0 are assumed to be Slater determinants. The excited-state wave functions ψ_i are approximated by linear combinations of singly-excited determinants, $\psi_\lambda \equiv \psi_{a-a'}$, which have been obtained from ψ_0 by replacing the ath occupied spin-orbital, $\eta_a = \varphi_a\sigma_a$, by the a'th virtual spin-orbital, $\eta_{a'} = \varphi_{a'}\sigma_{a'}$,

$$\psi_i = \sum_\lambda c_{i\lambda}\psi_\lambda, \tag{2.161}$$

where the matrix $c_{i\lambda}$ is unitary because the sets $\{\psi_i\}$ and $\{\psi_\lambda\}$ are both orthonormal. Substituting (2.161) into the first term in the second line of (2.51) yields

$$\begin{aligned}\varepsilon_{\text{ind}(A)} &\equiv -\sum_i{}^{\prime\,(A)} \frac{\langle\psi_0^A\psi_0^B|V|\psi_i^A\psi_0^B\rangle^2}{E_i^A - E_0^A}\\ &= -\sum_\lambda{}' \sum_\mu V_\lambda^A V_\mu^A \Big(\sum_i{}' \frac{c_{i\lambda}c_{i\mu}}{E_i^A - E_0^A}\Big), \quad \text{where}\end{aligned} \tag{2.162}$$

$$V_\lambda^A = \langle\psi_0^A\psi_0^B|V|\psi_\lambda^A\psi_0^B\rangle, \tag{2.163}$$

and similarly for V_μ^A. (For the sake of simplicity, we assume that all the wave functions are real.) It is now convenient to express V_λ^A in terms of matrix elements of the charge density operator defined in (2.101–103). In general, using (2.105), we may write

$$\begin{aligned}\langle\psi_i^A\psi_j^B|V|\psi_k^A\psi_l^B\rangle &= \int\int r_{AB}^{-1}\langle\psi_i^A\psi_j^B|\varrho^A(r_A)\varrho^B(r_B)|\psi_k^A\psi_l^B\rangle dr_A dr_B\\ &= \int\int r_{AB}^{-1} f_{ik}^A(r_A) f_{jl}^B(r_B) dr_A\, dr_B,\end{aligned} \tag{2.164}$$

where f_{ik}^A and f_{jl}^B are the transition-state charge distributions defined by

$$f_{ik}^A(r_A) = \langle \psi_i^A | \varrho^A(r_A) | \psi_k^A \rangle, \tag{2.165}$$

$$f_{jl}^B(r_B) = \langle \psi_j^B | \varrho^B(r_B) | \psi_l^B \rangle. \tag{2.166}$$

Actually, the above definition is an extension of the definition given in (2.104) for the charge distribution of the ground state. It should be noted that $f_{00}^A(r_A) \equiv f^A(r_A)$, $f_{00}^B(r_B) \equiv f^B(r_B)$.

As follows from (2.163) and the definition of ψ_λ, the calculation of V_λ^A requires a knowledge of the ground-state charge distribution $f^B(r)$ and the transition-state charge distributions of the form $f_{0\lambda}^A\ (r)$ [which from now on will be denoted $f_{a-a'}^A(r)$]:

$$f_{a-a'}^A(r) = \langle \psi_0^A | \varrho^A(r) | \psi_\lambda^A \rangle. \tag{2.167}$$

It is easy to show that

$$f_{a-a'}^A(r) = -\varphi_a^A(r)\varphi_{a'}^A(r)\delta_{aa'}^{\text{spin}}, \tag{2.168}$$

where $\delta_{aa'}^{\text{spin}} = \langle \sigma_a^A | \sigma_{a'}^A \rangle$.

We shall now assume that the orbitals used to construct the ground- and excited-state wave functions are localized. In this case V_λ may be given by

$$\begin{aligned} V_\lambda^A &\equiv V_{a-a'}^A \\ &= \int\int r_{AB}^{-1} f_{a-a'}^A(r_A) f^B(r_B) dr_A dr_B \\ &\approx -\mathcal{E}_{a,t}^B \mu_{a-a',t}^A, \end{aligned} \tag{2.169}$$

where $\mu_{a-a',t}^A$ is the tth component of the dipole moment of the localized transition-state charge distribution and $\mathcal{E}_{a,t}^B$ is the tth component of the electric field due to molecule B at orbital a (as usual, the repeated index t denotes a sum).

Let us now consider the factor in round brackets in (2.162) [2.62]. Turning to (2.161) we note that the states ψ_λ of the same mean energy (e.g., the states corresponding to similar localized excitations on identical bonds) will yield a cluster of eigenstates ψ_i upon mixing. If ψ_λ and ψ_μ are of different energies and, hence, belong to different clusters, there is no state ψ_i for which both $c_{i\lambda}$ and $c_{i\mu}$ are important at the same time. In this case, the sum

$$e_{\lambda\mu} \equiv \sum_i c_{i\lambda} c_{i\mu} / (E_i^A - E_0^A), \tag{2.170}$$

will obviously be small. If ψ_λ and ψ_μ belong to the same cluster and are of almost equal energies

$$E_\lambda^A \approx E_\mu^A \approx E_i^A, \tag{2.171}$$

the sum in (2.170) reduces to

$$e_{\lambda\mu} \approx \frac{1}{E^A - E_0^A} \sum_i c_{i\lambda} c_{i\mu} = \delta_{\lambda\mu}/(E^A - E_0^A). \tag{2.172}$$

Thus, the equation for the induction energy becomes

$$\begin{aligned}
\varepsilon_{\text{ind}(A)} &\approx -\sum_\lambda \frac{|V_\lambda^A|^2}{E_\lambda^A - E_0^A} \\
&= -\sum_a \sum_{a'} \frac{|V_{a-a'}^A|^2}{E_{a-a'}^A - E_0^A} \\
&= -\frac{1}{2} \sum_a \mathcal{E}_{a,t}^B \left(2 \sum_{a'} \frac{\mu_{a-a',t}^A \mu_{a-a',u}^A}{E_{a-a'}^A - E_0^A} \right) \mathcal{E}_{a,u}^B .
\end{aligned} \tag{2.173}$$

Comparing this equation with (2.140) we note that the factor in parentheses is the dipole polarizability of the ath localized orbital, i.e.,

$$\alpha_{a,tu}^A = 2 \sum_{a'} \frac{\mu_{a-a',t}^A \mu_{a-a',u}^A}{E_{a-a'}^A - E_0^A}. \tag{2.174}$$

Equation (2.173) can be now rewritten as

$$\varepsilon_{\text{ind}(A)} \approx -\frac{1}{2} \sum_a \mathcal{E}_{a,t}^B \alpha_{a,tu}^A \mathcal{E}_{a,u}^B . \tag{2.175}$$

A similar equation is obviously valid for $\varepsilon_{\text{ind}(B)}$, which represents the interaction energy between the permanent charge distribution of molecule A and the induced charge distribution of molecule B. The total induction energy is therefore given by

$$\varepsilon_{\text{ind}} = -\frac{1}{2} \left(\sum_a^{(A)} \mathcal{E}_{a,t}^B \alpha_{a,tu}^A \mathcal{E}_{a,u}^B + \sum_b^{(B)} \mathcal{E}_{b,t}^A \alpha_{b,tu}^B \mathcal{E}_{b,u}^A \right). \tag{2.176}$$

It is essential that ε_{ind} cannot, in principle, be reduced to a sum of pairwise interactions between pairs of localized charge distributions. This already follows from (2.176) in which it is seen that ε_{ind} is a quadratic function of the electric fields. The same result can be obtained by expressing the fields in terms of localized charge distributions which create these fields. If we define the charge densities of the orbitals as in (2.115,117), the net orbital charges will be equal to zero and the molecular charge distribution given in (2.116) can be approximated by a set of permanent dipoles associated with the localized orbitals. In this case, the electric fields can be expressed explicitly in terms of these permanent dipoles, and (2.176) can be rewritten as

$$\varepsilon_{\text{ind}} = -\frac{1}{2}\Big(\sum_b{}^{(B)} \sum_c{}^{(B)} \sum_a{}^{(A)} T_{tv}^{[2]} T_{uw}^{[2]} \mu_{b,t}^B \mu_{c,u}^B \alpha_{a,vw}^A R_{ba}^{-3} R_{ca}^{-3}$$

$$+ \sum_a{}^{(A)} \sum_c{}^{(A)} \sum_b{}^{(B)} T_{tv}^{[2]} T_{uw}^{[2]} \mu_{a,t}^A \mu_{c,u}^A \alpha_{b,vw}^B R_{ab}^{-3} R_{cb}^{-3}\Big). \quad (2.177)$$

It can now be seen that the induction energy also involves three-orbital terms arising from the interaction between an induced dipole in one molecule and two permanent dipoles in the other molecule, not just pairwise contributions ($b = c$ in the first sum and $a = c$ in the second sum).

Actually, (2.177) leads us to a bond-bond representation for the induction energy. Indeed, the localized innershell orbitals exhibit negligible dipole moments and small polarizabilities, so that their contribution to ε_{ind} can surely be ignored. Thus, as with the electrostatic energy, the localized orbital description provides no arguments in favor of the atom-atom model. At this step, the latter is usually introduced in a somewhat artificial way by distributing each bond polarizability between the two atoms joined by this bond. One of the possible ways to do this will be discussed in the next chapter.

It is to be noted that the approximation given in (2.172), which was used to go from (2.162 to 2.177), is sufficient to reduce the total molecular polarizability to the sum of the bond polarizabilities defined in (2.174). Indeed, the molecular dipole polarizability is given by, see (2.138),

$$\alpha_{tu} = 2\sum_i{}' \frac{\langle\psi_0|\mu_t|\psi_i\rangle\langle\psi_i|\mu_u|\psi_0\rangle}{E_i - E_0}. \quad (2.178)$$

For the wave functions given in (2.161) this expression can be reduced to

$$\alpha_{tu} = 2\sum_\lambda{}' \sum_\mu{}' \mu_{0\lambda,t}\mu_{0\mu,u} \sum_i \frac{c_{i\lambda}c_{i\mu}}{E_i - E_0}, \quad \text{where} \quad (2.179)$$

$$\mu_{0\lambda,t} = \langle\psi_0|\mu_t|\psi_\lambda\rangle \equiv \mu_{a-a',t}. \quad (2.180)$$

The same is true for $\mu_{0\mu,u}$. Using the approximation given in (2.172) for the third sum in (2.179), we immediately obtain

$$\alpha_{tu} = 2\sum_\lambda{}' \frac{\mu_{0\lambda,t}\mu_{\lambda 0,u}}{E_\lambda - E_0}$$

$$= 2\sum_a\sum_{a'} \frac{\mu_{a-a',t}\mu_{a-a',u}}{E_{a-a'} - E_0} = \sum_a \alpha_{a,tu}. \quad (2.181)$$

Thus, the bond polarizabilities defined in (2.174) are equal in meaning to the empirical transferable bond polarizabilities that can be determined by fitting (2.181) to the experimental molecular polarizabilities for a large number of molecules (see, for example, the tables compiled by *Le Fevre* [2.91]).

The above analysis of the total static polarizability in terms of bond polarizabilities is also valid for the frequency-dependent polarizability. The frequency-dependent bond polarizabilities are now defined as, see (2.146,174),

$$\alpha_{a,tu}(\omega) = 2\sum_{a'} \frac{\mu_{a-a',t}\mu_{a-a',u}(E_{a-a'} - E_0)}{(E_{a-a'} - E_0)^2 - \omega^2}, \tag{2.182}$$

and the total frequency-dependent dipole polarizability is given by

$$\alpha_{tu}(\omega) = \sum_a \alpha_{a,tu}(\omega). \tag{2.183}$$

Substituting (2.183) into (2.147) immediately reduces the dispersion energy to a sum of dispersion interactions between localized orbitals:

$$\varepsilon_{\text{disp}} = \sum_a{}^{(A)} \sum_b{}^{(B)} \varepsilon_{\text{disp},ab}, \quad \text{where} \tag{2.184}$$

$$\varepsilon_{\text{disp},ab} = \frac{1}{2} R_{ab}^{-6} T_{tu}^{[2]} T_{vw}^{[2]} \int_0^\infty \alpha_{a,tv}^A(\mathrm{i}\omega)\alpha_{b,uw}^B(\mathrm{i}\omega)d\omega. \tag{2.185}$$

The last expression provides a good basis for the derivation of simplified equations for the bond-bond dispersion energies. Thus, using the same assumptions as required to proceed from (2.148 to 2.160), one easily obtains the London- or Slater-Kirkwood-type equations (2.143,160) in which all parameters retain their usual meaning but now refer to bonds. Other useful approximations to the bond-bond dispersion energy can be found in [2.62].

To conclude this subsection, it is to be noted that the calculation of the polarization energy in terms of bond polarizabilities suffers from the same disadvantage as the calculation of the electrostatic energy in terms of bond multipoles or atomic point charges. This disadvantage lies in the fact that the values of the bond polarizabilities, when computed from (2.174,182) or other approximate equations [2.74], prove to be markedly sensitive to the particular wave function used. This has been demonstrated by *Amos* and *Crispin* [2.74]for the water molecule. Thus, the longitudinal component of the OH-bond polarizability varies from 1.42 to 5.16 a.u., depending on the choice of the wave function.

To improve the theoretical estimates of the bond polarizabilities, *Amos* and *Crispin* [2.74]recommend scaling them by a factor which brings the calculated and observed values for total molecular polarizabilities into coincidence. A similar scaling procedure has been suggested for the bond multipole moments.

2.2.4 Exchange Energy

In this subsection we shall derive, following *Claverie* [2.62], simplified equations for the first-order exchange energy as defined by the second term in (2.79)

$$\varepsilon_{\text{exch}} = \frac{\langle\phi_0|V|\phi_0\rangle\langle\phi_0|A'_{\text{int}}|\phi_0\rangle - \langle\phi_0|VA'_{\text{int}}|\phi_0\rangle}{1 - \langle\phi_0|A'_{\text{int}}|\phi_0\rangle}. \tag{2.186}$$

As with the polarization and electrostatic energies, we shall attempt to express the exchange energy in terms of interactions between localized charge distributions. It is seen in (2.186) that we have to consider two types of matrix elements, i.e., $\langle\phi_0|A'_{\text{int}}|\phi_0\rangle$ and $\langle\phi_0|VA'_{\text{int}}|\phi_0\rangle$ (the matrix element $\langle\phi_0|V|\phi_0\rangle$ is nothing but the electrostatic energy).

In order to simplify (2.186) one can neglect all terms in A'_{int} which involve two, three or more intermolecular permutations of the electrons. To justify this approximation, we note that the permutation operators P_λ, see (2.178), in the matrix elements involving A'_{int} generate terms which vary like the 2^λth power of the overlap integral S between the orbitals of the interacting molecules. Close to the van der Waals minimum S is usually a rather small quantity, so that it is indeed legitimate to replace A'_{int} by P_1.

If ψ_0^A and ψ_0^B are Slater determinants, the action of P_1 upon their product may be written as

$$P_1(\psi_0^A\psi_0^B) = \sum_{a=1}^{N_A}\sum_{b=1}^{N_B}(\psi_0^A|b/a)(\psi_0^B|a/b), \tag{2.187}$$

where $(\psi_0^A|b/a)$ denotes the replacement of the ath spin-orbital in ψ_0^A by the bth spin-orbital of ψ_0^B and similarly for $(\psi_0^B|a/b)$ (N_A and N_B are the number of electrons in A and B, respectively).

A further transformation of (2.187) can be made by representing the bth spin-orbital of ψ_0^B as

$$\eta_b^B = \eta'^B_b + \sum_{c=1}^{N_A} S_{cb}\eta_c^A, \tag{2.188}$$

where η'^B_b is the component of η_b^B that is orthogonal to all of the η_c^A. Thus, $(\psi_0^A|b/a)$ can be written as

$$(\psi_0^A|b/a) = (\psi_0^A|b'/a) + \sum_{c=1}^{N_A} S_{cb}\psi_{a-c}^A, \tag{2.189}$$

where b' denotes η'^B_b and ψ_{a-c}^A denotes the replacement of the ath orbital in ψ_0^A by the cth orbital belonging to the same molecule A. In the sum over

c all terms with $c \neq a$ vanish because the determinant ψ^A_{a-c} is identically equal to zero when η^A_c appears in ψ^A_{a-c} twice. (In the preceding subsection this was not so with $\psi^A_{a-a'}$ because $\eta^A_{a'}$ was a virtual spin-orbital.) Similar considerations can obviously be applied to $(\psi^B_j|a/b)$, so that (2.187) becomes

$$P_1(\psi^A_0\psi^B_0) = \sum_{a=1}^{N_A}\sum_{b=1}^{N_B}\{(\psi^A_0|b'/a)(\psi^B_0|a'/b) + S_{ab}[\psi^A_0(\psi^B_0|a'/b) + \psi^B_0(\psi^A_0|b'/a)] + S^2_{ab}\psi^A_0\psi^B_0\}. \quad (2.190)$$

We are now in a position to write down the expressions for the matrix elements containing P_1. Substituting (2.190) into $\langle\psi^A_0\psi^B_0|V P_1|\psi^A_0\psi^B_0\rangle$ yields [2.62]

$$\langle\psi^A_0\psi^B_0|V P_1|\psi^A_0\psi^B_0\rangle = \sum_{a=1}^{N_A}\sum_{b=1}^{N_B}\{\langle\psi^A_0\psi^B_0|V|(\psi^A_0|b'/a)(\psi^B_0|a'/b)\rangle + S_{ab}[\langle\psi^A_0\psi^B_0|V|\psi^A_0(\psi^B_0|a'/b)\rangle + \langle\psi^A_0\psi^B_0|V|\psi^B_0(\psi^A_0|b'/a)\rangle] + S^2_{ab}\langle\psi^A_0\psi^B_0|V|\psi^A_0\psi^B_0\rangle\} \quad . \quad (2.191)$$

For the matrix elements involving P_1 alone, the corresponding expression can easily be derived from (2.191) by replacing V by the unity operator and by using the normalization relation $\langle\psi^X_0|\psi^X_0\rangle = 1$ ($X = A$ or B). The result is [2.62]:

$$\langle\psi^A_0\psi^B_0|P_1|\psi^A_0\psi^B_0\rangle = \sum_{a=1}^{N_A}\sum_{b=1}^{N_B}\{[\langle\psi^A_0|(\psi^A_0|b'/a)\rangle + S_{ab}] + [\langle\psi^B_0|(\psi^B_0|a'/b)\rangle + S_{ab}]\}. \quad (2.192)$$

In this expression the matrix elements of the type $\langle\psi^A_0|(\psi^A_0|b'/a)\rangle$ vanish because b' is, by definition, orthogonal to the spin-orbitals a. Thus, (2.192) assumes a particularly simple form,

$$\langle\psi^A_0\psi^B_0|P_1|\psi^A_0\psi^B_0\rangle = \sum_{a=1}^{N_A}\sum_{b=1}^{N_B} S^2_{ab}. \quad (2.193)$$

By analogy with the ground-state and transition-state charge distributions given in (2.104,165), it is again convenient to introduce the overlap charge densities,

$$g^A_{a-b}(r) = \langle\psi^A_0|\varrho^A(r)|(\psi^A_0|b'/a)\rangle \quad (2.194)$$

and similarly for g^B_{b-a}.

Expressed in terms of these formal charge densities, the matrix element given in (2.191) assumes the form

$$\langle \psi_0^A \psi_0^B | V P_1 | \psi_0^A \psi_0^B \rangle = \sum_{a=1}^{N_A} \sum_{b=1}^{N_B} \Big\{ \int \int r_{AB}^{-1} g_{a-b}^A (r_A) g_{b-a}^B (r_B) dr_A dr_B$$

$$+ S_{ab} \Big[\int V^A (r_B) g_{b-a}^B (r_B) dr_B + \int V^B (r_A) g_{a-b}^A (r_A) dr_A \Big]$$

$$+ S_{ab}^2 \langle \psi_0^A \psi_0^B | V | \psi_0^A \psi_0^B \rangle \Big\}, \tag{2.195}$$

where $V^A(r_B)$ denotes the electrostatic potential at r_B due to the charge density f^A,

$$V^A(r_B) = \int r_{AB}^{-1} f^A (r_A) dr_A. \tag{2.196}$$

The equation for V^B is similar.

With the above definitions, the expression for the exchange energy is

$$\varepsilon_{\text{exch}} = - \sum_{a=1}^{N_A} \sum_{b=1}^{N_B} \Big\{ \int \int r_{AB}^{-1} g_{a-b}^A (r_A) g_{b-a}^B (r_B) dr_A dr_B$$

$$+ S_{ab} \Big[\int V^A (r_B) g_{b-a}^B (r_B) dr_B + \int V^B (r_A) g_{a-b}^A (r_A) dr_A \Big] \Big\}. \tag{2.197}$$

The explicit expression for g_{a-b}^A is similar to (2.168), namely

$$g_{a-b}^A (r) = -\varphi_a^A (r) \varphi'^B_b (r) \delta_{ab}^{\text{spin}}. \tag{2.198}$$

The expression for V^A is readily derived from (2.104,107,196). The result is

$$V^A(r_B) = \sum_\alpha Z_\alpha r_{\alpha B}^{-1} - \sum_{a=1}^{N_A} \int r_{AB}^{-1} [\varphi_a^A (r_A)]^2 dr_A. \tag{2.199}$$

Equation (2.197) is very convenient for analyzing $\varepsilon_{\text{exch}}$ in terms of localized charge distributions. Combining (2.189,194,198), the overlap charge density can be written as

$$g_{a-b}^A = \Big(\varphi_a^A \varphi_b^A - \sum_{c=1}^{N_A} S_{cb} \varphi_a^A \varphi_c^A \Big) \delta_{ab}^{\text{spin}}$$

$$= [\varphi_a^A \varphi_b^B - S_{ab} (\varphi_a^A)^2] \delta_{ab}^{\text{spin}} - \sum_{c=1;(c \neq a)}^{N_A} S_{cb} \varphi_a^A \varphi_c^A \delta_{ab}^{\text{spin}}. \tag{2.200}$$

If the φ's are localized orbitals, (2.200) may be regarded as a decomposition of the overlap charge density g_{a-b}^A into local, g'^A_{a-b}, and non-local, g''^A_{a-b}, parts:

$$g^A_{a-b} = g'^A_{a-b} + g''^A_{a-b}, \tag{2.201}$$

$$g'^A_{a-b} = [\varphi^A_a \varphi^B_b - S_{ab}(\varphi^A_a)^2]\delta^{\text{spin}}_{ab}, \tag{2.202}$$

$$g''^A_{a-b} = -\left(\sum_{c=1;(c\neq a)}^{N_A} S_{cb}\varphi^A_a \varphi^A_c \right)\delta^{\text{spin}}_{ab}. \tag{2.203}$$

A similar decomposition can obviously also be written for g^B_{b-a}.

It can be seen from (2.202) that the local part of g^A_{a-b} only depends on the orbitals φ^A_a and φ^B_b. The same is true for g^B_{b-a}. For this reason, a neglect of the non-local terms, g''^A_{a-b} and g''^B_{b-a}, in (2.201) and in the similar equation for g^B_{b-a} would result in the additivity of the first term of the exchange energy given in (2.197) with respect to molecular units (i.e., the inner shells, lone pairs and bonds corresponding to localized orbitals).

Intuitively it seems that the non-local parts of the overlap charge densities should be much smaller than the corresponding local terms. This follows from the product of different orbitals φ^A_a and φ^A_c in (2.203), which rapidly decreases with distance. If the orbitals a and b are the closest ones for a given configuration, the fact that the overlap integral S_{ab} in (2.202) is much larger than the integrals S_{cb} in (2.203) provides one more argument to neglect g'' compared to g'. Considering that the closest pair makes the largest contribution to the exchange energy, we conclude that the first term in (2.197) does represent a sum of nearly pairwise contributions from pairs of different localized orbitals.

The desired additive representation of the remaining two terms in (2.197) can only be obtained if the potentials V^A and V^B are replaced by the corresponding local terms, V^A_a and V^B_b, created by the orbitals φ^A_a and φ^B_b, respectively [the nuclear charge contributions to V^A and V^B may be distributed among the orbitals φ^A_a and φ^B_b using (2.117)]. The above substitution is, of course, justified if the orbitals a and b are the closest ones for the given configuration. For the remaining (non-closest) pairs such a substitution is senseless. It is, however, expected that the error introduded by this substitution is not very large because the contribution to the exchange energy from these remaining pairs is of less importance.

Thus, it is hoped that the exchange energy given in (2.197) can be represented with reasonable accuracy by a sum of terms $\varepsilon_{\text{exch},ab}$; each only dependent on the two orbitals indicated by the subscripts

$$\varepsilon_{\text{exch}} = \sum_a^{(A)} \sum_b^{(B)} \varepsilon_{\text{exch},ab}, \tag{2.204}$$

where

$$\varepsilon_{\mathrm{exch},ab} = -\int\int r_{AB}^{-1} g'^{A}_{a-b}(r_A) g'^{B}_{b-a}(r_B) dr_A dr_B$$
$$+ S_{ab}\left[\int V_a^A(r_B) g'^{B}_{b-a}(r_B) dr_B + \int V_b^B(r_A) g'^{A}_{a-b}(r_A) dr_A\right]. \qquad (2.205)$$

Turning to the physical meaning of the localized orbitals, we note that (2.204) involves the pairwise interactions between the inner shells, lone pairs and bonds corresponding to the localized orbitals φ_a^A and φ_b^B. As with the electrostatic and polarization energies, we can again neglect the contributions from the inner shells. This is now justified by the fact that $\varepsilon_{\mathrm{exch},ab}$ is nearly proportional to the square of the overlap integral S_{ab}, which is very small for the compact inner shell orbitals.

It is essential that the remaining (mostly, bond-bond) energies cannot be exactly reduced to a sum of atom-atom interactions. This immediately follows from the appearance of S_{ab}^2 in $\varepsilon_{\mathrm{exch},ab}$. Let us express the two orbitals a and b in terms of atomic orbitals [2.62,92],

$$\varphi_a = c_{1a}\chi_{1a} + c_{2a}\chi_{2a}, \qquad (2.206)$$

$$\varphi_b = c_{1b}\chi_{1b} + c_{2b}\chi_{2b}. \qquad (2.207)$$

In this case the overlap integral is given by

$$S_{ab} = c_{1a}c_{1b}\langle\chi_{1a}|\chi_{1b}\rangle + c_{1a}c_{2b}\langle\chi_{1a}|\chi_{2b}\rangle$$
$$+ c_{2a}c_{1b}\langle\chi_{2a}|\chi_{1b}\rangle + c_{2a}c_{2b}\langle\chi_{2a}|\chi_{2b}\rangle. \qquad (2.208)$$

It is now seen that, although the overlap integral itself reduces to a sum of atom-atom terms, the square does not.

Claverie [2.62]quoted the example of two hydrogen molecules, originally presented by *Salem* [2.92], to illustrate the inadequacy of the atom-atom representation for $\varepsilon_{\mathrm{exch}}$. Three configurations of the dimer are considered:

1. a linear configuration in which the closest atoms are a distance R apart;
2. a parallel configuration in which the molecules are a distance R apart;
3. a crossed configuration in which the four interatomic distances are all equal to R.

Using (2.208) with $c_{1a} = c_{2a} = c_{1b} = c_{2b}$ and $\chi_{1a} = \chi_{2a} = \chi_{1b} = \chi_{2b} = 1s$, it can easily be shown that the squares of the overlap integrals for these configurations are in the ratio,

$$S_{(1)}^2 : S_{(2)}^2 : S_{(3)}^2 \approx \frac{1}{4} : 1 : 4. \qquad (2.209)$$

On the other hand, if one represent S^2 by a sum of atom-atom functions $f(R)$, one gets

$$S_{(1)}^2 : S_{(2)}^2 : S_{(3)}^2 = \frac{1}{2} : 1 : 2. \qquad (2.210)$$

This means that if we choose $f(R)$ so as to reproduce $S^2_{(2)}$ exactly, $S^2_{(1)}$ will be overestimated and $S^2_{(3)}$ underestimated by a factor of two.

This apparent shortcoming of the atom-atom representation for S^2 is not, however, as serious as it appears at first sight. Inasmuch as S^2 is a very steep function of the intermolecular distance, the factor of two can easily be compensated by a small variation in R. In other words, if we choose $f(R)$ so as to reproduce $S^2_{(2)}$ exactly, the configurations for which the atom-atom estimates of S^2 satisfy the desired ratio given in (2.209) will not deviate much from the configurations (1) and (3).

3. The Atom-Atom Potential Method

In this chapter we shall give a general outline of the atom-atom potential method and discuss the ways most frequently used to derive the atom-atom-potential parameters. A large number of parameter sets, which covers the most important atomic species encountered in organic compounds, is presented. In all cases we have indicated the particular compounds and physical properties used in fitting the parameters of a potential. We hope that this material will assist the reader in choosing the set best suited to the particular problem he is interested in.

3.1 General Remarks

The preceding chapter has shown that an ab initio calculation of the intermolecular interaction energy for an intermediate separation is non-trivial using either the supermolecule or the perturbation method. Although the formal aspects of the problem have been developed fairly well, the computational difficulties, ultimately associated with the evaluation of a large number of two-electron integrals, seriously restrict the application of the ab initio methods to large and medium-sized molecules.

The prohibitive cost of ab initio calculations has led to the development of various semiempirical and empirical methods. All these methods can be roughly classified into two types. The first type utilizes explicitly the basic equations of the preceding chapter (e.g., the Roothaan equations) but approximates the time-consuming integrals by some parametrization or completely neglects some of the integrals. Descriptions and discussions of such methods, as applied to MO theory, may be found in [3.1]. For semiempirical versions of the Heitler-London and perturbation treatments, the reader is referred to [3.2–5].

The other type deals directly with the intermolecular potential; i.e. it simulates the potential using certain analytical functions. The parameters of these analytical functions are either treated empirically or approximated from theory.

The choice between the two approaches is a matter of taste since both have their advantages and disadvantages.

The major advantage of an analytical method lies in its extreme computational simplicity. For calculations involving repeated evaluations of the interaction potential for a large number of configurations, this advantage is of decisive importance. To illustrate this point we note that a calculation of the thermodynamic functions of a crystal using the cell model (Chap. 6) requires the evaluation of the lattice energy for 10^3–10^4 crystal configurations. Each evaluation of the lattice energy, in turn, involves an evaluation of the interaction potential for 15–20 distinct molecular pairs. It is clear that such calculations can only be performed with very simple analytical potentials. The same is even true for relatively easy calculations such as the calculations of the crystal structure of minimum energy or the calculations of force-constant matrices.

The central problem of an analytical method involves the choice of an adequate analytical representation for the interaction potential. The representations based on single-center multipole expansions show poor convergence even with such comparatively small molecules as ethylene and azabenzenes. This leads to the idea of partitioning the interacting molecules into small molecular fragments, i.e., bonds or atoms. The aim of such a partitioning is two-fold. First, it is expected that interactions between small fragments can be accurately described by one or two leading terms in the multipole expansions. Second, it is hoped that the properties (such as the multipole moments and polarizabilities) of the fragments will show good transferability between different molecules.

There are a number of arguments in favor of the approximation of transferable fragments. These are, first of all, the observations made in descriptive chemistry, which make it possible to identify particular atoms or bonds by their characteristic properties. Physical observations have also indicated that atoms and bonds retain their individuality, to a large extent, within the broad classes of organic molecules (e.g., within the aliphatic hydrocarbons). A representative example is due to vibrational spectroscopy which reveals the existence of characteristic frequencies attributable to particular bonds.

Further arguments are provided by the success of various additive schemes which interpret the molecular properties (such as atomization energy, dipole moment, polarizability, refractive index, etc.) in terms of bond or atomic increments.

There have been numerous attempts to justify theoretically the partitioning of molecules into small fragments and to define the "best transferable fragment". A comprehensive discussion of this subject and a number of relevant examples may be found in [3.6].

The principal difficulty involved in a theoretical treatment of the problem is the lack of a unique and physically-sound criterion to partition the molecular charge distribution into regions of localized charge. This leads to a diversity of the partitioning schemes which we shall now illustrate.

In the preceding chapter it has been mentioned that the canonical Hartree-Fock orbitals can be transformed, by a unitary transformation, to a set of localized orbitals corresponding to inner electronic shells, bonds and lone pairs. Numerical calculations [3.6] have shown a good transferability of the localized orbitals between chemically related molecules. Applied to the problem of intermolecular forces (Sect. 2.2.2–4), the localized-orbital description reduces the interaction potential to a sum of interactions between all bonds and lone pairs. The contribution of inner shells can be neglected.

The localized-orbital description, however, can hardly be regarded as a conceptual justification of the partitioning of a molecule into bonds and lone pairs since the very notion of an orbital does not have a clear physical meaning. It originates from a particular (Hartree-Fock) representation of the wave function.

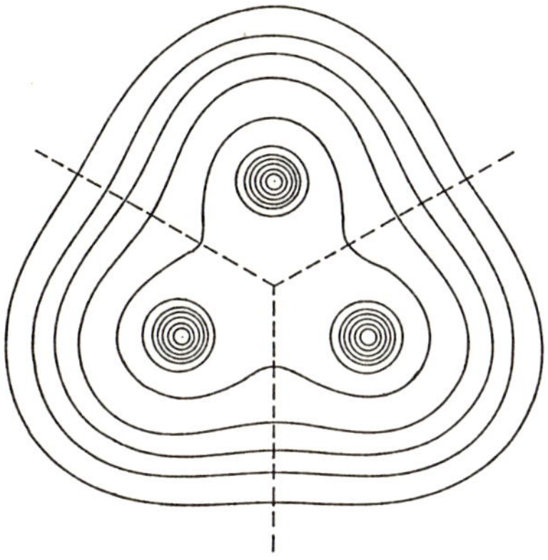

Fig. 3.1. Contour plot of the electron density in the plane of the carbon nuclei in cyclopropane. Partitioning surfaces are denoted by (- - -)

It is possible to solve the problem by dealing directly with the molecular charge distribution; a well-defined and observable property. This has been done, in particular, by *Bader* et al. [3.7–12] whose partitioning scheme is illustrated in Fig. 3.1. The molecular charge distribution is partitioned by the surface $S(\boldsymbol{r})$, defined by the gradient of the charge density $\nabla f(\boldsymbol{r})$, that passes through the point of minimum density between a pair of adjacent nuclei. A remarkable property of fragments bounded by such "zero-flux" surfaces is that they possess many theoretical properties and definitions in common with a total isolated system. In particular, the kinetic and potential energies of *Bader*'s fragments (and of the entire molecule as well) are well-defined and satisfy the virial theorem.

In *Bader*'s opinion, those fragments are the most transferable pieces of a molecule. This is because the zero-flux surface – by the nature of its definition in terms of $f(\boldsymbol{r})$ – maximizes the extent to which the charge distributions of the fragments are transferable between different systems. In principle, it is possible to describe the charge distributions in terms of multipole moments and polarizabilities and then use these charge distributions to calculate the intermolecular interaction energy.

Figure 3.1 shows that *Bader*'s virial fragments are not "bond-like" but rather "atom-like" in nature. This is just the result opposite to the localized-orbital description. Indeed, this is an example of how different partitioning schemes can lead to quite different results.

Besides the ambiguity involved in the definition of the best partitioning scheme, the theoretical treatment of the problem involves a further difficulty. We here imply the fact that the present theory is not yet capable of providing a reliable practical means for calculating the properties of molecular fragments, be they bond-like or atom-like. This has been demonstrated in the previous chapter for the bond multipole moments and polarizabilities. These are markedly sensitive to the particular wave function used in the calculation.

In such a situation it seems quite natural to adopt the simplest representation of the interaction potential and to treat the parameters of this representation in a purely empirical way, i.e., by fitting them to some selected properties of the system or to an "exact" potential calculated by an ab initio method. In such an empirical approach, the difference between the bond-bond and atom-atom representations of the interaction potential seems to be of minor importance since the interactions between the bonds can be simulated well by the interactions between the corresponding atoms (and vice versa).

From the computational point of view, the atom-atom representation of the interaction potential is obviously more convenient. This representation is closely related to the close-packing principle, so that it is natural to use it as a basis for calculating the properties of organic solids.

3.2 Formulation of the Atom-Atom Method

Formally, the atom-atom potential method may be regarded as an attempt to expand the interaction energy of two polyatomic molecules, ε^{AB}, over some (incomplete) set of two-center functions φ_{ab}, with the force centers placed on the atomic nuclei:

$$\varepsilon^{AB} = \sum_{a,b} \varphi_{ab}, \tag{3.1}$$

where the subscripts a and b run over all atoms in molecules A and B, respectively.[1]

There are at least three "levels of simplicity" that may be tried for the atom-atom potentials φ_{ab}. In the simplest model the potential is assumed to be isotropic and the parameters are assumed to depend only on the chemical nature of the interacting atoms (i.e., on the atomic numbers). If

[1] It is more convenient to henceforth label the nuclei using the Latin letters a, b, etc., thus departing from the earlier convention of labelling the nuclei with Greek letters.

this model is not flexible enough, we may assume that the parameters also depend on the valence state of the interacting atoms. And finally, in the most complicated model we may introduce anisotropic functions.

Numerous applications of the atom-atom potential method to simulate the interaction energy between organic molecules have shown that even the simplest model is already flexible enough to reproduce the properties of organic solids. Anisotropic potentials are now used very rarely; mainly to describe the hydrogen bond. But even in these rare cases the introduction of anisotropic potentials seems to be superfluous.

In the "isotropic" models the atoms a and b are usally assumed to behave like the atoms of rare gases via a (6-exp) Buckingham potential

$$\varphi_{ab} = -A_{ab}r_{ab}^{-6} + B_{ab}\exp(-C_{ab}r_{ab}), \tag{3.2}$$

or a (6-m) Lennard-Jones potential,

$$\varphi_{ab} = -A_{ab}r_{ab}^{-6} + B_{ab}r_{ab}^{-m} \quad (m = 8 - 14), \tag{3.3}$$

where r_{ab} is the interatomic distance.

More flexible potential functions are obtained by introducing a Coulombic term, $q_a q_b r_{ab}^{-1}$, into (3.2 or 3), with q_a and q_b formally treated as effective atomic charges. The resulting potentials are usually referred to as (6-exp-1) and (6-m-1) potentials, respectively.

Clearly, the value of a method is the higher, the smaller the number of adjustable parameters. In the simplest version of the atom-atom potential method it is assumed that A_{ab}, B_{ab} and C_{ab} only depend on the specific atoms involved in the interaction, regardless of their valence state or their molecular environment. In this version the total number of empirical parameters to be determined is rather small. Thus, if we use (6-exp) potentials, the interactions between hydrocarbon molecules will be described by only nine distinct parameters: $A_{ss'}$, $B_{ss'}$, and $C_{ss'}$ (s, s' = H or C). To further reduce the number of independent parameters, various combining rules for cross atomic interactions are usually applied. The geometric-mean combining rule for A and B

$$X_{ss'} = (X_{ss}X_{s's'})^{1/2}, \quad (X = A \text{ or } B), \tag{3.4}$$

and the arithmetic-mean combining rule for C

$$C_{ss'} = (C_{ss} + C_{s's'})/2, \tag{3.5}$$

are most frequently used. Applying these combining rules to hydrocarbons reduces the total number of independent empirical parameters to six.

In a more flexible version of the atom-atom potential method the atoms of the interacting molecules not only differ in their chemical nature but also differ in their surroundings in the molecules, or by some quantitative criteria characterizing the electron density and energy of the atoms. Thus, instead of using the same potential for all carbon atoms, one may, for example, distinguish between aliphatic and aromatic carbons and treat them as distinct atomic species. In this case, the number of parameters increases, so that the number of observations needed to determine these parameters must increase as well.

As mentioned in Sect. 3.1, the empirical parameters can be derived by either fitting them to selected properties of the system or to the "exact" intermolecular potential energy hypersurface calculated by an ab initio method. In the former case, one takes an arbitrary set of parameters, computes, for example, the lattice energy and equilibrium structure of a crystal, and then varies the parameters of the potential so as to achieve the best agreement between theory and experiment. In the latter case, one first calculates the ab initio interaction energy for some selected configurations, computes the corresponding model values for a trial set of parameters, and then varies the parameters so as to minimize the squares of the differences between the "exact" and model energies.

Thus, in both cases the atom-atom potential method represents a variational treatment using the atom-atom potentials given in (3.2,3) as trial functions, and $A_{ss'}$, $B_{ss'}$, $C_{ss'}$ as variational parameters. Clearly, there is no reason to attach any physical meaning to the parameters resulting from this variation treatment. Accordingly, there is no reason to regard the sum of the $-Ar^{-6}$ terms as the dispersion energy, the sum of the $B\exp(-Cr)$ terms as the exchange energy and the sum of the $qq'r^{-1}$ terms as the electrostatic energy. Since the atom-atom potential model is adjusted as a whole and it is not an exact representation of the true potential, all energy components become mixed after the fitting. The result is that the correspondence between the above atom-atom sums and the individual energy components is lost. An important consequence is that the use of (6-exp) or (6-m) potentials without Coulombic terms does not mean that the electrostatic component of the interaction energy has been completely neglected. If we adjust the potential as a whole, the electrostatic interactions will obviously be reflected in the resulting model, even though no explicit electrostatic terms appear in φ_{ab}.

Of course, the inclusion of explicit Coulombic terms should make the model more flexible and improve the description of the electrostatic interaction energy. However, the appearance of a Coulombic term is always displeasing from the computational point of view because of the poor convergence of Coulombic lattice sums. Also, the addition of new terms to (3.2,3) necessarily introduces new empirical parameters, which reduces the value of the model. Under these circumstances, it is always desirable, if possible, to do it without Coulombic terms.

The above speculations concerning the physical meaning of the atom-atom potentials do not refer to the case in which the fitting of (3.1) to the "exact" potential-energy hypersurface is carried out for each individual energy component separately. Thus, one may attempt to fit the dispersion energy by the sum of the $-A_{ab}r_{ab}^{-6}$ terms, the electrostatic energy by the sum of the $q_a q_b r_{ab}^{-1}$ terms, etc. This is probably the only case in which each of the above sums acquires a physical meaning. (Note that even then it is hardly justifiable to ascribe a physical meaning to the parameters themselves.)

We shall now make some preliminary remarks concerning the two purely empirical versions of the atom-atom potential method. We first consider the case in which the model potential has been fitted directly to the potential-energy hypersurface $\varepsilon(\boldsymbol{p})$, as calculated by a more or less accurate ab initio method. Figure 3.2a shows a hypothetical $\varepsilon(\boldsymbol{p})$ describing the interaction of two identical molecules A (solid line). For the sake of simplicity, the six coordinates p_i specifying the relative positions and orientations of the molecules are presented as a single parameter p. Thus, any point p on the abscissa corresponds to a particular configuration and energy $\varepsilon(p)$ of the interacting molecules. Let $\boldsymbol{x} = \{x_k\}$ be a vector whose components are the parameters $A_{ss'}$, $B_{ss'}$ and $C_{ss'}$. Let $\varepsilon'(p, \boldsymbol{x})$ denote the approximate inter-molecular potential as given by the model potential with the parameters $\boldsymbol{x}$. The fitting of $\varepsilon'(p, \boldsymbol{x})$ to $\varepsilon(p)$ generally involves the sampling of a set of configurations $p^{(i)}$ (at random or on a grid), the computation of $\varepsilon[p^{(i)}]$ and then the minimization of the quantity

$$\sum_i \{\varepsilon[p^{(i)}] - \varepsilon'[p^{(i)}, \boldsymbol{x}]\}^2, \quad \text{or} \tag{3.6}$$

$$\sum_i \{\varepsilon[p^{(i)}] - \varepsilon'[p^{(i)}, \boldsymbol{x}]\}^2 / \varepsilon[p^{(i)}]^2, \tag{3.7}$$

with respect to $\boldsymbol{x}$.

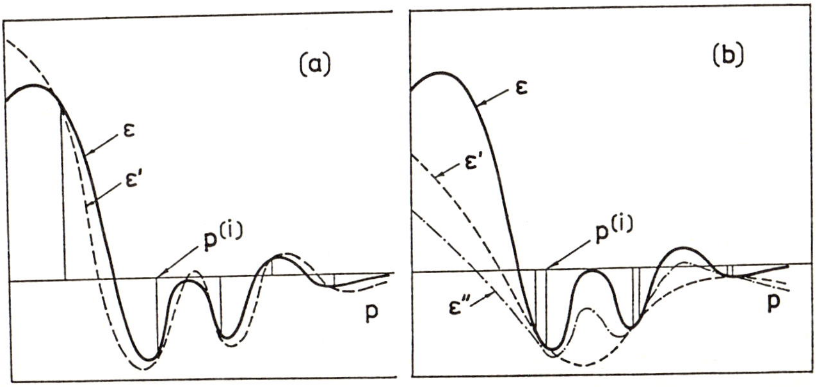

Fig. 3.2a,b. Schematic illustration of the fitting of a potential energy hypersurface ε by the model potentials ε' and ε'' : **(a)** direct fitting; **(b)** indirect fitting using selected crystal properties. (Arbitrary units)

A model potential obtained in such a way is shown in Fig. 3.2a by the broken line. The greater the number of sampled configurations $p^{(i)}$, the closer is the model potential $\varepsilon'(p, \boldsymbol{x})$ to the "exact" potential $\varepsilon(p)$, provided, of course, that the former is sufficiently flexible to reproduce the latter.

Let $\boldsymbol{x}^A$ denote the set of parameters that best describes the interaction of two molecules A. An important question that now arises is whether or not the set $\boldsymbol{x}^A$ is transferable to molecules similar to A (i.e., containing the same atoms, bonds and functional groupings). It is difficult to deduce an answer from theoretical considerations but practice has shown that such a transfer can be made. Direct evidence to support this point has been reported by *Clementi* et al. [3.13]. The results will be discussed in Sect. 3.5.

The observed transferability of atom-atom potentials is sometimes considered to be an indication that the potentials describe the inherent physical properties of the atoms involved. It is not unlikely, however, that the atom-atom potentials merely *simulate* the interactions of bonds or larger groups of atoms. Clearly, if we consider a series of molecules containing identical structural groups, the atom-atom potentials describing the interactions of these groups should be approximately the same for all molecules.

We now turn to the case in which the parameters $\boldsymbol{x}$ of the potential are determined by fitting them to selected crystal properties. The solid line in Fig. 3.2b represents the potential-energy hypersurface $\varepsilon(p)$ for two molecules A. The fitting to experimental crystal properties differs from that discussed previously mainly in that we are now unable to sample the trial configurations $p^{(i)}$ arbitrarily and, what is even more important, to increase the number of sampled points $p^{(i)}$. The "trial" configurations $p^{(i)}$ are now completely predetermined by the crystal structure. This number is very limited. To explain the latter, we note that the latttice energy and all potential-dependent properties of an organic crystal are mainly governed by the interaction of a molecule with its nearest neighbors. The number of nearest neighbors is usually 12–14. Due to crystal symmetry, the number of non-equivalent pairs formed by the central molecule and its nearest neighbors is even smaller. Thus, in a crystal of formic acid only 7 non-equivalent configurations make a significant contribution to the lattice energy. In general, the higher the crystal symmetry, the smaller the number of non-equivalent configurations that determine the lattice energy and other crystal properties.

In Fig. 3.2b the "important" configurations $p^{(i)}$ are shown by vertical lines. It is clear that any model potential $\varepsilon'(p, \boldsymbol{x})$ which fits $\varepsilon(p)$ and its derivatives *just at the points* $p^{(i)}$,

$$\varepsilon'[p^{(i)}, \boldsymbol{x}] \approx \varepsilon[p^{(i)}], \tag{3.8}$$

$$\frac{\partial^n \varepsilon'}{\partial p^n}[p^{(i)}, \boldsymbol{x}] \approx \frac{\partial^n \varepsilon}{\partial p^n}[p^{(i)}], \tag{3.9}$$

will correctly reproduce the lattice energy and all of the crystal properties sensitive to molecular displacements. If, for example, we are interested in the equilibrium structural parameters, only the first derivatives of $\varepsilon(p)$ must be fitted [$n = 1$ in (3.9)]. For lattice vibrations or elastic properties, the second derivatives must be fitted, too.

In fact, the constraints imposed upon the parameters $\boldsymbol{x}$ are less restrictive than the ones given in (3.8,9). Thus, in order to reproduce the lattice energy ϕ, it is not necessary to reproduce each energy $\varepsilon[p^{(i)}]$ but only the sum

$$\sum_i \varepsilon'[p^{(i)}, \boldsymbol{x}] = \phi. \tag{3.10}$$

Similar conditions can be formulated for other crystal properties y_m dependent on the derivatives of $\varepsilon(p)$:

$$f_m\left\{\frac{\partial^n \varepsilon'}{\partial p^n}[p^{(i)}, \boldsymbol{x}]\right\} = y_m, \tag{3.11}$$

where f_m represents a functional which relates the property y_m to the derivatives $\partial^n \varepsilon / \partial p^n$.

In general there is, of course, no garantee that a model potential $\varepsilon'(p, \boldsymbol{x})$ satisfying (3.10,11) will simultaneously satisfy (3.8,9). However, if the number of crystal properties involved in the fitting is sufficiently large, it is very likely that this is indeed so. An argument to support such a conclusion is that a model fitted to a sufficiently large set of crystal properties is usually capable of predicting crystal properties not included in the original fitting. This suggests that the model under consideration does reflect the behavior of the interaction potential in the vicinity of the important configurations. It is indeed very unlikely that a correct prediction of other properties is always due to chance.

We now suppose that by fitting the parameters $\boldsymbol{x}$ to the crystal properties we have obtained a model potential $\varepsilon'(p, \boldsymbol{x})$ which satisfies (3.8,9). As can be seen in Fig. 3.2b, where $\varepsilon'(p, \boldsymbol{x})$ is shown by a broken line, this does not necessarily mean that the model potential describes the entire potential-energy hypersurface $\varepsilon(p)$ equally well. Since the number of configurations $p^{(i)}$ governing the properties of the given crystal is very small (usually, 6–7), the model potential $\varepsilon'(p, \boldsymbol{x})$ may give erroneous results for configurations that are not encountered in the crystal. There are numerous examples to illustrate this point. For example, it has been convincingly demonstrated that many of the physical properties of the benzene crystal, such as the lattice energy, equilibrium structure, elastic constants, phonon frequencies and thermodynamic functions, can be satisfactorily reproduced using (6-exp) atom-atom potentials. However, the same potentials, when applied to the benzene dimer, always yield a sandwich-like arrangement as the most stable configuration. This is in conflict with experiment and ab initio calculations [3.14]. This apparent contradiction can be explained by the simple fact that the sandwich-like arrangement of molecules is not typical of the nearest neighbors in the benzene crystal. In other words, among the important points $p^{(i)}$ there is no point which describes the sandwich-like arrangement. For this reason, it is indeed possible to have a model potential which describes the crystal properties very well but, simultaneously, yields wrong results for a sandwich-like arrangement of the dimer.

A further conclusion that can be drawn from an analysis of Fig. 3.2b is that the potential-dependent properties of a crystal may be reproduced by more than one model potential. Since the part of the potential-energy hypersurface that is responsible for the crystal properties is very limited, it can be fitted with a number of distinct models including,of course, the model which fits the entire potential-energy hypersurface $\varepsilon(p)$. A possible model potential $\varepsilon''(p, \boldsymbol{x})$ which fits $\varepsilon(p)$ and its derivatives at $p^{(i)}$, but does not coincide with $\varepsilon'(p, \boldsymbol{x})$, is shown in Fig. 3.2b by a dash-dotted line.

If the number of crystals involved in the fitting is increased, the number of restrictive conditions given in (3.8,9) increases as well. Hence, the parameters of the potential can be determined with greater certainty. It is hoped that by increasing the number of crystals and crystal properties included in the fitting, the potenial parameter set approaches that which would be obtained by fitting the corresponding potential-energy hypersurface of the dimer.

For the sake of completeness, we must consider another version of the method, which will hereafter be called semiempirical. In this version some of the parameters are given a physical meaning. More frequently, these are atomic charges q_a which are not included in the fitting procedure but are determined, for example, from Mulliken populations. Although such an approach is aesthetically displeasing and is not quite consistent, we shall nevertheless pay some attention to it in Sect. 3.6.

We conclude this subsection by noting that the atom-atom potential method is sometimes used to fit not the entire interaction potential but only a part of it, e.g., the exchange and the dispersion contributions. The other contributions are evaluated using either a non-analytical method, a single-center multipole expansion, or a bond-bond model. Such a mixed approach seems to be justified if the interaction potential involves contributions which are a priori known to be poorly described by a normal atom-atom model. This is the case for strong electrostatic interactions which are poorly described by (6-exp-1) potentials, to say nothing of (6-exp) potentials. The addition of new, wisely-chosen terms to the model potential results in a marked improvement in the fit. Several relevant examples will be discussed in this and the following chapters.

3.3 Determination of Atom-Atom Potentials from Crystal Data

3.3.1 Fitting Procedures

To formulate the problem of finding the parameters of the potential that best reproduces the crystal properties, let us introduce two column vectors, $\boldsymbol{y}^0 = \{y_m^0\}$ and $\boldsymbol{y} = \{y_m\}$, whose components are the observed and calculated crystal properties, respectively, (the subscript m runs over all properties of all crystals involved in the fitting). Let $\boldsymbol{x} = \{x_k\}$ be a vector whose components are all unknown parameters $A_{ss'}$, $B_{ss'}$, etc. It will be assumed that $\boldsymbol{x}$ contains only independent variable parameters, while the remaining parameters are kept fixed or can be determined from the independent ones by applying combining rules such the ones given in (3.4,5). In general, the y_m are nonlinear functions of $\boldsymbol{x}$, i.e., $\boldsymbol{y} = \boldsymbol{y}(\boldsymbol{x})$. The problem of deriving the optimum parameters is thus to find a vector $\boldsymbol{x}^{\text{opt}}$ which provides as close a fit of $\boldsymbol{y}(\boldsymbol{x})$ to $\boldsymbol{y}^0$ as possible.

Before considering the ways of solving this problem, we shall briefly discuss the particular crystal properties most frequently used as components of $\boldsymbol{y}$ and $\boldsymbol{y}^0$. The aim of this preliminary discussion is to define the accuracy that may be regarded as acceptable in the fitting of y_m to y_m^0.

Lattice Energy. In most of the studies concerned with the determination of parameters the intermolecular part of the lattice energy, ϕ, is fitted directly to the observed enthalpy of sublimation, ΔH_s. The validity of equating ϕ to ΔH_s can be estimated from the definition of ΔH_s, by assuming the gaseous phase to be ideal:

$$\begin{aligned} H_{\text{gas}} &= PV_{\text{gas}} + 3RT + H_{\text{gas,intra}} \\ &= 4RT + \phi_{\text{gas,intra}} + E_{\text{gas,intra vib}}, \end{aligned} \tag{3.12}$$

where $\phi_{\text{gas,intra}}$ and $E_{\text{gas,intra vib}}$ denote the intramolecular potential and vibrational energies, respectively. As to H_{solid}, it may be represented as

$$\begin{aligned} H_{\text{solid}} =& \phi_{\text{inter}} + E_{\text{solid, inter vib}} \\ &+ \phi_{\text{solid, intra}} + E_{\text{solid, intra vib}}, \end{aligned} \tag{3.13}$$

where the meaning of the last three terms is obvious.

Subtracting (3.13) from (3.12), one obtains

$$\begin{aligned} \Delta H_s =& 4RT + \Delta\phi_{\text{intra}} + \Delta E_{\text{intra vib}} \\ &- (\phi_{\text{inter}} + E_{\text{solid, inter vib}}), \end{aligned} \tag{3.14}$$

where Δ denotes the difference between the gaseous and solid states.

The intramolecular energy difference $\Delta\phi_{\text{intra}}$ usually contributes very little to ΔH_s. For typical organic crystals made up of "rigid" molecules,

$\Delta\phi_{\text{intra}}$ is of the order of 10^{-3}kcal/mol [3.15]. It seems that $\Delta\phi_{\text{intra}}$ is not very significant even for flexible molecules whose conformation in the gaseous phase may markedly differ from that in the crystal state. Indeed, the weak crystal field may appreciably affect the molecular conformation only when the intramolecular potential energy surface is very shallow. This suggests that the intramolecular energy differences cannot be very large even though the molecular conformations may differ significantly.

The term $\Delta E_{\text{intra vib}}$ is usually more important than $\Delta\phi_{\text{intra}}$. In each particular case it can be readily estimated by comparing the vibrational frequencies in the gaseous and crystal states. In general, the significance of $\Delta E_{\text{intra vib}}$ increases with stronger intermolecular forces and weaker intramolecular forces. The contribution of $\Delta E_{\text{intra vib}}$ to ΔH_s may be important even at absolute zero. Thus, for n-hexane and n-octane $\Delta E_{\text{intra vib}}$ amounts to 5% of the corresponding enthalpies of sublimation at 0 K [3.15].

The magnitude of the last term in (3.14) is approximately 2–3 kcal/mol for most organic crystals at room temperature. According to *Warshel* and *Lifson* [3.15], $E_{\text{solid,inter vib}}$ is an important factor even at absolute zero. About 5% of the zero-point enthalpy of sublimation of n-hexane is accounted for by this term.

Rough estimates of ϕ from ΔH_s can usually be made by neglecting $\Delta\phi_{\text{intra}}$ and $\Delta E_{\text{intra vib}}$ and by assuming that $E_{\text{solid,inter vib}}$ is equal to $6RT$. This reduces (3.14) to

$$\Delta H_s \approx -2RT - \phi_{\text{inter}}. \tag{3.15}$$

The term $2RT$, about 1.2 kcal/mol at room temperature, is frequently ignored.

Considering the assumptions made in passing from (3.14 to 15) as well as the inaccuracy involved in the experimental determination of ΔH_s (1–3 kcal/mol), it becomes clear that discrepancies up to 3–4 kcal/mol between the calculated and observed enthalpies of sublimation should not cause any concern when judging the quality of the parameters of a potential.

From the computational point of view, an estimation of the lattice energy is reduced to a calculation of the lattice sums. Thus, for Buckingham-type potentials the energy is computed as

$$\phi = \frac{1}{2}\sum_{ss'}\Big[-A_{ss'}\sum_{(ss')} r^{-6} + B_{ss'}\sum_{(ss')}\exp(-C_{ss'}r)\Big], \tag{3.16}$$

where the subscripts s and s' run over all atomic species in the crystal. The summation in the second and third sums is carried out over all atomic pairs of the interaction type specified by the subscripts s and s'. For the lattice sums we shall use the notation,

$$S^{(m)} = \sum r^{-m}, \quad S^{(\exp)} = \sum \exp(-Cr). \tag{3.17}$$

Explicit expressions for the lattice sums will be presented in Chap. 4. Here we only note that it is the computation of the lattice sums and their derivatives that is the most time-consuming step in the calculations of crystal properties. This circumstance will be taken into account when comparing the various fitting procedures.

Parameters of the Equilibrium Crystal Structure. In most applications of the atom-atom potential method the equilibrium structure of a crystal is assumed to be the structure which corresponds to a minimum of the lattice energy, ϕ. The latter is considered to be a function of a column vector $\boldsymbol{p} = \{p_i\}$ whose components are the structural parameters needed to completely specify the crystal structure. Thus, the equilibrium parameters $\bar{\boldsymbol{p}}$ are defined to be those which satisfy the equation,

$$\nabla_p \phi(\bar{\boldsymbol{p}}) = 0, \tag{3.18}$$

where $\nabla_p \phi$ denotes the gradient of ϕ with respect to p_i,

$$\nabla_p \equiv \{\partial / \partial p_i\}. \tag{3.19}$$

The use of the lattice energy instead of the free energy in the determination of the equilibrium crystal structure involves two kinds of errors. The first kind involves the neglect of thermal expansion which usually changes the dimensions of the unit cell by values of the order of 0.1 Å and the molecular orientation by several degrees (in going from absolute zero to room temperature). Secondly, even at 0 K the crystal structure is not governed by the lattice energy alone since the structural parameters may be appreciably affected by zero-point vibrations. For example, the effect of the zero-point vibrations on the unit-cell dimensions of n-hexane and n-octane was found to vary from 0.06 to 0.16 Å, which is comparable to the effect of thermal expansion [3.15].

Thus, in fitting the equilibrium structural parameters, as defined by (3.18), to the corresponding observed values it does not make any sense to strive for an accuracy better than several tenths of an Angstrom in the unit-cell dimensions and several degrees in the molecular orientation. This also refers to the case when one deals with the experimental structural data extrapolated to absolute zero. Of course, in all cases the calculations should yield a structure somewhat denser than the one observed experimentally.

In deriving the optimum parameters it is sometimes more convenient to deal with the derivatives $f_i \equiv -\partial\phi / \partial p_i$, rather than the structural parameters themselves. These derivatives may be formally treated as generalized forces that act upon the crystal structure. If the f_i's are computed at the observed crystal configuration $\boldsymbol{p} = \boldsymbol{p}^0$, the respective components of $\boldsymbol{y}^0$ should all be taken equal to zero. The validity of using the equation,

$$\nabla_p \phi(\boldsymbol{p}^0) = 0, \tag{3.20}$$

can be discussed in the same terms that were used for (3.18). The tolerable deviations of $\partial\phi/\partial p_i$ from zero can be roughly estimated by expanding $\nabla_p\phi(\bar{\boldsymbol{p}})$ in a Taylor series about $\boldsymbol{p}^0$ and truncating the series after the second term,

$$\nabla_p\phi(\bar{\boldsymbol{p}}) = \nabla_p\phi(\boldsymbol{p}^0) + \underline{\underline{\phi_p}}(\boldsymbol{p}^0)(\bar{\boldsymbol{p}} - \boldsymbol{p}^0), \tag{3.21}$$

where $\underline{\underline{\phi_p}}$ is the matrix of the second derivatives

$$\underline{\underline{\phi_p}} = \{\partial^2\phi/\partial p_i\partial p_j\}. \tag{3.22}$$

Since $\nabla_p\phi(\bar{\boldsymbol{p}}) = 0$ (by the definition of $\bar{\boldsymbol{p}}$), (3.21) becomes

$$\nabla_p\phi(\boldsymbol{p}^0) = \underline{\underline{\phi_p}}(\boldsymbol{p}^0)(\boldsymbol{p}^0 - \bar{\boldsymbol{p}}). \tag{3.23}$$

The tolerable deviations from (3.20) may be estimated using the error thresholds of the structural parameters. In each particular case, however, the deviations depend on the slope of the potential at $\boldsymbol{p} = \boldsymbol{p}^0$, so that the matrix $\underline{\underline{\phi_p}}$ must be known in order to evaluate them.

Elastic Constants. In order to write down the equation for the elastic constants, the components of the vector $\boldsymbol{p}$ are divided into two groups: those describing the lattice deformations (unit-cell dimensions and angles) and those describing the internal deformations (molecular rotations and translations). The lattice deformations are usually specified by the deformation tensor ε_i whose components can be readily expressed in terms of the unit-cell parameters. The internal deformations u_k are implicitly dependent on the lattice deformations,

$$\partial\phi/\partial u_k = 0. \tag{3.24}$$

The elastic constants c_{ij} can be computed from the equation (Chap. 6),

$$c_{ij} = \left(\frac{\partial^2\phi}{\partial\varepsilon_i\partial\varepsilon_j} + \sum_k \frac{\partial^2\phi}{\partial\varepsilon_i\partial u_k}\frac{\partial u_k}{\partial\varepsilon_j}\right)/V_c, \tag{3.25}$$

where V_c is the volume of the unit cell. Again we imply that all of the contributions to the free energy of the crystal may be neglected except for the static lattice energy ϕ.

The elastic constants usually exhibit a pronounced temperature dependence. Thus, the magnitude of c_{33} of the naphthalene crystal, extrapolated to absolute zero, is about twice the value observed at room temperature. It is therefore clear that an inclusion of elastic constants in a fit only makes sense when the available experimental data allows an extrapolation of the observed constants to absolute zero. Unfortunately, low-temperature data on the elastic properties of organic crystals are meager.

Phonon Frequencies. The inclusion of phonon frequencies in a fit is always very desirable. Unlike the lattice energy and equilibrium structural parameters, the phonon frequencies are related to the second derivatives of the potential and thus allow a more thorough examination of the potential energy hypersurface to be made.

Experimental phonon frequencies, as determined by IR and Raman spectroscopy or the coherent inelastic scattering of thermal neutrons, are usually accurate to within 1 to 2 cm^{-1} [3.16]. The accuracy of the calculated frequencies depends to a great extent on the model that is adopted in solving the lattice dynamical problem. It is a standard practice of many researchers to assume that the molecules in a crystal move as rigid units. The reliability of this assumption has been examined by *Pawley* and *Cyvin* [3.17] for naphthalene and by *Dolling* et al. [3.18] for hexamethylenetetramine. It has been found that the differences between the values for the rigid and deformable-molecule models may be as much as 7 to 9 cm^{-1}. For more flexible molecules the differences may be even larger.

A further source of inaccuracy in the calculated phonon frequencies lies in the neglect of the anharmonicity of the lattice vibrations. An attempt to estimate the anharmonic contributions to the lattice frequencies has been made by *Warshel* [3.19] for the librational frequencies of the n-hexane crystal. They prove to be very large, ranging up to 40% of the frequency computed in the harmonic approximation. It seems, however, that the effect of the anharmonicity should not, in fact, be so serious. The numerous calculations of the phonon frequencies using the atom-atom potential method and the harmonic approximation reproduce the experimental data with a much better accuracy. It is not unlikely that the atom-atom potentials, when fitted to the observed (anharmonic) frequencies using the harmonic model, partly absorb the anharmonic effects and then effectively reproduce these effects in the lattice dynamical calculations.

Considering the above remarks, it seems reasonable to adopt a value of 10 cm^{-1} as a tolerable difference between the observed phonon frequencies and those calculated using the harmonic rigid-molecule approximation.

We now proceed to a discussion of the methods of fitting $\boldsymbol{y}(\boldsymbol{x})$ to $\boldsymbol{y}^0$. The strategy used to find the best parameters $\boldsymbol{x}^{\mathrm{opt}}$ depends a great deal on whether or not the parameters $\bar{\boldsymbol{p}}$ of the equilibrium crystal structure are explicitly among the components of $\boldsymbol{y}$. We first consider the case in which the equilibrium crystal structure is fitted explicitly, i.e., via a direct calculation of $\bar{\boldsymbol{p}}$ but not $\nabla_p \phi(\boldsymbol{p})$. For the sake of convenience, all of the crystal properties other than $\bar{p}_i$ shall be denoted by y'_m. Thus, we first consider the case in which $\boldsymbol{y} = \{\bar{p}_i,\ y'_m(\bar{\boldsymbol{p}})\}$.

The simplest and most accurate, but probably most tedious, method of deriving the best set of parameters consists in the following. For a trial parameter set $\boldsymbol{x}$ one first of all minimizes the lattice energy as a function of $\boldsymbol{p}$. The resulting structural parameters $\bar{\boldsymbol{p}}$ and other properties y'_m (computed at $\bar{\boldsymbol{p}}$) are then compared with the experimental values. The trial set $\boldsymbol{x}$ is corrected and the energy minimization repeated. Such a method, usually called the direct minimization method, is typical of the early applications of the atom-atom potentials. For example, it was applied by *Kitaigorodsky* and *Mirskaya* [3.20,21] to derive the parameters of the potentials for hydrocarbons. Essentially the same approach was followed by *Giglio* et al. [3.22–28], with the exception that they did not try to optimize the parameters but tested the various available potentials suggested by other researchers.

The direct minimization method is the most reliable method of deriving the optimum parameters since it deals with the exact values for $\bar{\boldsymbol{p}}$ and implies that all other crystal properties y'_m are computed at $\bar{\boldsymbol{p}}$. However, this method suffers from two seriuous handicaps. First, the minimization of ϕ with respect to $\boldsymbol{p}$ requires much expensive computer time because it involves repeated calculations of the lattice sums at various $\boldsymbol{p}$'s. Secondly, the direct minimization method, at least in the form used by *Kitaigorodsky* and *Mirskaya* [3.20,21], does not provide a definite analytical rule to improve the trial parameter set after each minimization cycle. (Such a rule is difficult to obtain in an exact form because $\bar{p}_i$ depends on $\boldsymbol{x}$ in an implicit manner.)

An attempt to surmount the above two difficulties has been undertaken by *Warshel* and *Lifson* [3.15,29]. Let us introduce, following them, a column vector $\boldsymbol{\Delta y} = \{\Delta y_m\}$ whose components are the differences between the calculated and observed quantities included in the fit, i.e, $\Delta y_m = y_m - y_m^0$. Let $\underline{\underline{W}}$ be a matrix of weighting factors that takes into account the uncertainties in the determination of the observables y_m^0, as well as the inaccuracies in the calculations of y_m. The quality-of-fit parameter Q may then be defined as

$$Q = \widetilde{\boldsymbol{\Delta y}}\,\underline{\underline{W}}\,\boldsymbol{\Delta y}. \tag{3.26}$$

Usually, the matrix $\underline{\underline{W}}$ is assumed to be diagonal. Its elements are the inverse squares of the allowable errors discussed at the beginning of this subsection. It should be noted that the use of a diagonal matrix for $\underline{\underline{W}}$ neglects any possible correlations between the errors in y_m. This may seriously affect the least-squares solution, particularly the estimates of the standard deviations of the derived parametes. Hence, if any correlations are known, it would seem reasonable to include them in the form of a non-diagonal weight matrix.

Returning to (3.26), we note that both $\boldsymbol{\Delta y}$ and Q are functions of the parameters $\boldsymbol{x}$: $\boldsymbol{\Delta y} = \boldsymbol{\Delta y}(\boldsymbol{x})$, $Q = Q(\boldsymbol{x})$. Any displacement $\boldsymbol{\delta x}$ from the point $\boldsymbol{x}$ improves the agreement between theory and experiment if it produces a decrease in Q. The best displacement is determined from

$$\partial Q / \partial \delta x_k = 0. \tag{3.27}$$

Let us now expand $\boldsymbol{\Delta y}(\boldsymbol{x})$ in a Tailor series about $\boldsymbol{x}$, i.e.,

$$\boldsymbol{\Delta y}(\boldsymbol{x}+\boldsymbol{\delta x}) = \boldsymbol{\Delta y}(\boldsymbol{x}) + \underline{\underline{Z}}\boldsymbol{\delta x} + \ldots, \tag{3.28}$$

where $\underline{\underline{Z}}$ denotes a Jacobian matrix whose elements are $Z_{mk} = \partial y_m / \partial x_k$. For small displacements $\boldsymbol{\delta x}$ only terms linear in $\boldsymbol{\delta x}$ may be retained in (3.28). In this case an expression for the optimal displacement is readily deduced from (3.26–28)

$$\boldsymbol{\delta x} = -(\underline{\underline{\tilde{Z}}}\,\underline{\underline{W}}\,\underline{\underline{Z}})^{-1}\underline{\underline{\tilde{Z}}}\,\underline{\underline{W}}\boldsymbol{\Delta y}(\boldsymbol{x}). \tag{3.29}$$

When using (3.29), care should be taken to see that $\boldsymbol{\delta x}$ is small enough. The latter condition may be introduced by adding the product $\xi\boldsymbol{\delta x \delta x}$ to the left-hand side of (3.27), where ξ is a Lagrange multiplier (a sufficiently large number). This modifies the expression for the optimal displacement to the following form [3.15,29]

$$\boldsymbol{\delta x} = -(\underline{\underline{\tilde{Z}}}\,\underline{\underline{W}}\,\underline{\underline{Z}} + \xi\underline{\underline{I}})^{-1}\underline{\underline{\tilde{Z}}}\,\underline{\underline{W}}\boldsymbol{\Delta y}(\boldsymbol{x}), \tag{3.30}$$

where $\underline{\underline{I}}$ is the unit matrix.

The best parameter set can be found using an iterative procedure

$$\boldsymbol{x}^{(k+1)} = \boldsymbol{x}^{(k)} + \boldsymbol{\delta x}^{(k)}, \tag{3.31}$$

in which $\boldsymbol{\delta x}^{(k)}$ is determined at each given step of the process from (3.30).

The above procedure provides a means of successfully improving the parameters. This procedure does not, however, relief us from the necessity of knowing the equilibrium parameters $\bar{\boldsymbol{p}}$ at each iteration step. To avoid a direct minimization of the lattice energy, *Warshel* and *Lifson* [3.15] recommended a Newton-Rapshon estimate for $\bar{\boldsymbol{p}}$

$$\bar{\boldsymbol{p}} = \boldsymbol{p} - \underline{\underline{\phi}}_p^{-1}\nabla_p\phi, \tag{3.32}$$

where $\boldsymbol{p}$ is either equal to an estimate of the equilibrium parameters at the preceding iteration step, $\bar{\boldsymbol{p}}[\boldsymbol{x}^{(k-1)}]$, or the experimental parameters $\boldsymbol{p}^0$. Both $\underline{\underline{\phi}}_p$ and $\nabla_p\phi$ in (3.32) are to be computed at the point $\boldsymbol{p}$. It is clear, however, that the success of (3.32) will strongly depend on the closeness of the selected crystal configuration $\boldsymbol{p}$ to the equilibrium configuration $\bar{\boldsymbol{p}}$ corresponding to the running values of the parameters $\boldsymbol{x}^{(k)}$.

A further problem encountered in the use of (3.30,31) involves the evaluation of the Jacobian matrix $\underline{\underline{Z}}$. Let us consider the elements Z_{mk} corresponding to a property y'_m. Since y'_m is defined at the equilibrium crystal configuration, it depends on $\boldsymbol{x}$ both explicitly and implicitly via $\bar{\boldsymbol{p}}(\boldsymbol{x})$. Therefore,

$$Z_{mk} = \frac{\partial y'_m}{\partial x_k} + \sum_i \frac{\partial y'_m}{\partial p_i}(\bar{\boldsymbol{p}}) \frac{\partial \bar{p}_i}{\partial x_k}. \tag{3.33}$$

The sum in this equation only vanishes if $y'_m \equiv \phi$. In this case the evaluation of Z_{mk} reduces to a computation of the explicit terms, e.g.,

$$\frac{\partial \phi}{\partial A_{ss'}} = -\tfrac{1}{2} S^{(6)}_{ss'}, \tag{3.34}$$

$$\frac{\partial \phi}{\partial B_{ss'}} = \tfrac{1}{2} S^{(m)}_{ss'}. \tag{3.35}$$

In general, the implicit terms in (3.33) do not vanish, so that one also needs to evaluate the derivatives $\partial y'_m / \partial p_i$ and $\partial \bar{p}_i / \partial x_k$. The equations for the former are rather cumbersome but can nevertheless be derived in a closed form for each particular property y'_m. The derivatives $\partial \bar{p}_i / \partial x_k$ can be roughly estimated from (3.32)

$$\partial \bar{\boldsymbol{p}} / \partial x_k = -\underline{\underline{\phi}}_p^{-1} \partial \nabla_p \phi / \partial x_k - \partial \underline{\underline{\phi}}_p^{-1} / \partial x_k \nabla_p \phi. \tag{3.36}$$

Both $\partial \nabla_p \phi / \partial x_k$ and $\partial \underline{\underline{\phi}}_p^{-1} / \partial x_k$ can be easily expressed in terms of the lattice sums and their derivatives. We shall not write down the explicit expressions but refer the reader to [3.15].

In analyzing the fitting procedure suggested by *Warshel* and *Lifson* [3.15,29], it is important to note that even the Newton-Raphson estimate for $\bar{\boldsymbol{p}}$ does not allow us to neglect a recalculation of the lattice sums and their derivatives at each new iteration step. This follows from the fact that both y'_m and Z_{mk} have to be computed at $\boldsymbol{p} = \bar{\boldsymbol{p}}$ which varies, according to (3.32), from iteration to iteration. To minimize computer time, *Warshel* and *Lifson* [3.15] recommended a recalculation of the lattice sums every three to five iterations. This, however, introduces certain errors in the quality-of-fit parameter Q and the elements of $\underline{\underline{Z}}$. As a result, the iterative process may converge more slowly and it is not clear whether the time saved in computing the lattice sums will really reduce the entire time required for the iterative process.

The derivation of the optimum parameters is considerably simplified by excluding the $\bar{p}_i$ from the components of $\boldsymbol{y}$ and replacing them by the generalized forces $f_i = -\partial \phi / \partial p_i$. At $\boldsymbol{p} = \boldsymbol{p}^0$ these forces should vanish if the model for the intermolecular potential is correct. The chief advantage of using the $f_i(\boldsymbol{p}^0)$ instead of the $\bar{p}_i$ themselves lies in the fact that all of the lattice sums in the expressions for the f_i need only be computed once at $\boldsymbol{p} = \boldsymbol{p}^0$. For atom-atom potentials of the Lennard-Jones form we have

$$f_i(\boldsymbol{p}^0) = -\sum_{ss'} \left[-A_{ss'} \frac{\partial S^{(6)}_{ss'}}{\partial p_i}(\boldsymbol{p}^0) + B_{ss'} \frac{\partial S^{(m)}_{ss'}}{\partial p_i}(\boldsymbol{p}^0) \right]. \tag{3.37}$$

Once the derivatives $\partial S_{ss'}^{(6)}/\partial p_i$ and $\partial S_{ss'}^{(m)}/\partial p_i$ have been computed (at $\boldsymbol{p} = \boldsymbol{p}^0$), the subsequent evaluation of f_i merely requires the multiplication of these derivatives by the parameters $A_{ss'}$ and $B_{ss'}$. The same can be done with Buckingham-type potentials unless the exponents $C_{ss'}$ are treated as independent internal variables.

The above approach was first used by *Williams* [3.30] to derive the parameters of the potentials for hydrocarbons. Besides the generalized forces, the components of $\boldsymbol{y}$ included the lattice energies ϕ and the elastic constants c_{ij}. Both ϕ and c_{ij} were calculated at $\boldsymbol{p}^0$, so that their computation was just as simple as the computation of the f_i using (3.37). More recent studies [3.16,31–34] have more or less used the same approach. The only modification that was made included the phonon frequencies ω_k in the set of physical properties involved in the fitting.

The method suggested by Williams suffers from two obvious shortcomings. Strictly speaking, all of the physical quantities such as ϕ, c_{ij} and ω_k can only be compared with experimental results when computed at $\bar{\boldsymbol{p}}$ (not at $\boldsymbol{p}^0$). This follows from the definitions of these quantities which imply that the lattice is at equilibrium. This shortcoming becomes especially important in the initial steps of the optimization process in which the $\bar{\boldsymbol{p}}$ corresponding to the starting values of the parameters $\boldsymbol{x}$ may prove to be too far from $\boldsymbol{p}^0$. In this case the calculated terms $\phi(\boldsymbol{p}^0)$, $c_{ij}(\boldsymbol{p}^0)$ and $\omega_k(\boldsymbol{p}^0)$ may differ very much from the respective correct values $\phi(\bar{\boldsymbol{p}})$, $c_{ij}(\bar{\boldsymbol{p}})$ and $\omega_k(\bar{\boldsymbol{p}})$, with the result that the parameter Q, see (3.26), will not properly reflect the actual quality of fit. As a result, the optimization process may converge very slowly or not at all.

To remedy the above difficulty, the optimization may be performed in two steps. In the first step one only takes into account the generalized forces. The minimization of Q with $y_m = f_i(\boldsymbol{p}^0)$ and $y_m^0 = 0$ yields parameters that ensure that $\bar{\boldsymbol{p}} \approx \boldsymbol{p}^0$. Once these parameters have been found, one can continue the optimization process considering all of the properties involved in the fitting.

The second shortcoming of the Williams method lies in the fact that the generalized forces f_i are a rather poor criterion for the assessment of the correspondence between the calculated and the experimental crystal structure. This follows from (3.23) which can be rewritten as

$$\boldsymbol{p}^0 - \bar{\boldsymbol{p}} = \underline{\underline{\phi}}_p^{-1}(\boldsymbol{p}^0)\nabla_p\phi(\boldsymbol{p}^0). \tag{3.38}$$

It can be seen that the deviations of the calculated equilibrium structural parameters from the observed ones do not only depend on the gradient $\nabla_p\phi(\boldsymbol{p}^0)$ but also depend on the magnitudes of the second derivatives of the lattice energy at $\boldsymbol{p}^0$. *Hagler* and *Lifson* [3.35] have shown that the difference $p_i^0 - \bar{p}_i$ may be appreciable in spite of the fact that the corresponding derivatives $f_i(\bar{\boldsymbol{p}})$ are very small. A more suitable criterion to assess the quality of the parameters that describe the equilibrium crystal configuration seems to be the quantity [3.35]

$$|\boldsymbol{p}^0 - \bar{\boldsymbol{p}}| = |\underline{\underline{\phi}}_p^{-1}(\boldsymbol{p}^0)\nabla_p\phi(\boldsymbol{p}^0)|. \tag{3.39}$$

The use of (3.39) instead of $\nabla_p\phi(\boldsymbol{p}^0)$ does not complicate the fitting procedure since all of the lattice sums that appear on the right-hand side of (3.39) need only be computed once at $\boldsymbol{p} = \boldsymbol{p}^0$. The criterion given in (3.39) will probably fail if the starting parameters are very poor. In this case $\bar{\boldsymbol{p}}$ is too far from $\boldsymbol{p}^0$, so that (3.21) is no longer valid. A more correct expression for $\boldsymbol{p}^0 - \bar{\boldsymbol{p}}$ can be obtained by including the third derivatives into the Taylor expansion in (3.21). Although this considerably complicates the calculations, the estimate for the quality-of-fit becomes more reliable and the fitting process converges very quickly [3.35].

We conclude this section by noting that the conditions

$$|\nabla_p\phi(\boldsymbol{p}^0)| = \text{minimum} \quad \text{and} \tag{3.40}$$

$$|\underline{\underline{\phi}}_p^{-1}(\boldsymbol{p}^0)\nabla_p\phi(\boldsymbol{p}^0)| = \text{minimum} \tag{3.41}$$

alone are insufficient to derive the parameters if $A_{ss'}$ and $B_{ss'}$ are all treated as independent variables. Indeed, since the derivatives $\partial\phi/\partial p_i$ are linear in $A_{ss'}$ and $B_{ss'}$, see (3.37), the minimization in (3.40) or (3.41) may yield the trivial solution $A_{ss'} = B_{ss'} = 0$. In order to avoid such a situation, some of these parameters should be kept fixed during the fitting process or a penalty function $w[\phi(\boldsymbol{p}) - \Delta H_s]^2$ should be added to (3.40,41).

A comparison of some of the optimization methods discussed in this subsection has been made by *Leh-Yeh Hsu* and *Williams* [3.36] in their treatment of perchlorohydrocarbons. The results of this comparison will be presented in Sect. 3.3.4.

3.3.2 Hydrocarbons

The simplest atom-atom potentials used to calculate the properties of hydrocarbons were suggested by *Kitaigorodsky* [3.37,38] as early as 1961. These were the "universal" potentials for H...H, H...C and C...C nonbonded interactions.

Let us write down the Buckingham-type potential in the conventional "reduced" notation

$$\varphi^*(z) = \frac{6}{\alpha - 6}\exp[-\alpha(z-1)] - \frac{\alpha}{\alpha - 6}z^{-6}, \tag{3.42}$$

where $\varphi^* = \varphi/\varepsilon^0$, $z = r/r^0$, ε^0 is the depth of the potential well and r^0 is the position of the minimum of the potential curve. The parameter α

in (3.42) characterizes the reduced curvature of the potential curve at the minimum:

$$(\partial^2\varphi^*/\partial z^2)_{z=1} = 6\alpha(\alpha - 7)(\alpha - 6). \tag{3.43}$$

In deriving the universal curves the parameters ε^0 and α were assumed to be the same for the three non-bonded interactions. The r^0's were assigned values somewhat greater than the sums of the respective intermolecular radii ($r^0_{CC} = 3.8$, $r^0_{CH} = 3.6$, $r^0_{HH} = 2.6$ Å). This was done in order to take into account the slight compression the molecules experience when they are packed into their crystal. The parameters ε^0 and α were taken as 0.075 kcal/mol and 13, respectively. (For the C...C interaction this yields a potential curve nearly coincident with the one derived by *Crowell* [3.39] from the properties of graphite.) The universal parameters provided quite satisfactory results for the lattice energies of methane, cyclopentane and benzene, as well as for the lattice periods and compressibility of methane. In [3.20c, 40–42] the parameters of the universal potential curves were corrected to values which are listed in Table 3.1 ("universal").

Table 3.1. Parameters of atom-atom (6-exp) potentials for hydrogen and carbon [kcal/mol, Å]

Set label	Interaction	A	B	C	ε^0	r^0	α	Reference
Universal	H...H	57	42000	4.86	0.067	2.80	13.6	[3.20c]
	H...C	154	42000	4.12	0.067	3.30	13.6	
	C...C	358	42000	3.58	0.067	3.80	13.6	
MKB74	H...H	29	49000	4.29	0.030	2.80	12.0	[3.47]
	H...C	118	18600	3.94	0.049	3.30	13.0	
	C...C	421	71600	3.68	0.079	3.80	14.0	
MPD77	H...H	51	31651	4.78	0.057	2.81	13.4	[3.48]
	H...C	145	31692	3.93	0.045	3.49	13.7	
	C...C	369	31735	3.33	0.041	4.16	13.9	
W66	H...H	36	4000	3.74	0.011	3.46	12.9	[3.30]
	H...C	139	9411	3.67	0.056	3.28	12.0	
	C...C	535	74460	3.60	0.093	3.85	13.9	
W67	H...H	27.3	2654	3.74	0.010	3.37	12.6	[3.49]
	H...C	125	8766	3.67	0.049	3.30	12.1	
	C...C	568	83630	3.60	0.095	3.88	14.0	
WS77[a]	H...H	32.5	2790	3.74	0.013	3.30	12.3	[3.50]
	H...C	137	15640	3.67	0.034	3.60	13.2	
	C...C	577	87700	3.60	0.094	3.90	14.0	
PDB78	H...H	-26.5	1264	3.74	no minimum			[3.16]
	H...C	330	15475	3.67	0.197	3.03	11.1	
	C...C	10	41800	3.60	0.0003	5.48	19.7	
TW66	H...H	26.5	2260	3.74	0.011	3.29	12.3	[3.52]
	H...C	128	8810	3.67	0.051	3.29	12.1	
	C...C	567	78659	3.61	0.102	3.83	13.8	
TW67	H...H	44.1	3430	3.79	0.023	3.14	11.9	[3.52]
	H...C	150	9773	3.67	0.063	3.26	12.0	
	C...C	443	65821	3.61	0.076	3.87	14.0	

[a] Complemented by a Coulombic term with $q_H = -q_C = 0.153\,e$

Strange as it may seem, the simple universal potentials yielded acceptable results not only for the equilibrium crystal structures, which is not very surprising in view of the choice of the r^0's, but also for the enthalpies of sublimation of benzene [3.40], naphthalene [3.41], anthracene [3.41], biphenyl and dibenzyl [3.43]. Moreover, the universal potentials have proven to be rather successful in the calculation of the spectra of intermolecular vibrations in crystals of benzene, naphthalene and anthracene [3.44,45].

Disadvantages of the universal potentials became apparent as soon as the calculations were extended to other crystals and different crystal properties. Thus, the calculated lattice energy of coronene was 6 kcal/mol lower than the experimental enthalpy of sublimation [3.46]. Poor agreement was also observed for the elastic constants of benzene, naphthalene and anthracene.

These failures led *Mirskaya* [3.21] to abandon the assumption that ε^0 and α are the same for all types of interactions. In deriving new potentials all of the parameters for the H...H and C...C potential curves were treated as independent variables, while those for the H...C curves were determined from the combining rules given in (3.4,5). The optimum parameters were found using the direct minimization method described in the preceding subsection. The elastic constants were fitted in addition to the crystal structure and the energy [3.21,47]. Probably, the best parameter set derived in such a manner was the one reported by *Mirskaya, Kozlova,* and *Bereznitskaya* in 1974 [3.47]. Hereafter, it will be referred to as the MKB74 set. The final refinement of the MKB74 set was performed by fitting the enthalpy of sublimation and the elastic constants extrapolated to absolute zero for the naphthalene crystal. The parameters of the MKB74 set are listed in Table 3.1.

Another attempt to improve the universal potential curves was made by *Mackenzie* et al. [3.48] who used these curves as a starting approximation in a fitting of selected phonon frequencies of the naphthalene crystal. By treating the six parameters of the H...H and C...C interactions as variables and applying the combining rules given in (3.4,5) to cross interactions, *Mackenzie* et al. [3.48] minimized the quantity

$$\chi^2 = \sum_i [\omega_i(\boldsymbol{p}^0) - \omega_i^0]^2 / \sigma_i^2 (N - 6), \tag{3.44}$$

where N is the number of frequencies included in the fit and σ_i is the experimental uncertainty. During the minimization of χ^2 the parameters exhibited a high degree of correlation, which led to computational instabilities when more than one parameter was allowed to vary at a time. The best potential set derived by *Mackenzie* et al. [3.48] (the MPD77 set in Table 3.1) was obtained by allowing the six parameters to vary independently. The MPD77 potentials lead to an even slightly better correspondence with experiment with respect to the reproducibility of the crystal structure of naphthalene than the starting universal potentials.

Williams [3.30,49] has suggested several sets of parameters based on the heats of sublimation, equilibrium crystal structures and the elastic constants of a large number of hydrocarbons. In order to linearize the calculated quantities with respect to the unknown parameters, the exponents C_{HH}, C_{HC} and C_{CC} were not adjusted but were deduced from other considerations. Thus, C_{CC} was taken as $3.60\,Å^{-1}$ from the paper by *Crowell* [3.39], C_{HH} was derived from a quantum-mechanical calculation of the repulsion of two H_2 molecules ($C_{HH} = 3.74\,Å^{-1}$), while C_{CH} was taken as the mean of the two. By analogy with two hydrogen H_2 molecules whose repulsion has been satisfactorily described by a dumbbell model in which the repulsion centers have been shifted 0.07 Å into the H–H bond, the hydrogen force centers in hydrocarbons are also assumed to be shifted from the normal hydrogen positions by the same distance towards the carbon atom. This is equivalent to assuming that the C–H bond is 0.07 Å shorter than its normal length.

The six unknown parameters $A_{ss'}$ and $B_{ss'}$ (s, s' = H or C) were first fitted using weighted least-squares to 77 observational equations involving the crystal structures, heats of sublimation and elastic constants of nine aromatic compounds. The number of independent variables was varied from six to three using the geometric mean combining rules given in (3.4) and by treating B_{HH} as an externally or internally adjustable parameter. In all cases the parameters exhibited a high degree of correlation. This can be clearly seen in Table 3.2 in which the correlation matrix for the best parameter set derived in [3.30] is shown (set W66 in Table 3.1).

Table 3.2. Correlation matrix for the independent variable parameters of set W66 [3.30]

	A_{CC}	B_{CC}	B_{CH}	A_{CH}
A_{CC}	1	-0.87	0.43	-0.83
B_{CC}		1	-0.37	0.60
B_{CH}			1	-0.80
A_{CH}				1

In the following paper by *Williams* [3.49] the 77 observational equations for the nine aromatic crystals were complemented by 31 equations involving the crystal properties of n-pentane, n-hexane, n-octane, cubane, adamantane, congressane and polyethylene. The parameter set (W67) that yields the best fit is given in Table 3.1. In deriving this set A_{CH} was computed from (3.4) while B_{CH} was treated as an independent variable.

The reliability of the W67 set was verified independently by using it to find the minimum-energy structure for several aromatic and non-aromatic compounds. The deviations from experiment did not exceed 0.2 Å for the lattice constants and 4.4% for the heats of sublimation.

Williams and *Starr* [3.50] have modified the atom-atom potentials for hydrocarbons to include a Coulombic term $q_a q_b / r_{ab}$. The number of adjustable parameters was only increased by one since the charges on the carbon and hydrogen atoms are assumed to satisfy the following equations:

$$\begin{aligned} &q_\mathrm{H} + q_\mathrm{C} = 0, \quad q_\mathrm{H} - q_\mathrm{C} = \Delta q; \quad \text{(for a CH group)} \\ &2q_\mathrm{H} + q_\mathrm{C} = 0, \quad q_\mathrm{H} - q_\mathrm{C} = \Delta q; \quad \text{(for a CH}_2\text{ group)} \\ &3q_\mathrm{H} + q_\mathrm{C} = 0, \quad q_\mathrm{H} - q_\mathrm{C} = \Delta q, \quad \text{(for a CH}_3\text{ group)} \end{aligned} \tag{3.45}$$

i.e., the three groups are assumed to be electrically neutral and the charge separation Δq in the C–H bond to be a constant. The only new parameter that appears in the above equations is Δq. Treating Δq in complete analogy to $A_{ss'}$ and $B_{ss'}$, *Williams* and *Starr* [3.50] have derived a new parameter set which is also presented in Table 3.1 (set WS77).

To assess the quality of a potential set in predicting the equilibrium crystal structure, *Williams* [3.51] used the discrepancy factor,

$$R = \left[\sum w_i f_i^2(\boldsymbol{p}^0) / \sum w_i\right]^{1/2}. \tag{3.46}$$

The inclusion of Coulombic terms into the atom-atom potentials resulted in a marked decrease in R. Thus, in the least-squares treatment of 118 observational equations for nine aromatic and nine saturated hydrocarbons [3.51] the discrepancy factor fell from 5.21 to 3.26 when the Coulombic terms were included.

Powell et al. [3.16] have attempted to improve the W67 potential set by using the data on the phonon frequencies, heats of sublimation and equilibrium structure of deuterated benzene. The fitting procedure is similar to the one described by (3.30,31), with $y_m = \phi(\boldsymbol{p}^0)$, $f_i(\boldsymbol{p}^0)$, $\omega_k(\boldsymbol{p}^0)$. Only the generalized forces f_i relating to molecular rotations were considered, while the stability of the crystal lattice was not examined. It was found that the parameters $B_{ss'}$ and $C_{ss'}$ were highly correlated and that the elements of the Jacobian matrix $\partial y_m/\partial B_{ss'}$ and $\partial y_m/\partial C_{ss'}$, corresponding to the same property y_m and the same type of interaction ss', were nearly proportional to each other. To overcome this difficulty, the parameters $C_{ss'}$ were not varied during the optimization process.

As can be seen in Table 3.1, in which the improved parameter set is presented (set PDB78), the resulting parameters are quite different from the starting ones. In particular, A_HH has changed its sign and A_HH has been reduced by a factor of about fifty. In spite of these changes, the agreement between observed and calculated phonon frequencies was improved only slightly (by $0.15\,\mathrm{cm}^{-1}$ on the average), so that both the starting W67 set and the improved set can be regarded as being nearly equally successful.

Another attempt to refine the W67 parameter set was made by *Taddei* et al. [3.52] who fitted the optically active phonon frequencies, heat of sublimation and generalized forces for the benzene crystal. The resulting TW67 set along with the TW66 set, derived from the same data base but starting from the W66 parameter set, are given in Table 3.1.

Table 3.1 illustrates the diversity of the parameters related to the same type of interaction but derived from different crystals or different properties. The origin of this diversity has been discussed in Sect. 3.2.

It is interesting to note that not only do the constants A, B and C of different parameter sets differ markedly but also the parameters ε^0 and r^0, which are frequently believed to have more physical significance than A, B and C. Moreover, it is seen that the atom-atom potential may contain a negative parameter A, which is meaningless if the $-Ar^{-6}$ term is associated with the dispersion attraction. Thus, the data presented in Table 3.1 support the conclusion that the individual parameters of the atom-atom potentials cannot be given a physical meaning.

The W67 potentials seem to be most reliable of the (6-exp) potentials listed in Table 3.1. Indeed, they were fitted to the largest number of observations. In the subsequent chapters of this book we shall demonstrate the applicability of these potentials to a great variety of problems encountered in the physics and chemistry of organic solids.

3.3.3 Atom-Atom Potentials for Interactions Involving Sulfur, Selenium, Oxygen and Nitrogen Atoms

The compound most suitable for a derivation of the parameters for S...S interactions is obviously crystalline sulfur, in which only one type of nonbonded interaction occurs. *Nauchitel* and *Mirskaya* [3.53] have obtained S...S parameters proceeding from a Buckingham-type potential written in the reduced form given in (3.42). The steepness parameter α was kept fixed at the "universal" value of 13.6. To derive a value for r^0, the fact that the equilibrium structural parameters of crystals with a single type of interaction are independent of ε^0 was used. Assuming an arbitrary value for the depth, ε', of the potential well, r^0 was varied until the minimum-energy structure of orthorhombic sulfur, as found by a direct minimization of the lattice energy, was as close to the experimental structure as possible. This yielded a value of 3.9 Å for r^0. The parameter ε^0 could then be readily found by scaling ε' by a factor

$$k = \Delta H_s / \phi(\overline{p}). \tag{3.47}$$

The resulting parameters are presented in Table 3.3 (set NM71). These parameters, when combined with the "universal" potential for C...C interactions, yield acceptable values for the equilibrium unit-cell parameters of CS_2. The deviations from the experimental values are within 0.2 Å.

In more recent studies *Rinaldi* and *Pawley* [3.54,55] have shown that the reproducibility of the heat of sublimation and the crystal structure of orthorhombic sulfur is rather insensitive to the choice of the parameter C. This conclusion was drawn from the following considerations. Let us write

Table 3.3. Parameters of atom-atom (6-exp) potential sets involving sulfur, selenium, oxygen and nitrogen atoms [kcal/mol, Å]

Set label	Interaction	A	B	C	ε^0	r^0	α	Reference
NM71[a]	S...S	2346	235000	3.49	0.378	3.90	13.6	[3.53]
	S...C	847	99400	3.54	0.141	3.88	13.6	
RP75	S...S	2761	41182	2.90	0.299	4.07	11.8	[3.55]
G79	Se...Se	3550	277288	3.35	0.442	4.06	13.6	[3.56]
	Se...Se	2757	82632	3.35	0.834	3.41	11.4	
KMN69[a]	O...O	259	77700	4.18	0.122	3.25	13.6	[3.57]
	O...C	313	57100	3.85	0.090	3.53	13.6	
MN72[b]	N...N	259	42000	3.78	0.067	3.60	13.6	[3.58]
	N...O	256	57000	3.98	0.090	3.42	13.6	
G74[c]	N...N	760	105400	3.60	0.133	3.85	13.9	[3.60]
	N...C	375	11480	3.60	0.298	2.84	10.2	
	N...H	143	4833	3.67	0.131	2.77	10.1	
LAM79	N...N	762	105600	3.46	0.083	4.14	14.3	[3.61]
	N...C	374	11340	3.46	0.173	3.12	10.8	
	C...C	568	83630	3.46	0.063	4.18	14.4	
R73[c,d]	N...N	350	65000	3.64	0.055	3.93	14.3	[3.62]
	N...C	450	74000	3.62	0.073	3.90	14.1	
	N...H	100	6800	3.69	0.043	3.24	12.0	
GB80[c]	N...N	-798	26997	3.60	no minimum			[3.33]
	N...C	467	76699	3.60	0.072	3.94	14.2	
	N...H	590	13824	3.67	1.022	2.38	8.7	
GB81[e]	N...N	430	2700	3.91	no minimum			[3.64]
	N...C	400	222600	3.91	0.062	3.98	15.6	
	N...H	82	7600	3.70	0.027	3.42	12.7	
	H...H	248	7130	3.75	0.042	2.42	9.1	
	H...C	35	11080	3.70	0.005	4.09	15.1	
	C...C	170	30200	3.91	0.059	3.41	13.3	

[a] Combined with universal potentials of Table 3.1;
[b] Combined with O...O potential of this table;
[c] Combined with W67 potentials of Table 3.1;
[d] Supplemented by the electrostatic dipole-dipole interaction between lone pair on a nitrogen atom and the nearest C–H bond ($\mu_N\mu_{CH} = 0.6\,D^2$);
[e] Supplemented by a many-center electrostatic dipole-dipole model with $\mu_{CH} = 0.1625\,e$ Å, $d_{CH} = 0.7$ Å from C, $\mu_N = -0.3579\,e$ Å, $d_N = 0.012$ Å from N and a dielectric constant of 1.59

the lattice energy in the form

$$\phi = -A\sum_i r_i^{-6} + B\sum_i \exp(-Cr_i), \tag{3.48}$$

and then consider all possible homogeneous deformations of the crystal

$$r_i \to r_i + \delta r_i, \quad \text{where} \quad \delta r_i = \gamma r_i. \tag{3.49}$$

The change in the lattice energy produced by the deformation δr_i is given by

$$\delta\phi = A\sum_i 6r_i^{-7}\delta r_i - BC\sum_i \exp(-Cr_i)\delta r_i$$

$$= \gamma\left[6A\sum_i r_i^{-6} - BC\sum_i r_i \exp(-Cr_i\right] \quad . \tag{3.50}$$

At equilibrium, the following condition must be satisfied:

$$\lim_{\gamma\to 0} \frac{\delta\phi}{\gamma} = 0. \tag{3.51}$$

Substituting $\delta\phi$ into (3.51) yields

$$6A\sum_i r_i^{-6} = BC\sum_i r_i \exp(-Cr_i). \tag{3.52}$$

The equations (3.48,52) impose two constraints upon the three parameters A, B and C :

$$B = \frac{\phi}{\sum_i \exp(-Cr_i) - C\sum_i r_i \exp(-Cr_i)}, \tag{3.53}$$

$$A = \frac{B\sum_i \exp(-Cr_i) - \phi}{\sum_i r_i^{-6}}. \tag{3.54}$$

In (3.53, 54) the lattice sums should be computed at the observed crystal configuration, and ϕ should be equated to the observed lattice energy.

For each given C one can thus find A and B, which ensure the coincidence of the calculated lattice energy with the observed value, as well as the stability of the crystal structure with respect to a homogeneous deformation. Of course, (3.53,54) only represent the necessary but not sufficient conditions for a minimum lattice energy. Actually all possible lattice distortions, besides homogeneous deformations, should be taken into consideration. Nevertheless, the use of (3.53,54) considerably simplifies the search for parameters since this search is now reduced to a one-parameter problem.

Rinaldi and *Pawley* [3.54] have examined several values for the adjustable parameter C from 2.8 to 4.0 Å^{-1}. (The corresponding variations in A and B, as calculated from (3.53,54), were 1191 to 1005 kcal Å^6/mol and 10100 to 533829 kcal/mol, respectively.) For each given parameter set the lattice energy of the crystal was minimized with respect to all structural distortions consistent with the crystal symmetry. It turned out that the variation of the parameters within the ranges indicated above has a negligible effect on the resulting equilibrium structure and lattice energy. Thus, the changes in the equilibrium lattice constants, the molecular orientation and the lattice energy were 0.03 Å, 0.1° and 0.04 kcal/mol, respectively. In our opinion these results are of great sifnificance because they clearly show that even for a crystal with a single type of interaction a knowledge of the lattice energy and the equilibrium structure is not enough to derive all three parameters of the (6-exp) potential unambiguously.

The final choice of parameters for S...S interactions was made by *Rinaldi* and *Pawley* [3.55] on the basis of lattice dynamical calculations. When the sum of the squares of the differences between the calculated and observed phonon frequencies was plotted against C, the curve exhibits a well-defined minimum near $C = 2.9\,\text{Å}^{-1}$. The same value of C afforded the best agreement between the calculated and observed unit-cell volumes.

The best parameters found by *Rinaldi* and *Pawley* [3.55] for S...S interactions are listed in Table 3.3 (set RP75).

Crystalline selenium Se_8 is just as convenient a compound as crystalline sulfur for the derivation of parameters. Several functions intended to describe the non-bonded interaction in the Se_8 crystal have been reported by *Govers* [3.56]. The parameters of two functions, which provide a slightly better agreement with experiment than the other ones, are listed in Table 3.3 (sets G79). The first parameter set was fitted to the heat of sublimation and crystal structure of the α-modification of Se_8, following a procedure similar to the one applied by *Nauchitel* and *Mirskaya* [3.53] to crystalline sulfur. The second parameter set was obtained by fitting 21 observational equations which take into account the generalized forces for the α- and β-phase crystal structures and the heat of sublimation. The parameter C of this set was not varied but kept fixed at the value found previously. It is interesting to note that the Se...Se potential curves suggested by *Govers* [3.56] are very similar to the curve obtained by *Nauchitel* and *Mirskaya* [3.53] for sulfur.

Parameters for the interactions involving oxygen atoms were first derived by *Kitaigorodsky* et al. [3.57] using the crystal structure and the heat of sublimation of CO_2 extrapolated to absolute zero. The interaction energy of two CO_2 molecules was assumed to be the sum of (6-exp) atom-atom potentials and a quadrupole-quadrupole term, as computed from the observed molecular quadrupole moment. The C...C interactions were treated using the "universal" potential parameters, while the C...O interactions were calculated from the combining rules given in (3.4,5). The steepness parameter α of the O...O interaction curve was kept fixed at the universal value of 13.6. The parameters r^0 and ε^0 were adjusted to yield the best agreement between the calculated and observed lattice energies and equilibrium crystal structures. The resulting parameter set is given in Table 3.3 (set KMN69).

In a more recent study by *Mirskaya* et al. [3.58] this set was employed to derive the parameters for molecular nitrogen. The calculations were performed with a crystal of N_2O using the same experimental quantities and exactly the same assumptions, as described above for CO_2. It is interesting to note that the N...N interactions in a N_2O crystal can be satisfactorily described by the universal curve ($\varepsilon^0 = 0.067$ kcal/mol, $\alpha = 13.6$) with r^0 equal to 3.6 Å (Table 3.3). The reliability of the N...N parameters has been verified independently by computing the equilibrium crystal structure and the heat of sublimation of α-nitrogen and hexamethylenetetramine. In the former case a quadrupole-quadrupole term was again added to the intermolecular potential. The results exhibit an acceptable agreement with experiment. The deviations are 0.2–0.3 Å for the lattice constants and about 10% for the lattice energy. The derived parameter set MN72 is given in Table 3.3.

Another parameter set for N...N interactions has been derived by *Govers* [3.59,60] following the Williams method described in Sect. 3.3.1. The crystal data for cyanogen, dicyanoethylene, tetracyanoethylene, s-tetrazine, pteridine and pyridazino[4,5-d]pyridazine and the heats of sublimation of the first two were used in the calculations. *Govers* [3.59,60] assumed that the W67 parameter set could be applied to the interactions involving carbon and hydrogen atoms. The N...N and N...H exponents were taken as being equal to C_{CC} and C_{CH} of the W67 parameter set (3.60 and 3.67 Å^{-1}, respectively). The best parameter set for the N...N, N...C and N...H interactions G74 is given in Table 3.3. It is to be noted that the generalized forces $f_i(\boldsymbol{p}^0)$ obtained with the G74 parameter set for the six crystals listed above were rather far from zero. Thus, the generalized forces corresponding to the unit-cell dimensions were sometimes as large as 4 kcal/(mol Å).

The G74 parameter set was thoroughly examined by *Luty* et al. [3.61] in a calculation of the equilibrium structure and the phonon frequencies of the monoclinic modification of tetracyanoethylene. (The latter was also used by *Govers* [3.59,60] in his fitting.) The results reported by *Luty* et al. [3.61] were far from satisfactory. Thus, the calculated monoclinic angle exceeded the observed value by 8° and the calculated and observed phonon frequencies sometimes differed by as much as 30 cm^{-1}. The calculated lattice periods and molecular orientation were on the whole in acceptable agreement with experiment, except perhaps the lattice period a which was about 0.6 Å smaller than the observed value.

In order to improve the Govers-Williams parameter set, *Luty* et al. [3.61] calculated the sum of the weighted squares of the differences between the observed and calculated structural parameters and phonon frequencies for selected values of the steepness parameter C. It was assumed that $C_{CC} = C_{CN} = C_{NN}$. The other parameters were kept fixed at the Govers-Williams values. The best agreement with experiment was obtained with $C = 3.46$ Å^{-1}; a value markedly lower than the initial one ($C = 3.60$ Å^{-1}). The resulting parameter set is presented in Table 3.3 (set LAM79). The parameters A and B of this set differ slightly from those of the G74 parameter set. This is because we have considered set c in *Govers'* thesis [3.60], to be the best Govers-Williams parameter set while *Luty* et al. [3.61] refer to the set reported by *Govers* [3.59] (the latter set coincides with set *b* in [3.60]).

The correction introduced by *Luty* et al. [3.61] for the C...C, C...N and N...N exponents has appreciably improved the calculated results for the monoclinic modification of tetracyanoethylene. This is especially true of the phonon frequencies whose maximum deviations from the experimental values are reduced to $10 - 13\,\mathrm{cm}^{-1}$. The corrected parameters were also tested by calculating the equilibrium structure of the cubic modification of tetracyanoethylene. The result was rather discouraging: the equilibrium lattice period proved to be ~0.4 Å larger than the observed value at room temperature.

A different parameter set for interactions involving nitrogen atoms has been derived by *Reynolds* [3.62] from the elastic constants and selected phonon frequencies of the pyrazine crystal. The interactions of two pyrazine molecules was treated within the usual scheme using Buckingham-type potentials. The only difference was that the N... H–C bond was treated as a weak hydrogen bond and the N...C and N...H interactions in it were supplemented by the electrostatic interaction of two dipoles placed on the lone pair of nitrogen (μ_N) amd the H–C bond (μ_{CH}). Only three adjustable parameters were involved in the fitting, i.e., A_{NN}, B_{NH} and the product $\mu_N\mu_{CH}$ (the orientation of the dipoles was kept fixed according to the orientation of the lone pairs and the C–H bonds in the crystal). The interactions involving H and C were treated using the W67 parameter set, while the remaining constants, except C_{NN}, were determined from combining rules similar to the ones given in (3.4,5). The exponent C_{NN} was interpolated from the W67 value for C_{CC} and the repulsion exponent of two neon atoms. To assess the quality-of-fit, Reynolds used the quantity

$$\chi^2 = \frac{1}{N-M} \sum_{n=1}^{N} [(\omega_n^2 - \omega_n^{0^2})/(2\Delta\omega_n^0)(\omega_n^0)]^2, \tag{3.55}$$

where N is the number of observed phonon frequencies, M is the number of adjustable parameters and $\Delta\omega_n^0$ is the experimental error in ω_n^0(~2%). The best parameters found by *Reynolds* [3.62] are listed in Table 3.3 (set R73). For this set $\chi = 3.4$ which corresponds to an average frequency deviation of about 7%. The parameter set was tested independently by calculating the equilibrium structure and heat of sublimation of the crystal. The calculated parameters of the crystal structure showed an acceptable agreement with experiment, while the lattice energy was appreciably higher than the observed value (15.2 and 10.9 kcal/mol, respectively).

It is worth noting that the formal electrostatic dipole-dipole term included by Reynolds in his model potential proves to be rather important. About 30% of the lattice energy is due to this term. Setting $\mu_N\mu_{CH} = 0$ and adjusting the other parameters alone gives a very poor fit ($\chi \approx 6.3$).

A further attempt to describe the interactions of nitrogen-containing molecules in terms of (6-exp) atom-atom potentials has been made by *Gamba* and *Bonadeo* [3.33]. They included the crystal structure data for pyrazine, pyridazino[4,5-d]pyridazine and s-tetrazine, the heats of sublimation of pyrazine and the α-modification of s-triazine, as well as several lattice frequencies of pyrazine and s-triazine in their observational equations. They also compared the calculated and observed heats of transition of s-triazine. The fitting procedure that was used was essentially the one described by (3.30,31) with $y_m = \phi(\boldsymbol{p}^0)$, $f_i(\boldsymbol{p}^0)$, $\omega(\boldsymbol{p}^0)$. The generalized forces f_i were only computed for molecular rotations and translations, while the stability of the crystal structures with respect to lattice deformations was not examined. The C...C, C...H, and H...H parameters were again taken from the W67 set and were not varied in the course of the fitting process. As starting approximations to the parameters describing the interactions between nitrogen atoms, *Gamba* and *Bonadeo* [3.33] used the parameters of *Govers* [3.59], *Besnainou* and *Cummings* [3.63] and the W67 parameter set in which nitrogen is considered to have the same parameters as carbon.

After a refinement all three starting parameter sets yielded almost identical potentials. The parameters obtained from the W67 set are listed in Table 3.3 (set GB80). The fit achieved with these parameters was rather poor. Although the heats of sublimation and equilibrium structures were reproduced with an acceptable accuracy, the pattern of lattice vibrations was appreciably distorted. Furthermore, the calculated lattice energies of the high- and low-temperature phase of s-triazine proved to be in the reverse order to that expected from the stabilities of these phases.

Two attempts to improve the fit were undertaken. First of all, *Gamba* and *Bonadeo* [3.33] considered the possibility of the formation of weak hydrogen bonds in the pyrazine and pyridazino[4,5-d]pyridazine crystals in which a few shortened non-bonding contacts N...H (~2.5 Å) occur. This was done by refining the parameters for these contacts independently. However, a substantial improvement in the fit was not achieved. The second attempt was made by adding an electrostatic quadrupole-quadrupole term to the intermolecular potential. The molecular quadrupole moments were roughly estimated using bond population moments and a dipole moment ascribed to the lone pair of the nitrogen atom. Unfortunately, *Gamba* and *Bonadeo* [3.33] failed to refine the parameters of this new model potential and used the old parameter set which does not include a quadrupole-quadrupole term. It is, therefore, again not very surprising that an improvement in the fit was not achieved.

After the publication of a paper by *Mulder* et al. [3.14], in which quite reliable ab initio data for the multipole moments of seven azabenzene molecules was reported, *Gamba* and *Bonadeo* [3.64] made a new attempt at improving their model potential. The idea was the same as in [3.33], i.e., to add an explicit electrostatic term to the sum of the (6-exp) atom-atom potentials.

The electrostatic charge distributions of the azabenzene molecules were simulated by placing a dipole on the line joining the N atoms to the molecular centers and another dipole on the line defined by the C–H bonds. Thus, the electrostatic part of the model only involved four parameters. These were chosen to be the magnitudes of the two dipoles, μ_N and μ_{CH}, and their distances to the atoms of the molecular ring, d_N and d_C. The four unknown parameters were found by fitting to the multipoles (up to the hexadecapole) of all seven azabenzenes including benzene. Such a simple model for the molecular charge distributions proved to be strikingly good. It gave a reasonable agreement even for the higher-order multipoles not included in the fit. The best values for the four parameters are given in the footnote to Table 3.3. These values were kept fixed in the subsequent refinement of the model potential.

The electrostatic model derived at this stage was first tested in combination with the usual (6-exp) atom-atom potentials by computing the lattice energy, equilibrium conditions and the limiting lattice frequencies of pyrazine. The parameters of the N...N, N...H and N...C interactions were treated as adjustable parameters, while those of the H...H, H...C and C...C interactions were first taken from the parameter set of *Califano* et al. [3.65a] derived from the crystal properties of benzene and naphthalene. The model potential of *Califano* et al. [3.65a] combined (6-exp) atom-atom potentials with a quadrupole-quadrupole term. The quadrupoles were placed on the molecular centers. This model potential will be discussed in Chap. 5.

It was found that it was impossible to adequately fit the experimental data without modifying the H...H, H...C and C...C parameters. When these parameters were allowed to vary, an excellent fit to the observed properties was achieved. Another important finding was that no reasonable fit could be obtained using only atom-atom interactions. Only a moderate improvement was achieved by taking into account quadrupole-quadrupole interactions.

In the second step of their calculations *Gamba* and *Bonadeo* [3.64] extended the set of observations to include the crystal properties of non-polar azabenzenes. The polar pyrimidine was left to check the transferability of the model potential. The complete set of observations that was used is given in Table 3.4 under the heading "Observed".

At first only the parameters of the atom-atom potential were refined (except for the parameters C which were strongly correlated with B). The resulting fit, although satisfactory, could be further improved by introducing a common multiplicative factor for the electrostatic interactions. This is equivalent to introducing a dielectric constant to account for a weakening of the electrostatic forces by the dielectric medium. The simultaneous refinement of all of the parameters resulted in the values listed in Table 3.3 (set GB81).

Table 3.4. Static and dynamical properties of benzene and five azabenzene crystals [kcal/mol, cm^{-1}, Å, deg.] (The atom-atom, quadrupole-quadrupole and higher-order electrostatic contributions to the calculated properties are given in relative units, such that atom-atom + quadr.-quadr. + higher terms = 1) [3.64]

Property		Obs.	Calc.	Relative contributions		
				Atom-atom	Quadr.-quadr.	Higher terms
Benzene						
Heat of sublimation		10.7				
Lattice energy		-11.4	-12.2	0.73	0.26	0.01
Molecular rotation						
	ΔR_x	0	0.1	9.84	-0.70	-8.14
	ΔR_y	0	- 0.9	0.84	-0.03	0.19
	ΔR_z	0	0.8	3.42	0.00	-2.42
Lattice frequencies						
	A_u	...	97	0.94	0.01	0.05
		...	65	0.95	0.05	0.00
		...	53	1.02	0.01	-0.03
	B_{1u}	85	84	0.97	0.00	0.03
		70	71	0.94	-0.07	0.13
	B_{2u}	94	97	0.98	0.04	-0.02
		53	56	1.13	-0.01	-0.12
	B_{3u}	94	95	1.07	-0.04	-0.03
		53	50	1.14	0.01	-0.15
	A_g	92	93	0.87	0.03	0.10
		79	72	0.62	0.28	0.10
		51	48	1.26	0.04	-0.30
	B_{1g}	128	131	0.74	0.08	0.18
		100	96	0.68	0.64	-0.32
		57	52	1.42	0.11	-0.53
	B_{2g}	...	101	0.70	0.54	-0.24
		90	92	0.98	0.05	-0.03
		79	80	0.79	0.38	-0.17
	B_{3g}	128	128	0.77	0.15	0.08
		92	86	0.91	0.06	0.03
		61	64	0.45	0.27	0.28
Pyrazine d_4 at 300 K						
Heat of sublimation		12.8				
Lattice energy		-13.7	-13.0	0.62	0.05	0.33
Molecular rotation						
	ΔR_y	0	0.2	15.17	-36.80	22.63
Lattice frequencies						
	A_u	...	103	1.14	-0.15	0.01
	B_{1u}	68	66	0.49	0.36	0.15
	B_{2u}	48	49	1.30	0.00	-0.30
	A_g	94	94	0.62	0.27	0.11
	B_{3g}	105	101	0.59	0.28	0.13
	B_{1g}	87	89	-0.28	0.12	1.16
		42	42	0.24	-1.80	2.56
	B_{2g}	67	65	1.51	-0.72	0.21
		42	38	-1.66	-1.65	4.31

Table 3.4 (continued)

Property	Obs.	Calc.	Relative contributions		
			Atom-atom	Quadr.-quadr.	Higher terms
Pyrazine at 180 K					
Lattice energy	-14.5	-13.5	0.58	0.05	0.37
Molecular rotation					
ΔR_y	0	0.9	3.86	-7.24	4.38
Lattice frequencies					
A_u	...	127	1.11	-0.11	0.00
B_{1u}	75	75	0.56	0.30	0.14
B_{2u}	52	58	1.25	0.01	-0.26
A_g	114	114	0.64	0.30	0.06
B_{3g}	123	124	0.62	0.29	0.09
B_{1g}	93	102	-0.29	0.10	1.19
	47	50	0.28	-1.64	2.36
B_{2g}	77	80	1.46	-0.54	0.08
	47	42	-2.18	-1.80	4.80
α-*s-Triazine*					
Heat of sublimation	10.3				
Lattice energy	-11.3	-11.7	0.71	-0.07	0.36
Lattice frequencies					
A_{2g}	...	100	0.21	0.01	0.78
	...	84	0.79	0.14	0.07
A_{2u}	...	56	1.75	0.00	-0.75
E_g	91.4	93	1.03	-0.06	0.03
	67.5	66	0.29	-0.30	1.01
E_u	84	80	0.71	-0.04	0.33
β-*s-Triazine*					
Lattice energy	-11.4	-12.0	0.71	-0.07	0.36
Molecular translation					
ΔT_y	0	- 0.004	11.95	0.00	-10.95
Molecular rotation					
ΔR_y	0	0.6	3.01	3.67	- 5.68
Lattice frequencies					
A_g	101	100	1.06	-0.06	0.00
	75	73	0.83	-0.20	0.37
A_u	85	97	0.78	-0.03	0.25
B_g	...	111	0.86	0.02	0.12
	104	100	0.72	0.06	0.22
	...	98	0.45	-0.03	0.58
	80	76	0.05	-0.22	1.17
B_u	106	100	0.83	-0.02	0.19
	70	60	1.69	0.00	-0.69

Table 3.4 (continued)

Property	Obs.	Calc.	Relative contributions		
			Atom-atom	Quadr.-quadr.	Higher terms
s-Tetrazine					
Lattice energy	...	-19.6	0.46	0.50	0.04
Molecular rotations					
ΔR_x	0	2.4	-0.54	4.82	-3.28
ΔR_y	0	0.8	2.15	-1.56	0.41
ΔR_z	0	- 1.5	-1.52	-0.37	2.89
Lattice frequencies					
A_u	...	96	0.85	0.12	0.03
	...	64	1.90	-0.62	-0.28
B_u	...	116	0.83	0.11	0.06
A_g	...	107	0.71	-0.04	0.33
	...	56	1.21	0.73	-0.94
	...	43	0.21	1.26	-0.47
B_g	...	126	0.88	-0.05	0.17
	...	96	-0.51	2.72	-1.21
	...	87	0.09	0.59	0.32

The calculated crystal properties of the azabenzenes are compared with the experimental properties in Table 3.4. The agreement is generally very good.

Besides the total values for the crystal properties we have also included in Table 3.4 the contributions to these properties from the atom-atom and electrostatic terms. The electrostatic contributions are also further decomposed into contributions from quadrupole-quadrupole and higher-order interactions.

The partitioning of the crystal properties in Table 3.4 seems to have some physical meaning since the electrostatic part of the model potential was adjusted separately (proceeding from the inherent properties of the molecular charge distributions). This is quite different from the usual situation in which the model potential is adjusted as a whole and the individual components of the interaction potential become indistinguishable in the resulting model.

Several important conclusions can be drawn from Table 3.4. For example, the dominant contribution in the benzene crystal is always made by the atom-atom terms. This explains the success of the simple atom-atom model in this particular crystal. A very different situation is found in the other crystals. Although the atom-atom interactions still make a major contribution to the lattice energies, the electrostatic interactions dominate many phonon frequencies. This is a good example of how the electrostatic interactions may largely cancel in the lattice energy, but are still important for the properties governed by the derivatives of the interaction potential.

A further conclusion that can be drawn from Table 3.4 is that the higher-order contributions to the electrostatic energy are not at all negligible. On the contrary, they even dominate in many cases. This is not very surprising in light of the findings of *Mulder* et al. [3.14,65b] discussed in Sect. 2.2.2.

The model potential derived by *Gamba* and *Bonadeo* [3.64] from the properties of benzene and five non-polar azabenzenes proved to be well transferable to the polar pyrimidine molecule. The fit provided by the model for the crystal properties of pyrimidine was of nearly the same quality as observed in Table 3.4.

A methodology much similar to the one used by *Gamba* and *Bonadeo* [3.64] was followed by *Williams* and *Cox* [3.65c] in deriving potential parameters for azahydrocarbons. Again, the electrostatic terms of the potential model were calibrated independently of the other terms, using ab initio calculations. In contrast, the electrostatic interactions were simulated by a point-charge model, with the force centers placed at the atomic nuclei and the lone pairs of the aromatic nitrogen atoms. Furthermore, the electrostatic parameters were fitted not to the molecular multipole moments, as in [3.64], but directly to the ab initio molecular electrostatic potential surrounding the molecules (the quality of such a fitting will be discussed separately in Sect. 3.5).

The remaining parameters of the model potential were determined by fitting the generalized forces for 10 azahydrocarbon and 17 hydrocarbon crystals (a total of 146 observational equations). The proper scaling of the potential parameters was achieved by fitting the heats of sublimation for benzene, α-nitrogen and n-hexane. The starting values for the H...H, H...C and C...C parameters were those of the WS77 set in Table 3.1. The exponents C_{CC}, C_{HH} and C_{NN} were kept fixed in the fitting (the C_{NN} was taken to be $3.78\,Å^{-1}$).

It is important that the parameters describing the interactions of the H and C atoms have remained, after the fitting, practically unchanged. This indicates a good transferability of the WS77 parameters into the azahydrocarbon crystals. For the nitrogen parameters the following values have been obtained: $A_{NN} = 1378.4\,Å^6$kcal/mol, $B_{NN} = 254529$ kcal/mol.

The resulting model potential was tested by using it to calculate the equilibrium crystal structures of the 27 compounds in the data base and four more azahydrocarbons which were not used in the fitting. In all cases the derived potential yielded more accurate predictions than the other potentials available in the literature.

For most of the azahydrocarbon crystals the Coulombic contribution to the lattice energy proved to be very important, ranging up to a maximum contribution of 59% in the case of 1,1,2,2-cyclopropanetetracarbonitrile.

For the sake of completeness, we should also mention a six-site model potential suggested by *Price* and *Stone* [3.65d] for azabenzenes. This potential includes anisotropic potential functions on the carbon and nitrogen atoms to represent the associated hydrogen atoms and the nitrogen's lone pairs. The potential predicts the crystal structure parameters of benzene, pyridine, pyrazine, pyrimidine, s-triazine and s-tetrazine within a few percent.

3.3.4 Halogenated Hydrocarbons

Crystals of halogenated hydrocarbons are, generally speaking, very convenient objects to examine the usefulness of the atom-atom potential method. The structures as well as the heats of sublimation of these crystals have been for the most part determined. The molecules of halogenated hydrocarbons are rigid enough to neglect the interplay between intra- and intermolecular interactions. The intramolecular force fields are known which is important in lattice dynamical calculations. Many halogenated hydrocarbons also exhibit phase transitions in the solid state. A calculation of these transitions is a good test for the quality of a model potential.

Bonadeo and *D'Alessio* [3.31,32] have derived parameters for Cl...Cl interactions by fitting 54 observables for the crystals of C_6Cl_6, 1,3,5-$C_6H_3Cl_3$ and *p*-$C_6H_4Cl_2$. The heats of sublimation, generalized forces for molecular rotations and translations, and some lattice vibrations were among the observables that were used. The interactions involving carbon and hydrogen atoms were treated using the parameters of the W67 potential. As a starting point for the Cl...Cl parameters, *Bonadeo* and *D'Alessio* [3.31,32] used the parameters employed by *Kitaigorodsky* and *Dashevsky* [3.66] in a conformational analysis of chlorosubstituted benzenes. A refinement of these parameters, as well as of those describing Cl...C and Cl...H interactions, was made using an iterative process given in (3.30,31).

As in the case of deuterated benzene which was discussed in Sect. 3.3.2, the rows of the Jacobian $\underline{\underline{Z}}$, corresponding to $B_{ss'}$ and $C_{ss'}$, were nearly proportional. Throughout the refinement process this proportionality was preserved. This again suggests that the $B_{ss'}$ and $C_{ss'}$ are not independent. These correlations caused *Bonadeo* and *D'Alessio* [3.31,32] to keep $C_{ss'}$ fixed and to vary $A_{ss'}$ and $B_{ss'}$ only.

The resulting potential set (set BA73 in Table 3.5) reproduced the observed quantities very well – to within $5\,\mathrm{cm}^{-1}$ for the phonon frequencies, 1° and 0.01 Å for molecular rotations and translations, respectively, and about 2 kcal/mol for the sublimation energies.

An interesting observation reported in [3.32] was that the uniqueness of the potential set was mainly due to the heats of sublimation. When the latter were not included in the fit, the refinement process became unstable, in the sense that there existed a number of different parameter sets which reproduced the remaining observables almost equally well.

Table 3.5. Parameters of atom-atom (6-exp-1) potentials of chlorinated hydrocarbons [kcal/mol, Å, e]

Set label	Interaction	A	B	C	ε^0	r^0	α	$q_{Cl}=-q_C$	Reference
BA73[a]	Cl...Cl	3650	263000	3.51	0.827	3.65	12.8	0.0	[3.31]
	Cl...C	-631	44200	3.65	no minimum				
	Cl...H	1005	33300	3.62	0.784	2.85	10.3		
RKW74[a]	Cl...Cl	2300	426000	3.65	0.374	3.91	14.3	0.0	[3.67]
	Cl...C	1140	180000	3.62	0.191	3.88	14.0		
BB74a[a]	Cl...Cl	1495	185710	3.52	0.224	3.95	13.9	0.0	[3.68]
	Cl...C	923	124391	3.56	0.146	3.92	14.0		
BB74b[a]	Cl...Cl	1744	212432	3.52	0.265	3.94	13.9	0.1058	[3.68]
	Cl...C	995	133288	3.56	0.158	3.91	13.9		
MC78[b]	Cl...Cl	2980	4580	2.26	0.197	4.21	9.5	0.0	[3.69]
	Cl...C	1055	16700	2.94	0.127	3.99	11.7		
	Cl...H	322	4560	3.07	0.077	3.50	10.8		
LW80a[c]	Cl...Cl	1850	221003	3.51	0.277	3.95	13.9	0.1090	[3.36]
	Cl...C	1033	139276	3.56	0.164	3.91	13.9		
LW80b[c]	Cl...Cl	1995	248028	3.51	0.289	3.97	13.9	0.1022	[3.36]
	Cl...C	1072	147546	3.56	0.167	3.93	14.0		
LW80c[c]	Cl...Cl	1794	207250	3.51	0.275	3.93	13.8	0.0842	[3.36]
	Cl...C	1017	134873	3.56	0.163	3.90	13.9		
LW80d[c]	Cl...Cl	1897	227158	3.51	0.283	3.95	13.9	0.1019	[3.36]
	Cl...C	1046	141202	3.56	0.165	3.91	13.9		

[a] Combined with W67 potentials of Table 3.1;
[b] Combined with MKB74 potentials of Table 3.1;
[c] Combined with WS77 potentials of Table 3.1

Quite different parameters for interactions involving chlorine have been derived by *Reynolds* et al. [3.67] who partly proceeded from the same observations as used by *Bonadeo* and *D'Alessio* [3.31,32]. The model potential adopted by *Reynolds* et al. [3.67] involved four independent parameters; A_{ClCl}, B_{ClCl}, A_{ClC} and B_{ClC}. For C...C, C...H and H...H interactions the W67 parameter set was used. Considering the correlations usually occurring between B and C, the parameter C_{ClCl} was not adjusted but interpolated from the repulsion exponents of argon and neon. The remaining parameters were determined from the combining rules given in (3.4,5).

The four adjustable parameters were fitted to 34 observations including the crystal structure data for 1,2,4,5-$C_6H_2Cl_4$, C_6Cl_6 and the α-phase of p-$C_6H_4Cl_2$ and the IR and Raman lattice frequencies and heats of sublimation of the last two. The resulting parameter set (labeled RKW74 in Table 3.5) afforded a fit within 0.05 Å for the unit-cell dimensions, 2° for the unit-cell angles, 2 kcal/mol for the heats of sublimation and 12 cm^{-1} for the phonon frequencies.

The RKW74 parameter set was tested independently in a calculation of the lattice dynamics, equilibrium crystal structures and heats of sublimation of the β-phase of p-$C_6H_4Cl_2$ and p-$C_6D_4Cl_2$. For the equilibrium structures and heats of sublimation the agreement with experiment was practically the same as in the original fitting. For the lattice vibrations the results were much worse.

Reynolds et al. [3.67] have attempted to improve the model potential by assigning an additional (anisotropic) polarizability to the molecules and placing dipoles on the C–Cl and C–H (or C–D) bonds. However, a marked improvement in the fit was not observed.

Bates and *Busing* [3.68] have compared the RKW74 and BA73 potential sets with their own set developed for the C_6Cl_6 crystal. They used, as observations, the heat of sublimation, lattice parameters, molecular orientation and optically active external-mode frequencies. Two functions of the (6-exp) and (6-exp-1) form were tested and two parameter sets were derived (sets BB74a and BB74b). The Buckingham potential (3.2) was written in reduced notation differing somewhat from the form given in (3.42),

$$\varphi(r) = \varepsilon^0(1 - 6\beta/r^0)^{-1}\{-(r^0/r)^6 + 6\beta/r^0 \exp[(r^0 - r)/\beta]\}, \quad (3.56)$$

where ε^0 and r^0 have the same meaning as in (3.42). For set BB74b a Coulombic term was added to (3.56). By comparing (3.56) to (3.42) it is easily seen that the softness parameter β is related to the steepness parameter α

$$\beta = r^0/\alpha. \quad (3.57)$$

Altogether, there were ten parameters to be determined for C_6Cl_6: $\varepsilon^0_{ss'}$, $r^0_{ss'}$, $\beta_{ss'}$ (s, s' = C,Cl), and $q_{Cl} = -q_C$. To reduce this number, all of the C...C parameters were fixed at the values for the W67 potential set, while β_{ClCl} was fixed at the value for the BA73 potential set. For cross atom-atom interactions, the following combining rules were assumed to be valid:

$$\varepsilon^0_{ss'} = (\varepsilon^0_{ss}\varepsilon^0_{s's'})^{1/2}, \quad (3.58)$$

$$r^0_{ss'} = (r^0_{ss} + r^0_{s's'})/2, \quad (3.59)$$

$$\beta_{ss'} = (\beta_{ss} + \beta_{s's'})/2. \quad (3.60)$$

When the respective parameters of the ss and $s's'$ potentials do not differ very significantly from each other, the combining rules given in (3.4,5 and 58–60) yield almost identical curves for the cross interactions.

Using the above assumptions the number of parameters was reduced to two for the set BB74a (ε^0_{ClCl} and r^0_{ClCl}) and three for the set BB74b (ε^0_{ClCl}, r^0_{ClCl} and $q_{Cl} = -q_C$). A refinement of these parameters was made in three steps. In the first step the parameters were varied until the best least-squares fit between the calculated and observed structural parameters and lattice energy was obtained. In the second step the structural parameters were varied to yield a minimum energy while the parameters determined

Table 3.6. The lattice energy, structural parameters and lattice frequencies of hexachlorobenzene calculated with different parameter sets [3.68]

Property		Observed	Calculated Parameter set BA73	RKW74	BB74a	BB74b
Lattice energy [kcal/mol]		-23.6	-23.5	-30.9	-22.3	-22.0
Lattice periods [Å]	a	8.048	8.538	8.282	8.242	7.974
	b	3.836	3.649	3.689	3.679	3.806
	c	16.599	16.707	17.194	17.124	16.711
Monoclinic angle [deg]	β	116.9	121.5	122.4	120.9	116.7
Molecular rotations [deg]	ΔR_x	0.0	14.8	17.6	13.6	-0.2
	ΔR_y	0.0	7.1	2.2	-0.4	1.5
	ΔR_z	0.0	-2.6	0.5	0.3	0.2
Lattice frequencies [cm^{-1}]	A_g	56	62	64	51	59
	B_g	54	55	61	49	58
	A_g	45	31	52	44	46
	B_g	38	42	50	40	41
	A_g	21	22	31	21	22
	B_g	25	35	30	26	29
	A_u	51	65	62	50	52
	A_u	...	42	50	41	40
	B_u	...	40	35	27	23

in the first step were kept constant. Finally, in the third step the lattice frequencies were computed using the parameters found in the first step and the equilibrium structure determined in the second step. Thus, the lattice frequencies were not used explicitly in the least-squares fitting.

The resulting parameters of the BB74a and BB74b sets are listed in Table 3.5. A comparison of the calculated and observed properties of the C_6Cl_6 crystal is given in Table 3.6 [3.68]. Apart from the data based on the BB74a and BB74b parameter sets, we present the properties calculated using the BA73 and RKW74 parameter sets. In all cases the equilibrium structural parameters were determined via a direct minimization of the lattice energy with respect to all of the structural parameters whose variation was allowed by the crystal symmetry. The lattice frequencies were all computed at the equilibrium configuration.

In comparing the results obtained with the parameter sets BB74a and BB74b, it is seen that both sets predict nearly the same lattice energy and lattice frequencies. With respect to the structural parameters, the parameter set BB74b provides a better agreement with experiment than BB74a, especially for β, c and R_x.

A further inspection of Table 3.6 shows that the parameter sets BA73 and RKW74 yield markedly worse results for the equilibrium structural parameters than those based on the parameter sets BB74a and BB74b. A substantial deviation from experiment is observed for the RKW74 value of the lattice energy. In their original paper *Reynolds* et al. [3.67] reported a lattice energy of 20.5 kcal/mol, which is 10.4 kcal/mol lower than the value obtained by *Bates* and *Busing* [3.68] using the same (RKW74) parameter set. Our own calculation of the lattice energy of hexachlorobenzene using the RKW74 potential confirmed the results of *Bates* and *Busing*. We found a lattice energy of 30.2 kcal/mol at $\boldsymbol{p} = \boldsymbol{p}^0$. *Bates* and *Busing* [3.68] proposed that the above discrepancy was due to the failure of *Reynolds* et al. [3.67] to carry the lattice energy summation to convergence. Such an explanation is, however, improbable for the summation radii adopted by *Reynolds* et al. [3.67]. It seems more likely that the discrepancy stems from an error in the computer program used by *Reynolds* et al. [3.67].

Turning to the lattice frequencies listed in Table 3.6, one can see that all of the potential sets examined provide an acceptable agreement with experiment; except for the lower pairs of A_g and B_g frequencies whose sequence, as determined from the BA73 potential, differs from that determined experimentally.

The potential sets discussed above have been criticized by *Mirsky* and *Cohen* [3.69,70], and *Wheeler* and *Colson* [3.71] because these sets do not reproduce the shortened Cl...Cl contacts (~3.4 Å) found in some chlorinated benzene crystals. However, such a shortcoming does not seem to be very dramatic if we are only interested in the structural parameters not in the particular details of the crystal structure. For example, an increase in the Cl...Cl distance (3.4 Å) in the p-dichlorobenzene crystal to its normal value (~3.7–3.8 Å) only requires a slight rotation of the molecules (by 2–3°) or a very small distortion of the unit cell. This means that the same model potential may fail to reproduce the short contacts but nevertheless be quite suitable for calculating the parameters of the equilibrium crystal structure.

Mirsky and *Cohen* [3.69] have derived a new (6-exp) potential for Cl...Cl interactions starting from the crystal structure and heat of sublimation (extrapolated to absolute zero) of C_6Cl_6. In Table 3.5 this potential is labeled MC78.

Very recently, *Leh-Yeh Hsu* and *Williams* [3.36] have reported four new potentials for Cl...Cl interactions (potentials LW80a to LW80d in Table 3.5). The potential was taken to be of the (6-exp-1) type and was combined with the WS77 C...C potential of Table 3.1. Cross atom-atom interactions were treated using the combining laws given in (3.4,5). Taking into account the strong correlations between $B_{ss'}$ and $C_{ss'}$, *Leh-Yeh Hsu* and *Williams* [3.36] did not vary C_{ClCl} but kept it fixed at the value used by *Bonadeo* and *D'Alessio* [3.31,32], and *Bates* and *Busing* [3.68]. In all there were three independent parameters to be optimized (A_{ClCl}, B_{ClCl} and $q_{\mathrm{Cl}} = -q_{\mathrm{C}}$). Note that the carbon atoms not bonded to chlorine were assumed to have a zero net charge. The three unknown parameters were fitted to 38 observations including the crystal structure data for hexachlorobenzene, octachloronaphthalene, octachloropentafulvalene, decachlorophenanthrene and decachloropyrene, and the heat of sublimation of hexachlorobenzene.

The paper by *Leh-Yeh Hsu* and *Williams* [3.36] is of particular interest because a comparison of the various methods of optimizing parameters was attempted. The potentials LW80a to LW80b given in Table 3.4 were obtained using the following methods:

LW80a. Direct Minimization. The equilibrium structural parameters $\bar{p}_i$ were treated explicitly, i.e., the $\bar{p}_i$ of each trial parameter set $\boldsymbol{x}$ were determined directly by minimizing the lattice energies with respect to the corresponding structural parameters. To judge the quality of $\boldsymbol{x}$, *Ley-Yeh Hsu* and *Williams* [3.36] used the discrepancy index

$$Q(\boldsymbol{x}) = \left\{N^{-1}\sum_i w_i[\bar{p}_i(\boldsymbol{x}) - p_i^0]^2\right\}^{1/2}, \tag{3.61}$$

where N is the total number of observations. The weights w_i are the inverse squares of the error thresholds and are equal to 1% for the lattice constants, 1° for the cell angles, 2° for the molecular rotation and 0.1 Å for the molecular translation. The parameters that optimize (3.61) were found by varying $\boldsymbol{x}$ externally on a grid.

LW80b. The Method of Hagler and Lifson [3.35]. The quality of fit was again determined using (3.61) but the differences $\bar{p}_i - p_i^0$ were estimated from (3.38). The best parameter set was found by minimizing Q with respect to $\boldsymbol{x}$.

LW80c. Force Minimization with a Diagonal-Weight Matrix. The best parameters were found by minimizing the sum of the weighted squares of the generalized forces f_i computed at the observed crystal configuration

$$R(\boldsymbol{x}) = \sum_i w_i f_i^2(\boldsymbol{p}^0). \tag{3.62}$$

The weights w_i are the diagonal elements of the matrix $\underline{\underline{W}}$ defined by [3.72],

$$\underline{\underline{W}} = [\tilde{\underline{\underline{\phi}}}_p(\boldsymbol{p}^0)\underline{\underline{V}}\,\underline{\underline{\phi}}_p(\boldsymbol{p}^0)]^{-1}, \tag{3.63}$$

where $\underline{\underline{V}}$ is a diagonal matrix whose elements are the error thresholds assigned to the structural parameters [i.e., the elements of $\underline{\underline{V}}$ coincide with the w_i in (3.61)]. The minimization of (3.62) was performed using the iterative process described by (3.30,31).

LW80d. Force Minimization with a Full-Weight Matrix. The parameters were derived in complete analogy to the above method with the exception that cross terms were also included in (3.62) with the appropriate weights w_{ij} defined by (3.63). Instead of minimizing (3.62), *Ley-Yeh Hsu* and *Williams* [3.36] minimized the quantity,

$$R(x) = \sum_{i,j} w_{ij} f_i(p^0) f_j(p^0). \qquad (3.64)$$

As seen from Table 3.5, the parameters found by the four different methods do not differ very significantly. Moreover, the values of the parameters are all rather similar to the values derived by *Bates* and *Busing* [3.68] from the properties of hexachlorobenzene alone (set BB74b).

Leh-Yeh Hsu and *Williams* [3.36] have tested the five potential sets (LW80a–d and BB74b) by finding the minimum-energy structures for five perchlorohydrocarbons. As expected, the best results were observed with the LW80a potential; a potential whose derivation was based on the exact values of the equilibrium structural parameters. A very striking fact was that the BB74b potential was only slightly worse, even though it was derived from the crystal properties of hexachlorobenzene alone.

The LW80b and LW80c parameter sets provided as nearly a good fit to the observed crystal structures as did the LW80a set. Thus, it is doubtful whether the extra computer time required for the direct minimization method is justifiable. The fit observed with the LW80c potential was not satisfactory, which is unexpected in view of the success achieved with the same method (force minimization with a diagonal-weight matrix) in deriving the parameters for hydrocarbon crystals [3.49,50].

Thus, we have listed almost all of the potentials suggested so far to describe Cl...Cl interactions in chlorohydrocarbons. There have also been a few other Cl...Cl potentials derived from the properties of solid and gaseous chlorine or from the conformational analyses of chlorine-containing molecules ([3.73] and references in [3.69]). However, these potentials were not intended to be used to calculate the properties of organic crystals and will not be discussed here.

With respect to halogens other than chlorine, most of the calculations available in the literature were carried out using unrefined parameters transferred from the pair interaction potentials of rare gas atoms [3.22]. The only attempt to refine these parameters especially for crystals has been undertaken by *Burgos* and *Bonadeo* [3.34] for brominated benzenes. The refinement process was exactly the same as the one applied by *Bonadeo* et al. to deuterated benzene [3.16], nitrogen-containing heterocyclic compounds [3.33] and chlorinated benzenes [3.31,32]. The parameters were derived by fitting them to 28 optically active lattice frequencies of 1,3,5-$C_6Br_3H_3$, *p*-$C_6H_4Br_2$ and *p*-$C_6D_4Br_2$, structural data of the same crystals and two crystal modifications of 1,2,4,5-$C_6H_2Br_4$, and also the heat of sublimation of *p*-$C_6H_4Br_2$. The H...H, H...C and C...C parameters were again fixed at the corresponding values of the W67 parameter set. Two potential sets were

Table 3.7. Parameters of atom-atom (6-exp) potentials of brominated benzenes [kcal/mol, Å]. (Combined with the W67 parameter set of Table 3.1) [3.34]

Set	Interaction	A	B	C	ε^0	r^0	α
I	Br...Br	4580	149000	3.14	0.595	3.98	12.5
	Br...C	730	78500	3.37	0.077	4.20	14.1
	Br...H	555	18050	3.44	0.219	3.25	11.2
II	Br...Br	7830	64300	2.78	0.968	3.92	10.9
	Br...C	421	118000	3.57	0.041	4.29	15.3
	Br...H	401	20900	3.55	0.141	3.35	11.9

obtained corresponding to two different starting points. Despite appreciable differences in the individual parameters of the two sets (Table 3.7), nearly equivalent fits to the experimental data were obtained. All deviations from experiment were within the tolerable limits discussed in Sect. 3.3.1.

3.3.5 Amides and Carboxylic Acids

The crystals of amides and carboxylic acids have received much attention in recent years, which partly stems from the view that the intermolecular potentials inferred from the properties of these crystals may be transferable to systems of biological importance [3.74]. A characteristic feature of crystals of amides and carboxylic acids, compared to the systems considered thus far, is the presence of strong hydrogen bonds (N-H...O=C bonds in amides and O-H...O=C bonds in carboxylic acids). It is obvious that a simple combination of the O...O and H...H non-bonded potentials discussed in the preceding subsections will fail to describe the interactions between the oxygen and hydrogen atoms involved in a hydrogen bond and that a special potential will be necessary. As to the analytical form of this potential, it is not clear a priori whether the usual analytical expressions such as (3.2,3) will be suitable or that another function will be required. *Hugler* et al. [3.75] have derived parameters for crystals of amides using the usual (6-m-1) potential for all atomic species except the N, H and O atoms involved in a N-H...O hydrogen bond. The latter were treated using the following function

$$\begin{aligned}\varphi = &(-A_{\mathrm{NO}}r_{\mathrm{NO}}^{-6} + B_{\mathrm{NO}}r_{\mathrm{NO}}^{-m} - A_{\mathrm{HO}}r_{\mathrm{HO}}^{-6} + B_{\mathrm{HO}}r_{\mathrm{HO}}^{-m})[1 - f(\theta)]\\ &+ q_{\mathrm{N}}q_{\mathrm{O}}r_{\mathrm{NO}}^{-1} + q_{\mathrm{H}}q_{\mathrm{O}}r_{\mathrm{HO}}^{-1}\\ &+ D\{\exp[-2\alpha(r_{\mathrm{OH}} - r^0)] - 2\exp[-\alpha(r_{\mathrm{OH}} - r^0)]\}f(\theta). \quad (3.65)\end{aligned}$$

The terms in the first line of (3.65) correspond to interactions N...O and H...O of the Lennard-Jones type attenuated by an angular dependence $[1 - f(\theta)]$. Two analytical forms were examined for $f(\theta)$; i.e.,

$$f(\theta) = \exp[-(\theta/\theta_0)^2], \quad (3.66)$$

where θ is the supplement of the N-H...O angle, and

$$f(\theta) = \exp[-(\theta_1/\theta_0)^2] + \exp[-(\theta_2/\theta_0)^2], \tag{3.67}$$

where θ_1 and θ_2 are the angles between the H...O vector and the two lone pairs on the oxygen atom.

The second line of (3.65) contains Coulombic terms while the third line represents a conventional Morse potential attenuated by an angular dependence $f(\theta)$.

The unknown parameters in (3.65–67) as well as those needed to describe all non-bonded interactions were fitted to the crystal structure data and heats of sublimation of eight and six amides, respectively. The empirical atomic charges were also fitted to the dipole moments of three amides. The number of independent adjustable parameters was substantially reduced by using the parameters for the alkyl groups CH_n of alkanes [3.15] and by applying the geometric-mean combining rule to all cross interactions. In addition, the groups CO, NH, NH_2 and CH_n were assumed to be electrically neutral, which immediately imposes additional constraints upon the atomic charges.

The optimization of the parameters was performed using the iterative process described by (3.30,31). The vector $\boldsymbol{\Delta y}$ contains all differences between the calculated and observed structural parameters $\boldsymbol{\Delta p}$, as given by (3.38), as well as the differences in the lattice energies and dipole moments. The optimum parameters derived in this manner were then used to find the exact equilibrium configurations of the crystals by directly minimizing the lattice energy with respect to the structural parameters.

A very interesting result was that the Morse potential in (3.65) is not needed to describe the interactions between the atoms involved in the hydrogen bond. The standard deviations of all relevant parameters (except that of r^0) exceeded the parameters themselves and the potential well D was found to be rather small. A similar situation was observed with the *Lippincott-Schroeder* [3.76,77] and *Stockmayer* [3.78a] potentials. Thus, the intermolecular interactions in amides can be sufficiently described by a set of usual (6-m-1) atom-atom potentials. A special function is not needed to take into account the hydrogen bond.

The parameters derived by *Hagler* et al. [3.75] are listed in Table 3.8 for two distinct values of the exponent m ($m = 9$ and 12). In both parameter sets the hydrogen atom involved in a hydrogen bond is associated with the zero-potential parameters ε^0 and r^0. In other words, the intermolecular potential ignores the presence of this hydrogen atom in the molecule (except for its Coulombic interactions). This is in agreement with the fact that a hydrogen atom bonded to an electronegative atom affects the electron density distribution around the latter only slightly, so that the resulting distribution remains almost spherical [3.75].

Table 3.8. Parameters of atom-atom potentials of amides [3.75] [kcal/mol, Å, e]

Interaction	A	B	ε^0	r^0	q
(6-12-1) potentials					
$H_{CH_n} \cdots H_{CH_n}$	32.9	7150	0.038	2.75	0.10
$O \cdots O$	502	275000	0.228	3.21	-0.38
$N_{NH} \cdots N_{NH}$	1230	2271000	0.167	3.93	-0.28
$N_{NH_2} \cdots N_{NH_2}$	1230	2271000	0.167	3.93	-0.83
$C_{CH_n} \cdots C_{CH_n}$	532	1811000	0.039	4.35	-n·0.10
$C_{CO} \cdots C_{CO}$	1340	3022000	0.148	4.06	0.38
$H_{NH} \cdots H_{NH}$	0	0			0.28
$H_{NH_2} \cdots H_{NH_2}$	0	0			0.41
(6-9-1) potentials					
$H_{CH_n} \cdots H_{CH_n}$	15.0	445	0.0025	3.54	0.11
$O \cdots O$	1410	45800	0.198	3.65	-0.46
$N_{NH} \cdots N_{NH}$	2020	86900	0.161	4.01	-0.26
$N_{NH_2} \cdots N_{NH_2}$	2020	86900	0.161	4.01	-0.82
$C_{CH_n} \cdots C_{CH_n}$	1230	38900	0.184	3.62	-n·0.11
$C_{CO} \cdots C_{CO}$	355	12500	0.042	3.75	0.46
$H_{NH} \cdots H_{NH}$	0	0			0.26
$H_{NH_2} \cdots H_{NH_2}$	0	0			0.41

The correlation matrix as well as the standard deviations reported by *Hagler* et al. [3.75] for the parameters of Table 3.8 indicated that considerable difficulties were involved in the determination of the parameters from the data base used. It is worth noting that the highest correlations and standard deviations were found for the parameters of the amide carbon. *Hagler* et al. [3.75] attributed this to the fact that in all of the crystals studied the amide carbon atom is only in contact with a few neighboring molecules and, hence, contributes very little to the total intermolecular potential. As a result, the calculated crystal properties and the quality of fit are almost insensitive to the parameters of the potential of the amide carbon atom. Appreciable correlations between the parameters $A_{ss'}$ and $B_{ss'}$ of the same type (ss') of interaction can also be found. The correlation coefficients vary from 0.7 to 0.9. In view of the strong correlations and large standard deviations, it seems rather surprising that the absolute values of the particular parameters have received so much attention in [3.75].

The force field developed for amides by *Hagler* et al. [3.75] can also be applied to carboxylic acids. *Lifson* et al. [3.74] have shown that the only new parameter needed to reproduce the crystal structures, heats of sublimation and dipole moments of 14 carboxylic acids (a total of 110 observations) was the net charge on the hydroxyl hydrogen atom. The parameters of the carbonyl oxygen atom in amides can be used for both oxygen atoms of the carboxyl group COOH. The van der Waals parameters of the carboxyl carbon atom were assumed to be equal to those of the amide carbon atom, while the charge was determined from the electroneutrality of the COOH group. Optimization of the adjustable parameter, the charge on the hydroxyl hydrogen atom yielded values of 0.41 and 0.35*e* with the (6-9-1) and (6-12-1) potential sets, respectively.

Recently the force field suggested by *Hagler* et al. [3.75] has been modified slightly to include the atomic dipole and quadrupole moments into the electrostatic part of the potential [3.78b]. The values of the dipole and quadrupole moments were derived from experimental electron density distributions. An application of this model potential will be discussed in Sect. 4.3.

Another attempt to derive parameters for the crystals of carboxylic acids has been undertaken by *Derissen* and *Smit* [3.79]. A set of 70 equations was drawn up using the crystal structure and heat of sublimation of acetic, α- and β-oxalic, α- and β-fumaric and isophthalic acid. The heats of dimerization and dimeric structures of formic, acetic and propionic acid were also fitted. The non-bonded interactions were described by (6-exp-1) potentials except for the atoms involved in hydrogen bonds. The latter were treated using a Lippincott-Schroeder potential supplemented by Coulombic terms

$$\begin{aligned} V_{\mathrm{HB}} = D\Big[1 &- \exp\Big(-\frac{n(r_{\mathrm{OH}} - r^0_{\mathrm{OH}})^2}{2r_{\mathrm{OH}}}\Big) \\ &- \frac{1}{g}\exp\Big(-\frac{ng(r_{\mathrm{H\ldots O}} - r^0_{\mathrm{H\ldots O}})^2}{2r_{\mathrm{H\ldots O}}}\Big)\Big] + A\exp(-br_{\mathrm{O\ldots O}}) \\ &- Br^{-m}_{\mathrm{O\ldots O}} + q_{\mathrm{H}}q_{\mathrm{O}}r^{-1}_{\mathrm{H\ldots O}} + q^2_{\mathrm{O}}r^{-1}_{\mathrm{O\ldots O}}. \end{aligned} \tag{3.68}$$

Of the eleven unknown parameters of (3.68), four were taken from [3.77] ($r^0_{\mathrm{OH}} = r^0_{\mathrm{H\ldots O}} = 0.97$ Å, $g = 1.45$ and $m = 1$). For non-bonded cross interactions the combining rules given in (3.4,5) were used. The number of adjustable parameters was further reduced by fixing the atomic charges at the values obtained in a CNDO/2 Mulliken population analysis (i.e., the adopted model potential was not, strictly speaking, completely empirical). The starting values for the non-bonded parameters were taken from the W67 set for C and H and the KMN69 set for O. For the parameters of the potential of the hydrogen bond, *Derissen* and *Smit* [3.79] started with the Lippincott-Schroeder values [3.77]. However, *A* and *B* were estimated under the condition that the calculated dimerization energy of formic acid is equal to the observed value.

The optimization of the parameters was performed using three different methods similar to those applied by *Leh-Yeh Hsu* and *Williams* [3.36] to derive the parameter sets LW80b, LW80c and LW80d. A slightly better overall agreement with experiment was obtained for the set derived by the diagonal-weight matrix method, although all three optimized sets and

Table 3.9. Parameters of the potentials of carboxylic acids [kcal/mol, Å]. (Given in parentheses are the standard deviations) [3.79]

Interaction	A	B	C	ε^0	r^0	α
H...H	30.8(10.0)	824(860)	3.965(0.35)	no minimum		
C...C	613.5(82.0)	68644(32250)	3.609(0.15)	0.130	3.71	13.4
O...O	247.9(45.0)	99422(31250)	4.070(0.11)	0.071	3.56	14.5

Hydrogen bonding parameters in (3.68)

D	n	A	b	B
113.4(8.9)	9.192(0.32)	$44.08(3.15)\cdot 10^5$	4.806(0.03)	19.0(1.8)

even the starting one reproduced the crystal structure data fairly well. The best parameters for carboxylic acids, found by *Derissen* and *Smit* [3.79], are listed in Table 3.9 together with their standard deviations. It is seen that the repulsion parameters B for all types of interactions cannot be accurately determined by the data base used. This is especially true for the repulsion parameter B_{HH} whose standard deviation exceeds the parameter itself. Since the parameter A_{HH} is also determined with a very large uncertainty, the resulting H...H non-bonded potential is unreliable as a whole and appears to be superfluous. In any case, the values reported by *Derissen* and *Smit* [3.79] for the H...H parameters are not realistic. Indeed, the H...H potential curve corresponding to the parameters of Table 3.9 does not have a minimum at intermediate distances and increases rapidly as the hydrogen atoms come closer together. At a distance of 1 Å, for example, the H...H interaction energy is as low as −15 kcal/mol; a value which is not compensated for by the electrostatic repulsion of the hydrogen atoms and the interactions of the carbon atoms bonded to these hydrogen atoms. Such an inherent shortcoming should obviously result in H...H contacts that are too short and an intermolecular energy that is too large when the interacting molecules approach one another so that their C–H bonds lie on a straight line. In the dimer and crystal structures examined by *Derissen* and *Smit* [3.79] such configurations have not however been encountered. That is why the above shortcoming of this potential field does not manifest itself and is of no practical significance.

The model potential derived by *Derissen* and *Smit* [3.79] is a good example of a model which correctly describes the important points on the potential energy hypersurface, but yields absurd results for the unimportant configurations.

To complete our discussion of the atom-atom potentials derived from crystal data, we present in Table 3.10 an extensive potential set used by *Giglio* et al. [3.22–28] in their calculations of crystal structures. As already mentioned in Sect. 3.3.1, *Giglio* et al. [3.22–28] did not optimize the parameters but merely tested the various available potentials, in particular,

Table 3.10. Parameters of atom-atom potentials used by *Giglio* et al. [3.22–28] in crystal-structure calculations [kcal/mol, Å]. (Me represents a CH_3 group and Q represents Cl and S; the potentials are implied to have the form $-Ar^{-6} + Br^{-m}\exp(-Cr)$)

Pair	A	$B\cdot10^{-3}$	C	m	Pair	A	$B\cdot10^{-3}$	C	m
H...H	49.2	6.6	4.08	0	N...Q	711.5	288.6	1.811	6
H...C	125.0	44.8	2.04	6	N...Br	983.6	165.2	1.517	6
H...N	132.0	52.1	2.04	6	N...I	1415.6	272.7	1.500	6
H...O	132.7	42.0	2.04	6	O...O	358.0	259.0	0	12
H...Me	380.5	49.1	3.705	0	O...Me	1026.3	272.7	1.665	6
H...Q	265.2	40.5	3.851	0	O...Q	715.5	239.2	1.811	6
H...Br	366.7	25.6	3.557	0	O...Br	989.3	140.9	1.517	6
H...I	527.8	43.4	3.540	0	O...I	1423.6	232.6	1.500	6
C...C	327.2	301.2	0	12	Me...Me	2942.0	273.9	3.329	0
C...N	340.0	340.0	0	12	Me...Q	2051.1	251.6	3.475	0
C...O	342.3	278.7	0	12	Me...Br	2836.0	133.5	3.181	0
C...Me	981.1	291.1	1.665	6	Me...I	4081.0	219.9	3.164	0
C...Q	684.0	255.4	1.811	6	Q...Q	1430.0	220.8	3.621	0
C...Br	945.8	149.0	1.517	6	Q...Br	1977.2	123.0	3.327	0
C...I	1361.0	246.0	1.500	6	Q...I	2845.2	206.8	3.310	0
N...N	354.0	387.0	0	12	Br...Br	2733.8	65.8	3.033	0
N...O	356.0	316.2	0	12	Br...I	3934.0	107.6	3.016	0
N...Me	1020.5	325.9	1.665	6	I...I	5661.0	174.2	2.999	0

the interaction potentials of rare gas atoms and the potentials used in conformational analyses. The test for the quality of a potential set generally consisted in a computation of the minimum-energy crystal structure and a comparison of this structure with experiment. Crystal properties other than the equilibrium structural parameters were not examined. Hence, the potential set listed in Table 3.10 seems to be only suitable for calculations of crystal structures. Several applications of this potential set will be discussed in Chap. 4.

3.4 The Use of Molecular Data in Deriving the Parameters of Potentials

Calculations of the internal energy of an organic molecule using empirical and semiempirical approaches have been discussed in [3.80] and are beyond the scope of this book. Since the first applications of atom-atom potentials to calculations of conformational energies [3.81–86], there has been considerable progress in both theoretical and computational quantum chemistry. As a result a number of problems previously treated by empirical methods have become solvable on the basis of a much more rigorous approach. At least this is the case for small and medium-sized organic molecules. However, purely empirical methods have still to be used for large molecules and macromolecules.

As far as the majority of empirical conformational-energy models treat non-bonded interactions in exactly the same way as the intermolecular-energy models, it is of interest to clear up two questions:

i) can the intermolecular atom-atom potentials be used to calculate the conformational energy; and
ii) can the use of molecular data in fitting potential parameters help us to select the best atom-atom potentials?

Before discussing these interrelated questions, it is reasonable to give a brief outline of the principal ideas used in empirical conformational-energy calculations.

In most empirical calculations the conformational energy of an organic molecule is expressed as a sum of interactions between bonded atoms (1–2 interactions), between atoms bonded to the same atom (1–3 interactions), between atoms separated by two atoms (1–4 interactions), etc.

$$\phi_{\text{intra}} = \sum f^{(1-2)} + \sum f^{(1-3)} + \sum f^{(1-4)} + \text{higher-order interactions.} \tag{3.69}$$

The first two sums can be approximated by their quadratic expansions because the mutual displacements of atoms involved in (1–2) and (1–3) interactions are usually small. This is not the case for the other terms in (3.69), which require more complicated analytical expressions.

The success of an empirical conformational-energy model depends a great deal on the choice of internal molecular coordinates. It is actually this choice that ultimately determines the transferability of empirical parameters in (3.69). For the first sum this choice can be made uniquely by assuming that $f^{(1-2)}$ is a function of the corresponding bond length l. The function $f^{(1-2)}$ may thus be associated with the bond stretching energy and take the following form,

$$f^{(1-2)} = f_{\text{str}}(l) = \frac{1}{2}K_l(l - l^0)^2, \tag{3.70}$$

where K_l and l^0 are adjustable parameters (the force constant and the ideal bond length, respectively).

Let us consider a (1–3) interaction between the atoms 1 and 3 bonded to the same atom 2. A convenient internal coordinate for the function $f^{(1-3)}$ is the bond angle β between the bonds 1–2 and 2–3. The distance between the atoms 1 and 3 is indeed mainly determined by the magnitude of β, while the bond lengths are of lesser importance. The function $f^{(1-3)}$ may thus be regarded as characterizing the bond-angle bending energy and, considering the smallness of the variations in β, can be written as

$$f^{(1-3)}(\beta) = f_{\text{bend}}(\beta) = \frac{1}{2}K_\beta(\beta - \beta^0)^2, \tag{3.71}$$

where K_β and β^0 are the bond-angle bending force constant and the ideal bond angle, respectively.

The (1–4) and higher-order interactions in (3.69) are usually treated using (6-exp) or (6-m) non-bonded potentials f_{nb}, although a Coulombic term is occasionally included as well. An additional term is also needed for (1–4) interactions to reproduce the internal rotation barriers. This term is referred to as a torsional potential f_{tors} and is considered to be a function of the dihedral angle τ formed by the planes 123 and 234 in the atomic sequence 1234 (thus, τ describes the internal rotation about the bond 2–3). The torsional potential $f_{\text{tors}}(\tau)$ is a periodic function of τ and can therefore be described by the leading terms in a Fourier expansion of the potential

$$f_{\text{tors}}(\tau) = \frac{1}{2}K_\tau(1 - \cos n\tau), \tag{3.72}$$

where the integer n is determined by the symmetry of the rotation.

The intramolecular hydrogen bonds are ordinarily treated in exactly the same way as described in the preceding section for intermolecular interactions.

Finally, the intramolecular energy is represented as

$$\phi_{\text{intra}} = \sum_l f_{\text{str}}(l) + \sum_\beta f_{\text{bend}}(\beta) + \sum_\tau f_{\text{tors}}(\tau) + \sum_r f_{nb}(r) + \sum_{\text{H-bonds}} f_{\text{H-bond}}(h), \tag{3.73}$$

where l, β, τ, and r run over all bonds, bond angles, torsional angles and non-bonded (1–4) and higher-order pairs of atoms, respectively, and h denotes the internal parameters of a hydrogen bond.

It should be noted that for the majority of problems encountered in the physics and chemistry of organic solids the bonds and bond angles may be considered to be rigid, so that the first two sums in (3.73) may be neglected. This is justified by the fact that the bond and bond-angle deformation energies are generally 2–4 orders of magnitude greater than the intermolecular energy. Hence, the interplay between the bond and bond-angle deformations and the displacements of the molecule as a whole may be neglected. This is not so for the third and fourth sums in (3.73) which may allow a substantial variation in τ and r without appreciable changes in the conformational energy.

Equation (3.73) represents the simplest empirical model for ϕ_{intra}. Additional terms are frequently included in the model to make it more flexible. This mainly refers to various cross terms which describe the interplay between different internal coordinates. For example, a stretch-bending cross term may be introduced [3.87–90],

$$f_{\text{str-bend}} = \frac{1}{2}K_{\beta l}\Delta l(\beta - \beta^0), \tag{3.74}$$

where $K_{\beta l}$ is the stretch-bending force constant, $\Delta l = |l_1 - l_1^0| + |l_2 - l_2^0|$ and β is the angle between the adjacent bonds 1–2 and 2–3. Another analytical form for the stretch-bending potential was used by *Warshel* and *Lifson* [3.15,29],

$$f_{\text{str-bend}} = \frac{1}{2} K_r (r - r^0)^2, \tag{3.75}$$

where r is the distance between two atoms bonded to a common atom. Although the bond lengths and bond angle do not appear in (3.75) explicitly, an implicit dependence of $f_{\text{str-bend}}$ on these coordinates exists since r implicitly depends on them.

A further cross term appears in the above model [3.15,29] which describes the coupling between the bond-angle bending and torsional contributions and has the form

$$f_{\text{bend-tors}} = -K_{\beta\tau}(\beta_1 - \beta_1^0)(\beta_2 - \beta_2^0) \cos\tau. \tag{3.76}$$

Here β_1 and β_2 are the bond angles on both sides of a bond and τ is the torsional angle describing the rotation about this bond. The cross term in (3.76) is found to be important in reproducing the correct splitting of certain internal mode vibrations of different symmetry [3.15].

For aromatic systems additional energy terms describing the non-planar deformations of the aromatic rings and the out-of-plane displacements of the peripheral bonds adjacent to the aromatic nuclei are needed. *Kitaigorodsky* and *Dashevsky* [3.91,92] have suggested the following potential to account for the non-planar deformation energy,

$$f_{\text{non-pl}} = \frac{1}{2} K_\gamma \sin^2\gamma \approx \frac{1}{2} K_\gamma \gamma^2, \tag{3.77}$$

where γ is either the angle between a bond adjacent to the aromatic ring and the plane of the two nearest bonds of the ring, or the angle formed by the planes of two successive bond pairs of the same ring, e.g., the angle between the planes C1C2C3 and C2C3C4. Several alternative approaches to the description of the non-planar deformation energy may be found in [3.80].

Increasing the number of different energy terms in a model potential clearly brings about a closer fit to experiment. It should, however, be realized that the inclusion of new terms in (3.73) increases the number of adjustable parameters; thereby, reducing the value of the model and increasing the number of observations needed to determine the unknown parameters. In each particular case a judicious compromise should therefore be made between the number of energy terms in ϕ_{intra} and the accuracy of the calculated quantities.

The fitting of empirical parameters in (3.70–77) is usually done using exactly the same procedures, as described in Sect. 3.3.1. Among the experimental quantities most frequently employed in a fitting are the molecular conformations at equilibrium, vibrational frequencies, rotational barriers and energy differences between particular conformers.

The experience gained in applying the atom-atom approximation to the calculations of intra- and intermolecular energies has shown that non-bonded potential functions derived from molecular data alone are, as a rule, unsuitable for calculating crystal properties. This is not surprising because the atom-atom potentials f_{nb} in (3.73) appear as an integral part of the total empirical force field. Hence, they depend on the other parameters which, in turn, are not in any way related to the intermolecular energy. In general, conformational energy calculations are not a good basis for the determination of non-bonded parameters. Indeed, the non-bonded contribution to ϕ_{intra} is only comparable to the bond stretching or bond-angle bending energy at small interatomic distances where a strong repulsion between the atoms dominates. The remaining (attractive) part of the non-bonded potential proves to be of little significance in most cases, so that it is very difficult to determine the three parameters of the potential with certainty.

Thus, it seems that in computing the total energy of a crystal it is reasonable to use different parameters for the same non-bonded interactions, according to the atomic pair, intermolecular or intramolecular, under consideration. Such an approach, though not quite consistent, should yield more reliable results than those based on a parameter set derived from either crystal or molecular data separately.

In spite of these obvious considerations, there have been many attempts made to derive a consistent force field suited to both intra- and intermolecular energy calculations simultaneously. The first attempt of such a kind was undertaken by *Warshel* and *Lifson* [3.15] for crystals of alkanes. The intermolecular-energy contribution to the total energy of a crystal was calculated using (6-m-1) potentials. The same interaction potentials were assumed for all non-bonded atoms (separated by three and more bonds) within a molecule. The intramolecular contribution to the total energy also included the harmonic potentials (3.70,71,75), the torsional energies (3.72) and the bend-torsional cross terms (3.76).

The empirical parameters of the force field were fitted to an extensive set of crystal and molecular data including: the unit-cell dimensions and heats of sublimation of n-hexane and n-octane, the enthalpy differences between gauche- and trans-butanes and between axial and equatorial methylcyclohexanes, the excess enthalpies for a series of cycloalkanes and the molecular conformations and vibrations of a number of cyclo- and n-alkanes.

At first *Warshel* and *Lifson* [3.15] set $m = 12$ in their (6-m-1) non-bonded potentials. As might be expected, the (6-12-1) potentials previously fitted to the molecular data alone [3.29] could not adequately reproduce the crystal properties. An attempt to re-adjust the parameters of the (6-12-1) potentials by fitting them to the molecular and crystal data simultaneously failed. An acceptable agreement with experiment could only be achieved for either the molecular or the crystal data. The potentials derived from the molecular data gave values of the intermolecular energy which were too low, while those derived from the crystal data exaggerated the conformational energy. This circumstance led *Warshel* and *Lifson* [3.15] to test softer potential functions, with $m = 10$, 9, and 8. The best overall agreement with experiment was obtained with $m = 9$. The derived parameters are listed in Table 3.8.

Those results [3.15] show that it is, in principle, possible to find a set of non-bonded potentials suitable for calculating both the intra- and intermolecular energies of a selected class of compounds. At the same time, it is seen that an adequate description of intramolecular interactions places severe constraints upon the non-bonded potential. These constraints should evidently reduce the flexibility of the non-bonded potentials in the calculations of the crystal properties. Thus, it is felt that the requirements made by the conformational energy upon the softness of the non-bonded potentials will interfere with the fitting of these potentials to crystal properties sensitive to the second derivatives of the potentials (e.g., lattice vibrations and elastic constants).

Warshel and *Lifson* [3.15] argued that the simultaneous use of both molecular and crystal data in fitting non-bonded potentials was justified due to the necessity of simultaneously examining both the short- and long-range parts of a potential. Indeed, the molecular properties are mainly sensitive to the short-range part of a potential, while the crystal properties are determined by the intermediate and long-range parts. Therefore, the simultaneous fitting of a non-bonded potential to molecular and crystal data might be thought to allow one to examine the entire potential and hence to determine the parameters with greater accuracy. This would be the case if the atom-atom potentials were real constituents of the actual intra- and intermolecular energies and had the same physical meaning in both cases. In fact, we are dealing with two quite different empirical models, each of them involving different assumptions and imposing its own requirements upon the non-bonded potentials. Hence, the information inferred from molecular data can hardly be regarded to complement the information derived from crystal properties.

The empirical model described by (3.70–77) is sometimes used for much more sophisticated force fields. Thus, *Warshel* and *Kaplus* [3.93] followed the empirical scheme in (3.73) to describe the σ-electron contribution to the conformational energy of conjugated molecules. Their model was based on a formal separation of the σ- and π-electron energies. The latter was represented by a semiempirical LCAO-MO model of the Pariser-Parr-Pople type [3.94] that was corrected for the overlap of nearest-neighbor orbitals. The force field involved a total of 67 adjustable parameters (11 for the π-electron integrals and 56 for the empirical functions) which were fitted to an extensive set of molecular and crystal data for ethylene, benzene, butadiene and propylene in their ground and excited electronic states.

It is worth noting that the non-bonded parameters of the force field were adjusted independently of the other parameters, using only quantities governed by the intermolecular interactions. The resulting non-bonded potentials were kept fixed during the subsequent fitting of the remaining parameters to the molecular data. Thus, the undesirable dependence of the non-bonded potentials on the assumptions and parameters of the conformational-energy model was avoided. The model suggested by *Warshel* and *Kaplus* [3.93] is aesthetically displeasing as it combines two approaches based on quite different principles and assumptions to calculate the same quantity (ϕ_{intra}). However, the use of such models seems to be unavoidable in treating large conjugated molecules. Neither the empirical scheme given in (3.73) nor the semiempirical approach alone is sufficient in this case. Another advantage of this model lies in the possibility of calculating the molecular and crystal properties in the excited electronic states. Such calculations are hardly possible with a pure empirical approach because the relevant experimental information is usually insufficient to refine all of the additional parameters needed to describe the excited states.

To conclude this section we shall briefly outline the EPEN (Empirical Potential using Electrons and Nuclei) model suggested by *Scheraga* et al. [3.95–97]. Similar to the models of *Warshel* and *Lifson,* and *Warshel* and *Kaplus,* the EPEN model was also designed to simulate both intra- and intermolecular interactions in organic systems. Being a purely empirical model, it however differs in many respects from others. We shall restrict ourselves to a revised version of the EPEN model, i.e., to the EPEN/2 model.[2]

The basic assumptions of the EPEN/2 model are given below. A molecule is considered to be composed of interaction centers which carry charge. Positive charges are located on the atomic nuclei and are equal to $Z - 2$ (except for hydrogen which carries a charge of $+1$). Centers of negative charge or electrons are not located off the nuclei. The bonding electrons are located along the line joining two bonded atoms, while the lone-pair electrons or π electrons are located off this line. All bond lengths and bond angles are kept fixed. The energy of a molecule or a crystal is taken to be the sum of Coulombic interactions between all of the point charges and Buckingham-type (6-exp) interactions between the electrons. The parameters of the (6-exp) potential are not assumed to be the same for all electrons

[2] A non-empirical variant of the EPEN/2 model, in which the model parameters are determined by fitting to ab initio interaction energies, was put forward by *Marchese* et al. [3.98]

but are assumed to be dependent on the chemical nature of a particular electron. Thus, one distinguishes between the carbon sp^3 σ-electrons, the sp^2 σ-electrons and the π-electrons, as well as between the bonding and lone-pair electrons of oxygen atoms. Moreover, special potentials are introduced to describe the interactions between electrons separated by four heavy atoms and the interactions between π electrons ($\pi \ldots \pi$ interactions). The geometrical parameters determining the location of the electrons are treated as adjustable parameters similar to the A, B and C parameters of the (6-exp) potentials. As a result, the number of adjustable parameters in the EPEN/2 model is comparable with the number of parameters to be fitted using the usual atom-atom approach. An advantage of this model is that it does not require any special functions to describe the torsional energy and the energy of the hydrogen bonds. These energies are well described by Coulombic and (6-exp) non-bonded potentials.

Actually, the EPEN/2 model represents a version of a bond-bond model potential in which the empirical (6-exp-1) potentials are used to simulate the interactions between all localized bond and lone-pair orbitals.

The ability of the EPEN/2 model to reproduce crystal properties has not yet been firmly established. *Nemenoff* et al. [3.97a] have compared the calculated and observed lattice constants and energies for several saturated compounds including water (ice), pentane, ammonia, hexane, octane, methanol, ethanol, glucose and methylamine. In all of the cases, except glucose whose calculated lattice energy of −54.7 kcal/mol seems to be too low, the agreement with experiment was satisfactory. The differences in the lattice periods and lattice energies were of the order of 0.1 Å and 3 kcal/mol, respectively. The agreement was worse for carboxylic acids [3.97b], amides and peptides [3.97c]. Thus, the calculated lattice energy of succinic acid was −23.6 kcal/mol compared to an experimental value of −28.1 kcal/mol; for oxamide it was −17.5 instead of −27.1 kcal/mol. The EPEN/2 estimate of the lattice energy of L-alanine (−45 kcal/mol) also seems to be unrealistic although no relevant experimental evidence is available to verify this result. Serious discrepancies were observed for the lattice constants. Thus, the calculated unit cell parameter b of L-alanine was more than 1 Å greater than the experimental value. For oxamide, the calculated angles in the unit cell deviated 7–8 degrees from the observed ones.

Even poorer results were obtained with the classical objects of the atom-atom potential method – the crystals of benzene and naphthalene – where the discrepancies between the calculated and observed lattice periods amounted to 1.4 Å [3.95]. In spite of these failures, it is difficult to draw any general conclusions with respect to the use of the model in predicting lattice energies and crystal structures. This difficulty stems from the fact that the parameters of the EPEN/2 model were not only fitted to crystal properties but also to a large set of molecular data including dipole moments, molecular conformations, rotational barriers and energy differences between particular molecular conformers. As a result, it is not clear whether the EPEN/2 model would be capable of adequately reproducing crystal properties if all the parameters were only fitted to crystal data.

3.5 Ab Initio Atom-Atom Potentials

Historically, the atom-atom potential method first appeared in its experimental form discussed in Sect. 3.3. Extensive attempts to use the potentials to simulate the results of rigorous quantum-mechanical calculations were undertaken much later and were mainly stimulated by the success of the experimental version. The principal objectives of these studies were: (1) to clarify the nature of the atom-atom potentials, the origin of their success and their transferability, and, (2) to develop easily computable model potentials that can be used in repeated calculations of the interaction energy.

In this section we shall discuss the most interesting studies of this kind, – placing more emphasis on the studies involving organic crystals. *Wasiutynski* et al. [3.99] attempted to describe the ab initio interaction potential of two ethylene molecules using an atom-atom model and then to see if the latter reproduces the static and dynamic properties of the ethylene crystal. The ab initio interaction energy between two ethylene molecules was taken to be the sum of the first- and second-order contributions, i.e.,

$$\varepsilon = \varepsilon_1 + \varepsilon_2 . \tag{3.78}$$

The first-order interaction energy ε_1 takes into account the electrostatic and exchange energies:

$$\varepsilon_1 = \langle A\psi_0^A\psi_0^B|H|A\psi_0^A\psi_0^B\rangle - \langle\psi_0^A|H^A|\psi_0^A\rangle - \langle\psi_0^B|H^B|\psi_0^B\rangle , \tag{3.79}$$

i.e., ε_1 is nothing but the Heitler-London interaction energy defined by (2.38). The basis set used was (9,5/4) contracted to [4,2/2] (double zeta).

The second-order interaction energy ε_2 was evaluated using a multipole expansion that was truncated after the two leading terms. The coefficients C_6^{disp} and C_8^{disp} were determined using a basis of polarized quality. The induction contribution to ε_2 was neglected.

The error estimates were 10% for ε_1 and 20% for ε_2 [3.99].

The atom-atom potential adopted by *Wasiutynski* et al. [3.99] was of the usual (6-exp-1) form. All of the parameters were treated empirically. It involves three more adjustable parameters, x_{H}, y_{H} and y_{C} (atomic coordinates), which allow the point charges on H and C to be shifted from the nuclei while the molecular symmetry remains unchanged.

An attempt to adapt all of the adjustable parameters at once, by fitting them to the total interaction energy, failed because of the strong correlations between the parameters. Hence, the parameters were adjusted separately for each individual energy component. The charges $q_C = -2q_H$ and the coordinates x_H, y_H and y_C were derived by fitting the Coulombic part of the atom-atom potential to the long-range part of the electrostatic energy ε_{mult}. The fitting was carried out for 64 configurations of the dimer. The standard deviation was 3.6%. An adjustment of the point charges placed on the atomic nuclei resulted in a markedly worse fit ($\sigma = 23\%$).

The parameters of the formal exchange terms of the atom-atom potentials $B_{ss'}$ and $C_{ss'}$ were fitted to the "short-range" energy defined by

$$\varepsilon_{sh.r.} = \varepsilon_1 - \varepsilon_{p.ch.}, \tag{3.80}$$

where $\varepsilon_{p.ch.}$ is the electrostatic contribution to the potential model computed from the point charges determined in the previous fitting. The simultaneous variation of all nine parameters $B_{ss'}$ and $C_{ss'}$ (s, $s' =$ H or C) without any restrictive conditions yielded a fairly good fit to $\varepsilon_{sh.r.}$ ($\sigma = 13\%$). The resulting parameters, however, prove to be unsuitable for calculating the equilibrium crystal structure and phonon frequencies. In our opinion, this was due to the fact that the set of configurations for the dimer, sampled in the fitting of $\varepsilon_{sh.r.}$, was too restrictive (16 configurations) and did not include all important configurations of the ethylene crystal (Sect. 3.2).

To remedy the above difficulties with $B_{ss'}$ and $C_{ss'}$, the following procedure was used. The parameters B_{HH} and C_{HH} were determined independently of the other parameters from those configurations whose repulsion energy was mainly governed by the H...H repulsion. Thereafter the parameters B_{CC} and C_{CC} were determined by using the combining rules given in (3.4,5) for B_{HC} and C_{HC}, and keeping B_{HH} and C_{HH} fixed at the previously derived values. The fit to $\varepsilon_{sh.r.}$ was naturally appreciably worse with the parameters obtained in this way ($\sigma = 33\%$).

The parameters $A_{ss'}$ were fitted to the long-range dispersion energy ε_2. Again, the three parameters A_{HH}, A_{HC} and A_{CC} exhibited a strong correlation when fitted to ε_2 without any restrictive conditions. This was avoided by applying the combining rule given in (3.4) to A_{HC}. The resulting fit to ε_2 was very good. The relative standard deviation was 7.1%.

The quality of the fittings of the various components of the ab initio potential is illustrated in Figs. 3.3–5. It is clearly seen that the adopted empirical model is flexible enough to reproduce all essential features of the interaction between two ethylene molecules.

The values of the parameters derived by *Wasiutynski* et al. [3.99] are listed in Table 3.11. These values were used to calculate the heat of sublimation, the equilibrium crystal structure and nine lattice vibrations of ethylene. Similar calculations were carried out using the W67 (6-exp) and the (6-exp-1) parameter set [3.100] (the latter does not differ significantly from the WS77 set in Table 3.1). For the equilibrium dimensions of the unit cell, the molecular orientation and the heat of sublimation, both the ab initio and the experimental atom-atom parameter sets yielded an acceptable

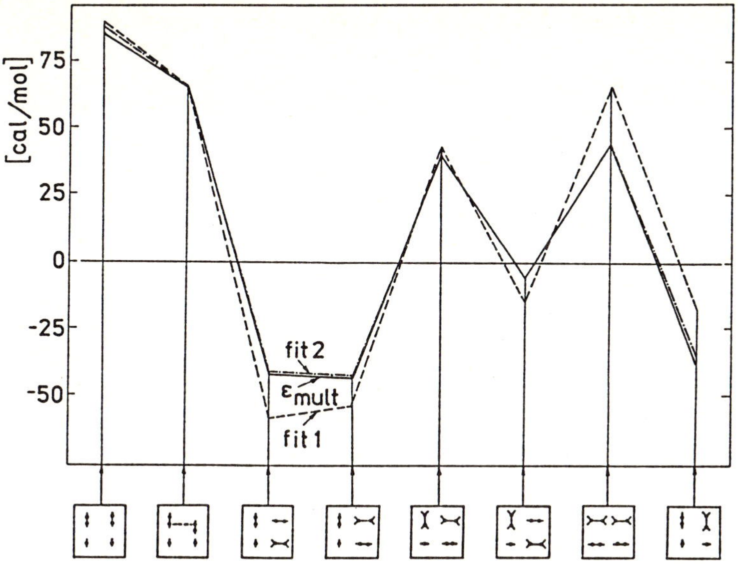

Fig. 3.3. Orientation dependence of the electrostatic interaction in the ethylene dimer at $R = 12$ a.u. [3.99]. $\varepsilon_{\mathrm{mult}}$: ab initio results (multipole expanded); fit 1: point charges on the atomic nuclei; fit 2: point charges shifted from the nuclei

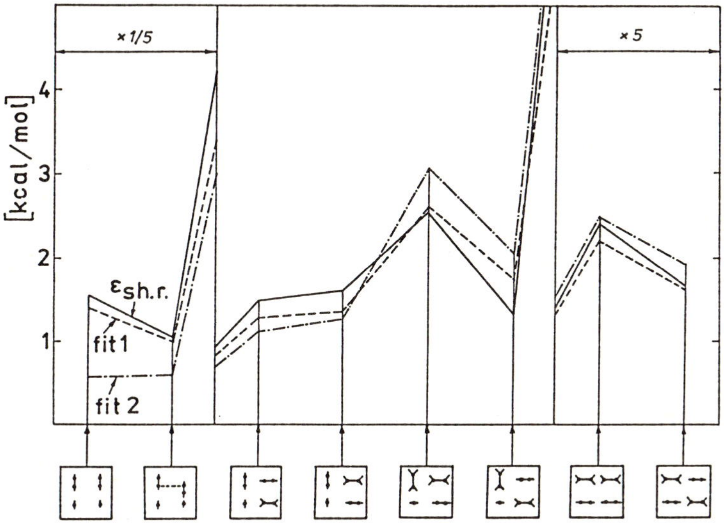

Fig. 3.4. Orientation dependence of the short-range repulsion in the ethylene dimer at $R = 8$ a.u. [3.99]. $\varepsilon_{\mathrm{sh.r.}}$: ab initio results calculated from (3.80); fit 1: atom-atom potentials without combining constraints for C...H interactions; fit 2: atom-atom potentials with combining constraints for C...H interactions

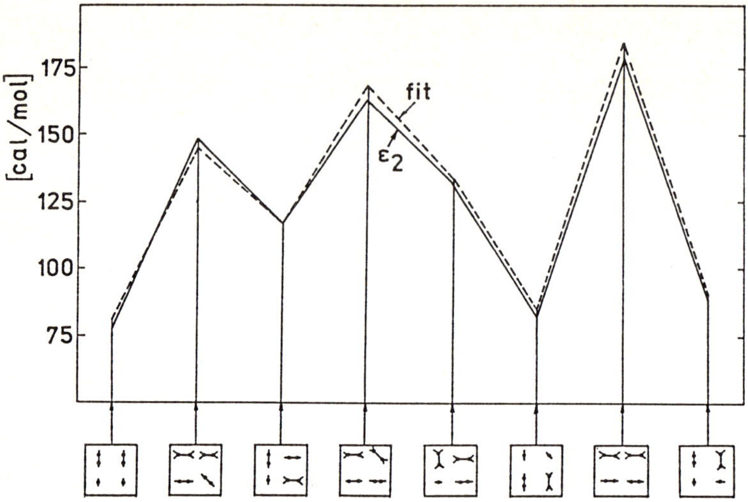

Fig.3.5. Orientation dependence of the long-range dispersion interaction in the ethylene dimer at $R = 12$ a.u. [3.99]. ε_2 : ab initio results; fit: atom-atom potentials with combining constraints for C...H interactions

Table 3.11. Parameters of the (6-exp-1) potentials fitted to the ab initio interaction energy of the ethylene dimer [kcal/mol, Å, e] [3.99]

Interaction	A	B	C	ε^0	r^0	α
H...H	20	1500	3.70	0.008	3.29	12.2
H...C	132	6378	3.43	0.033	3.56	12.2
C...C	876	27110	3.16	0.128	3.90	12.3

Point charges and their coordinates[a]

$q_C = -2q_H = -0.5274,$

$x_C = 0,\ y_C = \pm 0.5549;\ x_H = \pm 0.8308,\ y_H = 1.0095$

[a]The molecular coordinate axis x is chosen to lie in the molecular plane, perpendicular to the C=C bond; the y axis is along the C=C bond

fit. Much worse results were obtained for the lattice vibrations. For the ab initio set, the average deviation from the observed frequencies was $28\,\mathrm{cm}^{-1}$ (the largest was $72\,\mathrm{cm}^{-1}$). The experimental sets proved to be more successful. Thus, the average deviation for the (6-exp-1) potential set [3.100] was $11\,\mathrm{cm}^{-1}$ (the largest was $35\,\mathrm{cm}^{-1}$).

Wasiutynsky et al. [3.99] attempted to play down the success of the experimental parameter sets by arguing that the latter were adapted to crystals of related hydrocarbons and to properties intimately related to those examined for ethylene. In other words, it was implied that the parametrization of the experimental sets had implicitly involved properties of the ethylene crystal. This is not however so. Thus, the magnitudes of the phonon frequencies are governed by the second derivatives of the atom-atom potentials, while the experimental sets were fitted to the lattice energies and equilibrium crystal structures, i.e., to properties almost insensitive to these derivatives.

The two experimental parameter sets mentioned above were also tested with respect to their ability to reproduce the interaction energy of the dimer [3.99]. Unfortunately, the comparison of the ab initio and empirical values was carried out for each individual energy component separately. Thus, the sum $\Sigma - AR^{-6}$ was compared with ε_2, $\Sigma qq'r^{-1}$ with $\varepsilon_{\mathrm{mult}}$ and so on. In our opinion, such a comparison is hardly justifiable because the sums of the individual atom-atom terms do not have an exact physical meaning and should not generally reproduce particular energy components (Sect. 3.2). It is not very surprising, therefore, that the above sums exhibited considerable deviations from the respective ab initio results when computed from the experimental parameter sets: $\sigma = 62\%$ for $\varepsilon_{\mathrm{mult}}$, 53% for $\varepsilon_{\mathrm{sh.r.}}$ and 21% for ε_2 (all values refer to the (6-exp-1) parameter set [3.100]).

The fitting procedure [3.99] is very similar to the one applied by *Smit* et al. [3.101] to derive the atom-atom parameters for carboxylic acids. Again, in the latter study the model potential was not adjusted as a whole but was adjusted for each individual energy component separately. The ab initio part of this study has already been discussed in Sect. 2.1. A total of 12 configurations for the formic-acid dimer were generated, including seven non-equivalent configurations in the first coordination sphere of the crystal. In addition, sixteen configurations for the methanol/formaldehyde complex were generated. The interaction energy of the complex was calculated in exactly the same way as described in Sect. 2.1 for the formic-acid dimer.

The model potential adopted by *Smit* et al. [3.101] was of the (6-exp-1) form but included additional attractive exponential terms $-D\exp(-Fr_{\mathrm{HO}})$ to improve the description of the hydrogen bond. The point charges q_i were obtained from the point-charge model which best reproduced the electrostatic potential created by the six lowest molecular multipoles. It is essential that the point charges derived in such a way differ markedly from those obtained from Mulliken populations (Table 2.7) and accurately reproduce both the ab initio multipole moments and $\varepsilon_{\mathrm{mult}}$. The values of q_i for methanol and formic acid are listed in Table 3.12 [3.101]. The point charges for formic acid were additionally corrected for the difference between the ab initio and the experimental quadrupole moments. The corrected point charges, listed in the last column of Table 3.12, were obtained by fitting the point-charge model to the electrostatic potential resulting from the experimental quadrupole moment and higher ab initio multipole moments (up to the 2^6-th pole).

The parameters $B_{ss'}$ and $C_{ss'}$ (s, s' = H, C or O) were fitted to the short-range repulsion energy as in [3.99]. Cross interactions were treated using the combining rules given in (3.4,5) for $B_{ss'}$ and $C_{ss'}$. The resulting

Table 3.12. Point-charge models for methanol and formic acid, fitted to the ab initio electrostatic potentials [e] [3.101]

Methanol		
C	0.431	
O	-0.787	
H (in OH)	0.465	
H (in CH_3, trans to OH)	-0.001	
H (in CH_3, gauche to OH)	-0.054	
Formic acid		
		a
C	0.945	0.523
O (in C=O)	-0.659	-0.467
O (in OH)	-0.775	-0.521
H (in CH)	-0.027	0.057
H (in OH)	0.516	0.408

[a]Corrected for the difference between the ab initio and the experimental quadrupole moment

model potential for the first-order interaction energy ($\varepsilon_1 = \varepsilon_{\text{elst}} + \varepsilon_{\text{exch}}$) gave a fairly good fit to the corresponding ab initio energies [the average deviation was ~14% for the formic-acid dimer (12 configurations) and 10% for the methanol/formaldehyde complex (16 configurations)].

The attractive part of the atom-atom potentials, $-Ar^{-6} - D\exp(-Fr)$ ($D \neq 0$ only for H and O atoms involved in a hydrogen bond), was fitted to the second-order interaction energy which included the induction, charge-transfer and dispersion energies. The parameters $A_{ss'}$ were again treated using (3.4). For the best parameters the average deviation from the ab initio second-order energies was 11% for the formic-acid dimer and 5% for the methanol/formaldehyde complex.

The complete list of parameters derived by *Smit* et al. [3.101] is given in Table 3.13. The parameter set A was obtained as described above. It provided an acceptable fit for the total interaction energies (the average deviation was about 20% for the formic-acid dimer and 3% for the methanol/formaldehyde complex). In order to improve the fit, the parameters of set A were

Table 3.13. Atom-atom parameters fitted to the ab initio interaction energies of the formic acid dimer and the methanol/formaldehyde complex [kcal/mol, Å]. (Given in parentheses are the standard deviations. Set A: refined on ε_1 and ε_2 separately; set B: refined on ε using set A as starting values) [3.101]

	Set A	Set B
A_{OO}	246.4(113)	247.5(130)
B_{OO}	46965(18127)	50184(16652)
C_{OO}	4.36(0.23)	4.35(0.21)
A_{CC}	24.7(120)	37.0(247)
B_{CC}	54976(84000)	66604(72137)
C_{CC}	3.47(0.51)	3.45(0.39)
A_{HH}	35.6(20)	45.5(27)
B_{HH}	1652(670)	1689(589)
C_{HH}	3.98(0.25)	4.04(0.22)
D_{HH}	147.8(58)	138.4(60)
F	2.053(0.20)	2.10(0.23)

taken as starting values and then all of the parameters were simultaneously fitted to the total interaction energies. In our opinion, this was hardly reasonable because the physical meaning of the starting potential (set A) was obviously lost after such a procedure. In Table 3.13 the improved parameter set is labelled B. This set afforded an only slightly better fit than that achieved with set A.

To examine the quality of their parameters *Smit* et al. [3.101] have calculated the energy and equilibrium structure of the dimers and crystals of formic and acetic acids, and also the crystals of α- and β-oxalic acids. Actually, this was also a test for the transferability of the derived parameters because acetic and oxalic acids were not included in the original fitting. For a comparison, all of the calculations were also performed with the experimental parameter set derived by *Derissen* and *Smit* [3.79] (Table 3.9).

Both the ab initio and the experimental potential sets were quite successful in predicting the equilibrium structural parameters. As might be expected, a slightly better agreement was observed with the experimental set. With respect to the dimer and lattice energies, the experimental set still showed an acceptable agreement (within 2 kcal/mol), while the results based on the ab initio set were, in some cases, far from satisfactory. Thus, the dimerization energy of formic acid was overestimated by 5.8 kcal/mol, while the lattice energy of β-oxalic acid was underestimated by 8.7 kcal/mol. We are, however, not inclined to ascribe these discrepancies to flaws in the atom-atom potential method. Instead, it seems that the model developed by *Smit* et al. [3.101] merely reproduces inaccuracies in the ab initio calculations (see the error estimates discussed in Sect. 2.1 for the formic-acid dimer and crystal).

An attempt to fit the ab initio results to an atom-atom model was also undertaken by *Huiszoon* and *Mulder* [3.102] for seven azabenzenes.

The ab initio part of their calculations has already been discussed in Sect.2.2.3. They [3.102] restricted themselves to the dispersion part of the long-range interaction energy. The ab initio dispersion energies were computed for a large number of configurations (1200 for each dimer) and fitted by a sum of $-Ar^{-6}$ atom-atom terms.

From a theoretical point of view, this approximation should not yield a good representation of the longe-range dispersion energy as given by the multipole expansion in (2.135). To verify this, let us express the sum,

$$\varepsilon_{\text{disp}}^{(\text{at-at})} = -\sum_{a,b} A_{ab} r_{ab}^{-6}, \tag{3.81}$$

as a power series in R^{-1} [3.102]. The result may be written in the form

$$\varepsilon_{\text{disp}}^{(\text{at-at})} = -\sum_{l_A, l_B = 0} C_{l_A+l_B+6}^{(\text{at-at})} R^{-(l_A+l_B+6)}, \quad \text{where} \tag{3.82}$$

$$C^{(\text{at-at})}_{l_A+l_B+6} = P^{[l_A+l_B]}_{(t);(u)} G^{[l_A+l_B]}_{(t);(u)}, \quad \text{and} \tag{3.83}$$

$$P^{[l_A+l_B]}_{(t);(u)} = \sum_{a,b} A_{ab} r_{t_1}(a) r_{t_2}(a) \dots r_{t_{l_A}}(a) r_{u_1}(b) r_{u_2}(b) \dots r_{u_{l_B}}(b), \tag{3.84}$$

$$G^{[l_A+l_B]}_{(t);(u)} = R^{(l_A+l_B+6)} \frac{(-1)^{l_A}}{l_A! l_B!} (\nabla_{t_1} \nabla_{t_2} \dots \nabla_{t_{l_A}} \times \nabla_{u_1} \nabla_{u_2} \dots \nabla_{u_{l_B}}) R^{-6}. \tag{3.85}$$

All of the indices have the same meaning as in (2.82,83,87) and the summation over repeated indices is again implicit.

In analyzing (3.83–85) it is easily seen that the coefficient of the leading term in (3.83),

$$C^{(\text{at-at})}_6 = \sum_{a,b} A_{ab},$$

is independent of the relative orientation of the interacting molecules (unlike the usual C_6 dispersion coefficient). For this reason, (3.81) should not, in principle, give good results at long distances where the $C_6 R^{-6}$ contribution is dominant.

In spite of this apparent disadvantage, (3.81), in which A_{ab} only depends on the chemical nature of the interacting atoms, is flexible enough to be adapted to the exact dispersion-energy hypersurface. In fitting (3.81) to the "exact" energy of each dimer separately, the standard deviations were as low as 4–5%. The optimum parameters derived from these individual fittings were different for different dimers. Thus, the parameter A_{HH} varied from 25 to 55, A_{CC} from 375–690 and A_{NN} from 765–1000 kcal · Å^6/mol. Moreover, some of the parameters for s-triazine and s-tetrazine were negative. It is, however, hardly reasonable to attach any great significance to these results in view of the strong correlations between the individual parameters. To reduce the correlations, the geometric-mean combining rule given in (3.4) was applied to cross interactions. Nevertheless, the correlation problem was not completely resolved.

A very important result obtained by *Huiszoon* and *Mulder* [3.102] was that despite the marked differences between the specific potential sets (i.e., between those derived for each dimer separately), there existed a common potential set applicable to each dimer. This common set was obtained by fitting the ab initio dispersion energies of all of the dimers simultaneously. The standard deviation characterizing the quality of the common set proved to be 6.8%, i.e., it is comparable to the standard deviations obtained with the specific sets (4–5%). The parameters of the common potential set are the following: $A_{CC} = 650$, $A_{NN} = 504$ and $A_{HH} = 27$ kcal · Å^6/mol.

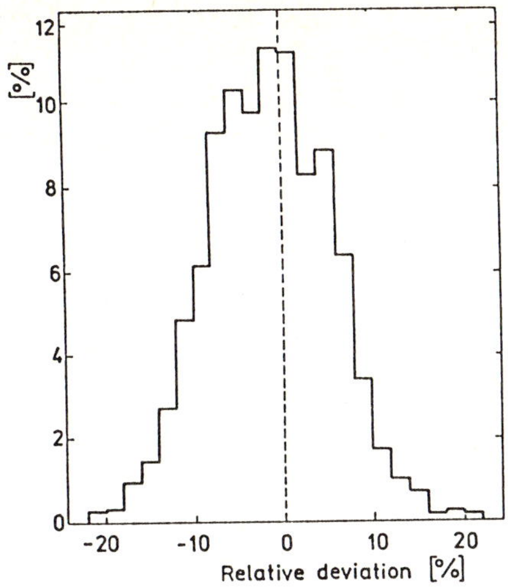

Fig. 3.6. The relative deviations of atom-atom dispersion energies from ab initio results for seven azabenzenes [3.102]

The distribution of the relative deviations between the fitted and ab initio dispersion energies for the seven azabenzenes is shown in Fig. 3.6. This distribution, as well as the low magnitude of the standard deviation, provides convincing evidence for the transferability of the common potential set between the azabenzenes.

Huiszoon and *Mulder* [3.102] also examined the ability of three experimental potential sets to reproduce the ab initio long-range dispersion energy. The three sets that were tested were G74, MN72 (Table 3.3) and a set by *Momany* et al. [3.103] to be discussed in the next section. At large intermolecular distances, the exponential terms of the atom-atom potentials were neglected. Hence, the long-range dispersion energy was directly compared with the sum of the Ar^{-6} terms. For the last two sets the agreement with the ab initio values was unsatisfactory, while the G74 set yielded a fairly good fit ($\sigma = 10.5$). The latter result was unexpected because atom-atom potentials derived from crystal data generally simulate the total interaction potential but not the dispersion part alone (see discussion in Sect. 3.2). It seems, therefore, that the good fit obtained with the G74 potential set is most likely fortuitous.

Very recently, the above calculations on the azabenzenes have been extended to cover the electrostatic part of the long-range interaction potential [3.104]. The calculations were based on the ab initio molecular multipole moments (up to the 2^6th pole) evaluated by *Mulder* et al. [3.14]. Several types of point-charge models were tried and several fitting procedures were tested. The simplest model (M1) contains point charges on the atoms, with the constraint of molecular electrical neutrality. An atomic charge in this model was assumed to only depend on the chemical nature of a particular atom, regardless of its chemical surroundings. Model M2 also contains point charges on the lone pairs of the nitrogen atoms (1 a.u. from the nitrogen atom). Model M3 is similar to M2, except that the electroneutrality constraint was applied, individually, to all of the C–H bonds and the nitrogen/lone-pair fragments (lone pair and nitrogen itself). Model M4 is similar to M3 but the charges on the lone pairs are allowed to shift. Finally, model M5 is similar to M3, except that the atoms are distinguished by their chemical surroundings.

Four different fitting procedures were used (F1–F4). In F1 the parameters were determined by a least-squares fitting of the multipole moments, as calculated from an assumed point-charge distribution, to the corresponding ab initio results. In F2 the best parameters were found by minimizing the orientationally-averaged square of the difference between the electrostatic potentials predicted by the model and the ab initio multipoles 10 a.u. from the molecular center. In F3 the parameters were determined similarly but with respect to the interaction energy of two molecules 15 a.u. apart.

Procedure F4 is similar to F2 but utilizes the same set of point charges for all of the molecules. The results of F2–F4 are rather insensitive to the choice of a particular intermolecular distance provided that the distance is large enough to ensure a convergence of the multipole expansion.

The results for model M1 fitted by procedure F1 were completely unsatisfactory. The deviations for the electrostatic potential range from 20% (benzene) to 74% (s-tetrazine). Also, the model parameters resulting from (F1, M1) showed a high degree of correlation (up to 95%) and no transferability.

The use of procedure F2 somewhat improved the fit but the parameters remained highly correlated and non-transferable. A real improvement only resulted when point charges were introduced to the lone pairs of the nitrogen atoms (model M2). This can be seen from Table 3.14 in which the results for (F2, M1) and (F2, M2) are compared.

An important result of model M2 was the approximate electroneutrality of the C–H bonds and the nitrogen/lone-pair fragments ($q_C \approx -q_H$, $q_N \approx -q_{lp}$). This result, together with the strong correlation between q_C and q_H and between q_N and q_{lp}, justified the use of separate neutrality constraints for the next model M3.

Model M3 was the most attractive since it only involved two adjustable parameters for each molecule (one for benzene). As can be seen from Table 3.14, M3 yields quite an acceptable approximation to the electrostatic potential and exhibits a good transferability of its parameters.

In Fig. 3.7 the behavior of the electrostatic potential of benzene along the molecular axes x and z is shown [3.104]. The three curves presented correspond to the point-charge model M3, an ab initio multipole expansion and the exact (non-expanded) results. It is seen that the point-charge model breaks down at nearly the same intermolecular distance as the multipole expansion, from which the model was derived. At larger distances the agreement is very good.

Table 3.14. Point-charge models [e] for seven azabenzenes, fitted to the ab initio multipole expanded results [3.104]

Code fit	Molecule[a]	q_H	q_C	q_N	q_{lp}	σ^b %
(F2, M1)	1	0.162	-0.162	-	-	4.7
	2	0.14	-0.08	-0.28	-	16.6
	3	0.11	0.02	-0.25	-	12.1
	4	0.10	0.02	-0.26	-	24.3
	5	0.07	0.22	-0.59	-	10.4
	6	-0.01	1.18	-1.17	-	6.5
	7	-0.16	1.2	-0.52	-	16.6
(F2, M2)[c]	2	0.170	-0.156	0.52	-0.59	3.2
	3	0.183	-0.149	0.48	-0.55	1.0
	4	0.178	-0.15	0.59	-0.61	4.8
	5	0.19	-0.18	0.6	-0.63	2.4
	6		Ill-conditioned problem			
	7	0.23	-0.18	0.53	-0.55	1.8
(F2, M3)[c]	2	0.181	$q_C=-q_H$	0.76	$q_{lp}=-q_N$	5.8
	3	0.219		0.74		4.4
	4	0.201		0.73		7.1
	5	0.191		0.64		2.4
	6	0.222		0.69		3.8
	7	0.255		0.58		2.0

[a] 1: benzene, 2: pyridine, 3: pyridazine, 4: pyrimidine, 5: pyrazine, 6: s-triazine, 7: s-tetrazine;
[b] σ is the standard relative deviation for the fit on the electrostatic potential;
[c] The fits (F2, M2) and (F2, M3) for benzene are identical to (F2, M1)

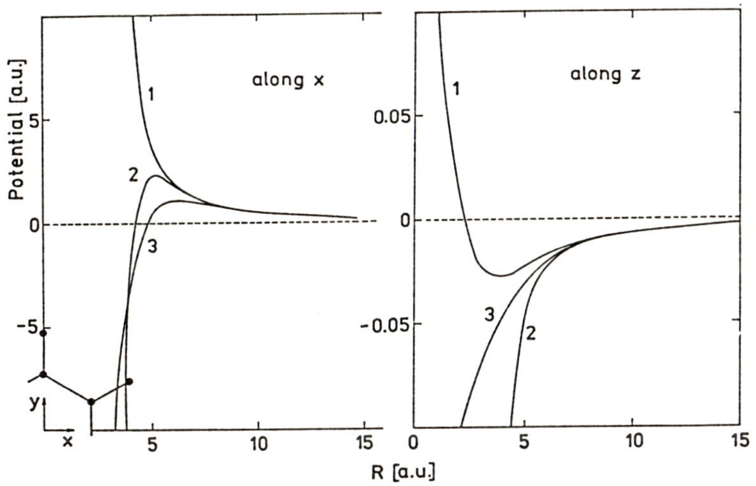

Fig. 3.7. Dependence of the electrostatic potential of benzene on the distance along the x and z axes [3.104]. *1*: exact potential; *2*: multipole expanded potential; *3*: potential predicted by point-charge model

The application of the fitting procedure F3 to the model M3 gave very similar point charges, but the resulting quality of fit was nearly twice as poor as that achieved in F2 for the electrostatic potential. This indicates that it is more difficult to fit the interaction energy than the electrostatic potential. A reasonable representation of the latter does not imply an equivalent representation of the former.

When the lone-pair charges are allowed to shift (model M4), the charges and their distances to the nitrogen atoms are highly correlated. In general, somewhat better fits are obtained, although the lone-pair dipole moment is almost the same as in model M3.

The use of model M5, in which the atoms are also distinguished by their chemical environment, had only a minor effect on the quality-of-fit compared to model M3.

Finally, when all of the molecules are assigned the same set of charges (combination F4, M3), the fit was, as expected, somewhat worse but nevertheless remained quite acceptable (~12% for the electrostatic potential). The best transferable charges were $q_H = -q_C = 0.19$ a.u., $q_N = -q_{lp} = 0.68$ a.u.

A point-charge model similar to M3 was also employed by *van der Linden* and *van Duijneveld* [3.105a] to fit the non-expanded ab initio potentials of the azabenzenes. The wave functions used to compute the electrostatic potentials were exactly the same as the ones employed by *Mulder* et al. [3.14] to compute the multipole moments. The resulting point charges are listed in Table 3.15. A comparison with the values listed in Table 3.14 (combination F2, M3) shows a fair agreement. The small observed differences may be partially ascribed to penetration effects.

The atom-atom representation of the intermolecular potential has also proven to be flexible enough to simulate the interactions in benzene-benzene and water-benzene dimers [3.105b]. The ab initio interaction energy was calculated in an SCF CI approximation using a partitioning scheme similar to that suggested by *Kitaura* and *Morokuma* [3.105c]. In calculating ε_{elst}, a many-center multipole expansion was used, while ε_{ind} and ε_{disp} were evaluated using a local partitioning scheme.

Table 3.15. Point-charge models [e] for seven azabenzenes, fitted to the ab initio (non-expanded) electrostatic potential [3.105a]. (Key to molecules is given in footnote a of Table 3.14; the point charges are distributed as in model M3 of Table 3.14: $q_C = -q_H$, $q_{lp} = -q_N$)

Molecule	q_H	q_N
1	0.121	-
2	0.157	0.721
3	0.198	0.743
4	0.185	0.701
5	0.150	0.502
6	0.178	0.642
7	0.300	0.567

The ab initio potential was fitted by atom-atom analytical functions of the (1-4-6-9-12) form. In these functions the coefficients at r^{-1} were chosen so as to reproduce the experimental dipole and quadrupole moments.

The resulting model potential reproduced the ab initio calculations remarkably well. In particular, the model correctly predicted a perpendicular configuration of the benzene dimer to be more stable then a parallel configuration. When used to calculate the second virial coefficient of benzene, the model potential reproduced the experimental data with an accuracy of 25%. This certainly is an acceptable accuracy considering that it corresponds to an average error in the interaction potential of about 0.1 kcal/mol [3.105b].

A significant contribution to the ab initio version of the atom-atom potential method has been made by *Clementi* et al. in their studies of the interactions between water molecules [3.106–108], between water and simple ions [3.109–112], and between water and ions, on the one hand, and the basic components of proteins and nucleic acids, on the other hand [3.13,113–116]. The analytical potentials that were obtained were intended for use in Monte-Carlo simulations aimed at elucidating the structural organization of aqueous solutions containing ions and biologically important molecules [3.117–122a]. For an application of the potentials to crystallographical problems the reader is referred to a Monte-Carlo study of the network of water molecules hydrating a 2 : 1 complex of proflavine and deoxycytidylyl-3′,5′-guanosine [3.122b]. (The latter work will be briefly discussed in Sect. 4.5.)

Already the first applications of atom-atom potentials to simulate the interaction energy between two water molecules showed that the model is flexible enough to reproduce the fine structure of the potential-energy hypersurface. This is well illustrated by the improvement in the calculated properties of liquid water, produced by successive improvements in the ab initio calculations.

The best analytical potential to simulate the interaction of two water molecules is probably the one derived by *Matsuoka* et al. [3.106b] from very accurate CI calculations on the water dimer. This potential is of the form

$$\varepsilon = \sum_{a,b} [q_a q_b r_{ab}^{-1} + B_{ab} \exp(-C_{ab} r_{ab}) - F_{ab} \exp(-D_{ab} r_{ab})]. \qquad (3.86)$$

The point charges q_H of the hydrogen atoms were placed on the atomic nuclei, while q_O was shifted ~0.226 Å along the bisector of the HOH angle. All other force centers (both the repulsive and attractive ones) were placed on the atomic nuclei. The last term in (3.86) was intended to take into account correlation effects and was assumed to be non-zero for O...H interactions only. In deriving the best parameters for (3.86) a partitioning of the interaction energy into its individual components was not undertaken, so that a physical meaning cannot be given to the numerical values obtained for the model parameters (Table 3.16).

Table 3.16. Parameters [a.u.] of analytical $H_2O \ldots H_2O$ interaction potentials fitted to an ab initio CI potential surface [3.106b]

Interaction	B	C	F	D
H...H	0.662712	1.299982	0	0
H...O	2.684452	1.439787	0.675342	1.141494
O...O	1864.271482	2.753110	0	0
$q_H = -\frac{1}{2} q_O = 0.565117^a$				

[a] q_O is assumed to have been shifted 0.4877 a.u. from the oxygen nucleus along the bisector of the HOH angle

In a more recent work *Clementi* and *Habitz* [3.122c] have attempted to improve the water-water potential using a more flexible basis set, more extended CI interactions and a larger number of dimer configurations. The resulting model potential proved to be less repulsive and showed a shift to larger distances. Unfortunately, the second virial coefficients calculated with the new potential were about a factor of two off the corresponding experimental values. In this respect the old potential of *Matsuoka* et al. [3.106b] was substantially better.

In studies of biologically important systems *Clementi* et al. [3.107–116] employed a more traditional model. In it all of the force centers were placed on the atomic nuclei and the potential was of the usual (6-12-1) form

$$\varphi_{ab}(r_{ab}) = -A_{ab}r_{ab}^{-6} + B_{ab}r_{ab}^{-12} + G_{ab}q_a q_b r_{ab}^{-1}. \tag{3.87}$$

Although the point charges q_a and q_b were not adjusted but were calculated from gross Mulliken populations, this is of little significance because the constants G_{ab}, as well as A_{ab} and B_{ab}, were treated as adjustable parameters.

The distinctive feature of above model potential, as opposed to the models discussed previously, is a much finer classification of the interacting atoms into "atomic species". Besides the chemical nature of an atom, *Clementi* et al. [3.115] also took into consideration its net charge, q_a, and the energy of its molecular orbital valency state (MOVS) defined as

$$\mathrm{MOVS}_a = \sum_i (e_{ia} + I_{ia}), \tag{3.88}$$

where e_{ia} and I_{ia} are the one-center contributions of atom a to the orbital energy e_i, see (2.20), and the one-electron orbital energy I_i, see (2.9), respectively.

The above two quantities, q_a and MOVS_a, are measures of the electron density and the energy of atom a in a molecule. It is important that q_a and MOVS_a are well correlated with the chemical environment of atom a.

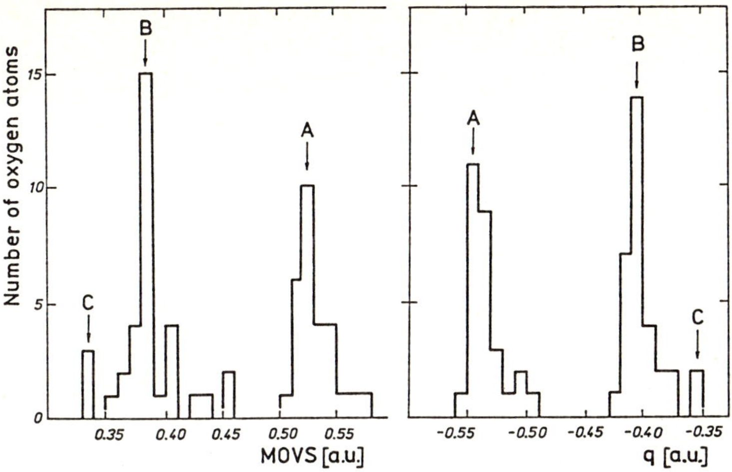

Fig. 3.8. Distribution of the oxygen atoms in the 22 naturally occurring amino acids and four bases of DNA [3.115a]. The three peaks marked by arrows correspond to the oxygen atoms in a OH group (peak A), the CO unit of a COOH group (B) and a C=O group (C)

The distributions of oxygen atoms over their net charges and MOVS's for 22 naturally occurring amino acids and four bases of DNA are shown in Fig. 3.8 [3.115a]. One can easily observe three well-defined peaks in the distributions, indicating that all oxygen atoms in the systems can be grouped into three classes on the basis of their q_a's and MOVS$_a$'s. An analysis of particular contributions to the peaks has shown that the three classes correspond exactly to oxygen atoms which appear in either the OH group (class A), the CO unit of the COOH group (class B) or the C=O group (class C).

In deriving the atom-atom parameters, atoms belonging to different classes were treated as distinct atomic species, even though they may be chemially identical (i.e., have the same atomic number). A final choice of classes was made after the fitting procedure: if the parameter set of any two classes was nearly identical, the two classes were grouped together into a single class.

In Sect. 3.2 we have noted that a model potential may reproduce the interaction potential by simulating the interactions between individual molecular fragments. In this respect, the classification of "atomic species" suggested by *Clementi* et al. [3.115] seems to be very judicious because it is intimately related to the chemical environment of a given atom in a molecule. It is clear that by assigning distinct potentials to chemically identical atoms in distinct atomic groups, we can simulate the interactions between the given groups more accurately than in the case in which the identical atoms are treated with the same potentials. (In the later case we are actually forced to employ a mean potential which represents a compromise between the potentials best suited to the particular groups.)

The idea to distinguish between atoms similar in the chemical sense but belonging to distinct molecular fragments is, of course, not new and has already been exploited in experimental versions of the atom-atom potential method. However, in the latter case this could not be done properly because of the shortage of observations needed to uniquely determine all of the associated parameters. The ab initio version of the atom-atom potential method is free of this shortcoming because the number of observations (i.e., the number of sampled configurations) that can be taken is, in principle, unlimited.

The success of the classification suggested by *Clementi* et al. [3.115a] is evidenced by the good transferability of their atom-atom potentials. This transferability can already be judged by the very existence of a potential set suited simultaneously to a large number of molecules. An example of such a set is the one describing the interactions of water with 21 naturally occurring amino acids and four bases of DNA [3.115a,b]. This set involves 56 distinct potentials and reproduces the interaction energies of 2212 configurations with a standard deviation of ~0.6 kcal/mol. A more critical test of the transferability has been reported in [3.115c] in which the aforementioned set was applied to the water/phenylalanine complex not explicitly included in the original fitting. The results of this test proved to be encouraging: for 55 configurations with interaction energies ranging from −4.09 to 3.18 kcal/mol the common potential set from [3.115a,b] gave a standard deviation of ~0.4 kcal/mol from the ab initio values. That is to say, the examined potential set was found to be applicable to the new system almost as well as to the systems involved in the original fitting.

Before concluding this subsection we note that there may be a mixed version of the atom-atom potential method, in which the potential parameters are determined by combining ab initio and experimental results. Properly speaking, the model potentials of *Gamba* and *Bonadeo* [3.64] and *Williams* and *Cox* [3.65c] discussed in Sect. 3.3.3 have already been such mixed models since their electrostatic parameters have been calibrated using ab initio data.

More emphasis on ab initio results has been placed in a mixed model suggested by *Böhm* et al. [3.122d] for CH_4, CH_3F, CHF_3, CH_3Cl, CH_2Cl_2, CH_3CN and CO_2. The model was essentially based on the decomposition of the interaction potential given in (3.78). The first-order interaction energy ε_1 was obtained in ab initio SCF calculations and then fitted by pairwise (exp-1) potentials. For CH_4, CH_3F, CHF_3 and CH_3CN the force centers were located at the nuclei, while for the other three compounds a slight shift (within 0.1Å) off the nuclei was required to attain a satisfactory fit.

The second-order interaction energy ε_2 was represented as a sum of atom-atom dispersion interactions. The parameters of the latter were fitted to the experimental second virial coefficients. The agreement of the model predictions with the available experimental and theoretical results proved to be satisfactory, except for CH_3CN.

3.6 Semiempirical Atom-Atom Potentials

As indicated in Sect 3.2, all or, more frequently, a part of the parameters in semiempirical versions of the atom-atom potential method are not adjusted but are calculated from some kind of partitioning scheme. In order to do so, one should necessarily assume that the parameters and hence the atom-atom potentials themselves have definite physical meanings. In this sense the semi-empirical methods differ from the pure empirical ones which merely exploit the atom-atom potentials as convenient basis functions to expand the interaction energy.

In deriving the parameters of semiempirical potentials the interaction energy between two atoms is usually represented as a sum of electrostatic, induction, dispersion and exchange repulsion contributions

$$\varphi = \varphi_{\text{elst}} + \varphi_{\text{ind}} + \varphi_{\text{disp}} + \varphi_{\text{exch}}. \tag{3.89}$$

We shall now briefly discuss three representative calculation schemes used to estimate the individual terms of (3.89).

The first of these schemes is the one used by *Poltev*, *Sukhorukov* et al. [3.123,124]. The dispersion contribution to the interaction between atoms a and b is calculated by the London formula

$$\varphi_{\text{disp},ab} = -\frac{3}{2}\alpha_a\alpha_b\frac{I_a I_b}{I_a + I_b}r_{ab}^{-6}, \tag{3.90}$$

where α and I are the atomic dipole polarizabilities and ionization potentials, respectively. The polarizabilities are deduced from empirical atomic refractive indices as used in the additive atomic representations of molecular refractive indices [3.125]. The ionization potentials are assumed to depend on the valency state of a given atom and are taken from [3.126] (Table 3.17).

The electrostatic component of the potential given in (3.89) is evaluated using the monopole-monopole approximation with atomic charges obtained by the Del Re method for σ electrons [3.127] and by the Hückel method for π electrons [3.128].

The induction energy due to the polarization of atom a by the atomic charge on b is given by

$$\varphi_{\text{ind},\,b\to a} = -\frac{1}{2}\alpha_a\mathcal{E}_{b\to a}^2, \tag{3.91}$$

where $\mathcal{E}_{b\to a}$ is the strength of the electric field induced by atom b at atom a. Note that the induction contribution to the interaction energy is not additive with respect to $\varphi_{\text{ind},b\to a}$. Hence, in computing the lattice induction energy one should determine the total electric field at the ath atom and then perform the summation over a.

Table 3.17. Atomic polarizabilities, ionization potentials, intermolecular radii and parameters of atom-atom potentials used by *Poltev* et al. [3.123,124] [kcal/mol, Å]

Atom type	α	I	r^0	A	B	C
H	0.42	313.7	2.70	41.5	11000	4.54
H in H-bond	0.42	313.7	0.80	41.5	9900	4.54
>C<	0.94	336.8	3.76	223	828000	4.58
>C=	1.25	334.5	3.72	392	1310000	4.58
(H)(H)>N–(C)	0.89	382.9	3.44	228	373000	4.59
(C)(C)>N–(H)	0.94	382.9	3.44	254	416000	4.59
(C)(C)>N–(C)	1.06	382.9	3.44	322	528000	4.59
-N=	1.42	445.2	3.40	673	996000	4.59
O=	0.84	581.3	3.20	307	284000	4.60
-O-	0.64	553.6	3.24	159	162000	4.60

The exchange contribution is represented by a conventional exponential dependence,

$$\varphi_{\text{exch},ab} = B_{ab} \exp\left(-C_{ab} r_{ab}\right). \qquad (3.92)$$

The exponents C_{ab} are taken from a plot of C versus the atomic number, constructed by *Scott* and *Scheraga* [3.129], who used the data on the scattering of high-velocity atomic beams. For cross interactions, the geometric-mean combining rule given in (3.4) is used.

The parameter B_{ab} in (3.92) is estimated from the equilibrium distance r^0_{ab} defined as a solution of

$$\partial(\varphi_{\text{disp},ab} + \varphi_{\text{exch},ab})/\partial r_{ab} = 0. \qquad (3.93)$$

By substituting (3.90,92) into (3.93) one finds that

$$B_{ab} = 6A_{ab} \frac{\exp\left(C_{ab} r^0_{ab}\right)}{(r^0_{ab})^7 C_{ab}}, \qquad (3.94)$$

where A_{ab} has its usual meaning and is defined by (3.90).

The set of equilibrium distances adopted in [3.123,124] is presented in Table 3.17 together with all other parameters of the calculation scheme. It is worth noting that the equilibrium distances of Table 3.17 do not correspond

Table 3.18. Components of the lattice energy and the heat of sublimation [kcal/mol] of five aromatic compounds [3.123]

	Φ_{elst}	Φ_{ind}	Φ_{disp}	Φ_{exch}	Φ	ΔH_{subl} [a]
Naphthalene	-0.3	-0.1	-22.8	5.0	-18.1	17.0 to 17.3
Anthracene	-0.5	-0.2	-31.1	7.3	-24.4	23.4 to 24.4
p-Benzoquinone	-2.4	-0.4	-17.1	5.8	-14.1	15.0
Pyrazine	-1.3	-0.3	-17.7	4.8	-14.4	13.5
Imidazole	-8.1	-0.5	-19.6	8.1	-20.2	20.4

[a] The heats of sublimation are given as quoted in [3.123]

to any standard set of intermolecular or van der Waals distances. It seems, therefore, that the r^0's were varied somewhat, although the original papers do not give a relevant explanation.

The parameters of Table 3.17 were tested with respect to the reproducibility of the sublimation energy of several organic crystals. The results listed in Table 3.18 exhibit a fairly good agreement with experiment. This agreement seems very surprising because the calculation scheme does not involve any adjustable parameter (except r^0_{ab}), and treats the components of the interaction energy in a too crude way. The latter statement is particularly true of the electrostatic energy, whose calculated values, obtained using the point-charge approximation with semi-empirical atomic charges, usually do not agree at all with the experimental values.

The second calculation scheme, which may be classified semiempirical, was introduced by *Momany* et al. [3.103]. To estimate the dispersion energy the Slater-Kirkwood equation was used

$$\varphi_{\mathrm{disp},ab} = -\frac{3}{2}\frac{\alpha_a\alpha_b}{(\alpha_a/N_a)^{1/2}+(\alpha_b/N_b)^{1/2}}r_{ab}^{-6}, \tag{3.95}$$

where N_a and N_b are the effective number of outer-shell electrons in atoms a and b, respectively, as suggested by *Pitzer* [3.130]. The values of N were interpolated from a plot of the effective number of outer-shell electrons versus the atomic number for rare-gas atoms. The atomic polarizabilities were taken from *Dalgarno* [3.131] (Table 3.19).

The induction contribution to φ_{ab} was neglected while the electrostatic energy was computed from the atomic point charges estimated from CNDO/2 Mulliken populations. The dielectric constant in Coulomb's law was taken to be equal to 2.

The exchange contribution to φ_{ab} was written in the form

$$\varphi_{\mathrm{exch},ab} = B_{ab}r_{ab}^{-12}, \tag{3.96}$$

where B_{ab} is treated as an adjustable parameter.

Exactly the same terms were included in the interaction potential φ_{ab} for hydrogen and proton-acceptor atoms involved in a hydrogen bond, except that the attractive term was assumed to be proportional to r^{-10} (in-

Table 3.19. Parameters of the semi-empirical atom-atom potentials used by *Momany* et al. [3.103] [kcal/mol, Å]

Atom type	Code	α	N	A	B	ε^0	r^0
Aliphatic H	H_1	0.42	0.85	45.5	14100	0.037	2.92
Primary and secondary amine or amide H	H_2	0.42	0.85	45.5	8430	0.062	2.68
Aromatic H	H_3	0.42	0.85	45.5	14390	0.036	2.93
Hydroxyl or carboxylic acid H	H_4	0.42	0.85	45.5	11690	0.044	2.83
Aliphatic C	C_6	0.93	5.20	370.5	906030	0.038	4.12
Carbonyl, carboxylic acid, or peptide bond C	C_7	1.51	5.20	766.6	1048980	0.141	3.74
Aromatic C	C_8	0.93	5.20	370.5	475300	0.073	3.70
Primary or secondary amide N	N_{13}	0.87	6.10	363.1	732540	0.045	3.99
Uncharged primary or secondary amine N; or primary or secondary amide N as proton acceptor in a hydrogen bond	N_{14}	0.93	6.10	401.0	374940	0.107	3.51
Carbonyl or carboxylic acid (C =O) O	O_{17}	0.84	7.00	369.0	170190	0.200	3.12
Hydroxyl or carboxylic acid (C-O-H) O	O_{18}	0.59	7.00	217.2	125630	0.094	3.24
Sulfur in sulfur-containing heterocyclic systems, sulfohydryl groups and thio ethers	S_{20}	0.34	14.8	249.0	363180	0.043	3.78
Hydrogen-bonding parameters							
$H_2 \cdots O_{17}$				4014	120400	1.11	1.90
$H_4 \cdots O_{17}$				5783	133440	5.92	1.66
$H_4 \cdots O_{18}$				4610	130330	1.71	1.84
$H_2 \cdots N_{14}$				8244	328970	0.55	2.19

stead of r^{-6}). The coefficients of the r^{-10} terms were taken from an earlier study [3.132] in which the (10–12) hydrogen-bond potential was fitted to the experimental dimerization energy, dimer structure and CNDO/2 results for several hydrogen-bonded dimers.

All of the repulsive non-bonded parameters B_{ab} were determined in a purely empirical way, by fitting to the crystal structures of several hydrocarbons, carboxylic acids, amines and amides. The resulting parameter set is presented in Table 3.19.

It is to be noted that in some cases the minimum-energy structures and lattice energies found by *Momany* et al. [3.103] using the parameters of Table 3.19 showed marked deviations from experiment. Thus, the calculated lattice energy of sebacic acid was −33.7 kcal/mol versus an experimental value of –38.4 kcal/mol, –11.3 kcal/mol versus –17.0 kcal/mol for formamide, and –19.5 kcal/mol versus –27.1 kcal/mol for oxamine. The discrepancies in the lattice periods were sometimes as much as 0.62 Å (pyrazine), 0.63 Å (formic acid) and 0.82 Å (oxalic acid). These discrepancies, as well as those observed for the lattice energies, are perceptibly greater than those observed using pure empirical model potentials.

As is the practice of many researchers dealing with semi-empirical atom-atom models, *Momany* et al. [3.103] paid too much attention to the particular components of the lattice energy and their role in stabilizing a crystal structure. This is hardly justifiable in view of the existence of pure empirical terms in the model. Clearly, the empirical parameters B_{ab}, when fitted to the observed crystal structures, should ultimately absorb all components of the total potential because neither of the remaining semiempirical terms reproduces the corresponding energy contributions correctly. *Derissen* and *Smit* [3.79] have illustrated this point by noting that, in the case of formic acid, for example, the point-charge model based on CNDO/2 calculations makes a contribution to the lattice energy, which is three to four times smaller than the contribution calculated from ab initio point charges. The use of the latter point charges instead of the CNDO/2 ones would thus substantially affect the adjustable part of the lattice energy and, hence, yield quite a different distribution of the lattice energy among its components.

We now turn to the semiempirical model potential developed by *Caillet* et al. [3.133]. As in the two models discussed above, the electrostatic contribution to φ_{ab} was calculated using the monopole approximation with semiempirical LCAO-MO charges. To evaluate the induction contribution they [3.133] used (3.91), as did *Poltev* [3.123], but followed a different calculation scheme in finding the atomic polarizabilities. Believing that atomic polarizabilities are less transferable than bond polarizabilities, they preferred to employ a set of bond polarizibility increments α_u, as given by *Le Fevre* [3.134], and to deduce α_a from

$$\alpha_a = \sum_u \frac{n_{u,a}}{n_{u,a} + n_{u,b}} \alpha_u, \tag{3.97}$$

where u runs over all of the bonds sharing atom a; b labels the atom bonded to the ath atom by the uth bond; and $n_{u,a}$ and $n_{u,b}$ are the effective numbers that characterize the participation of the ath and bth atoms in the uth bond. The values for $n_{u,a}$ were taken from

$$n_{u,a} = N_u/2 + N_a^c/\nu_a, \tag{3.98}$$

where N_u is the number of electrons shared by the ath and bth atoms of the uth bond (e.g., 2 for a C–C bond, 4 for a C=C bond, 3 for a $\mathrm{C\underline{\;---\;}C}$ bond in benzene, etc.); and N_a^c is the number of electrons of the ath atom, not involved in a chemical bond (except the 1s electrons). For the elements of the first row N_a^c coincides with the number of lone pair electrons. It was assumed that the N_a^c electrons are equally distributed among the ν_a bonds sharing the ath atom. A somewhat more complicated scheme was suggested to evaluate N_u for conjugated systems, and N_a^c for the elements beyond the first row. In some cases *Caillet* et al. [3.133] preferred to determine the induction energy directly from the bond polarizabilities and atomic charges instead of reallocating the former among the atoms [3.135].

To calculate the dispersion and exchange contribution to φ_{ab}, a universal potential of the form

$$\varphi_{\text{disp},ab} + \varphi_{\text{exch},ab} = k_a k_b[-A(r_{ab}/r_{ab}^{\text{W}})^{-6} + (1 - q_a/N_a^{\text{val}}) \times (1 - q_b/N_b^{\text{val}})B \exp(-\alpha r_{ab}/r_{ab}^{\text{W}})], \tag{3.99}$$

was used [3.133], where q_a and q_b are the net atomic charges, also employed to determine $\varphi_{\text{elst},ab}$; N_a^{val} and N_b^{val} are the number of valence electrons of atoms a and b; the terms $(1 - q_a/N_a^{\text{val}})$ and $(1 - q_b/N_b^{\text{val}})$ are intended to take into account the effects of the electronic populations on the atom-atom exchange repulsion; and r_{ab}^{W} is defined as $2(r_a^{\text{W}} r_b^{\text{W}})^{1/2}$ where r_a^{W} and r_b^{W} are the van der Waals atomic radii ($r_{\text{H}}^{\text{W}} = 1.2\,\text{Å}$, $r_{\text{C(aliphatic)}}^{\text{W}} = 1.7\,\text{Å}$, $r_{\text{C(aromatic)}}^{\text{W}} = 1.77\,\text{Å}$, $r_{\text{N}}^{\text{W}} = 1.6\,\text{Å}$, $r_{\text{O}}^{\text{W}} = 1.5\,\text{Å}$, $r_{\text{P}}^{\text{W}} = 1.85\,\text{Å}$, $r_{\text{S}}^{\text{W}} = 1.8\,\text{Å}$) [3.136]. The empirical parameters A, B and α were assumed to be independent of the atomic species and equal to 0.214 kcal/mol, 47 000 kcal/mol and 12.35, respectively. The ratio A/B was chosen so that the minimum of (3.99) occurred at $r_{ab} = 13/11\, r_{ab}^{\text{W}}$. The parameters k_a and k_b were introduced to allow the well depth of the potential in (3.99) to take on different values according to the atomic species involved in a particular interaction ($k_{\text{H}} = k_{\text{C}} = 1$, $k_{\text{N}} = 1.10$, $k_{\text{O}} = 1.36$, $k_{\text{P}} = 2.10$, $k_{\text{S}} = 2.40$).[3]

The interaction between hydrogen and proton-acceptor atoms involved in a hydrogen bond has the same analytical form as (3.99). However, the parameters A, B and α were assigned distinct values according to the interatomic distance. Two characteristic distances, r^m and r^M, were introduced ($r^m < r^M$). For $r > r^M$ the same parameters as the ones listed above for non-bonded interactions were taken. The following values, $A' = A/5$, $B' = B/2.7$ and $\alpha' = 13.8$, were adopted for $r < r^m$. At intermediate distances, $r^m < r < r^M$, *Caillet* et al. [3.133] used interpolated values defined by

$$k(x) = \frac{K + K'}{2} + \frac{-K - K'}{2}(0.375x^5 - 1.25x^3 + 1.875x), \tag{3.100}$$

where

[3] The parameter set first reported in [3.133a] has been somewhat extended and corrected according to a private communication by J. Claverie.

$$x = \left(r - \frac{r^m + r^M}{2}\right) \Big/ \left(\frac{r^M - r^m}{2}\right), \tag{3.101}$$

and K is one of the three parameters A, B and α.

The polynomial $0.375x^5 - 1.25x^3 + 1.875x$ was chosen so that the first and second derivatives vanish at $x = \pm 1$ and $K(1) = K$ and $K(-1) = K'$. The values adopted for r^m and r^M were 1.8 and 2.6 Å, respectively.

Recently, *Caillet* et al. [3.137] have extended their model to include the interactions of several monovalent ions (Na^+, K^+, F^-, Cl^- and Br^-). For each ionic species, only two parameters, r_a^W and k_a, were treated as adjustable parameters, while the others were held fixed at the values listed above. The parameters r_a^W and k_a were determined by fitting the well depths and equilibrium distances of the dispersion plus repulsion curves in (3.99) to the respective theoretical values for the interactions between rare gas atoms and the above ions [3.138]. Such a fitting yielded only the products $k_{ion}k_{atom}$ and $r_{ion}^W r_{atom}^W$ so that a determination of k_{ion} and r_{ion}^W required a knowledge of k_{atom} and r_{atom}^W. These latter parameters were derived from the theoretical curves for the interactions between the rare gas atoms themselves [3.138]. It is worth noting that k_{ion} and r_{ion}^W varied appreciably depending on which particular rare gas was used to derive these parameters. Thus, $r_{Na^+}^W$ varied from 1.172 to 1.578 Å, $r_{F^-}^W$ from 1.578 to 1.903 Å and $r_{Cl^-}^W$ from 1.785 to 2.515 Å. These variations indicate that the multiplication rules for k_a and r_a^W and probably the (6-exp) potential itself cannot be used to describe the interactions between ions and rare gas atoms. In spite of this fact it has been found justifiable to use the results of their fitting as a basis for the construction of their potential set [3.137]. The final values adopted for k_{ion} and r_{ion}^W were: $k_{ion} = 1.4$ (Na^+), 2.9 (K^+), 0.92 (F^-), 1.1 (Cl^-), 1.3 (Br^-); $r_{ion}^W = 1.2$ (Na^+), 1.46 (K^+), 1.9 (F^-), 2.5 (Cl^-), 2.8 Å (Br^-).

The intermolecular-energy model developed by *Caillet* et al. [3.133a, 135,137] suffers from the same disadvantages exhibited by the two semiempirical models discussed above. This refers, first of all, to the representation of the electrostatic energy which is evaluated using the point-charge approximation with semempirical LCAO-MO atomic charges. The use of semiempirical point charges also calls in question the validity of the estimates of the induction energy. Considering the possible inaccuracy in ε_{ind}, as well as the smallness of the contribution of ε_{ind} to the lattice energy (of the order of 0.1 kcal/mol [3.133a]), it seems that the inclusion of the induction terms in the calculation scheme is a superfluous complication of the model.

Despite the apparent shortcomings of this model, it has provided quite reliable results for lattice energies and equilibrium crystal structures. This is not very surprising because the model involves terms evaluated on a pure empirical basis and the adopted equilibrium distances correlate well with the respective van der Waals radii.

In conclusion, it is to be noted that the list of parameters presented in the tables is, of course, not exhaustive. As the atom-atom potential method is finding ever increasing applications, it would be almost impossible to discuss all of the parameter sets suggested so far. Thus, we have confined ourselves to the most frequently used potentials and the most representative procedures of deriving their parameters. In the calculations to be discussed in the subsequent chapters several other parameter sets will appear. All necessary information concerning the manner in which these sets were derived will be presented while discussing the relevant calculations.

4. Lattice Statics

The ability to evaluate, in a very simple way, the intermolecular interaction energy provides a powerful tool for handling a wide range of problems that arise in solid-state sciences of organic materials. This chapter is devoted to problems which can be treated, to a reasonable approximation, in terms of the mechanical stability of a crystal lattice. In Sect. 4.1 we shall derive the general conditions of mechanical stability, suitable for dealing with a lattice of rigid or partially rigid molecules. Section 4.2 describes a procedure for finding the stable crystal configuration within the framework of a symmetry-constrained model. The explicit expressions for the lattice energy and its derivatives are presented, adapted to the specific form of the intermolecular interaction energy. Also discussed in this section are the convergence properties of the lattice sums and several techniques for the global and local minimization of the lattice energy.

Sections 4.3–6 review the applications of static lattice-energy calculations, including predictions of the stable crystal configuration, the studies on the effect of crystal forces on molecular conformation, and the use of atom-atom potentials in the crystallographic analysis of monomeric and polymeric substances.

4.1 The Lattice at Equilibrium

It is quite natural to begin a chapter devoted to lattice statics with a discussion of the conditions that must be satisfied for a lattice to be stable. These conditions were first formulated by *Born* and *Huang* [4.1] for the general case of a crystal to which the Born-Oppenheimer approximation may be applied. For molecular crystals, however, this formalism proves to be inconvenient. We shall now illustrate this point.

Let us consider an arbitrary crystal containing N unit cells. Let $\boldsymbol{a}$, $\boldsymbol{b}$ and $\boldsymbol{c}$ denote the three basis vectors specifying the edges of a representative (reference) unit cell. Each unit cell in the crystal may be uniquely specified by a vector $\boldsymbol{l}$ whose components l_1, l_2 and l_3 are integer numbers defining the position of the origin of the given unit cell relative to the origin of the reference (zero-th) unit cell:

$$\boldsymbol{r}^l = l_1\boldsymbol{a} + l_2\boldsymbol{b} + l_3\boldsymbol{c}. \tag{4.1}$$

Let $\boldsymbol{r}^{al}$ denote a vector whose components r_α^{al} are the Cartesian coordinates of the ath atom in the lth unit cell. The potential energy of a crystal is a function of all atomic coordinates and can be expanded in power series of the atomic displacements. The equilibrium conditions can then be readily derived by requiring the linear coefficients of this expansion to vanish.

According to *Born* and *Huang* [4.1], only those displacements, $\Delta \boldsymbol{r}^{al} \equiv \boldsymbol{u}^{al}$, are to be considered which leave the lattice ideal. The most general form for such displacements is given by

$$u_\alpha^{al} = u_\alpha^a + \sum_\beta u_{\alpha\beta} r_\beta^{al}. \tag{4.2}$$

The symmetric part of $u_{\alpha\beta}$ is nothing but the strain tensor, as defined in elasticity theory. The deformations corresponding to $u_{\alpha\beta}$ and u_α^a are usually known as external and internal deformations, respectively.

An essential feature of organic molecular crystals is that their constituent molecules, or at least certain molecular fragments, can be considered to be rigid. In developing a theory describing the lattice statics of organic crystals, we would, therefore, want to have a formalism which would be easily adaptable to the rigid-molecule or partially-rigid-molecule approximation. A description of lattice deformations in the form of (4.2) is unsuitable in this respect. This is quite obvious because deformations described by (4.2) always affect the molecular shape.

A special formalism which is general enough but, simultaneously, applicable to rigid or partially-rigid molecules has been suggested by *Chandrasekharan* and *Walmsley* [4.2]. In the subsequent discussion we shall follow this formalism, except for some minor changes in notation. To describe the crystal configuration, we shall use the vectors $\boldsymbol{R}^{kl} = \{R_\alpha^{kl}\}$ (each of them specifying the translation, orientation and conformation of a particular molecule in a particular unit cell) instead of the Cartesian atomic coordinates $\boldsymbol{r}^{al}$. In the notation R_α^{kl} l has the same meaning as in (4.1), while k labels the molecules in the unit cell. Thus, each molecule in the crystal will be labeled by a pair of indices, kl, except for the molecules in the unit cell at the origin, for which the index $l = 0$ will be omitted. It will be assumed that the first three components of the vector $\boldsymbol{R}^{kl}$ are the center-of-mass coordinates of the molecule kl. The next three components ($\alpha = 4,5,6$) describe the orientation of the molecule, while the remaining components ($\alpha > 6$) are internal coordinates specifying the molecular conformation. In the rigid-molecule approximation the internal coordinates are lacking and $\boldsymbol{R}^{kl}$ represent a six-dimensional translation-rotation vector. The other limiting case is obtained when we allow independent displacements for all of the individual atoms. In this case the number of internal components in $\boldsymbol{R}^{kl}$ is equal to $3n - 6$, where n is the number of atoms in the molecule.

With the above definitions, any displacement $\Delta \boldsymbol{R}^{kl} \equiv \boldsymbol{U}^{kl}$, which leaves the lattice ideal may be represented as

$$U_\alpha^{kl} = U_\alpha^k + \sum_\beta u_{\alpha\beta} R_\beta^{kl}, \quad \alpha,\beta = 1,2,3,$$

$$U_\alpha^{kl} = U_\alpha^k, \quad \alpha>3. \tag{4.3}$$

We now consider the changes in the potential energy produced by displacements of the form given in (4.3). In the pair approximation the total potential energy of a crystal has the following form

$$\phi = \sum_{kl} E^{kl} + \frac{1}{2} \sum_{kl} \sum_{k'l'}{}' \varepsilon^{kl,k'l'}, \tag{4.4}$$

where E^{kl} and $\varepsilon^{kl,k'l'}$ represent the internal energy of molecule kl and the interaction energy of molecules kl and $k'l'$, respectively. The prime on the sum over $k'l'$ means that if $k = k'$, the term with $l = l'$ should be omitted from the summation. Since all unit cells in the crystal are identical and remain so after any lattice deformation described by (4.3), the potential energy ϕ may be rewritten as

$$\phi = N\Big(\sum_k E^k + \frac{1}{2} \sum_k \sum_{k'l'}{}' \varepsilon^{k,k'l'}\Big), \tag{4.5}$$

where k runs over all of the molecules in the reference unit cell. The index $l = 0$ labeling the reference unit cell is again dropped.

It is now convenient to introduce the energy density defined by

$$\phi_c \equiv \phi/N = \sum_k E^k + \frac{1}{2} \sum_k \sum_{k'l'} \varepsilon^{k,k'l'}. \tag{4.6}$$

The change in ϕ_c, produced by small displacements U_α^k and $u_{\alpha\beta}$, may be expanded in a Taylor series in powers of U_α^k and $u_{\alpha\beta}$,

$$\Delta\phi_c = \sum_{\alpha k} \frac{\partial \phi_c}{\partial U_\alpha^k} U_\alpha^k + \sum_{\alpha\beta} \frac{\partial \phi_c}{\partial u_{\alpha\beta}} u_{\alpha\beta} + \text{higher-order terms.} \tag{4.7}$$

If we expand ϕ_c about the equilibrium configuration, the coefficients of U_α^k and $u_{\alpha\beta}$ should vanish, i.e.,

$$\frac{\partial \phi_c}{\partial U_\alpha^k} = 0, \tag{4.8}$$

$$\frac{\partial \phi_c}{\partial u_{\alpha\beta}} = 0. \tag{4.9}$$

Before discussing the explicit expressions for the above derivatives, it is worth noting that the derivatives $\partial\phi_c/\partial u_{\alpha\beta}$, from now on denoted as $S_{\alpha\beta}$, are directly related to the components of the stress tensor $\sigma_{\alpha\beta}$:

$$S_{\alpha\beta} = V_c \sigma_{\alpha\beta}, \tag{4.10}$$

where V_c is the unit-cell volume. This can be easily seen by comparing the change in energy produced by a pure lattice deformation,

$$\Delta\phi_c = \sum_{\alpha\beta} S_{\alpha\beta} u_{\alpha\beta}, \tag{4.11}$$

with the corresponding expression in elasticity theory. Note that $S_{\alpha\beta}$, as well as $\sigma_{\alpha\beta}$, is symmetric with respect to α and β. This follows from the fact that a displacement described by $u_{\alpha\beta} = -u_{\beta\alpha}$ $(\alpha \neq \beta)$ represents a pure rotation of the lattice as a whole and should therefore leave the energy density unchanged.

The explicit expressions for $\partial\phi_c/\partial U_\alpha^k$ and $S_{\alpha\beta}$ are readily derived from (4.6). To avoid any confusion concerning the indices k in (4.6) and U_α^k, we shall replace k in (4.6) by k''. Inasmuch as $E_{k''}$ and $\varepsilon^{k'',k'l'}$ depend on U_α^k implicitly, via the dependence of $\boldsymbol{R}_{k''}$ and $\boldsymbol{R}_{l'k'}$ on U_α^k, we may write

$$\frac{\partial\phi_c}{\partial U_\alpha^k} = \sum_{k''}\sum_{\gamma} \frac{\partial E^{k''}}{\partial R_\gamma^{k''}} \frac{\partial R_\gamma^{k''}}{\partial U_\alpha^k} + \frac{1}{2}\sum_{k''}\sum_{l'k'}\left(\sum_{\gamma} \frac{\partial\varepsilon^{k'',k'l'}}{\partial R_\gamma^{k''}} \frac{\partial R_\gamma^{k''}}{\partial U_\alpha^k} + \sum_{\gamma} \frac{\partial\varepsilon^{k'',k'l'}}{\partial R_\gamma^{k'l'}} \frac{\partial R_\gamma^{k'l'}}{\partial U_\alpha^k}\right). \tag{4.12}$$

For lattice deformations of (4.3) the derivatives $\partial R_\gamma^{k'l'}/\partial U_\alpha^k$ are given by

$$\frac{\partial R_\gamma^{k'l'}}{\partial U_\alpha^k} = \delta_{k'k}\delta_{\alpha\gamma}, \tag{4.13}$$

where $\delta_{k'k}$ and $\delta_{\alpha\gamma}$ are the Kronecker deltas. Substituting this expression and a similar expression for $\partial R_\gamma^{k''}/\partial U_\alpha^k$ into (4.12) yields

$$\frac{\partial\phi_c}{\partial U_\alpha^k} = \frac{\partial E^k}{\partial R_\alpha^k} + \frac{1}{2}\left(\sum_{k'l'}{}' \frac{\partial\varepsilon^{k,k'l'}}{\partial R_\alpha^k} + \sum_{k''l'}{}' \frac{\partial\varepsilon^{k'',kl'}}{\partial R_\alpha^{kl'}}\right). \tag{4.14}$$

The resulting equation can be further simplified by noting that the two sums are equal. To prove this we make use of the relation

$$\frac{\partial\varepsilon^{k'',kl'}}{\partial R_\alpha^{kl'}} = \frac{\partial\varepsilon^{k''(-l'),k0}}{\partial R_\alpha^k}. \tag{4.15}$$

This relation follows from the fact that the derivative $\partial\varepsilon^{kl,k'l'}/\partial R_\alpha^{k'l'}$ is invariant with respect to any translation of the lattice as a whole; i.e.,

$$\frac{\partial\varepsilon^{kl,k'l'}}{\partial R_\alpha^{k'l'}} = \frac{\partial\varepsilon^{k(l+l''),k'(l'+l'')}}{\partial R_\alpha^{k'(l'+l'')}}. \tag{4.16}$$

Considering (4.15), the last sum in (4.14) may be rewritten as

$$\sum_{k''l'}{}' \frac{\partial \varepsilon^{k'',kl'}}{\partial R_\alpha^{kl'}} = \sum_{k''l'}{}' \frac{\partial \varepsilon^{k''(-l'),k0}}{\partial R_\alpha^k} = \sum_{k''(-l')}{}' \frac{\partial \varepsilon^{k,k''l'}}{\partial R_\alpha^k}, \tag{4.17}$$

which is identical to the next to last sum because the summation over l' is carried out over all lattice vectors. Finally, (4.14) becomes

$$\frac{\partial \phi_c}{\partial U_\alpha^k} = \frac{\partial}{\partial R_\alpha^k}(E^k + \psi^k), \tag{4.18}$$

where ψ^k denotes the interaction energy of molecule k and its surrounding molecules:

$$\psi^k = \sum_{l'k'}{}' \varepsilon^{k,k'l'}. \tag{4.19}$$

It is now seen that (4.8) represents the condition that molecule k is force-free. Considering that E^k only depends on the internal coordinates, these force-free conditions may be rewritten as

$$\frac{\partial \psi^k}{\partial R_\alpha^k} = 0, \quad \alpha = 1,\ldots,6, \tag{4.20}$$

$$\frac{\partial}{\partial R_\alpha^k}(E^k + \psi^k) = 0, \quad \alpha > 6. \tag{4.21}$$

The maximum number of force-free conditions is $3n \cdot N_d$, where N_d is the number of crystallographically distinct molecules in the unit cell. If the molecules in a crystal retain some of their symmetry elements, the number of force-free conditions is reduced. In general, for each molecule k the number of independent conditions coincides with the number of totally symmetric site group coordinates [4.2]. In the rigid- or partially-rigid-molecule approximation the number of force-free conditions is further reduced. In each particular case, this number can easily be determined by considering all of the possible independent coordinates whose variation leaves the molecular site group unchanged and does not affect the conformation of the rigid fragments.

We now turn to a consideration of $S_{\alpha\beta}$. Differentiation of (4.6) with respect to $u_{\alpha\beta}$ yields

$$S_{\alpha\beta} = \frac{1}{2}\sum_k \sum_{k'l'}{}' \Big(\sum_\gamma \frac{\partial \varepsilon^{k,k'l'}}{\partial R_\gamma^k}\frac{\partial R_\gamma^k}{\partial u_{\alpha\beta}} + \sum_\gamma \frac{\partial \varepsilon^{k,k'l'}}{\partial R_\gamma^{k'l'}}\frac{\partial R_\gamma^{k'l'}}{\partial u_{\alpha\beta}}\Big). \tag{4.22}$$

If the displacements of $R_\gamma^{k'l'}$ are described by (4.3), the derivatives $\partial R_\gamma^{kl}/\partial u_{\alpha\beta}$

are given by

$$\frac{\partial R_\gamma^{kl}}{\partial u_{\alpha\beta}} = R_\beta^{kl}\delta_{\alpha\gamma}, \quad \alpha,\, \beta,\, \gamma = 1,2,3. \tag{4.23}$$

Thus, (4.22) reduces to

$$S_{\alpha\beta} = \frac{1}{2}\sum_k \sum_{k'l'} \Big(\frac{\partial \varepsilon^{k,k'l'}}{\partial R_\alpha^k} R_\beta^k + \frac{\partial \varepsilon^{k,k'l'}}{\partial R_\alpha^{k'l'}} R_\beta^{k'l'} \Big). \tag{4.24}$$

Considering that for $\alpha = 1,2,3$,

$$\frac{\partial \varepsilon^{k,k'l'}}{\partial R_\alpha^{k'l'}} = -\frac{\partial \varepsilon^{k,k'l'}}{\partial R_\alpha^k}, \tag{4.25}$$

we may rewrite (4.24) as

$$S_{\alpha\beta} = \frac{1}{2}\sum_k \sum_{k'l'}{}' \frac{\partial \varepsilon^{k,k'l'}}{\partial R_\alpha^k} R_\beta^{k,k'l'}, \quad \text{where} \tag{4.26}$$

$$R_\beta^{k,k'l'} \equiv R_\beta^k - R_\beta^{k'l'} \quad (\beta = 1,2,3). \tag{4.27}$$

The stress-free conditions for a crystal are thus formulated as follows

$$S_{\alpha\beta} = \frac{1}{2}\sum_k \sum_{k'l'}{}' \frac{\partial \varepsilon^{k,k'l'}}{\partial R_\alpha^k} R_\beta^{k,k'l'} = 0 \quad \text{with } \alpha,\beta = 1,2,3; \quad \alpha \le \beta. \tag{4.28}$$

The number of independent stress-free conditions to be fulfilled is determined by the crystal symmetry. In cubic crystals the diagonal elements of $S_{\alpha\beta}$ are all equal, while the off-diagonal elements automatically vanish. Thus, in cubic crystals only one independent stree-free condition is to be satisfied. In order to determine if a crystal configuration is stress-free, it is sufficient to check whether one of the $S'_{\alpha\alpha}$s vanishes.

For crystals of hexagonal, tetragonal and trigonal symmetry, two independent conditions are to be fulfilled: $S_{33} = 0$ and either $S_{11} = 0$ or $S_{22} = 0$ (it is assumed that the Cartesian coordinate axis 3 is chosen parallel to the highest symmetry axis). For orthorhombic crystals, the number of independent stress-free conditions increases to three. In this case all three $S'_{\alpha\alpha}$s must be separately set equal to zero for equilibrium.

Non-zero off-diagonal elements $S_{\alpha\beta}(\alpha \neq \beta)$ only appear in monoclinic and triclinic crystals. In monoclinic crystals, there is only one such element, S_{13}, if we choose the Cartesian coordinate axis 2 to be parallel to the unique symmetry direction in the crystals. In triclinic crystals, all of the $S_{\alpha\beta}$ may be distinct from zero, so that all six stress-free conditions must be explicitly satisfied at equilibrium.

If the unit cell of a crystal contains crystallographically equivalent molecules, the expressions for the energy density and its derivatives may be represented in a form more convenient for practical calculations. Let us introduce a new index m, which will only label crystallographically distinct molecules in the reference unit cell. Let N_m denote the number of molecules in the unit cell crystallographically equivalent to molecule m ($\sum_m N_m = N_c$, the total number of molecules in the unit cell). The energy density given in (4.6) may then be rewritten as

$$\phi_c = \sum_m N_m \left(E^m + \frac{1}{2}\psi^m\right), \tag{4.29}$$

where ψ^m, defined by (4.19), represents the interaction energy of molecule m and its surrounding molecules. With the above convention for m, the definition of ψ^m may be rewritten as

$$\psi^m = \sum_{m's'}{}' \varepsilon^{m,m's'}, \tag{4.30}$$

where s' runs over all symmetry operations, including the lattice translations. The pair $m's'$ labels the molecule generated from molecule m' by applying the symmetry operation s'. The prime on the sum means that for $m' = m$ the identity operations ($s' = 1$) should be omitted from the summation.

In the expression for $S_{\alpha\beta}$, the summation over k, k' and l' may also be replaced by a summation over m, m' and s', provided that each contribution (m, m', s') is given a weight N_m :

$$S_{\alpha\beta} = \frac{1}{2}\sum_m N_m \sum_{m's'}{}' \frac{\partial\varepsilon^{m,m's'}}{\partial R_\alpha^m} R_\beta^{m,m's'}. \tag{4.31}$$

It is to be noted that the calculation of the tensor $S_{\alpha\beta}$ using (4.31) only yields correct results for its significant components, i.e., for those components which do not automatically vanish by symmetry. To exemplify, let us consider a monoclinic crystal with two crystallographically equivalent molecules in the unit cell, labeled 1 and 2 and related by the the symmetry element s_0. Let $S_{\alpha\beta}^{1,n}$ denote a representative contribution to $S_{\alpha\beta}$ resulting from the interaction of molecule 1 with an arbitrary molecule n :

$$S_{\alpha\beta}^{1,n} = \frac{\partial\varepsilon^{1,n}}{\partial R_\alpha^1}(R_\beta^1 - R_\beta^n). \tag{4.32}$$

By applying the symmetry operation s_0 to molecules 1 and n, we obtain the contribution,

$$S_{\alpha\beta}^{2,ns_0} = \frac{\partial\varepsilon^{2,ns_0}}{\partial R_\alpha^2}(R_\beta^2 - R_\beta^{ns_0}). \tag{4.33}$$

This contribution is always explicitly present in $S_{\alpha\beta}$ because for the adopted crystal symmetry we may rewirte (4.26) as

$$S_{\alpha\beta} = \frac{1}{2}\sum_{n\neq 1}(S_{\alpha\beta}^{1,n} + S_{\alpha\beta}^{2,ns_0}), \tag{4.34}$$

where n runs over all of the molecules in the crystal, except molecule 1.

To find a relationship between $S_{\alpha\beta}^{1,n}$ and $S_{\alpha\beta}^{2,ns_0}$, let us go from the initial reference coordinate system $\{x, y, z\}$ to a new coordinate system, $\{x', y', z'\}$, whose center and axes are generated from those of $\{x, y, z\}$ by the symmetry operation s_0. In this new coordinate system the pair $2, ns_0$ is identical to the pair $1, n$ in the old system $\{x, y, z\}$, and hence

$$\frac{\partial\varepsilon^{2,ns_0}}{\partial R'^2_\alpha}(R'^2_\beta - R'^{ns_0}_\beta) = S_{\alpha\beta}^{1,s_0}. \tag{4.35}$$

In monoclinic crystals the action of any allowable symmetry operation on the coordinates R_α $(\alpha = 1, 2, 3)$ may be represented as

$$R'_\alpha = B_\alpha R_\alpha + T_\alpha, \tag{4.36}$$

where $B_\alpha = \pm 1$. The inverse transformation is given by

$$R_\alpha = B_\alpha R'_\alpha - T_\alpha. \tag{4.37}$$

Using the two last equations, the left-hand side of (4.35) may be transformed,

$$\frac{\partial\varepsilon^{2,ns_0}}{\partial R'^2_\alpha}(R'^2_\beta - R'^{ns_0}_\beta) = \frac{\partial\varepsilon^{2,ns_0}}{\partial R^2_\alpha}(R^2_\beta - R^{ns_0}_\beta)B_\alpha B_\beta. \tag{4.38}$$

Consequently,

$$S_{\alpha\beta}^{1,n} = S^{2,ns_0}B_\alpha B_\beta, \tag{4.39}$$

and, using (4.34), we find

$$S_{\alpha\beta} = \tfrac{1}{2}(1 + B_\alpha B_\beta)\sum_{n\neq 1} S_{\alpha\beta}^{1,n}. \tag{4.40}$$

We can now compare this result for $S_{\alpha\beta}$ with the result obtained using (4.31). In our particular case of a crystal with two crystallographically equivalent molecules in the unit cell, (4.31) may be rewritten as

$$S_{\alpha\beta} = \frac{1}{2} \cdot 2 \sum_{n \neq 1} S_{\alpha\beta}^{1,n} . \tag{4.41}$$

It is seen that (4.40,41) are identical for $\alpha = \beta$. This, however, is not always so for $\alpha \neq \beta$. Let us assume that the unique symmetry direction in the crystal is parallel to the Cartesian coordinate axis 2. In this case the vector $\{B_\alpha\}$ is equal to either $\{1,-1,1\}$ or $\{-1,1,-1\}$. Hence, (4.40 and 41) are identical only for S_{13}, i.e., for the only off-diagonal component of $S_{\alpha\beta}$, which may be distinct from zero in monoclinic crystals. As concerns S_{12} and S_{23}, these are identically equal to zero when determined from (4.40), but are not, when determined from (4.41).

An explicit computation of S_{12} and S_{23} is, of course, senseless for any crystal configuration of monoclinic symmetry. However, if we nevertheless want to determine them, we would have to consider all of the molecules in the unit cell to obtain the correct result: $S_{12} = S_{23} = 0$.

The above considerations have shown that (4.31) cannot serve as a definition for the tensor $S_{\alpha\beta}$, but can nevertheless be used to compute its significant components.

4.2 Determination of Equilibrium Crystal Configurations Using a Symmetry-Constrained Model

4.2.1 Equilibrium Conditions

The most typical situation encountered in practical calculations of the equilibrium crystal structure is the one in which the space group of a crystal, the number of molecules in the unit cell and, frequently, even the unit-cell dimensions are already known from experiment. In this case the problem is reduced to a search for the energy minimum in the subspace of those structural parameters whose variation leaves the crystal symmetry unchanged.

Let us consider a crystal with N_d crystallographically distinct molecules in the unit cell, each having n degrees of freedom. If the crystal symmetry is left unchanged during the search for the equilibrium structure, any crystal configuration is uniquely specified by $6 + nN_\mathrm{d}$ structural parameters $\boldsymbol{p} = \{p_i\}$. Six parameters describe the geometry of the unit cell while the other nN_d describe the translation, orientation and conformation of each of the crystallographically distinct molecules. In this symmetry-constrained model all molecules in the crystal move in concert with the displacements of the N_d basis molecules, so that the crystal symmetry is always left unchanged. The coordinates of the molecules surrounding the basis molecules in the crystal are not independent variables but are functions of the coordinates of the basis molecules and of the unit-cell parameters:

$$\boldsymbol{R}^{ms} = \boldsymbol{R}^{ms}(\underline{\underline{L}}, \boldsymbol{R}^{m}), \tag{4.42}$$

where $\underline{\underline{L}}$ represents the unit-cell parameters, $\boldsymbol{R}^m$ the coordinates of the mth basis molecule, and $\boldsymbol{R}^{ms}$ the coordinates of the molecule generated from molecule m by the symmetry operation s. The dependence of $\boldsymbol{R}^{ms}$ on $\underline{\underline{L}}$ is a result of the definition of the symmetry operations in the unit-cell space. For the moment, we shall not give any explicit definition for $\underline{\underline{L}}$ and $\boldsymbol{R}^m$, so that these may represent any parameters uniquely specifying the geometry of the unit cell, and the translation, orientation and conformation of the basis molecules.

The lattice energy per unit cell is

$$\begin{aligned}\phi_c &= \sum_m N_m \left[E^m(\boldsymbol{R}^m) + \frac{1}{2}\sum_{m's'}{}' \varepsilon^{m,m's'}(\boldsymbol{R}^m, \boldsymbol{R}^{m's'})\right] \\ &= \sum_m N_m \left\{E^m(\boldsymbol{R}^m) + \frac{1}{2}\sum_{m's'}{}' \varepsilon^{m,m's'}[\boldsymbol{R}^m, \boldsymbol{R}^{m's'}(\underline{\underline{L}}, \boldsymbol{R}^{m'})]\right\} \\ &= \phi_c(\underline{\underline{L}}, \boldsymbol{R}^1, \ldots, \boldsymbol{R}^{N_\mathrm{d}}),\end{aligned} \tag{4.43}$$

i.e., the lattice energy is also a function of the unit-cell parameters and the coordinates of the N_d basis molecules.

In the symmetry-constrained model the equilibrium crystal configuration is defined to be a solution of the equations,

$$\nabla_L \phi_c = 0, \tag{4.44}$$

$$\nabla_{R^m} \phi_c = 0, \quad m = 1, \ldots, N_\mathrm{d}, \tag{4.45}$$

where ∇_L and ∇_{R^m} denote the gradients with respect to the components of $\underline{\underline{L}}$ and $\boldsymbol{R}^m$. (Note that $\nabla_L + \nabla_{R^1} + \ldots + \nabla_{R^{N_\mathrm{d}}} \equiv \nabla_p$).

It can be readily shown that any crystal configuration satisfying (4.45) will simultaneously satisfy the force-free conditions given in (4.20,21), regardless of the particular choice of the coordinates $\boldsymbol{R}^m$ in (4.45). Equation (4.44) leads to the stress-free condition. In the latter case, however, the choice of $\underline{\underline{L}}$ and $\boldsymbol{R}^m$ is important because it is desirable that the relationship between the derivatives $\partial\phi_c/\partial L_{\alpha\beta}$ and the components of the tensor $S_{\alpha\beta}$ be as simple as possible. (This will be particularly important in the search for the equilibrium crystal configuration at a non-zero external stress $\sigma_{\alpha\beta}$.)

Let us define the unit-cell parameters $\underline{\underline{L}}$ as the components of the unit-cell vectors $\boldsymbol{a}$, $\boldsymbol{b}$ and $\boldsymbol{c}$ in a reference Cartesian coordinate system $\{\boldsymbol{x}, \boldsymbol{y}, \boldsymbol{z}\}$. If we choose the $\boldsymbol{x}$ axis to be coincident with $\boldsymbol{a}$ and $\boldsymbol{z}$ to lie in the plane defined by $\boldsymbol{a}$ and $\boldsymbol{c}$, the components of $\underline{\underline{L}}$ will be given by

$$\underline{\underline{L}} = \begin{pmatrix} a_x & b_x & c_x \\ 0 & b_y & 0 \\ 0 & b_z & c_z \end{pmatrix}, \tag{4.46}$$

where a_x denotes the projection of $\boldsymbol{a}$ on the axis x, and so forth.

It is worthwhile to write down the relations between $L_{\alpha\beta}$ and the lengths and angles of the unit cell usually used to specify the geometry of the unit cell;

$$L_{11} = a, \tag{4.47}$$

$$L_{12} = b \cos\gamma, \tag{4.48}$$

$$L_{13} = c \cos\beta, \tag{4.49}$$

$$L_{22} = \frac{b}{\sin\beta}(1 - \cos^2\alpha - \cos^2\beta - \cos^2\gamma + 2\cos\alpha\cos\beta\cos\gamma)^{1/2}, \tag{4.50}$$

$$L_{32} = \frac{b}{\sin\beta}(\cos\alpha - \cos\beta\cos\gamma), \tag{4.51}$$

$$L_{33} = c \sin\beta. \tag{4.52}$$

Let us now choose the orientational and conformational variables R_α^m ($\alpha>3$) so that they are independent of $L_{\alpha\beta}$. This will always be the case if the R_α^m ($\alpha>6$) specify the molecular conformation with respect to some molecule-fixed coordinate frame $\{\boldsymbol{\xi}^m, \boldsymbol{\eta}^m, \boldsymbol{\varsigma}^m\}$, while R_4^m, R_5^m, R_6^m describe the orientation of the latter with respect to the reference coordinate system $\{\boldsymbol{x},\boldsymbol{y},\boldsymbol{z}\}$. With these choices, the dependence of ϕ_c on $L_{\alpha\beta}$ will be completely determined by the dependence of ϕ_c on the vectors, $\boldsymbol{r}^{m,m's'} \equiv r^m - r^{m's'}$, joining the molecular centers. We may therefore write

$$\begin{aligned} \frac{\partial\phi_c}{\partial L_{\alpha\beta}} &= \frac{1}{2}\sum_m N_m \sum_{m's'}\sum_\gamma \frac{\partial\varepsilon^{m,m's'}}{\partial r_\gamma^{m,m's'}} \frac{\partial r_\gamma^{m,m's'}}{\partial L_{\alpha\beta}} \\ &= \frac{1}{2}\sum_m N_m \sum_{m's'}\sum_\gamma \frac{\partial\varepsilon^{m,m's'}}{\partial r_\gamma^m} \frac{\partial r_\gamma^{m,m's'}}{\partial L_{\alpha\beta}}, \end{aligned} \tag{4.53}$$

where we have used the obvious relation

$$\frac{\partial\varepsilon^{m,m's'}}{\partial r_\gamma^{m,m's'}} = \frac{\partial\varepsilon^{m,m's'}}{\partial r_\gamma^m}. \tag{4.54}$$

A very simple formula for $\partial\phi_c/\partial L_{\alpha\beta}$ results if the translational parameters R_1^m, R_2^m and R_3^m are chosen to be the fractional unit-cell coordinates of the center of molecule m ($R_\alpha^m \equiv X_\alpha^m$, $\alpha = 1, 2, 3$). In this case the dependence of $r_\gamma^{m,m's'}$ on $L_{\alpha\beta}$ is given by

$$r_\gamma^{m,m's'} = \sum_\delta L_{\gamma\delta} X_\delta^{m,m's'}, \tag{4.55}$$

where $X_\delta^{m,m's'}$ denotes the difference $X_\delta^m - X_\delta^{m's'}$. Differentiation of this expression with respect to $L_{\alpha\beta}$ yields

$$\frac{\partial r_\gamma^{m,m's'}}{\partial L_{\alpha\beta}} = \delta_{\alpha\beta} X_\beta^{m,m's'}, \tag{4.56}$$

so that (4.53) assumes a particularly simple form

$$\frac{\partial\phi_c}{\partial L_{\alpha\beta}} = \frac{1}{2}\sum_m N_m \sum_{m's'} \frac{\partial\varepsilon^{m,m's'}}{\partial r_\alpha^m} X_\beta^{m,m's'}. \tag{4.57}$$

It is worth noting that if the translational coordinates R_α^m ($\alpha = 1, 2, 3$) had been chosen to be the Cartesian coordinates of the molecular center, the following expression would have been obtained,

$$\frac{\partial\phi_c}{\partial L_{\alpha\beta}} = \frac{1}{2}\sum_m N_m \sum_{m's'} \frac{\partial\varepsilon^{m,m's'}}{\partial r_\alpha^m} T_\beta^{s'}, \tag{4.58}$$

where $T_\beta^{s'}$ denotes the βth component of the translational part of the symmetry operation s'.

The advantage of (4.57) over (4.58) lies in the fact that the former is related in a very simple way to the components of the tensor $S_{\alpha\beta}$, as given by (4.31). Indeed, using (4.55,57) we may rewrite (4.31) as

$$\begin{aligned} S_{\alpha\beta} &= \frac{1}{2}\sum_m N_m \sum_{m's'} \frac{\partial\varepsilon^{m,m's'}}{\partial r_\alpha^m} \sum_\gamma L_{\beta\gamma} X_\gamma^{m,m's'} \\ &= \sum_\gamma L_{\beta\gamma} \frac{1}{2}\sum_m N_m \sum_{m's'} \frac{\partial\varepsilon^{m,m's'}}{\partial r_\alpha^m} X_\gamma^{m,m's'} \\ &= \sum_\gamma L_{\beta\gamma} \frac{\partial\phi_c}{\partial L_{\alpha\gamma}}. \end{aligned} \tag{4.59}$$

It is useful to rewrite the resulting expression using the explicit form of the matrix $L_{\alpha\beta}$, see (4.47–52):

$$S_{11} = a_x \frac{\partial \phi_c}{\partial a_x} + b_x \frac{\partial \phi_c}{\partial b_x} + c_x \frac{\partial \phi_c}{\partial c_x}, \tag{4.60}$$

$$S_{12} = b_y \frac{\partial \phi_c}{\partial b_x}, \tag{4.61}$$

$$S_{13} = b_z \frac{\partial \phi_c}{\partial b_x} + c_z \frac{\partial \phi_c}{\partial c_x}, \tag{4.62}$$

$$S_{22} = b_y \frac{\partial \phi_c}{\partial b_y}, \tag{4.63}$$

$$S_{23} = b_y \frac{\partial \phi_c}{\partial b_z}, \tag{4.64}$$

$$S_{33} = b_z \frac{\partial \phi_c}{\partial b_z} + c_z \frac{\partial \phi_c}{\partial c_z}. \tag{4.65}$$

It is now clearly seen that any crystal configuration satisfying the equilibrium conditions given in (4.44) is always stress-free.

The search for a crystal configuration which is stable at zero (atmospheric) pressure does not involve any difficulties. It is easily accomplished by minimizing the lattice energy with respect to $L_{\alpha\beta}$ and R_α^m. The problem becomes more complicated when the goal is to find a crystal configuration which is in equilibrium with a specified external stress $\sigma_{\alpha\beta}$. (For example, such a problem may be encountered in the studies of stress-induced polymorphic transitions). An obvious way of finding such a configuration is to minimize the quantity

$$\sum_{\beta \geq \alpha} \left(\sigma_{\alpha\beta} + V_c^{-1} \sum_\gamma L_{\beta\gamma} \frac{\partial \phi_c}{\partial L_{\alpha\gamma}} \right)^2 + \sum_\alpha \sum_m \left(\frac{\partial \phi_c}{\partial R_\alpha^m} \right)^2, \tag{4.66}$$

with respect to $L_{\alpha\beta}$ and R_α^m (note that $V_c = a_x b_y c_z = L_{11} L_{22} L_{33}$). Compared to the usual minimization of ϕ_c with respect to $L_{\alpha\beta}$ and R_α^m, the minimization of (4.66) is much more complicated because it involves a determination of higher-order energy derivatives.

A comparatively simple problem is to find a crystal configuration that is stable under a specified hydrostatic pressure P. In this case the components of the tensor $S_{\alpha\beta}$ must fulfill the equation

$$S_{\alpha\beta} = -P V_c \delta_{\alpha\beta}. \tag{4.67}$$

As can be seen from (4.60–65), this means that the derivatives of ϕ_c with respect to the diagonal components of $L_{\alpha\beta}$ must satisfy

$$a_x \frac{\partial \phi_c}{\partial a_x} = b_y \frac{\partial \phi_c}{\partial b_y} = c_z \frac{\partial \phi_c}{\partial c_z} = -P a_x b_y c_z, \tag{4.68}$$

while all of the off-diagonal derivatives as well as the derivatives with respect to R_α^m must vanish:

$$\frac{\partial \phi_c}{\partial p_i'} = 0, \tag{4.69}$$

where p_i' denotes all structural parameters other than a_x, b_y and c_z.

The required crystal configuration may be found by minimizing the static part of the Gibbs potential, $G_c^0 = \phi_c + PV_c$, with respect to $L_{\alpha\beta}$ and R_α^m. Considering that $V_c = L_{11}L_{22}L_{33}$, it can be easily shown that the conditions, $\partial G_c^0/\partial L_{\alpha\beta} = \partial G_c^0/\partial R_\alpha^m = 0$, are equivalent to (4.68,69).

In finding the crystal configuration at a non-zero pressure it is sometimes more convenient to replace the constant-pressure constraints by constant-volume ones. In this case the problem consists of finding the crystal configuration with the lowest lattice energy for a specified unit-cell volume V_c. In other words, we now have to find the structural parameters which minimize the function

$$f = \phi_c(p', a_x, b_y, c_z) \tag{4.70}$$

under the constraint

$$a_x b_y c_z = V_c. \tag{4.71}$$

In these equations V_c is considered to be a fixed quantity and p' represents, as in (4.69), all the structural parameters other than a_x, b_x and c_z. The restrictive condition given in (4.71) can be incorporated into (4.70) by expressing, for example, c_z in terms of a_x, b_y and V_c. This yields

$$f(p', a_x, b_y) = \phi_c[p', a_x, b_y, c_z(a_x, b_y)]. \tag{4.72}$$

Thus, the problem is reduced to a minimization of the lattice energy with respect to p', a_x and b_y.[4] The parameter c_z, needed for the actual computation of the lattice energy now represents a dependent variable; c_z is determined as $V_c/a_x b_y$ for any trial p_i', a_x and b_y.

The derivatives of the lattice energy under the constant-volume constraint of (4.71) are given by

$$\frac{\partial f}{\partial p_i'} = \frac{\partial \phi_c}{\partial p_i'}, \tag{4.73}$$

[4] Note that the dimension of this problem is smaller than that of a minimization of G_c^0.

$$
\begin{aligned}
\frac{\partial f}{\partial a_x} &= \frac{\partial \phi_c}{\partial a_x} + \frac{\partial \phi_c}{\partial c_z}\frac{\partial c_z}{\partial a_x} \\
&= \frac{\partial \phi_c}{\partial a_x} - \frac{\partial \phi_c}{\partial c_z}\frac{V_c}{a_x^2 b_y} \\
&= \frac{\partial \phi_c}{\partial a_x} - \frac{c_z}{a_x}\frac{\partial \phi_c}{\partial c_z},
\end{aligned} \tag{4.74}
$$

$$
\frac{\partial f}{\partial b_y} = \frac{\partial \phi_c}{\partial b_y} - \frac{c_z}{b_y}\frac{\partial \phi_c}{\partial c_z}. \tag{4.75}
$$

These derivatives vanish at the crystal configuration found by minimizing f with respect to p', a_x and b_y :

$$
\frac{\partial f}{\partial p'_i} = \frac{\partial f}{\partial a_x} = \frac{\partial f}{\partial b_y} = 0. \tag{4.76}
$$

This implies that the following equations are fulfilled

$$
\frac{\partial \phi_c}{\partial p'_i} = 0, \tag{4.77}
$$

$$
a_x \frac{\partial \phi_c}{\partial a_x} = b_y \frac{\partial \phi_c}{\partial b_y} = c_z \frac{\partial \phi_c}{\partial c_z}. \tag{4.78}
$$

By comparing these equations with (4.68,69), we see that the resulting crystal configuration is force-free and conforms to the hydrostatic compression conditions.

The only shortcoming of the above procedure is that the particular pressure which corresponds to the resulting configuration is not known beforehand, but has to be determined from (4.68) after completing the minimization. This shortcoming does not, however, seem to be very significant because generally an interest does not lie in the crystal properties at a particular pressure, but in the pressure *dependence* of these properties. The required dependence can easily be found by calculating the equilibrium crystal configuration for a selected set of unit-cell volumes.

The above procedure can be successfully applied to determine the transition point of a pressure-induced polymorphic transition I→II, provided of course that the unit-cell symmetries of phases I and II are known. The calculation amounts to a determination of the equilibrium structures of I and II for several trial unit-cell volumes. Having determined the equilibrium structures, one plots the pressure dependence of the thermodynamic potentials, $\phi_{\mathrm{I}} + PV_{\mathrm{I}}$ and $\phi_{\mathrm{II}} + PV_{\mathrm{II}}$, and then finds the intersection point of these two curves.

4.2.2 The Lattice Energy and Its Derivatives

So far, we have not been interested in the particular form of the interaction potential. The only assumption that has been made is that the lattice energy of a crystal is additive with respect to the interactions of its constituent molecules. In this section we go a little further and assume that the intermolecular interaction energies may, in their turn, be expressed in terms of pairwise interactions of the atoms that make up the interacting molecules. Based on this assumption, we represent the lattice energy as a function of structural parameters which are consistent with the parameters of the previous section. Also, we shall derive formulas for the first and second derivatives of the lattice energy, which are of great use in the practical computation of the equilibrium crystal configuration.

Explicitly, in the atom-atom approximation the energies $\varepsilon^{m,m's'}$ may be represented as

$$\varepsilon^{m,m's'} = \sum_{aa'} \varphi_{aa'}(r^{am,a'm's'}), \quad \text{where} \tag{4.79}$$

$$r^{am,a'm's'} \equiv |\boldsymbol{r}^{am,a'm's'}| = |\boldsymbol{r}^{am} - \boldsymbol{r}^{a'm's'}|, \tag{4.80}$$

and $\boldsymbol{r}^{a'm's'} = \{r_\alpha^{a'm's'}\}$ denotes the Cartesian coordinates of atom a' in molecule $m's'$ (similarly for $\boldsymbol{r}^{am}$). To simplify the notation, we shall from now on denote the potential $\varphi_{aa'}\,(r^{am,a'm's'})$ simply as $\varphi_{aa'}$, and its derivatives with respect to the interatomic distance as $\varphi'_{aa'}$ and $\varphi''_{aa'}$.

In order to insert (4.79) into the equations for the lattice energy and its derivatives, it is first of all necessary to express $\boldsymbol{r}^{a'm's'}$ in terms of the structural parameters $L_{\alpha\beta}$ and R_α^m chosen as indicated in the previous section. We remind the reader that a definite choice for the parameters R_α^m was only made for $\alpha = 1, 2$ and 3. That is, the matrix $L_{\alpha\beta}$ was defined by (4.47–52), while R_1^m, R_2^m and R_3^m were chosen to be the fractional coordinates of the origins of the molecule-fixed coordinate frames $\{\boldsymbol{\xi}^m, \boldsymbol{\eta}^m, \boldsymbol{\varsigma}^m\}$. With respect to the rotational and conformational parameters R_α^m ($\alpha>3$), it was only assumed that R_4^m, R_5^m and R_6^m describe the molecular orientation relative to the reference coordinate frame $\{\boldsymbol{x}, \boldsymbol{y}, \boldsymbol{z}\}$, while R_α^m ($\alpha>6$) specify the molecular conformation in the molecule-fixed coordinate system $\{\boldsymbol{\xi}^m, \boldsymbol{\eta}^m, \boldsymbol{\varsigma}^m\}$.

Let us now choose R_4^m, R_5^m and R_6^m to be the three Euler angles $\chi^m = \{\chi_1^m, \chi_2^m, \chi_3^m\}$ that describe the orientation of the molecule-fixed coordinate system with respect to the reference coordinate system. The Cartesian coordinates of atom a in basis molecule m, $\boldsymbol{r}^{am} = \{r_\alpha^{am}\}$, are then given by

$$\boldsymbol{r}^{am} = \underline{\underline{G}}(\chi^m)\,\boldsymbol{\varrho}^{am} + \underline{\underline{L}}\boldsymbol{X}^m, \tag{4.81}$$

where $\underline{\underline{G}}$ is the Euler matrix whose components are functions of the three Euler angles, i.e.,

$$\underline{\underline{G}}(\chi) = \begin{pmatrix} (\cos\chi_1 \cos\chi_2 & (-\sin\chi_1 \cos\chi_2 & (\sin\chi_2 \\ -\cos\chi_3 \sin\chi_1 & -\cos\chi_3 \cos\chi_1 & \times \sin\chi_3) \\ \times \sin\chi_2) & \times \sin\chi_2) & \\ (\sin\chi_2 \cos\chi_1 & (-\sin\chi_1 \sin\chi_2 & (-\sin\chi_3 \\ +\cos\chi_3 \cos\chi_2 & +\cos\chi_3 \cos\chi_2 & \times \cos\chi_2) \\ \times \sin\chi_1) & \times \cos\chi_1) & \\ (\sin\chi_1 \sin\chi_3) & (\cos\chi_1 \sin\chi_3) & \cos\chi_3 \end{pmatrix}. \tag{4.82}$$

The vector $\boldsymbol{\varrho}^{am}$ in (4.81) denotes the Cartesian coordinates of atom a in molecule m in the molecule-fixed coordinate frame $\{\boldsymbol{\xi}^m, \boldsymbol{\eta}^m, \boldsymbol{\varsigma}^m\}$. Note that $\boldsymbol{\varrho}^{am}$ only depends on the internal molecular coordinates R_α^m $(\alpha>6)$, from now on denoted as $\boldsymbol{g}^m = \{g_k^m\}$:

$$\boldsymbol{\varrho}^{am} = \boldsymbol{\varrho}^{am}(\boldsymbol{g}^m). \tag{4.83}$$

In the rigid-molecule approximation all ϱ_α^{am} are constants.

The fractional unit-cell coordinates of atom a in basis molecule m are given by

$$\boldsymbol{X}^{am} = \underline{\underline{L}}^{-1}\boldsymbol{r}^{am} = \underline{\underline{L}}^{-1}\underline{\underline{G}}^m \boldsymbol{\varrho}^{am} + \boldsymbol{X}^m. \tag{4.84}$$

The action of the symmetry operation s on these coordinates yields

$$X^{ams} = \underline{\underline{B}}^s \boldsymbol{X}^{am} + \boldsymbol{T}^s, \tag{4.85}$$

where $\underline{\underline{B}}^s$ and $\boldsymbol{T}^s$ denote the rotational and translational parts of the symmetry operation s.

Returning to the Cartesian coordinates, one finds that

$$\begin{aligned} \boldsymbol{r}^{ams} &= \underline{\underline{L}}\boldsymbol{X}^{ams} \\ &= \underline{\underline{L}}(\underline{\underline{B}}^s \underline{\underline{L}}^{-1}\underline{\underline{G}}^m \boldsymbol{\varrho}^{am} + \underline{\underline{B}}^s \boldsymbol{X}^m + \boldsymbol{T}^s) \\ &= \underline{\underline{B}}^s \underline{\underline{G}}^m \boldsymbol{\varrho}^{am} + \underline{\underline{L}}(\underline{\underline{B}}^s \boldsymbol{X}^m + \boldsymbol{T}^s), \end{aligned} \tag{4.86}$$

where we have used the obvious relation: $\underline{\underline{L}}\,\underline{\underline{B}}^s \underline{\underline{L}}^{-1} = \underline{\underline{B}}^s$. Finally, the vector $\boldsymbol{r}^{am,a'm's'}$ may be represented as follows

$$\begin{aligned} \boldsymbol{r}^{am,a'm's'} = &\underline{\underline{G}}(\chi^m)\boldsymbol{\varrho}^{am}(\boldsymbol{g}^m) - \underline{\underline{B}}^{s'}\underline{\underline{G}}(\chi^{m'})\boldsymbol{\varrho}^{a'm'}(\boldsymbol{g}^{m'}) \\ &+ \underline{\underline{L}}(\boldsymbol{X}^m \underline{\underline{B}}^{s'}\boldsymbol{X}^{m'} - \boldsymbol{T}^{s'}). \end{aligned} \tag{4.87}$$

In this expression we have explicitly indicated all of the parameters which will hereafter be treated as independent variables. These are

$$\underline{\underline{L}} = \{L_{\alpha\beta}\}, \quad \boldsymbol{X}^m = \{X_\alpha^m\}, \quad \chi^m = \{\chi_\alpha^m\}, \quad \boldsymbol{g}^m = \{g_k^m\} \quad . \tag{4.88}$$

Considering (4.43,79), the lattice energy per unit cell may now be written as

$$\phi_c = \sum_m N_m \Big[E^m + \frac{1}{2} \sum_{aa'm's'}{}' \varphi_{aa'}(r^{am,a'm's'})\Big], \tag{4.89}$$

where the summation over a and a' is carried out over all atoms of basis molecules m and m', respectively. If the site symmetry of molecule m is greater than 1, the sum over a may only be taken over the asymmetric part of the molecule. In this case each contribution $\varphi_{aa'}$ should be given a weight equal to the multiplicity of the crystallographic position occupied by atom a. The summation over a' may also be taken over the asymmetric part of molecule m', provided that the summation over s' also includes the symmetry operations which generate the entire molecule m' from its asymmetric part. Clearly, in the latter case we shall not save any computer time because all atoms of molecule m' should, in some way or another, appear in the sum.

The summation over a' may be carried out in a more economical way only for contributions with $m' = m$. For these contributions the sum over a' may only be taken for $a' \leq a$ (the $a' < a$ terms are given double weight). To prove that such a procedure is correct, we note that the sum over s' always includes the inverse element $\bar{s}'$ to any given s'. This means that the sum in (4.89) always contains, apart from the pair am, $a'ms'$, the pair $a'm$, $am\bar{s}'$. Considering that

$$r^{am,a'ms'} = r^{a'm,am\bar{s}'}, \tag{4.90}$$

it can be easily seen that the pairs am, $a'ms'$ and $a'm$, $am\bar{s}'$ contribute equally to the lattice energy, so that the above procedure is indeed legitimate.

The first derivative of the lattice energy given in (4.89) with respect to any parameter p_i of (4.88) is given by the expression,

$$\begin{aligned}\frac{\partial \phi_c}{\partial p_i} &= \sum_m N_m \Big(\frac{\partial E^m}{\partial p_i} + \frac{1}{2} \sum_{aa'm's'}{}' \varphi'_{aa'} \frac{\partial r^{am,a'm's'}}{\partial p_i}\Big)\\ &= \sum_m N_m \Big(\frac{\partial E^m}{\partial p_i} + \frac{1}{2} \sum_{aa'm's'} \varphi'_{aa'} \frac{1}{r^{am,a'm's'}}\\ &\quad \times \sum_\alpha r_\alpha^{am,a'm's'} \frac{\partial r_\alpha^{am,a'm's'}}{\partial p_i}\Big).\end{aligned} \tag{4.91}$$

In (4.91) the derivatives $\partial E^m/\partial p_i$ may differ from zero only when p_i is one of the conformational parameters g_k^m. The explicit form of the derivatives $\partial E^m/\partial g_k^m$ depends on the particular model adopted for E^m and will not be discussed here. For the empirical intramolecular-energy model described in Sect. 3.4, all expressions that are needed are readily obtained by differentiating (3.70–77).

For the intermolecular part of (4.91), the derivative φ' is given by

$$\frac{\partial \varphi}{\partial r} = \frac{6A}{r^7} - BC \exp(-Cr), \tag{4.92}$$

for a (6-exp) Buckingham-type potential; or by

$$\frac{\partial \varphi}{\partial r} = \frac{6A}{r^7} - \frac{mB}{r^{m+1}} \tag{4.93}$$

for a (6-m) Lennard-Jones potential. For (6-exp-1) and (6-m-1) potentials, φ' also includes the term $-qq'/r^2$.

The explicit expressions for $\partial r_\alpha^{am,a'm's'}/\partial p_i$ are readily derived by differentiating the following equation, see (4.87),

$$\begin{aligned} r_\alpha^{am,a'm's'} &= \sum_\beta G_{\alpha\beta}(\chi^m)\varrho_\beta^{am}(g^m) \\ &\quad - \sum_{\beta\gamma} B_{\alpha\beta}^{s'} G_{\beta\gamma}(\chi^{m'})\varrho_\gamma^{a'm'}(g^{m'}) \\ &\quad + \sum_\beta L_{\alpha\beta}(X_\beta^m - \sum_\gamma B_{\beta\gamma}^{s'} X_\gamma^{m'} - T_\beta^{s'}). \end{aligned} \tag{4.94}$$

Thus, the derivative of (4.94) with respect to a unit-cell parameter, $L_{\mu\nu}$, is given by

$$\frac{\partial r_\alpha^{am,a'm's'}}{\partial L_{\mu\nu}} = \delta_{\alpha\mu}\left(X_\nu^m - \sum_\gamma B_{\nu\gamma}^{s'} X_\gamma^{m'} - T_\nu^{s'}\right). \tag{4.95}$$

In this case the last sum in (4.91) is reduced to a single term,

$$\begin{aligned} &\sum_\alpha r_\alpha^{am,a'm's'} \frac{\partial r_\alpha^{am,a'm's'}}{\partial L_{\mu\nu}} \\ &\quad = r_\mu^{am,a'm's'}\left(X_\nu^m - \sum_\gamma B_{\nu\gamma}^{s'} X_\gamma^{m'} - T_\nu^{s'}\right). \end{aligned} \tag{4.96}$$

Differentiating (4.94) with respect to a translational parameter, $X_\gamma^{m''}$, yields

$$\frac{\partial r_\alpha^{am,a'm's'}}{\partial X_\gamma^{m''}} = \delta_{mm''} L_{\alpha\gamma} - \delta_{m'm''} \sum_\beta L_{\alpha\beta} B_{\beta\gamma}^{s'}. \tag{4.97}$$

Substituting this expression back into (4.91) results in a simplification of the formula for $\partial\phi_c/\partial X_\gamma^{m''}$ because of the presence of the Kronecker symbols in (4.97). In practical computations, however, it is more convenient to evaluate (4.97) numerically, substitute the result into (4.91), and perform the summation over m, a, a', m', and s' at the same time as the summation of the other derivatives.

The derivative of (4.94) with respect to a rotational parameter, $\chi_\sigma^{m''}$, is given by

$$\frac{\partial r_\alpha^{am,a'm's'}}{\partial \chi_\sigma^{m''}} = \delta_{mm''} \sum_\beta G_{\alpha\beta,\sigma}(\chi^m) \varrho_\beta^{am}(\boldsymbol{g}^m) - \delta_{m'm''} \sum_{\beta\gamma} B_{\alpha\beta}^{s'} G_{\beta\gamma,\sigma}(\chi^{m'}) \varrho_\gamma^{a'm'}(\boldsymbol{g}^{m'}), \tag{4.98}$$

where $G_{\alpha\beta,\sigma}$ represents a matrix whose elements are the derivatives of the respective elements of the Euler matrix given in (4.82) with respect to χ_σ ($G_{11,1} = -\sin\chi_1 \cos\chi_2 - \cos\chi_3 \cos\chi_1 \sin\chi_2$, and so forth).

Finally, differentiating (4.94) with respect to a conformational parameter, $g_k^{m''}$, we find

$$\frac{\partial r_\alpha^{am,a'm's'}}{\partial g_k^{m''}} = \delta_{mm''} \sum_\beta G_{\alpha\beta}(\chi^m) \frac{\partial \varrho_\beta^{am}}{\partial g_k^m} - \delta_{m'm''} \sum_{\beta\gamma} B_{\alpha\beta}^{s'} G_{\beta\gamma}(\chi^{m'}) \frac{\partial \varrho_\gamma^{a'm'}}{\partial g_k^{m'}}. \tag{4.99}$$

The explicit expression for the derivative $\partial\varrho_\beta^{am}/\partial g_k^m$ depends on the particular choice of the conformational parameter g_k^m and cannot be written in a general form. It is, however, worthwhile to consider a particularly important case, one in which g_k^m describes the rotation of a rigid molecular subgroup about the bond, $b - c$, joining this subgroup to the rest of the molecule (b belongs to the subgroup and c to the rest of the molecule). For any atom a of the rotating subgroup the required derivative is given by

$$\frac{\partial \varrho_\beta^{am}}{\partial g_k^m} = [\boldsymbol{e}_{bc} \times (\boldsymbol{\varrho}^{am} - \boldsymbol{\varrho}^{cm})]_\beta, \tag{4.100}$$

where $\boldsymbol{e}_{bc}$ is a unit vector in the direction of the bond $b - c$.

Thus, we have presented the most important equations needed for an analytical evaluation of the first derivatives of the lattice energy. Expressions for the second derivatives are readily derived by a double differentiation of (4.89). Retaining only the essential indices, the resulting equation may be represented as

$$\frac{\partial^2 \phi_c}{\partial p_i \partial p_j} = \sum_m N_m \left[\frac{\partial^2 E^m}{\partial p_i \partial p_j} + \frac{1}{2}\sum\left(\varphi'' \frac{\partial r}{\partial p_i}\frac{\partial r}{\partial p_j} + \varphi' \frac{\partial^2 r}{\partial p_i \partial p_j}\right)\right]$$

$$= \sum_m N_m \left\{\frac{\partial^2 E^m}{\partial p_i \partial p_j} + \frac{1}{2}\sum\left[\frac{1}{r^2}\left(\varphi'' - \frac{1}{r}\varphi'\right)\sum_\alpha r_\alpha \frac{\partial r_\alpha}{\partial p_i}\right.\right.$$

$$\left.\left.\times \sum_\alpha r_\alpha \frac{\partial r_\alpha}{\partial p_j} + \frac{1}{r}\varphi' \sum_\alpha \left(r_\alpha \frac{\partial^2 r_\alpha}{\partial p_i \partial p_j} + \frac{\partial r_\alpha}{\partial p_i}\frac{\partial r_\alpha}{\partial p_j}\right)\right]\right\}. \quad (4.101)$$

Since expressions for $\partial r_\alpha / \partial p_i$ are already available, it is only necessary to derive the explicit expressions for $\partial^2 r_\alpha / \partial p_i \partial p_j$. These are easily obtained by differentiating (4.95–100) with respect to each of the parameters listed in (4.88). Fortunately, only a few derivatives $\partial^2 r / \partial p_i \partial p_j$ are not identically equal to zero. These are

$$\frac{\partial^2 r_\alpha^{am,a'm's'}}{\partial X_\beta^{m''} \partial L_{\alpha\gamma}} = \delta_{mm''}\delta_{\gamma\beta} - \delta_{m'm''} B_{\gamma\beta}^{s'}, \quad (4.102)$$

$$\frac{\partial^2 r_\alpha^{am,a'm's'}}{\partial \chi_\sigma^{m''} \partial \chi_\tau^{m''}} = \delta_{mm''} \sum_\beta G_{\alpha\beta,\sigma\tau}(\chi^m) \varrho_\beta^{am}(\boldsymbol{g}^m)$$

$$- \delta_{m'm''} \sum_{\beta\gamma} B_{\alpha\beta}^{s'} G_{\beta\gamma,\sigma\tau}(\chi^{m'}) \varrho_\gamma^{a'm'}(\boldsymbol{g}^{m'}), \quad (4.103)$$

where

$$G_{\alpha\beta,\sigma\tau} = \frac{\partial^2 G_{\alpha\beta}}{\partial \chi_\sigma \partial \chi_\tau}, \quad (4.104)$$

$$\frac{\partial^2 r_\alpha^{am,a'm's'}}{\partial \chi_\sigma^{m''} \partial g_k^{m''}} = \delta_{mm''} \sum_\beta G_{\alpha\beta,\sigma}(\chi^m) \frac{\partial \varrho_\beta^{am}}{\partial g_k^m}$$

$$- \delta_{m'm''} \sum_{\beta\gamma} B_{\alpha\beta}^{s'} G_{\beta\gamma,\sigma}(\chi^{m'}) \frac{\partial \varrho_\gamma^{a'm'}}{\partial g_k^m}, \quad (4.105)$$

$$\frac{\partial^2 r_\alpha^{am,a'm's'}}{\partial g_k^{m''} \partial g_l^{m''}} = \delta_{mm''} \sum_\beta G_{\alpha\beta}(\chi^m) \frac{\partial^2 \varrho_\beta^{am}}{\partial g_k^m \partial g_l^m}$$

$$- \delta_{m'm''} \sum_{\beta\gamma} B_{\alpha\beta}^{s'} G_{\beta\gamma}(\chi^{m'}) \frac{\partial^2 \varrho_\gamma^{a'm'}}{\partial g_k^{m'} \partial g_l^{m'}}. \quad (4.106)$$

The above expressions for the components of the interatomic distance and their derivatives yield equations for evaluating the lattice energy and its derivatives when inserted into (4.89,91,101). It is to be stressed that the lattice energy appears in these formulas as a function of the structural parameters chosen in accordance with the conventions of Sect. 4.1. This means, in particular, that the substitution of the derivatives $\partial\phi_c/\partial L_{\alpha\beta}$ computed from (4.91,94,95) into (4.59) immediately yields the components of the stress tensor.

4.2.3 Minimization Techniques

In general, the search for the crystal structure at equilibrium consists of two steps. The first step, called a global search, involves a scanning of the space of the variable structural parameters with the aim of roughly locating the regions of low-energy crystal configurations. In the second step, the most favorable configurations are selected and then refined using a local minimization technique.

In most applications, the equilibrium crystal structure is approximately known from experiment and the aim of the lattice-energy calculation is merely to refine this structure. In this case a global search is unnecessary and any routine for local minimization may be used. There are, however, many situations in which the starting information consists of a knowledge of the crystal symmetry and, perhaps, of the unit-cell parameters. Moreover, one frequently deals with hypothetical molecular conformations and has no information whatsoever regarding the molecular packing. In such situations a great variety of crystal configurations have to be tried and the question arises as to how to do this in the most economical way.

The problem which has to be solved before proceeding to a global search involves the choice of the range of the structural parameters. Two obvious considerations should be taken into account to reduce the computational work. First of all a scanning of the parameter space should only cover crystallographically distinct configurations. Secondly, a priori poor (high-energy) regions of the parameter space should be rejected beforehand.

The smallest region to be scanned for the translational and orientational parameters, $\boldsymbol{X}^m$ and χ^m, is determined by the Cheshire group of the crystal and the molecular point group. The direct product of the two groups specifies the symmetry of the six-dimensional space of the translational and orientational parameters. The asymmetric unit in this space is exactly equal to the region to be scanned by $\boldsymbol{X}^m$ and χ^m. A comprehensive treatment of this point and numerous examples have been given by *Hirshfeld* [4.3].

The range for $\boldsymbol{X}^m$ may be further reduced by considering that the molecules in a crystal cannot approach closed symmetry elements, unless the molecule itself possesses that symmetry. In each particular case the regions forbidden for $\boldsymbol{X}^m$ can be approximately determined by a simple geometrical analysis, proceeding from the smallest molecular dimension. Such an analysis may also be used to choose the range of the unit-cell dimensions. In the latter case, however, both the smallest and largest molecular dimensions are essential, as well as the number of molecules in the unit cell and the crystal symmetry.

In practical computations, the range of the structural parameters p_i $(i = 1, \ldots, n)$ is generally represented by a parallelepiped in the n-dimensional parameter space defined by

$$P_n : p_i^{\min} \leq p_i \leq p_i^{\max}, \quad i = 1, \ldots, n. \tag{4.107}$$

Using the relations,

$$x_i = (p_i - p_i^{\min})/(p_i^{\max} - p_i^{\min}), \tag{4.108}$$

this parallelepiped is transformed into the unit cube

$$C_n : 0 \leq x_i \leq 1, \quad i = 1, \ldots, n. \tag{4.109}$$

The obvious way of scanning C_n is to use a regular grid. The coordinates of the points $\boldsymbol{x}^{(k)}$ describing a regular grid are given by

$$x_i^{(k)} = \left(j_i^{(k)} - \frac{1}{2}\right)/M, \quad j_i^{(k)} = 1, \ldots, M, \tag{4.110}$$

where the integer M determines the thickness of the grid. The total number of points in the grid is $N = M^n$.

Although regular grids are commonly used in practical computations, they are actually not a good means of scanning the parameter space. Indeed, let us suppose that the objective function $\phi(\boldsymbol{x})$ only depends on one parameter, say, x_1. By computing $\phi(\boldsymbol{x})$ on the grid described by (4.110), one only obtains M distinct values of ϕ. The lowest value of ϕ is selected from among these values, so that $N - M = N(1 - N^{1/n-1})$ sampled points prove to be superfluous. Clearly, almost the same situation is encountered when $\phi(\boldsymbol{x})$ is only slightly dependent on $x_2, x_3, \ldots, x_n$, i.e., when $\phi(\boldsymbol{x})$ may be represented as

$$\phi(\boldsymbol{x}) = \phi'(x_1, \ldots, x_n) + \phi''(x_1), \tag{4.111}$$

where $\phi'' \gg \phi'$.

In general, $\phi(\boldsymbol{x})$ may have m significant parameters,

$$\phi(\boldsymbol{x}) = \phi'(x_1, \ldots, x_n) + \phi''(x_1, \ldots, x_m), \tag{4.112}$$

where $m<n$ and $\phi'' \gg \phi'$. For such a function, the number of superfluous points is equal to $N - M^m = N(1 - N^{m/n-1})$.

The latter case is directly related to lattice-energy calculations because the lattice-energy hypersurface is often very shallow along some of the coordinate axes.

Instead of using regular grids, one may resort to uniform random grids. Requisite routines that generate a sequence of random points, $\gamma^{(k)} = \{\gamma_i^{(k)}\}$, uniformly distributed in the unit interval, $0 \le \gamma_i^{(k)} \le 1$, are available for any computer. The random grids are free from the above shortcoming of the regular grids because the probability of two random points having identical abscissae is equal to zero.

The advantage of random grids over regular grids is well illustrated in terms of the discrepancy D defined as follows. Let us choose an arbitrary point $\boldsymbol{X} = \{X_i\}$ in C_n. Let Π_n^X denote a parallelepiped with a diagonal $0X$ and faces parallel to the coordinate planes:

$$\Pi_n^X : 0 \le x_i < X_i, \quad i = 1, \ldots, n. \tag{4.113}$$

The discrepancy of a sequence of points $\boldsymbol{\xi}^{(k)}$ $(k = 1, \ldots, N)$ within the unit cube C_n is defined by [4.4]

$$D = \sup_{\boldsymbol{X} \in C_n} |S_N^X - NV^X|, \tag{4.114}$$

where V^X is the volume of Π_n^X and S_N^X is the number of points of $\boldsymbol{\xi}^{(k)}$, which fall into Π_n^X.

It can be seen from (4.114) that the discrepancy D decreases as the distribution of points $\boldsymbol{\xi}^{(k)}$ over C_n becomes more uniform. It can be shown that for regular grids the discrepancy increases with N as $N^{1-1/n}$, while for random grids it increases with N as $N^{1/2}$ [4.4]. This means that for $n \ge 3$ regular grids are less uniform and, hence, less suitable for scanning the parameter space.

The above conclusion should not be generalized to all non-random sequences. There are many deterministic sequences which are more uniform than the random ones. The best sequence of this kind is the $\Lambda\Pi_\tau$ sequence $\boldsymbol{q}^{(k)}$, whose discrepancy increases with N as $\ln^n N$ [4.4,5]. (Sequences with a slower growth of D are as yet unknown). It is essential that the $\Lambda\Pi_\tau$ sequence already ensures small discrepancies at small values of N and n. Another advantage of this sequence is that the projections of the points $\boldsymbol{q}^{(k)}$ onto the one-, two-, and three-dimensional faces of C_n also exhibit small discrepancies. The $\Lambda\Pi_\tau$ sequence is readily generated by a computer using a procedure described in [4.4,5].

The $\Lambda\Pi_\tau$ global search has been compared to the random search on various objective functions [4.5]. In all cases the former yielded much better results so that the computational work could be reduced by a factor of 2 to 4. Applications of the $\Lambda\Pi_\tau$ search to problems encountered in organic crystallography may be found in [4.6].

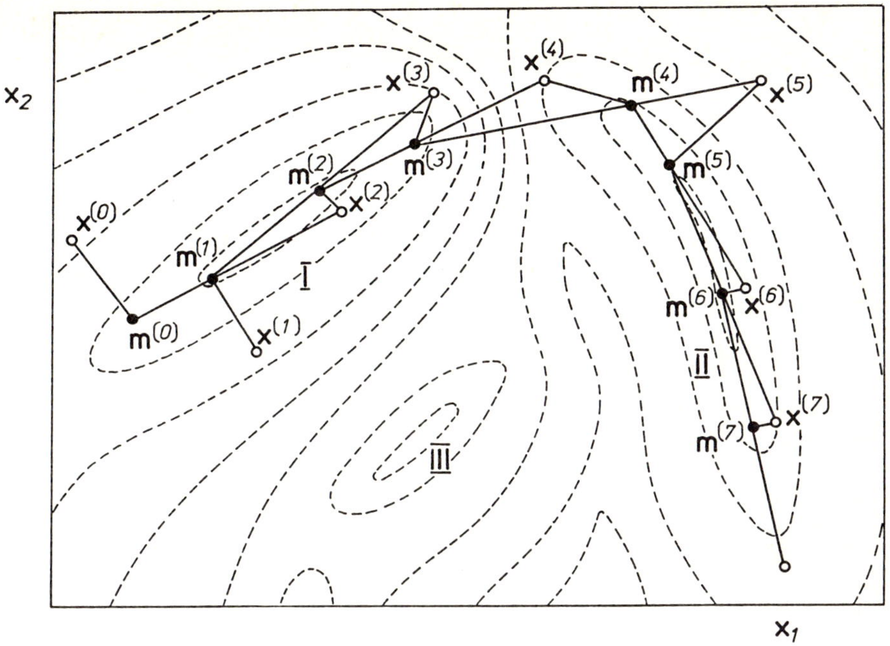

Fig. 4.1. A schematic illustration of the valley method for the case of a two-dimensional problem. (Arbitrary units)

An obvious shortcoming inherent to both the random and deterministic uniform grids, as applied to the search of good crystal configurations, lies in the fact that the overwhelming majority of sampled points proves to be of too high an energy and, hence, do not yield a reasonable molecular packing. There is, however, a possibility of scanning the parameter space over mostly low-energy regions. A way in which this can be done is illustrated in Fig. 4.1 for the case of a two-parameter problem. The search for a global minimum starts with a choice of two arbitrary points, $\boldsymbol{x}^{(0)}$ and $\boldsymbol{x}^{(1)}$, not too far apart. At these points one first computes the gradients, $\nabla\phi[\boldsymbol{x}^{(0)}]$ and $\nabla\phi[\boldsymbol{x}^{(1)}]$, and then finds the minima, $\boldsymbol{m}^{(0)}$ and $\boldsymbol{m}^{(1)}$, in the directions, $-\nabla\phi[\boldsymbol{x}^{(0)}]$ and $-\nabla\phi[\boldsymbol{x}^{(1)}]$, respectively. Thereafter one makes a step of length h in the direction, $\boldsymbol{m}^{(1)} - \boldsymbol{m}^{(0)}$, and thus determines a new point, $\boldsymbol{x}^{(2)}$,

$$\boldsymbol{x}^{(2)} = \boldsymbol{m}^{(1)} + \frac{\boldsymbol{m}^{(1)} - \boldsymbol{m}^{(0)}}{|\boldsymbol{m}^{(1)} - \boldsymbol{m}^{(0)}|} h. \tag{4.115}$$

At this new point one again computes the gradient and finds the minimum $\boldsymbol{m}^{(2)}$ in the direction $-\nabla\phi[\boldsymbol{x}^{(2)}]$. The process is a recurring one so that each new point $\boldsymbol{x}^{(k)}$ is determined from the minima of the two preceding interactions:

$$\boldsymbol{x}^{(k)} = \boldsymbol{m}^{(k-1)} + \frac{\boldsymbol{m}^{(k-1)} - \boldsymbol{m}^{(k-2)}}{|\boldsymbol{m}^{(k-1)} - \boldsymbol{m}^{(k-2)}|} h. \tag{4.116}$$

It can be seen from Fig. 4.1 that the points $\boldsymbol{m}^{(k)}$ move along the bottom of local minimum I closest to the starting points $\boldsymbol{x}^{(0)}$ and $\boldsymbol{x}^{(1)}$, surmount the barrier separating minima I and II, descend to minimum II, and so on. The success of such a search depends on two factors. The first is the shape of the lattice-energy hypersurface $\phi(\boldsymbol{x})$. The method works well when the hypersurface represents a valley in the parameter space, i.e., when a change in one or a few of the structural parameters affects the lattice energy only slightly. (In the Russian literature the method is usually referred to as the valley method [4.7].)

The second factor influencing the success of the global search is the choice of the length h. If the step is too large, the sequence of points $\boldsymbol{m}^{(k)}$ will not follow the direction of the valley but will actually represent a random sampling of the parameter space. On the other hand, if the step is too small, the points $m^{(k)}$ will not surmount the barriers separating the individual minima, but will move about a particular minimum.

We conclude the discussion of the strategy of global searches by noting that neither of the above procedures ensures that all of the minima on the lattice energy hypersurface will be revealed after a finite number of trials. Thus, in Fig. 4.1 we show the minimum III which is omitted when using the valley method. With uniform grids, there is also a probability of omitting the global minimum. Clearly, this probability increases if the global minimum is very narrow and the grid is too coarse. In organic crystals, very narrow and deep minima are generally associated with strong short-range attractive interactions such as hydrogen bonding. Fortunately, such minima can be approximately located beforehand from a simple geometrical analysis.

We now suppose that a global search has revealed some promising points. The next step is to find the local minima closest to these promising points. A great variety of methods have been developed for local minimization [4.8–11]. If the first and second derivatives of the objective function (the lattice energy) can be computed analytically, the minimization is best accomplished by the Newton-Raphson method. This method is based on the approximation of the objective function $\phi(\boldsymbol{p})$ by a quadratic function; the explicit expression of which is given by two leading terms of the Taylor expansion of $\phi(\boldsymbol{p})$. If the lattice energy were actually quadratic in $\boldsymbol{p}$, only one step would be required to reach the minimum-energy configuration from any starting configuration $\boldsymbol{p}^{(0)}$:

$$\bar{\boldsymbol{p}} = \boldsymbol{p}^{(0)} - \underline{\underline{\phi}}_{p}^{-1}[\boldsymbol{p}^{(0)}] \nabla_{p} \phi[\boldsymbol{p}^{(0)}]. \tag{4.117}$$

For a general objective function $\phi(\boldsymbol{p})$ this expression is not strictly corect. It only gives an approximation to $\bar{\boldsymbol{p}}$. In this case one has to make use of an iterative process in which the estimates of $\bar{\boldsymbol{p}}$ in the $(k+1)$-th and k-th step are related to each other as are $\bar{\boldsymbol{p}}$ and $\boldsymbol{p}^{(0)}$ in (4.117):

$$\boldsymbol{p}^{(k+1)} = \boldsymbol{p}^{(k)} - \underline{\underline{\phi}}_p^{-1}[\boldsymbol{p}^{(k)}]\nabla_p\phi[\boldsymbol{p}^{(k)}]. \tag{4.118}$$

This equation is usually written in the form,

$$\boldsymbol{p}^{(k+1)} = \boldsymbol{p}^{(k)} - \lambda^{(k)}\underline{\underline{\phi}}_p^{-1}[\boldsymbol{p}^{(k)}]\nabla_p\phi[\boldsymbol{p}^{(k)}], \tag{4.119}$$

where $\lambda^{(k)}$ is a scalar. This scalar may be either kept constant or adjusted at each iteration step so as to minimize the lattice energy in the search direction, $-\underline{\underline{\phi}}_p^{-1}[\boldsymbol{p}^{(k)}]\nabla_p\phi[\boldsymbol{p}^{(k)}]$.

It is essential that the value of the lattice energy is guaranteed to be improved at each iteration step only if the Hessian matrix $\underline{\underline{\phi}}_p[\boldsymbol{p}^{(k)}]$ is positive definite, i.e., if $\phi(\boldsymbol{p})$ is convex at each $\boldsymbol{p}^{(k)}$. Far away from the minimum, this may not be so and, hence, the search may fail. To overcome this difficulty, several special methods have been devised to force the Hessian matrix to be positive definite. A brief outline of these methods and the necessary references may be found in [4.8]. The simplest solution of the problem is to use the method of steepest descent in the early stages of minimization, i.e., to replace $\underline{\underline{\phi}}_p^{-1}$ in (4.119) by the unit matrix.

An analytical evaluation of both $\nabla_p\phi$ and $\underline{\underline{\phi}}_p$ using the equations of Sect. 4.2.2 presents no difficulties if one deals with rigid molecules. In more general cases, the analytical expressions for $\nabla_p\phi$ and, particularly, for $\underline{\underline{\phi}}_p$ are very cumbersome and it is easy to make an error in programming these expressions. Moreover, there are several cases in which it is practically impossible to write down an explicit expression for particular derivatives. (This refers, for example, to the derivatives of the lattice energy with respect to the parameters h and δ appearing in the variable virtual-bond method to be described in Sect. 4.6.)

The literature on nonlinear programming contains a wide variety of local minimization techniques which require an analytical evaluation of only ϕ and $\nabla_p\phi$ (first-derivative methods) or even ϕ alone (derivative-free methods). In our view, the most efficient of the first-derivative methods, when applied to an optimization of the lattice energy, are the variable-metric or quasi-Newton methods. These methods are all based on (4.119). They do not use the exact Hessian but some approximation of it, inferred from the first derivatives in the preceding stages of the search. To illustrate how this can be done, let us expand the gradient at $\boldsymbol{p}^{(k+1)}$ in a Taylor series about $\boldsymbol{p}^{(k)}$,

$$\nabla_p\phi[\boldsymbol{p}^{(k+1)}] = \nabla_p\phi[\boldsymbol{p}^{(k)}] + \underline{\underline{\phi}}_p[\boldsymbol{p}^{(k)}][\boldsymbol{p}^{(k+1)} - \boldsymbol{p}^{(k)}] + \dots . \tag{4.120}$$

Neglecting higher-order terms,

$$\boldsymbol{p}^{(k+1)} - \boldsymbol{p}^{(k)} = \underline{\underline{\phi}}_p^{-1}[\boldsymbol{p}^{(k)}]\{\nabla_p\phi[\boldsymbol{p}^{(k+1)}] - \nabla_p\phi[\boldsymbol{p}^{(k)}]\}, \quad \text{or} \tag{4.121}$$

$$\boldsymbol{\Delta p}^{(k)} = \underline{\underline{\phi}}_p^{-1}[\boldsymbol{p}^{(k)}]\boldsymbol{\Delta g}^{(k)} \quad , \tag{4.122}$$

where $\boldsymbol{\Delta p}^{(k)}$ and $\boldsymbol{\Delta g}^{(k)}$ are defined by (4.121).

In the variable-metric methods, the approximating matrix for $\underline{\underline{\phi}}_p^{-1}$, hereafter denoted as $\underline{\underline{f}}$, is usually given by

$$\underline{\underline{f}}^{(k+1)} = \underline{\underline{f}}^{(k)} + \underline{\underline{\Delta}}\,\underline{\underline{f}}^{(k)}. \tag{4.123}$$

At the stage $\boldsymbol{p}^{(k+1)}$ one knows $\boldsymbol{\Delta p}^{(k)}$, $\boldsymbol{\Delta g}^{(k)}$ and $\underline{\underline{f}}^{(k)}$ and wishes to compute the unknown $\underline{\underline{\Delta}}\,\underline{\underline{f}}^{(k)}$ so that the relation

$$\boldsymbol{\Delta p}^{(k)} = \underline{\underline{f}}^{(k+1)}\boldsymbol{\Delta g}^{(k)}, \tag{4.124}$$

is fulfilled. Substituting (4.123) into (4.124), one immediately obtains

$$\underline{\underline{\Delta}}\,\underline{\underline{f}}^{(k)}\boldsymbol{\Delta g}^{(k)} = \boldsymbol{\Delta p}^{(k)} - \underline{\underline{f}}^{(k)}\boldsymbol{\Delta g}^{(k)}, \tag{4.125}$$

which is to be solved for $\underline{\underline{\Delta}}\,\underline{\underline{f}}^{(k)}$.

Equation (4.125) has an infinite number of solutions which may all be written in the form

$$\underline{\underline{\Delta}}\,\underline{\underline{f}}^{(k)} = \frac{\boldsymbol{\Delta p}^{(k)}\tilde{\boldsymbol{V}}^{(k)}}{\tilde{\boldsymbol{V}}^{(k)}\boldsymbol{\Delta g}^{(k)}} - \frac{\underline{\underline{f}}^{(k)}\boldsymbol{\Delta g}^{(k)}\tilde{\boldsymbol{W}}^{(k)}}{\tilde{\boldsymbol{W}}^{(k)}\boldsymbol{\Delta g}^{(k)}}, \tag{4.126}$$

where $\boldsymbol{V}^{(k)}$ and $\boldsymbol{W}^{(k)}$ are two arbitrary vectors. In essence, the various quasi-Newton methods differ from one another in their choice of $\boldsymbol{V}^{(k)}$ and $\boldsymbol{W}^{(k)}$. In the most popular and, probably, most efficient Davidon-Fletcher-Powell algorithm [4.8] these two vectors are chosen to be

$$\boldsymbol{V}^{(k)} = \boldsymbol{\Delta p}^{(k)}, \tag{4.127}$$

$$\boldsymbol{W}^{(k)} = \underline{\underline{f}}^{(k)}\boldsymbol{\Delta g}^{(k)}. \tag{4.128}$$

With these choices the equation for $\underline{\underline{f}}^{(k+1)}$ becomes

$$\underline{\underline{f}}^{(k+1)} = \underline{\underline{f}}^{(k)} + \frac{\Delta p^{(k)} \widetilde{\Delta p}^{(k)}}{\widetilde{\Delta p}^{(k)} \Delta g^{(k)}} - \frac{\underline{\underline{f}}^{(k)} \Delta g^{(k)} \widetilde{\Delta g}^{(k)} \tilde{\underline{\underline{f}}}^{(k)}}{\widetilde{\Delta g}^{(k)} \underline{\underline{f}}^{(k)} \Delta g^{(k)}} . \tag{4.129}$$

The starting $\underline{\underline{f}}$ is usually the unit matrix, so that the starting direction in the minimization is that of steepest descent. As the minimization progresses, the search direction $-\underline{\underline{f}}\nabla_p\phi$ gradually changes from the gradient to the Newton direction.

The Davidon-Fletcher-Powell algorithm outlined above may, in principle, be used in the free-derivative form, in which the first derivatives are estimated by difference schemes [4.12]. The performance of such a modified algorithm in lattice-energy calculations is generally quite acceptable, except in the immediate vicinity of the minimum where the numerical errors in the derivatives may somewhat deteriorate the convergence. For high-accuracy calculations, alternative free-derivative algorithms are recommended, such as the well-known algorithms of *Powell* [4.13] and of *Rosenbrock* [4.14].

4.2.4 Convergence Properties of the Lattice Energy

In computing the intermolecular part of the lattice energy

$$\phi_c = \frac{1}{2} \sum_m N_m \sum_{aa'm's'} \varphi_{aa'}(r^{am,a'm's'}), \tag{4.130}$$

one faces a convergence problem since the summation in (4.130) is over an infinitely extended crystal. For atom-atom potentials of the (6-exp)- or (6-m)-type the sum converges rather rapidly. A fairly good estimate of ϕ_c may be obtained by merely truncating $\varphi_{aa'}$ at a sufficiently large distance, $r^{\text{tr}}_{aa'}$, and considering only those atomic pairs which fulfill the requirement[5],

$$r^{am,a'm's'} \leq r^{\text{tr}}_{aa'} . \tag{4.131}$$

The latter procedure amounts to a formal replacement of (4.130) by

$$\phi^{\text{tr}}_c(r^{\text{tr}}) = \frac{1}{2} \sum_m N_m \sum_{aa'm's'} \varphi_{aa'} H(r^{\text{tr}}_{aa'} - r^{am,a'm's'}), \tag{4.132}$$

where $H(x)$ is the Heaviside function defined by

$$H(x) = \begin{cases} 1, & x \geq 0, \\ 0, & x < 0 \end{cases} . \tag{4.133}$$

[5] To avoid the introduction of new indices, we shall label $\varphi_{aa'}$ and $r^{\text{tr}}_{aa'}$ by the same indices used to label the interacting atoms. In fact, both $\varphi_{aa'}$ and $r^{\text{tr}}_{aa'}$ are determined by the *species* of the interacting atoms.

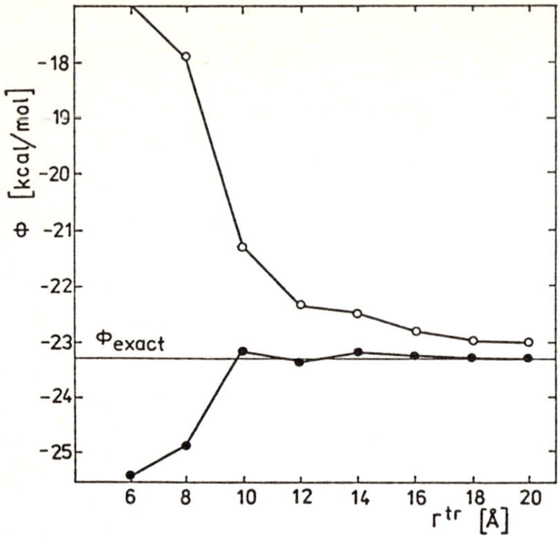

Fig. 4.2. The lattice energy of hexachlorobenzene as a function of truncation radius. (○ ○ ○) : calculated using the truncated sum in (4.132); (• • •) : corrected by the "smearing" formula (4.138)

Figure 4.2 shows the dependence of ϕ_c^{tr} on r^{tr} for a crystal of hexachlorobenzene, calculated using the MC78 potentials of Table 3.5 (open circles). The horizontal line represents the exact value of the lattice energy, obtained with a truncation radius so large that a further increase in r^{tr} does not produce any change in the lattice energy. The deviations of the open circles from the horizontal line provide the reader with an idea of the accuracy that can be attained with various truncation radii. Thus, an 8 Å truncation radius encompasses about 77% of ϕ_c, a 10 Å radius about 91% and a 16 Å radius about 98%.

The convergence of ϕ_c^{tr} to ϕ_c for (6-exp) and (6-m) potentials is obviously governed by the convergence of the Ar^{-6} terms, because the repulsive terms decrease very rapidly with increasing distance. For these potentials, the remainder of the lattice energy, which is suppressed in (4.132) by the Heaviside function, can be estimated by uniformly smearing the atoms outside the truncation sphere over the crystal and then replacing the summation by an integration. Let us consider the interaction of an atom a with all of the surrounding atoms a' within a spherical layer $(r, r + dr)$. If the surrounding atoms are assumed to be uniformly smeared over the crystal, the atomic density will be N_c/V_c and the number of atoms a' in the layer $(r, r + dr)$ will be equal to

$$n_{a'}(r, dr) = 4\pi r^2 dr N_c/V_c. \tag{4.134}$$

The interaction energy of atom a with all atoms a' in the layer is given by

$$\varepsilon^{aa'}(r, dr) = \varphi_{aa'}(r) n_{a'}(r, dr) = 4\pi \frac{N_c}{V_c} \varphi_{aa'} r^2 dr. \tag{4.135}$$

The total interaction energy of atom a with all atoms a' outside the sphere of radius $r^{\mathrm{tr}}_{aa'}$ can now be readily obtained by integrating (4.135) from $r = r^{\mathrm{tr}}_{aa'}$ to infinity, i.e.,

$$\varepsilon^{aa'}(r^{\mathrm{tr}}_{aa'}) = \frac{4\pi N_c}{V_c} \int_{r^{\mathrm{tr}}_{aa'}}^{\infty} \varphi_{aa'}(r) r^2 dr. \tag{4.136}$$

Neglecting the repulsive part of $\varphi_{aa'}$,

$$\varepsilon^{aa'}(r^{\mathrm{tr}}_{aa'}) = -\frac{4\pi N_c}{V_c} \int_{r^{\mathrm{tr}}_{aa'}}^{\infty} A_{aa'} r^{-6} r^2 dr = -\frac{4\pi N_c A_{aa'}}{3V_c (r^{\mathrm{tr}}_{aa'})^3}. \tag{4.137}$$

Since the number of atoms a in the unit cell is equal to N_c (one atom per molecule), the suppressed part of the lattice energy is given by

$$\begin{aligned}\phi_c^{\mathrm{suppr}}(r^{\mathrm{tr}}) &= \left[\frac{1}{2} N_c \sum_{aa'} \varepsilon_{aa'}(r^{\mathrm{tr}}_{aa'})\right] / N_c \\ &= -\frac{2}{3} \frac{\pi N_c}{V_c} \sum_{aa'} A_{aa'} (r^{\mathrm{tr}}_{aa'})^{-3}. \end{aligned} \tag{4.138}$$

Expressions for the derivatives of ϕ_c^{suppr} with respect to the structural parameters given in (4.88) are very easy to derive, considering that $V_c = L_{11} L_{22} L_{33}$.

The addition of (4.138) to (4.132) markedly improves the estimates of the lattice energy. This can be seen in Fig. 4.2 where the full circles correspond to these improved estimates.

The calculation of the lattice energy proves to be much more complicated when the atom-atom potentials involve Coulombic terms. In this case the sum given in (4.130) converges very slowly and conditionally, and special devices are required to force the lattice energy to convergence. The simple smearing procedure discussed above is completely impracticable. This immediately follows from (4.136) because the integral appearing in this equation diverges when $\varphi_{aa'}$ contains a Coulombic term.

We shall now briefly discuss two procedures, suitable for dealing with Coulombic interactions. These procedures are both based on the early work of *Ewald* [4.15] and *Bertaut* [4.16], and involve a transformation of the direct sum in (4.130), or a part of it, to a sum in the reciprocal space.

The first procedure has been developed by *Williams* [4.17] and is designed to evaluate lattice sums of the general form,

$$S^{(m)} = \frac{1}{2} \sum_{a' \neq a} q_a q_{a'} r_{aa'}^{-m}, \tag{4.139}$$

where a runs over the atoms of the reference (zero-th) unit cell and a' runs over the entire lattice. For $m = 1$ (4.139) yields the Coulombic lattice energy, including its intramolecular part. Since the latter can easily be evaluated term by term in a separate calculation and then subtracted from $S^{(m)}$, (4.139) can be readily adapted for an evaluation of the intermolecular part of the Coulombic lattice energy.

For $m = 6$, (4.139) describes the dispersion part of the lattice energy, provided that the dispersion coefficients can be represented as products $q_a q_{a'}$ (the latter requirement implies that the coefficients for cross interactions obey the geometric-mean combining law given in (3.4)). Again, the intramolecular part of the dispersion lattice energy can be evaluated separately and then subtracted from $S^{(m)}$.

In William's treatment of the series in (4.139) each term is first multiplied by a convergence function, $\eta(r)$, such that $\eta(0) = 1$ and $\eta(r)$ rapidly approaches zero as r increases. The Fourier transform of the remaining terms is then found and the transformed series is summed in reciprocal space.

For $\eta(r)$ Williams used the function

$$\eta_m(r) = \Gamma(m/2, \pi K_m^2 r^2)/\Gamma(m/2), \tag{4.140}$$

previously employed by *Nijboer* and *De Wette* [4.18] in a treatment of $S^{(m)}$ in which the sum is taken over a simple lattice. In (4.140) K_m is a constant and $\Gamma(n)$ and $\Gamma(n, x)$ are a gamma function and an incomplete gamma function, respectively.

The final expression for $S^{(m)}$ has the form

$$\begin{aligned} S^{(m)} = \frac{1}{2\Gamma(m/2)} \Bigg[& \sum_{a' \neq a} q_a q_{a'} r_{aa'}^{-m} \Gamma(m/2, \pi K_m^2 r_{aa'}^2) \\ & + V_c^{-1} \pi^{(m-3)/2} \sum_{\boldsymbol{h}_\lambda \neq 0} |F(\boldsymbol{h}_\lambda)|^2 h_\lambda^{m-3} \Gamma(-m/2 + 3/2, h_\lambda^2/K_m^2) \\ & + 2V_c^{-1} \pi^{m/2} K_m^{m-3}/(m-3) \Big(\sum_a q_a\Big)^2 \\ & - 2m^{-1} \pi^{m/2} K_m^m \sum_a q_a^2 \Bigg], \end{aligned} \tag{4.141}$$

where the sum over $\boldsymbol{h}_\lambda$ is taken over all of the points of the reciprocal lattice, except $\boldsymbol{h} = 0$; and $F(\boldsymbol{h})$ is the Fourier transform of the charge density $q_a \delta(\boldsymbol{r} - \boldsymbol{r}_a)$, obtained in exactly the same way as the crystallographic structure factor, but in which the atomic scattering factors are replaced by the charges q_a :

$$F(\boldsymbol{h}) = \sum_a q_a \exp(-2\pi i \boldsymbol{h} \cdot \boldsymbol{r}_a). \tag{4.142}$$

Equation (4.141) converges for any q_a, provided that $m>3$. If $\sum_a q_a = 0$ (the electroneutrality condition for the unit cell), then (4.141) converges for all $m>0$.

Williams [4.17] applied (4.141) to both the Coulombic and dispersion parts of the lattice energy. It is, however, very doubtful whether such a complicated approach is really needed for the dispersion energy which can be easily evaluated by combining (4.132 and 138). For this reason, we shall confine ourselves to a discussion of the Coulombic energy. An expression for $S^{(1)}$ is readily derived from (4.141) by setting $m = 1$ and assuming that $\sum_a q_a = 0$. The result is identical to the expression obtained by *Ewald* [4.15],

$$S^{(1)} = \frac{1}{2\pi V_c} \sum_{h_\lambda} |F(h_\lambda|^2 h_\lambda^{-2} \exp(-\pi h_\lambda^2/K_1^2) + \frac{1}{2} \sum_{a' \neq a} q_a q_{a'} r_{aa'}^{-1} [\mathrm{erfc}(\pi^{1/2} K_1 r_{aa'})] - K_1 \sum_a q_a^2, \qquad (4.143)$$

where $\mathrm{erfc}(x)$ is the complement of the error function, defined by

$$\mathrm{erfc}(x) = 1 - \frac{2}{\pi^{1/2}} \int_0^x \exp(-t)^2 dt. \qquad (4.144)$$

Comparing (4.143) with (4.139), we notice that the second sum in (4.143) is identical to the original sum in (4.139), except that each Coulombic term contains an extra factor $\mathrm{erfc}(\pi^{1/2} K_1 r)$, which suppresses the long-range interactions. The suppressed part of the original sum is recovered in the reciprocal sum and the constant term $-K_1 \sum_a q_a^2$.

For a small value of K_1 the major contribution to $S^{(1)}$ is due to the direct sum, whereas the reciprocal sum is of minor importance. As K_1 tends to zero, the reciprocal sum and the constant term vanish, so that (4.143) is now identical to (4.139). For a large value of K_1 the direct sum becomes small and the dominant contribution to $S^{(1)}$ is due to the reciprocal sum. Thus, the constant K_1 is a parameter that governs the distribution of $S^{(1)}$ among the direct and reciprocal sums.

In his early papers *Williams* [4.17,19] advised choosing a value of K_1 so as to make the reciprocal sum negligible. It is, however, hardly reasonable to follow this advice. As noted by *Leibfried* [4.20], the best convergence of $S^{(1)}$ should occur when the direct and reciprocal sums contribute nearly equally to $S^{(1)}$. This implies that it is, in principle, impossible to select K_1 so that the direct sum converges rapidly and, at the same time, the reciprocal sum is negligible.

The shortcoming of Williams' choice is immediately seen if we neglect the reciprocal sum in (4.143) and write

$$S^{(1)} \approx \frac{1}{2} \sum_{a' \neq a} q_a q_{a'} r_{aa'}^{-1} \operatorname{erfc}(\pi^{1/2} K_1 r_{aa'}) - K_1 \sum_a q_a^2. \tag{4.145}$$

Such a representation of $S^{(1)}$ implies that the suppressed part of the original direct sum, i.e.,

$$S^{(1),\mathrm{suppr}} = \frac{1}{2} \sum_{a' \neq a} q_a q_{a'} r_{aa'}^{-1} \operatorname{erf}(\pi^{1/2} K_1 r_{aa'}), \tag{4.146}$$

may be adequately recovered in the constant term $-K_1 \sum_a q_a^2$. It is, however, clear that there is in fact no correlation between $S^{(1),\mathrm{suppr}}$ and the constant term. Thus, $S^{(1),\mathrm{suppr}}$ may be either positive or negative, depending on K_1, whereas the constant term is always negative. Also, $S^{(1),\mathrm{suppr}}$ is obviously dependent on the crystal configuration, while the constant term is not.

To be sure, for each particular crystal it is always possible to select the parameter K_1 so that $S^{(1),\mathrm{suppr}}$ is exactly equal to the constant term. This equality, however, will no longer be valid if we change the crystal configuration or attempt to transfer the selected parameter K_1 to another crystal.

Thus, in computations of the Coulombic part of the lattice energy it seems prudent to only use the rigorous version of Williams' procedure, as given by (4.143). In this rigorous version one must evaluate both the direct and reciprocal sums and select K_1 so as to make the contributions from the two sums nearly equal to one another.

From the practical point of view, the necessity of dealing with both the direct and reciprocal sums is, of course, displeasing. In this respect, the method suggested by *Bertaut* [4.16] seems to be more attractive. Bertaut's method takes advantage of the well-known fact that the interaction energy of two point charges, q_1 at $\boldsymbol{r}_1$ and q_2 at $\boldsymbol{r}_2$, will remain unchanged if we replace the charges by spherically symmetrical charge distributions $q_1\sigma_1(|\boldsymbol{r} - \boldsymbol{r}_1|)$ and $q_2\sigma_2(|\boldsymbol{r} - \boldsymbol{r}_2|)$, with the only requirement that the distributions be non-overlapping and satisfy the normalization condition,

$$\int \sigma_i(|\boldsymbol{r} - \boldsymbol{r}_i|) dV = 1, \quad i = 1, 2. \tag{4.147}$$

Using the Fourier transform technique, *Bertaut* has derived the following representation for $S^{(1)}$,

$$S^{(1)} = \frac{1}{2\pi V_c} \sum_{\boldsymbol{h}_\lambda} |F(\boldsymbol{h}_\lambda)|^2 h_\lambda^{-2} |\psi(\boldsymbol{h}_\lambda)|^2 - \sum_a q_a^2 \int_{-\infty}^{\infty} |\psi(x)|^2 dx, \tag{4.148}$$

where $\psi(x)$ is the Fourier transform of the charge density distribution $\sigma(r)$.

Ewald's equation, (4.143), may be derived from (4.148) by substituting the normal Gaussian distribution for $\sigma(r)$ and then correcting the resulting expression for the penetration of the distributions on different atoms.

If the original point charges are assumed to be uniformly distributed over the volume of a sphere of radius R, i.e.,

$$\sigma(r) = \begin{cases} 3/4\pi R^3, & r \leq R \\ 0, & r > R \end{cases}, \tag{4.149}$$

the following equation results,

$$S^{(1)} = \frac{18\pi R^2}{V_c} \sum_{\boldsymbol{h}_\lambda} |F(\boldsymbol{h}_\lambda)|^2 (\sin\alpha_\lambda - \alpha_\lambda \cos\alpha_\lambda)^2 \alpha_\lambda^{-8} - \frac{3}{5R} \sum_a q_a^2, \tag{4.150}$$

where $\alpha_\lambda = 2\pi h_\lambda R$.

Due to the presence of the factor α_λ^{-8}, the reciprocal sum in (4.150) converges rather rapidly, so that a fairly good estimate of $S^{(1)}$ may be obtained by merely neglecting the terms with h_λ greater than a sufficiently large number, $h^{\max}$. Again, we may introduce a truncated sum, $S^{(1),\mathrm{tr}}(h^{\max})$, defined similarly to (4.132), using the Heaviside function $H(h^{\max} - h_\lambda)$ to suppress the long-range contributions to (4.150). The rate of convergence of $S^{(1),\mathrm{tr}}$ to $S^{(1)}$ strongly depends on the value adopted for R; the convergence is the better, the larger the radius R. Since R should not exceed one half of the shortest interatomic distance (the condition that the spheres are nonoverlapping), the maximum achievable rate of convergence is determined by the particular crystal structure.

In organic crystals, the maximum allowable value for R is limited by one half of the shortest bond length (~0.5 Å). Since the unit-cell volume is generally large in organic crystals, the density of "reflections" $\boldsymbol{h}_\lambda$ is also large, and it takes a rather large number of terms in (4.150) to force $S^{(1)}$ to convergence. This is illustrated in Fig. 4.3 in which the relative error in $S^{(1)}$ is plotted against $h^{\max}$ (open circles), as calculated by *Gramaccioli* and *Filippini* [4.21] for the benzene crystal with atomic point charges of 0.18 e. Shown in the same figure are the number of reflections $\boldsymbol{h}_\lambda$ that fall into the sphere of radius $h^{\max}$. It is seen that acceptable estimates of $S^{(1)}$ are obtained when the number of terms in (4.150) is of the order of 10^4.

Fortunately, there is a possibility of recovering, though approximately, the suppressed part of $S^{(1)}$, $S^{(1),\mathrm{suppr}} \equiv S^{(1)} - S^{(1),\mathrm{tr}}$. This can be done using a very simple smearing procedure similar to the one discussed for the dispersion part of the lattice energy. Such a procedure has been suggested by *Templeton* [4.22] and shall be discussed below.

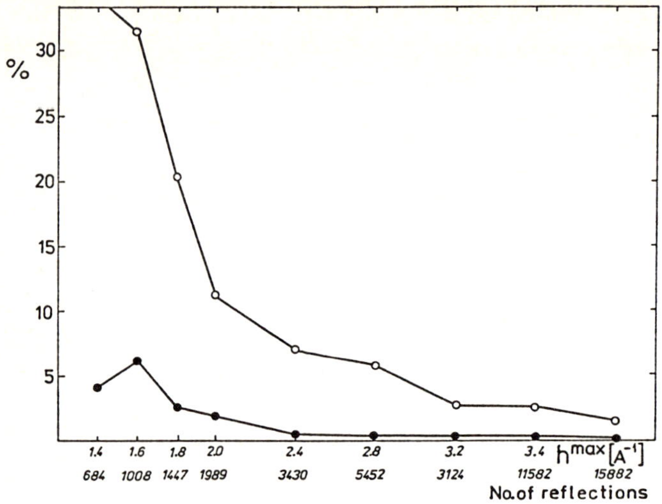

Fig. 4.3. Percentage error in the Coulombic lattice energy of benzene as a function of the truncation radius in the reciprocal space [4.21]. (∘∘∘) : calculated using the truncated sum in (4.150); (•••) : corrected according to (4.156)

Let us write down the explicit expression for $S^{(1),\mathrm{suppr}}$,

$$S^{(1),\mathrm{suppr}} = \frac{18\pi R^2}{V_c} \sum_{h_\lambda > h^{\max}} |F(\boldsymbol{h}_\lambda)|^2 (\sin\alpha_\lambda - \alpha_\lambda \cos\alpha_\lambda)^2 \alpha_\lambda^{-8}. \tag{4.151}$$

For reflections with large values of h_λ, the number of terms in a small interval of h_λ also becomes large so that $|F(\boldsymbol{h}_\lambda)|^2$ may be replaced by its average value, $\sum_a q_a^2$. Considering that the average density of terms in the reciprocal space is equal to V_c, $S^{(1),\mathrm{suppr}}$ may be approximated by the integral,

$$S^{(1),\mathrm{suppr}} \approx \frac{18\pi R^2}{V_c} \sum_a q_a^2 V_c \int_{h^{\max}}^{\infty} (\sin\alpha - \alpha\cos\alpha)^2 \times \alpha^{-8} 4\pi h^2 dh, \tag{4.152}$$

which can be reduced, after simple manipulations, to

$$S^{(1),\mathrm{suppr}} \approx \frac{3\sum q_a^2}{5\pi R} \left[\frac{3\sin^2\beta}{\beta^5} - \frac{6\sin\beta\cos\beta}{\beta^4} + \frac{2+\cos^2\beta}{\beta^3} - \frac{\sin\beta\cos\beta}{\beta^2} - \frac{1-2\sin^2\beta}{\beta} + \pi - 2Si(2\beta) \right], \tag{4.153}$$

where

$$\beta = 2\pi R h^{\max}, \quad \text{and} \tag{4.154}$$

$$Si(x) = \int_0^x \frac{\sin t}{t} dt. \tag{4.155}$$

Recently, *Gramaccioli* and *Filippini* [4.21] showed that for $\beta > \pi$ (4.153) may be approximated fairly well by a simple analytical expression,

$$S^{(1),\text{suppr}} \approx \frac{\sum q_a^2}{R} 0.016161 (2R h^{\max})^{-3.0381}. \tag{4.156}$$

This expression, when added to Bertaut's series terminated at $h_\lambda < h^{\max}$, see (4.150), provides a relatively simple and reliable procedure for calculating the Coulombic lattice energy. To illustrate the success of this procedure, Fig. 4.3 shows the relative error in the Coulombic lattice energy of the benzene crystal, corrected according to (4.156) (full circles) [4.21]. It can be seen that the introduced correction appreciably accelerates the convergence of the lattice energy. Thus, to attain an acceptable accuracy of the order of 2%, it is sufficient to take into account about 1500 independent reflections with $h_\lambda < 1.8\,\text{Å}^{-1}$. Similar results were also obtained for crystals of naphthalene and acetylene [4.21].

A good illustration of the importance of the convergence-acceleration methods in evaluating the Coulombic part of the lattice energy may be found in [4.23,24]. *Momany* et al. [4.23] have calculated the lattice energy of the three polymorphic modifications of glycine and found an unusually large difference of 13 kcal/mol between the lattice energy of λ-glycine and that of β-glycine due to the difference in the Coulombic lattice energy. The latter was evaluated by direct summation of the Coulombic interactions. To understand the origin of this unexpected large difference, *Derissen* et al. [4.24] recalculated the energies of the three glycine crystals using both the direct summation and Ewald's formula, (4.143). It turned out that the direct summation was only successful for α-glycine. In this case the direct sum already converged at $r^{\text{tr}} = 25\,\text{Å}$ to give an energy practically identical to that obtained by Ewald's formula. The convergence of the direct sum was somewhat poorer for β-glycine. The final direct estimate of the Coulombic energy was about 1.5 kcal/mol higher than the exact value. For γ-glycine, the direct sum oscillated, i.e., did not converge at all, whereas (4.143) converged very rapidly and yielded an energy close to that of α- and β-glycine.

The poor convergence of the direct Coulombic sum for β- and, particularly, γ-glycine can be satisfactorily explained in terms of the space group symmetry of the two glycines. Unlike α-glycine, the crystals of β- and γ-glycine are non-centrosymmetric and their unit cells have nonzero dipole moments. A CNDO/2 calculation of the unit-cell dipole moments yields a value of 4.1 D for β-glycine and 28.9 D for γ-glycine [4.24]. Since the dipole-dipole energy is proportional to the square of the dipole moment and decreases rather slowly with distance, the poor convergence of the Coulombic lattice energy of γ-glycine is well understood.

4.3 The Use of Atom-Atom Potentials in Predicting Stable Crystal Configurations

At the time of writing this book over two hundred papers which employ the atom-atom potential method to predict an equilibrium crystal configuration have appeared. A nearly complete review of this topic, covering the literature up to 1978, can be found in review articles by *Kitaigorodsky* [4.25], *Timofeeva* et al. [4.26] and *Ramdas* and *Thomas* [4.27]. Many of the studies not surveyed in these articles are referred to in the present monograph.

Most of the work reported thus far follows a standard procedure which starts from the observed crystal configuration and searches for the local energy minimum nearest to the starting point. The crystal symmetry is generally kept fixed during the energy minimization and the molecules are assumed rigid. Many calculations of this kind have been discussed in the preceding chapter. The calculations reproduced the crystal structure at equilibrium with a typical accuracy of the order of 0.1 Å for the lattice periods and molecular translations, and 1° for the molecular orientations.

Such an accuracy is in itself impressive but does not seem to be a very significant achievement. The fact is that the symmetry-constrained model severely restricts the motion of molecules in a crystal. For this reason, it is usually sufficient to reproduce a few intermolecular contacts to predict the entire crystal structure reasonably well.

Strictly speaking, the occurrence of a local energy minimum near the observed crystal configuration does not yet mean that the calculation correctly predicts the stable crystal structure. In fact, there are a variety of local minima on the potential-energy hypersurface of a crystal and, consequently, a variety of crystal packings satisfying the equilibrium conditions given in (4.44,45). Thus, a more stringent test for the reliability of the atom-atom potential method will be to see if the method is capable of predicting the observed equilibrium packing.

There are two ways of subjecting the atom-atom potential method to such a test. The first is to compare observed packing with some hypothetical (non-existing) packings and to see whether the calculation yields a lower potential energy for the former. The second way is to consider two polymorphs of the same substance and to check whether a calculation yields the potential energies in the correct sequence.

To remind the reader of the sequence of the lattice energies in the latter case, Fig. 4.4 gives an idealized thermodynamic picture of a temperature-

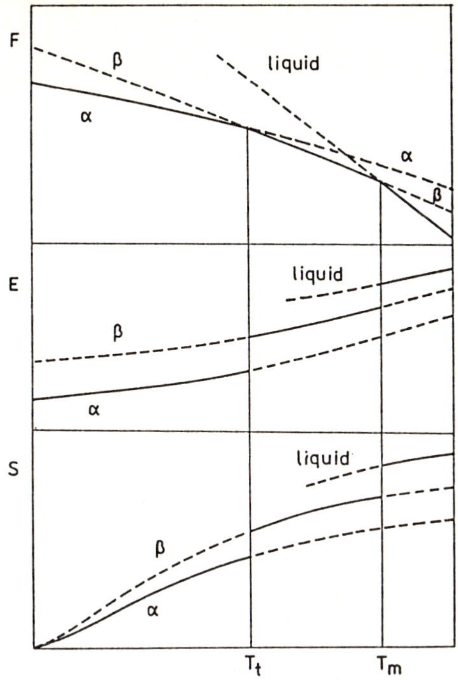

Fig. 4.4. An idealized picture of a temperature-induced polymorphic transition $\alpha \leftrightarrow \beta$ at T_t. (Arbitrary units). The solid and dashed lines represent the stable and metastable phases at a given temperature

induced polymorphic transition. The low-temperature polymorph is labeled α and the high-temperature one β. Solid lines correspond to the stable polymorph while the dashed lines correspond to the metastable one.

At the transition point, T_t, the free energies of the two polymorphs are equal while the internal energies and entropies are related as follows

$$E_\beta - E_\alpha = T_t(S_\beta - S_\alpha). \tag{4.157}$$

Since $S_\beta > S_\alpha$, the difference $E_\beta - E_\alpha$ should be positive, i.e., the low-temperature phase should have a lower (more negative) internal energy at $T = T_t$. For organic crystals, the changes in energy involved in polymorphic transitions are generally of the order of 1 kcal/mol.

Of course, the difference $E_\beta - E_\alpha$ depends on the temperature but is very unlikely to change its sign. Indeed, at temperatures below T_t, where the inequality

$$E_\beta - E_\alpha > T(S_\beta - S_\alpha) \tag{4.158}$$

is valid, the difference $E_\beta - E_\alpha$ must always remain positive. This can be immediately shown by setting $T = 0$ in (4.158). For temperatures above T_t, the range in which $E_\beta - E_\alpha$ remains positive should also be very wide. Thus, if we assume that $E_\beta - E_\alpha = 1$ kcal/mol and that the heat capacities of the two polymorphs differ by as great as 10 cal/(mol K) ($C_\beta < C_\alpha$), $E_\beta - E_\alpha$ will still remain positive up to $T = T_t + 100\,K$. In fact the heat capacities of different polymorphs differ very slightly. Hence, the difference $E_\beta - E_\alpha$ should generally remain positive at all temperatures and, furthermore, should not change significantly in absolute value.

Thus a calculation of the internal energies of two polymorphs should yield a lower energy for the polymorph being stable at lower temperatures. The same sequence should, as a rule, be valid for the static lattice energies because the vibrational contribution to E is comparatively small and, what is more important, can hardly be expected to differ significantly between different polymorphs. (Some representative figures will be presented in Chap. 6 devoted especially to the thermodynamics of organic crystals.)

A convenient object with which one can check the ability of the atom-atom potential method to distinguish between different polymorphs is the crystal of p-dichlorobenzene. This crystal is known to exist in three polymorphic forms: a high-temperature triclinic form (β) stable above 304.4 K, a room-temperature monoclinic form (α) stable between 304.4 and 273.8 K, and a low-temperature monoclinic form (γ) stable below 279.8 K [4.28]. The polymorphic transitions in the crystal are very sluggish, so that it is possible to keep all three polymorphs at the same temperature for a long time.

The crystal structure data necessary to compare the three different packings at the same temperature are available [4.29] and refer to a temperature of 100 K. From the observed sequence of the polymorphic transitions, the sequence of the lattice energies of the three polymorphs at 100 K should be $\phi_\gamma < \phi_\alpha < \phi_\beta$. An attempt to reproduce this sequence has been made by *Mirsky* and *Cohen* [4.30] using the MC78 parameter set. The results are given in Table 4.1. It is seen that the calculations fail to correctly predict the sequence of the lattice energies. The predicted sequence is the reverse of the expected one.

Suspecting the failure of the Mirsky and Cohen calculations to be due to defects in the MC78 parameter set, we have recalculated the lattice energies of the three polymorphs using two alternative parameter sets, i.e., the RKW74 and BB74a sets of Table 3.4. These results are also listed in Table 4.1. It is seen that both latter parameter sets yield lattice energies in the correct sequence.

Another comparison of high- and low-temperature polymorphs was made by *Ramdas* et al. [4.31] with the two modifications of the pyrene crys-

Table 4.1. Calculated lattice energies of the three polymorphs of p-dichlorobenzene at 100 K [kcal/mol]

Polymorph	Parameter set		
	MC78	RKW74	BB74a
β	-15.28	-18.38	-14.30
α	-15.14	-19.09	-14.49
γ	-15.12	-19.22	-14.93

tal. This work will be discussed in Sect.4.5. We only note here that the calculations correctly predicted a lower lattice energy for the low-temperature polymorph.

It is to be emphasized that a comparison of the potential energies of different polymorphs only makes sense when the sequence of the mutual transformations is known. Very often, however, the available information is limited to a knowledge of the relative stabilities of the polymorphs at a single temperature or in a limited temperature range. Thus *Caillet* and *Claverie* [4.32] have studied three polymorphs of 5,5-diethylbarbituric acid whose detailed phase diagram was unknown. Comparing the melting points of the three polymorphs, they concluded that the most stable is the hexagonal polymorph (I) melting at 463 K, then the monoclinic polymorph (II) melting at 456 K and finally the monoclinic polymorph (IV) melting at 449 K. Based on these data, the lattice energies of the three polymorphs were assumed to follow the sequence $\phi_I < \phi_{II} < \phi_{IV}$. Such an assumption, however, is not well substantiated. Indeed, from the observed relative stabilities of different polymorphs one can only deduce the sequence of the free energies, not of the lattice energies. It can be seen in Fig. 4.4, in which the melting points of the α and β polymorphs are given by the intersection points of the curve F_{liquid} with F_α and F_β, that the polymorph melting at a lower temperature may well have a lower internal energy.

We recall that the sequence of the lattice energies of different polymorphs is not determined by their relative stability, but by the fact which of the polymorphs is stable at lower and higher temperatures. Since the low-temperature part of the phase diagram of 5,5-diethylbarbituric acid is unknown, it can not be ruled out that the polymorphs metastable at room temperature are stable at lower temperatures.

Generally speaking, the sequence of the lattice energies of different polymorphs should coincide with that of the free energies only at temperatures below the lowest transition point. At other temperatures the two sequences are different. For example, the free energies of the three p-dichlorobenzene polymorphs at $T > 304.4$ K follow the sequence $F_\gamma > F_\alpha > F_\beta$, while the lattice energies follow the reverse order: $\phi_\gamma < \phi_\alpha < \phi_\beta$. This example clearly shows that the arrangement of the lattice energies of different polymorphs according to their relative stability at a particular temperature may be in error if possible polymorphic transformations at lower temperatures are not taken into account.

We now turn to a case in which an observed packing is compared with hypothetical (non-existing) packings. Under "non-existing packings" we shall imply packings which do not exist at any temperature, so that in all cases we shall expect the lowest lattice energy for the observed packings. Several comparisons of this kind can be found in the recent literature. Thus *Bernstein* and *Hagler* [4.33,34] have attempted to answer the question of why N-(p-chlorobenzylidene)-p-chloroaniline (CBCA) does not pack in the monoclinic $P2_1$ structure observed for its dimethyl analogue, N-(p-methylbenzylidene)-p-methylaniline (MBMA). The non-existence of this structure for CBCA is indeed unexpected because a chlorine atom is very close in size to a methyl group. Also, the molecules in the monoclinic crystal of MBMA assume a conformation in which the intramolecular energy is minimum. In the structures observed for CBCA the molecules adopt higher-energy conformations, so that it is even more surprising that CBCA does not crystallize in the monoclinic structure of MBMA.

Bernstein and *Hagler* [4.33,34] have compared three alternative packings for the CBCA molecules – two packings observed in the triclinic and orthorhombic polymorphs of CBCA and a hypothetical packing corresponding to the monoclinic form of MBMA. To avoid reaching conclusions which were dependent on potentials, the lattice-energy calculations were carried out with three different sets of potential: the (6-exp), (6-12) and (6-9) potentials for hydrogen and carbon atoms [4.35–37] and *Giglio*'s potential [4.38] for chlorine atoms (Tables 3.8,10). Electrostatic interactions were also taken into account using STO-3G point charges. In the minimization of the lattice energy the molecular conformational parameters were kept fixed at the values observed experimentally for the respective crystals. The results of the energy minimization are presented in Table 4.2. It is seen that all potential sets prefer the observed packings over the hypothetical one. The average difference in the intermolecular lattice energy between the hypothetical and orthorhombic packing is 1.6 kcal/mol, while that between the hypothetical and triclinic packing is 3.4 kcal/mol.

Table 4.2. Calculated lattice energies of the three polymorphs of p-(N-chlorobenzylidene)-p-chloroaniline [kcal/mol] [4.33,34]

Polymorph	Potential 6-exp	6-12	6-9
Monoclinic (hypothetical)	-20.75	-20.13	-41.23
Orthorhombic	-22.34	-21.68	-42.73
Triclinic	-23.99	-22.73	-45.71

Since the packings examined by *Bernstein* and *Hagler* [4.33] differ substantially in molecular conformation, the values listed in Table 4.2 should be corrected for the differences in the intramolecular energy. Ab initio calculations [4.34] have shown that the molecular conformation in the hypothetical packing is 0.7 kcal/mol more stable than that in the orthorhombic packing and ~2.1 kcal/mol more stable than that in the triclinic packing. By adding these values to the corresponding differences in the intermolecular lattice energy, one reaches the conclusion that the hypothetical packing remains somewhat less stable than the two observed packings.

The calculated lattice energies listed in Table 4.2 are markedly sensitive to the particular parameter set. In spite of this fact, all potentials used lead to the same sequence of lattice energies ($\phi_{\text{inter}}^{\text{monocl}} > \phi_{\text{inter}}^{\text{orthorh}} > \phi_{\text{inter}}^{\text{tricl}}$). Moreover, it is seen that the energy differences between the three packings do not vary substantially with the potential set. This implies that the errors arising in an application of the potential sets are largely systematic and are partly cancelled when comparing the different packings. It, therefore, seems very likely that the accuracy of the atom-atom potential method used to compare different packings is better than the accuracy of the calculations of the lattice energy itself.

In more recent studies [4.39,40] a comparison of observed and hypothetical packings has been extended to MBMA and "mixed" analogues of MBMA and CBCA – MBCA and CBMA. In all the cases the packings really observed proved to be more stable than the hypothetical ones, regardless of the model potential used.

We now turn to our early work [4.41] also concerned with observed and hypothetical packings. In this work we considered the crystals of acenaphthene, biphenylene and tolane. A common characteristic feature of these crystals is the occurrence of crystallographically distinct molecules in the unit cell. The observed crystal symmetries are $Pcm2_1$, $Z_1 = 2(m)$ and $Z_2 = 2(m)$ for acenaphthene, $P2_1/a$, $Z_1 = 2(\bar{1})$ and $Z_2 = 4(1)$ for biphenylene and $P2_1/a$, $Z_1 = 2(\bar{1})$ and $Z_2 = 2(\bar{1})$ for tolane. It was interesting to clear up the question of why the molecules of these compounds prefer these exotic packings over more common packings with one crystallographically distinct molecule in the unit cell. We tried monoclinic $P2_1/a$ packings with the lowest possible number of molecules in the unit cell as alternative (hypothetical) packings. In constructing these packings for biphenylene and tolane the molecules were placed on inversion centers, so that the number of molecules in the respective unit cells was equal to two [crystal symmetry $P2_1/a$, $Z = 2(\bar{1})$]. The hypothetical crystal of acenaphthene contained four symmetrically-related molecules in its unit cell with the basis molecule placed in a general position [crystal symmetry $P2_1/a$, $Z = 4(1)$]. That is, it was assumed that the acenaphthene molecules lose all of their symmetry elements in the crystal, while the biphenylene and tolane molecules retain their inversion centers. The above choice for the hypothetical crystal symmetries was governed by the fact that they are the most probable crystal symmetries for centrosymmetric and non-centrosymmetric molecules, respectively, based on the closest-packing principle and the actual distribution of organic molecular crystals over crystal symmetries [4.42].

In the search for the lowest-energy packing we varied all of the structural parameters whose variation was allowed by crystal symmetry. The observed packings were handled in the usual way, by finding the local energy minimum closest to the observed crystal configuration. The hypothetical packings were investigated with a global minimization technique, using

Table 4.3. Calculated lattice energies of acenaphthene, biphenylene and tolane in actual and hypothetical packings [kcal/mol] [4.41]

	Actual packing	Hypothetical packing
Acenaphthene	-19.1	-16.3
Biphenylene	-16.6	-16.2
Tolane	-17.9	-17.9

the valley method described in Sect. 4.2.3. The energy minima found in the global search were then refined using the Davidon-Fletcher-Powell algorithm. The results are given in Table 4.3. It is seen that the hypothetical packings of acenaphthene and biphenylene are 2.8 and 0.4 kcal/mol, respectively, less stable than the observed packings. For tolane, the calculations fail to distinguish between the real and hypothetical packings.

An unsuccessful attempt to predict an observed crystal structure can be found in recent papers by *Dzyabchenko* [4.43,44] concerned with crystalline benzene. Four alternative crystal symmetries ($Pbca, P2_1/c, P\bar{1}$ and $C2/c$) were tried. In each case the molecules were placed on the inversion centers. A thorough scanning of the configurational space has resulted in a variety of local minima. The deepest minima differed in energy by values of the order of 0.01 kcal/mol. Of the seven model potentials used, only the WS77 potential (Table 3.1) correctly predicted the observed packing ($Pbca$), while the other six yielded a global minimum corresponding to a $P2_1/c$ symmetry.

Other comparisons of real and hypothetical packings were made in [4.45–47]. *Hagler* and *Leizerowitz* [4.45] investigated the packing observed in the adipamide crystal and found it to be 0.5–1.0 kcal/mol more stable than two alternative hypothetical packings. *Derissen* and *Smit* [4.46] found that the observed chain structure of the acetic acid crystal was about 0.2 kcal/mol more stable than a hypothetical structure made up of hydrogen-bonded dimers.

Berkovitch-Yellin and *Leizerowitz* [4.47] compared the observed and hypothetical packings for the α- and β-polymorphs of oxalic acid and for formic acid. The hypothetical packings were generated by applying various symmetry operations to the hydrogen-bonded chains or layers really observed in crystals. In all the cases the observed packings proved to be substantially more stable than the alternative structures. Thus, in oxalic acid all hypothetical structures had energies more than 3 kcal/mol higher than the observed structures. For formic acid, all hypothetical structures were separated from the observed one by an energy gap of more than 5 kcal/mol.

In general, the work discussed in this section shows that the atom-atom potential method is, in most cases, capable of distinguishing between alternative packings. However, energy differences resulting from lattice-energy calculations are sometimes very small, and it is not yet clear whether these differences are indeed significant. We believe that it is too early to draw more definite conclusions because the number of pertinent examples is still very limited.

4.4 The Influence of Crystal Forces on the Molecular Conformation

In the current literature, the influence of crystal forces on the molecular conformation is being studied with an ever-increasing frequency. Practical interest in this problem mainly stems from the known difficulties involved in an experimental determination of the molecular conformation in gases, liquids and solutions. Considering the high accuracy and reliability of X-ray crystal-structure determinations, it would clearly be practical to have a means of predicting the molecular conformation in the gas phase or solution from the conformation in the crystal.

It can be seen from (4.21) that the equilibrium conformation of a molecule in a crystal is determined by a balance of intra- and intermolecular forces. If the conformational energy is expressed in an analytical form, as in the case of the empirical models discussed in Sect. 3.4, the determination of the molecular geometry in a crystal does not present any difficulties. It can easily be accomplished by minimizing the total (intermolecular and conformational) lattice energy, starting with the observed crystal configuration. The resulting molecular conformation is compared with the experimental conformation, and conclusions can then be drawn with respect to the reliability of the adopted model. Thereafter, the intermolecular force field is removed and the conformational energy is minimized to yield the free-molecule conformation.

Extensive calculations of crystal-state molecular conformations have been performed [4.19,48–56]. An analysis of these calculations has shown that empirical conformational-energy models combined with atom-atom model for the intermolecular potential provide a fairly reliable tool for predicting molecular conformations in crystals. To demonstrate the predictive power of these calculations, we shall briefly discuss some representative examples.

A classical example is the biphenyl molecule which possesses a soft internal degree of freedom (the twist angle τ between the phenyl rings). In the gas phase $\tau = 42^\circ$, while in the crystal the molecules are nearly planar ($\tau \approx 0$). The interplay between the intra- and intermolecular forces in the crystal was studied by *Casalone* et al. [4.57]. The conformational parameters that were treated as variables were the angle of twist τ, the bridge ($C_1 - C_1'$) bond length l and the ($C_1C_2H_2$) bond angle β [H_2 being the hydrogen atom closest to the bridge ($C_1 - C_1'$) bond]. The bond stretching and bond bending energies were evaluated from (3.70,71) with

$K_l = 2303.5\,\mathrm{kcal/(mol\,Å^2)}$, $l^0 = 1.483\,Å$, $K_\beta = 0.0377\,\mathrm{kcal/(mol\,degree^2)}$ and $\beta^0 = 120°$. The non-bonded interactions were treated with *Bartell*'s potentials [4.58]. The conformational energy also included a π-electron resonance contribution, which was evaluated by *Hückel*'s method [4.59].

For an isolated biphenyl molecule the energy was found to be a minimum at $\tau = 35°$. The crystal energy calculations were made by placing the molecular center of mass at the experimental position and varying only the internal parameters and orientations. At the minimum of the total lattice energy the molecule was planar, in full agreement with experiment.

It is interesting that the packing forces in a crystal of p-p′-bitolyl operate so that the relative orientation of the phenyl rings is close to that observed for an isolated biphenyl molecule ($\tau = 36$ and $40°$ for the two crystallographically distinct molecules in the unit cell [4.60]). Crystal-packing calculations similar to those for biphenyl have successfully reproduced this phenomenon to give $\tau = 35$ and $40°$ [4.60].

In a crystal of p-nitrobiphenyl, the crystal forces also affect the free-state orientation of the phenyl rings only slightly. In the minimum-energy crystal structure found by *Casalone* et al. [4.61] the angle of twist is equal to 32°, compared with the experimentally observed value of 33°.

Further, the crystal structure determination of 2-bromo-1,1-di-p-tolylethylene shows that the phenyl ring cis to the bromine atom is rotated 68° out of the plane of the ethylene, while the second phenyl is 24° out of the plane [4.62]. For an isolated molecule the minimum of the conformational energy occurs at 55° and 25°, respectively. However, when the molecule is placed in its crystal, the optimum angles of twist become 69° and 25°, respectively.

In general, comparisons of crystal- and free-state molecular conformations have shown, as expected, that crystal forces may pronouncedly affect molecular conformations only if a molecule possesses soft internal degrees of freedom. These are, almost exclusively, rotations about single bonds, which generally involve energy changes of the order of 1 kcal/mol. Bond stretching and bond-angle deformation are usually associated with much larger energy changes, which cannot be supplied by the weak intermolecular forces. Representative numerical values have been evaluated by *Warshel* et al. [4.63,64] for conjugated hydrocarbons. According to this work, the differences between the crystal-state and free-state molecular conformations are typically of the order of 0.001 Å for bond lengths, of 0.1° for bond angles, of 1° for rotations about double (C=C) bonds and of 10–20° for rotations about single bonds. Note that the distortions of torsional angles may, in principle, be even much larger because the crystal- and free-state conformations may correspond to different minima or, moreover, need not correspond to any minima of the torsional potential.

It is to be stressed that the principal aim of the work cited so far was to check the reliability of the adopted energy model and then to use this model to predict the gas-phase molecular conformation. In recent years, however, a few studies have been reported which attempt to explain, through model calculation, why the molecules in a crystal assume a particular conformation and do not assume some alternative conformation. This has been done, for example, by *Caillet* et al. [4.65] for the adrenaline tartrate crystal.

An essential conformational characteristic of the adrenaline molecule is the relative arrangement of the cationic head and the ring. This arrangement may be specified by two torsional angles, τ_1 and τ_2, shown in Fig. 4.5. The angle τ_1 describes the orientation of the plane of the side chain relative to that of the ring [$\tau_1 = 0$ if the two planes are parallel and the C_7-C_8 bond is cis with respect to C_1-C_6]. The other angle τ_2 describes the rotation of the cationic head about the C_7-C_8 bond.

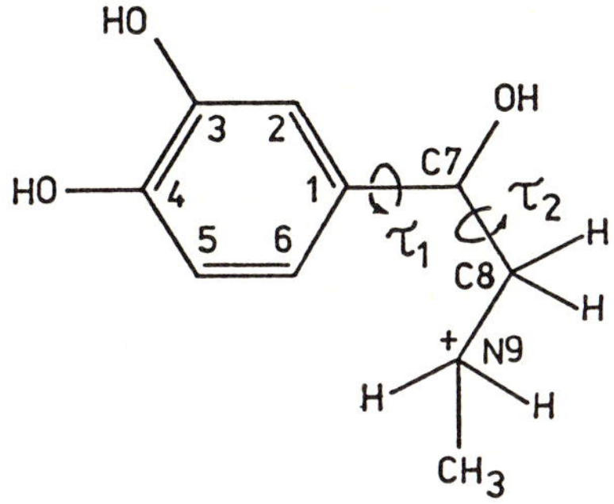

Fig. 4.5. Torsional angles τ_1 and τ_2 in adrenaline

A conformational energy map calculated by the PSILO method for the free adrenaline molecule [4.66] shows that the global minimum is at $\tau_1 = -90°$ and $\tau_2 = -60°$ (conformation G). There are also a few local minima, in particular, at $\tau_1 = 90°$ and $\tau_2 = 180°$ (conformation L) with an energy of about 3 kcal/mol higher than that of G.

In the crystal of adrenaline tartrate the adrenaline molecule is in a conformation with $\tau_1 = -3°$ and $\tau_2 = -179°$ (conformation C), which does not correspond to a local minimum on the conformational energy map (the energy of conformation C is 4 kcal/mol higher than that of G). To explain why adrenaline prefers, in a crystal, to assume the conformation C and not G or L, *Caillet* et al. [4.65] have resorted to crystal-packing calculations.

In the search for the most favorable packing of each of the three conformers they varied all of the crystal structure parameters, except the unit-cell angles and the number of molecules in the unit cell [4.65]. The latter were assumed to be equal to the experimentally observed values (conformer C). The starting positions of the tartrate and the aromatic ring of adrenaline were also taken from experiment, while the starting unit-cell lengths were increased somewhat to avoid too short contacts in the starting configuration. The intermolecular lattice energy was then minimized to give -213.4 kcal/mol for conformer C, -185.3 kcal/mol for G and -191.8 kcal/mol for L. By adding these values to the respective conformational energies, *Caillet* et al. [4.65] found the experimentally observed conformer C to be the most stable (the total energy of C is about 24 and 20 kcal/mol lower than those of G and L, respectively).

A critical reader can easily find two serious shortcomings in these calculations. The first is that the crystal configurations that were found by minimizing the intermolecular lattice energy cannot be considered to be equilibrium configurations because the soft internal degress of freedom, τ_1 and τ_2, were kept fixed during the minimization. For the particular crystal being considered, the fixation of the internal parameters is hardly justifiable. The values of the lattice energies obtained for the three conformers (−213.4, −185.3 and −191.8 kcal/mol) suggest that the intermolecular forces operating in the adrenaline tartrate crystal are strong and markedly sensitive to molecular conformation. What is even more important is that the differences between the calculated lattice energies are much greater than the height of the "barriers" separating the individual minima on the conformational energy map (~5 kcal/mol [4.66]). This implies that the intermolecular forces might well have significantly distorted the obtained configurations, to yield quite different structures and energies, had the internal degress of freedom been allowed to vary.

The second shortcoming inherent in the methodology adopted in [4.65] concerns the packings of the hypothetical conformers G and L. It is quite clear that the packings found for G and L, starting from the experimental structure of the adrenaline tartrate crystal, cannot be considered, with confidence, to be the lowest energy ones. To find out if this is indeed the case, it would have been necessary to try all possible crystal symmetries, find the global energy minimum for each and then select the deepest minimum. This is, of course, a hopeless task but for the moment no other approach can be suggested.

Thus, the results of *Caillet* et al. [4.65] cannot be used to explain the realization of conformer C in the crystal, but not conformers G and L. Similarly, it would be impossible to explain, using energy calculations, why biphenyl assumes a planar conformation in the crystal and does not assume the conformation corresponding to the minimum conformational energy. From the practical point of view, however, this is of minor importance because of actual interest is the question: "is a particular energy model reliable enough to predict the gas- or liquid-state molecular conformation?" A test for the reliability of the energy model is afforded by a simple comparison of the observed and calculated crystal-state molecular conformations, so that there is no real need of finding the optimum packings of hypothetical conformers.

4.5 The Atom-Atom Potential Method as an Aid in Determining Crystal Structures

In spite of the unquestionable success of the single-crystal diffraction methods in studying the structure of organic solids, there are many cases in which single-crystal work is impossible and another approach is needed. For instance, we refer to the studies of polymorphic transitions followed by shattering of single crystals of the original phase. Another important example is provided by polymeric substances, which rarely form sufficiently large and perfect single crystals. In such cases one has to be content with the raw data derived from powder or selected-area diffraction, infra-red absorption or Raman scattering. The atom-atom potential method may, in this case, prove to be of great utility, affording an independent criterion for assessing the reasonableness of a possible crystal configuration.

In this section we shall discuss some representative examples of how lattice-energy calculations can complement incomplete structural data on low-molecular-weight organic compounds (polymeric crystals will be discussed in the next section).

4.5.1 Determination of Crystal Structure from Unit-Cell Dimensions

We shall start our discussion with the low-temperature form of the pyrene crystal [4.31]. Crystals of pyrene, when cooled below the transition point at 120 K, generally shatter, which makes single-crystal diffraction work impossible. All experimental information about the the low-temperature phase (pyrene II) is actually reduced to a knowledge of the unit-cell dimensions and the space group of the crystal derived from X-ray powder and microelectron diffraction data [4.67]. According to these data, the observed transition produces only slight changes in the unit-cell parameters but no change in the space group $P2_1/a$. Some indirect information is provided by the fluorescence spectra of the crystal [4.68], indicating that the pyrene molecules remain grouped in pairs on going through the transition point.

In deriving the structure of pyrene II, *Ramdas* et al. [4.31] made the natural assumption that the molecular geometry did not change during the phase transition. The validity of this assumption for a crystal-structure calculation is beyond doubt in view of the high rigidity of the pyrene molecule. The equilibrium crystal configuration was found by minimizing the lattice energy with respect to the structural parameters whose variation was allowed by the crystal symmetry. During the minimization the unit-cell dimensions were kept fixed at the corresponding observed values.

An analysis of the minimum-energy crystal configuration has shown that the molecules in pyrene II are arranged in pairs, in agreement with the fluorescence spectra. Compared to the room-temperature phase (pyrene I), the overlap between paired molecules of pyrene II is greater, the separation of the molecular pairs being 3.53 Å in pyrene I and 3.44 Å in pyrene II. This also correlates with the fluorescence spectra which exhibit a shift in emission towards longer wavelengths upon cooling the sample below the transition point. A further argument in favor of the calculated crystal configuration is that the lattice energy of pyrene II is about 1 kcal/mol lower than that of pyrene I. This should indeed be so for a polymorphic modification stable at lower temperatures.

Thus, there are good reasons to believe that the calculated structure of pyrene II does correspond to reality.

Calculations very similar to those described above have been performed by *Halac* et al. [4.69] for the low-temperature phase (α) of 1,2,4,5-tetrachlorobenzene. Crystals of this phase are usually twinned, so that only the unit-cell dimensions and the number of molecules in the unit cell could be deduced from the available X-ray diffraction data [4.70]. A comparison of the unit-cell parameters of the α and the room-temperature β phase shows that the cells are very similar. It can therefore be concluded that the structures are closely related and may be derived from each other by a slight displacement of the molecules.

Three possible crystal symmetries compatible with the X-ray diffraction data were tried by *Halac* et al. [4.69] for the α phase. These were: (1) the $P1$ symmetry, with two crystallographically distinct molecules in the unit cell, (2a) the $P\bar{1}$ symmetry with two molecules occupying general positions related by an inversion center, and (2b) the $P\bar{1}$ symmetry, with two symmetrically distinct molecules occupying inversion centers.

The NQR spectra of the α phase, measured by *Monfils* [4.71], exhibit only four distinct peaks corresponding to four crystallographically distinct Cl atoms. This favors Models 2a and 2b because Model 1 involves eight symmetrically non-equivalent chlorine atoms.

The Raman spectra of the α phase [4.72] exhibit five bands in the external mode region. Considering that Model 1 predicts nine Raman-active external modes, while Models 2a and 2b predict only six, the centrosymmetrical Models 2a,b again seem more plausible.

The IR spectra of the α phase, measured by *Halac* et al. [4.69], do not exhibit any new bands in the internal mode region upon cooling the crystal through the $\beta \rightarrow \alpha$ transition point. The selection rules for the IR modes imply that this can only be so if the crystal symmetry is as in (2b).

Thus, it can be concluded from the spectroscopic evidence that the most probable crystal model for the α phase is (2b). The other two models cannot, however, be excluded because the absence of particular bands in the NQR, Raman and IR spectra may be the result of their extreme low intensity.

The final choice between the three possible crystal models was based on lattice-energy calculations. As in [4.31], the unit-cell dimensions were kept fixed at their observed values. The parameter sets that were used were the BA73 and RKW74 sets listed in Table 3.5.

The minimization of the lattice energy of Model 1 shifted the molecules to the inversion centers, thus giving Model 2b, in accordance with the spectroscopic evidence. Model 2a also proved to be uncomfortable for packing the molecules; the minimum energy of (2a) being about 12 kcal/mol higher than that of (2b).

One more argument in favor of Model 2b was obtained in a calculation of the Raman-active phonon frequencies for Models 2a and 2b. In both cases the crystal configurations used were those derived from the minimization of the lattice energy. The results for Model 2b were in remarkable agreement with experiment, while those for Model 2a showed appreciable discrepancies.

The above two applications of the atom-atom potential method refer to a comparatively simple case in which the unit-cell parameters of a crystal are known. The determination of the equilibrium crystal configuration does not involve any great difficulties because the fixation of the unit-cell dimensions at their observed values imposes rather severe constraints upon the possible molecular packings, and facilitates the search for the most probable packing. A more complicated situation is encountered with the triclinic modification of anthracene [4.73] in which only indirect and approximate knowledge concerning the geometry of the unit cell is available. The search for the unknown crystal structure was much more difficult and also involved a variation of the unit-cell parameters. We shall thoroughly discuss this case in Chap. 7, in connection with the solid-state photodimerization of anthracene.

A similar problem confronted *Cangeloni* and *Schettino* [4.74] in the determination of the structure of the low-temperature form of n-butane (n-butane III). It was only known with certainty, from the IR and Raman spectra, that the crystal possesses $P\bar{1}$ symmetry with one molecule per unit cell. Some indirect, though very important, evidence was provided by the intensity pattern of the three Raman-active librations, which was similar to the ones observed for the triclinic modifications of n-hexane and n-octane.

Considering that all triclinic even n-paraffin crystals possess very similar unit cells, except for the cell parameter c which varies regularly in increments of 2.4 Å in the successive members of the series, *Cangeloni* and *Schettino* [4.74] had good reasons to take the unit cell of n-hexane, with c shortened by 2.4 Å, as a trial unit cell of n-butane III. The subsequent search for the equilibrium structure of n-butane III was carried out in the standard way, by minimizing the lattice energy with respect to all of the structural parameters (including the unit-cell dimensions) whose variation was allowed by the observed crystal symmetry. The resulting crystal configuration had an energy of −7.4 kcal/mol, a value which correlates well with the experimental heat of sublimation. A further test for the validity of the derived configuration was made by using this configuration to calculate the lattice frequencies of the crystal. For the three Raman-active librations the agreement with experiment proved to be remarkably good, the largest deviation being $8\,\text{cm}^{-1}$.

4.5.2 Atom-Atom Potentials in Interpreting Single Crystal Diffraction Data

The atom-atom potential method has been very successful in determining crystal structures from very little experimental data. It is obviously of invaluable help in interpreting single-crystal diffraction data, both at the structure-solution and structure-refinement steps. Numerous supports for this view may be found in the literature [4.19,75–86].

We shall begin our discussion with the work by *Gavuzzo* et al. [4.78] on the crystal structure of coumarin. The experimental part of the work was carried out in a standard way using a single-crystal automatic diffractometer. A total of 750 independent reflections with intensities greater than 3σ were collected. The crystal structure was solved in three steps. In the first step an approximate molecular model was constructed, using standard bond lengths and angles and assuming that the molecule is planar. In the second step, the lattice energy hypersurface was scanned using 20° and 0.5 Å regular grids for the angular and translational parameters, respectively. In this step the unit-cell dimensions were kept fixed at their experimental values and the coumarine molecule was assumed rigid. The low-energy regions that were found were examined more thoroughly in the next step using finer grids for the structural parameters. Finally, two distinct minima noticeably deeper than the others were selected. Although the energies of these minima are almost equal, the calculation of the crystallographic discrepancy factors shows that one of the minima is much more preferable (the R values were 0.22 and 0.10).

The structure corresponding to the lower reliability factor was refined in a conventional manner to $R = 0.048$ for the 750 observed reflections. The refined crystal structure was in close agreement with the one found independently by *Kitaigorodsky* et al. [4.87] using the Patterson function.

Very similar calculations were performed by *Coiro* et al. [4.77] for 5α-androstan-3,17-dione. In solving the crystal structure the unit-cell parameters, crystal symmetry and molecular conformation were again kept unchanged, while the molecular rotations and translations were allowed to vary. A scanning of the parameter space revealed two satisfactory energy minima (minima I and II). An unexpected result was that the discrepancy factor of minimum I, the deepest energy minimum, is appreciably larger than that of minimum II ($R_{\rm I} = 0.31$, $R_{\rm II} = 0.22$). The reason for this became clear after the structure corresponding to II was refined and the energies of I and II were recalculated using the refined molecular conformation. With this new conformation, minimum II proved to be 2.5 kcal/mol deeper than minimum I; i.e., the wrong sequence of the original lattice energies can be attributed to errors in the starting molecular geometry.

4.5.3 The Case of Non-Rigid Molecule

Errors in the lattice energy, arising from the lack of an exact foreknowledge of the molecular conformation in a crystal, can be partly avoided by removing the assumption of molecular rigidity, at least for the softest intramolecular degrees of freedom. Here the empirical intramolecular-energy models discussed in Sect. 3.4 may be of great use. To illustrate this point, we turn to a crystal structure determination by *Adams* and *Ramdas* [4.84] of solution-grown 9,10-diphenylanthracene.

The 9,10-diphenylanthracene molecule cannot be treated as a rigid molecule because of the occurrence of two very soft internal degrees of freedom (the rotations of the phenyl groups about the C–C bonds). In the crystal, each half of the molecule is related to the other half by an inversion center, so that only one independent internal parameter had to be incorporated to specify the molecular conformation. Since the molecular translation was invariant by symmetry and the unit-cell dimensions were known from experiment, it only took a total of four parameters to specify any crystal configuration.

The search for the equilibrium crystal structure was carried out by minimizing the total crystal energy,

$$\phi = \phi_{\text{inter}} + \phi_{\text{nb mol}} + \phi_{\text{conj}}, \tag{4.159}$$

where ϕ_{inter} is the intermolecular lattice energy, $\phi_{\text{nb mol}}$ is the contribution from the non-bonded interactions between the phenyl rings and the anthracene fragment and ϕ_{conj} is an empirical potential intended to account for the loss in conjugation energy in going from a planar to a non-planar molecular conformation. The following analytical expression was adopted for ϕ_{conj}:

$$\phi_{\text{conj}} = -\phi_0 \cos^2 \tau, \tag{4.160}$$

where ϕ_0 is the conjugation energy of the planar molecule and τ is the dihedral angle between the planes of the anthracene fragment and the phenyl group. The conjugation energy ϕ_0 was assigned the value obtained by *Dewar* and *Harget* [4.88] in a SCF-MO treatment of biphenyl (8.6 kcal/mole).

For the observed space group $C2/c$ and four molecules in the unit cell there are two possible positions for the molecular centers within the cell: (1) at the crystallographic origin $(0,0,0)$, in which case the molecules are related by a diagonal glide; or (2) at (3/4, 3/4, 1/2), in which case the molecules are related by a simple c glide. Strictly speaking, the molecule may also be positioned on the crystallographic two-fold axis $(0, y, 1/4)$ with the molecular center at $y \approx 1/4$. However, such a possibility is very unlikely. Indeed, it would imply the retention of a two-fold axis and the loss of $\bar{1}$ molecular symmetry, which is never observed in organic crystals [4.42].

The best molecular packing was found using Model 2. The total potential energy is −64.5 kcal/mol, of which the intramolecular non-bonded contribution $\phi_{\mathrm{nb\,mol}}$ is 5.0 kcal/mol. The conjugation energy proved to be −1.1 kcal/mol, corresponding to $\tau \approx 70°$.

For the sake of comparison, the crystal structure of 9,10-diphenylanthracene was also solved using direct methods and then refined by least-squares in the conventional way. The basic structures obtained by these two methods are in close agreement. The differences can be attributed to the idealized molecular conformation adopted in the energy calculations.

4.5.4 Inclusion Compounds

The above studies by *Gavuzzo* et al. [4.78] and *Adams* and *Ramdas* [4.84] have shown that the atom-atom potential method may indeed be successfully used to solve a crystal structure. They cannot, however, convince the reader of *the necessity* to use the method. In both cases the crystal structure could also be determined as successfully applying traditional methods to interprete the single-crystal diffraction data.

Then again, many situations are encountered in which the traditional approach fails and only lattice-energy calculations can completely resolve the crystal structure. Very interesting examples of this can be found in the papers by *Giglio* et al. [4.79–81,83] concerned with the inclusion compounds of 3α, 12α-dihydroxy-5β-cholan-24-oic acid (DCA). In these compounds the host molecules of DCA form channels in which guest molecules are accommodated. The guest molecules occluded in these channels usually have one or more very soft external degree of freedom, so that the location of these molecules on the basis of diffraction data alone involves serious difficulties. Additional difficulties arise because the chains of the guest molecules generally possess a symmetry not coincident with the symmetry of the channel and, what is more, may also exhibit a substantial disorder.

A typical example is provided by the complex of DCA, ethanol and water [4.79]. The structure of this complex was first solved by direct methods from 300 reflections with $|E| \geq 1.42$. It gave all non-hydrogen atoms except one methyl carbon of DCA and all the atoms of ethanol and water. The space group was determined to be $P6_5$. Structure-factor calculations, with the unknown methyl carbon atom generated at the expected position, and a subsequent Fourier synthesis allowed the localization of only one more atom, hereafter denoted as O′, which may belong to either water or ethanol. Thereafter, the hydrogen atoms of DCA were generated at the expected positions and the coordinates of DCA and O′ were refined to $R = 0.09$. The subsequent difference synthesis revealed one new peak higher than $3\sigma(\varrho)$ (1.47 Å from O′), which could be attributed to the methylene carbon of ethanol.

The crystal structure derived at this step consisted of an array of parallel helices, each formed by the DCA molecules related by a 6_5 axis. The DCA molecules in a representative helix are mainly held together by hydrogen bonds, while the helices themselves are held together by van der Waals' forces. Each helix has a nearly cylindrical cavity, about 4 Å in diameter, in which ethanol and water molecules can be accommodated. One of the DCA oxygen atoms, located on the inner side of the helix, is about 2.7 Å from O′, a separation typical of a O–H...O hydrogen bond.

To determine the arrangement of the water and ethanol molecules inside the helices, potential-energy calculations were performed. Since the location of the oxygen atoms was already known (O′), it only remained to find the most favorable distribution of the water and ethanol oxygen atoms over the sequence of points generated from O′ by the 6_5 axis. For the water molecules, no rotational degrees of freedom appeared because the hydroxyl hydrogen atoms were neglected. The orientation of an ethanol molecule was only specified by the single variable φ, which describes the rotation of the methyl group about the bond joining O′ to the methylene carbon. (Recall that the coordinates of the latter were also known.) In all of the calculations the coordinates of the DCA molecules were kept fixed.

At first, a model containing one ethanol molecule per DCA molecule and no water molecules was investigated. This model, however, yielded unrealistically short contacts and was therefore rejected. Thereafter, several regular sequences with n ethanol molecules separated by a water molecule were tried. Only the sequences with $n \leq 5$ were considered, i.e., those sequences which resulted in no less than one water molecule per unit cell. Since the symmetry axis 6_5 could not be applied to these sequences, each ethanol molecule in the repeating unit had to be treated independently with its own rotational parameter φ_i.

The models with $n \geq 3$ had the same disadvantage as the model with one ethanol molecule per DCA molecule, i.e., all of the models were too tight and gave unreasonably short contacts. The best molecular packing was found with a model with $n = 2$, which corresponded to a DCA:ethanol:water ratio of 3:2:1. This finding was supported by gas-chromatographic measurements, which yielded a similar ratio, namely 3:1.95:0.95.

An acceptable molecular packing, though energetically less favorable, was also found for the sequence with $n = 1$. It was therefore concluded that the chains of the guest molecules may involve defects of the type ...E–E–W–E–E–W–(E–W)–E–E–W..., where E and W denote the ethanol and water molecules, respectively, and the defective section of the chain is enclosed in parentheses.

The best model determined by the potential-energy calculations was then used to complete the refinement of the crystal structure. The methylene carbon atoms of the ethanol molecules were introduced at the observed positions with an occupancy factor of 2/3. The methyl carbon atoms were given a weight of 1/3 and were placed at two distinct positions corresponding to the best φ_1 and φ_2 found in the potential-energy calculations. A least-squares refinement of the structure reduced the discrepancy factor to 0.08. Considering the complexity of the structure and the possible disorder, the resulting R-factor may be regarded as quite acceptable.

It is important to note that the application of the atom-atom potential method to inclusion compounds is not merely restricted to assisting the interpretation of diffraction data. Thus, potential-energy calculations may be successfully employed to predict what molecules can be accommodated in a given cavity and how these molecules will be arranged. For channel complexes, this is of particular importance in view of the ability of the guest molecules to undergo radiation-induced polymerization [4.89]. Hence, the atom-atom potential method opens up the way for the design of inclusion compounds which should, after irradiation and removal of the matrix compound, yield polymers with a prescribed structure.

Before concluding this section we should like to mention the recent work [4.90,91] which also deal with partially disordered structures. *Kim* and *Clementi* [4.90] have used the atom-atom potential method to locate the positions of water molecules in a 2:1 complex of proflavine and deoxycytidylyl-3′,5′-guanosine. Instead of the usual energy minimization, a Monte-Carlo method was used, simulating the thermal motion of the water molecules hydrating the complex. As in [4.73], not only have the positions of the water molecules been found but also the most probable number of water molecules in the unit cell.

Murthy and *Venkatesan* [4.91] investigated the crystal structure of 7-methyl-1(2)a,1(6)a,3(4)a-trihomocubane-1(6)a,3(4)a-dione. This structure exhibits an enantiomeric and positional disorder, with the result that the actually non-symmetric molecules occupy crystallographic centers of symmetry in the crystal. The aim of the lattice energy calculations was to see whether the diffraction experiment has revealed all possibilities of the disorder. The calculations of the lattice energies of the two enantiomers as a function of the molecular orientations have resulted in energy minima whose number and location are in good agreement with experiment.

4.6 Polymeric Crystals

4.6.1 Chain Conformation and Symmetry Constraints

The determination of the structure of crystalline regions in crystallizing polymers is one of the most important and promising applications of the atom-atom potential method. As has been mentioned earlier, large and perfect polymeric crystals suitable for single-crystal diffraction studies can only be prepared in a few cases, exclusively by the solid-state polymerization of single crystals of the parent monomers [4.89,92][6]. In all other cases one has to deal with rather imperfect crystallographic objects in which crystalline elements (crystallites) coexist with more or less amorphous regions. The fraction of the latter normally ranges from 20 to 80%. The crystallites are as a rule small, of the order of 10^2–10^3 Å in size, and much more defective than typical crystals of low-molecular weight compounds. These peculiarities inherent to polymers seriously complicate the solution of the structures.

The major source of information about the structure of crystallites is the wide-angle X-ray diffraction data of oriented polymeric samples (fibres or films). For the reasons indicated above, the diffraction patterns from such samples are of comparatively poor quality, which precludes the use of powerful crystallographic methods such as Fourier synthesis or the least-squares refinement of atomic parameters. The usual procedure in such cases has been to impose various stereochemical and packing constraints upon the crystal configuration and then to minimize the crystallographic discrepancy factor with respect to the remaining degrees of freedom.

The stereochemical constraints are generally incorporated by keeping the bond lengths and angles fixed at their standard values, although more sophisticated procedures are sometimes employed [4.94–99]. The use of stereochemical constraints significally reduces the number of variable structural parameters, thereby increasing the ratio of the observations to refinable parameters. Also, unlike the refinement of an atomic coordinate, the refinement of a conformational parameter within a stereochemically-constrained model usually involves the simultaneous displacement of more than one atom, which enhances the sensitivity of the R-factor to the variables of the model.

The packing constraints are ordinarily incorporated via the condition that the non-bonded contacts in a crystal configuration should not differ markedly from the respective sums of the van der Waals radii. This condition considerably reduces the volume of the parameter space to be scanned in the search for the optimum crystal configuration.

Unfortunately, even the incorporation of stereochemical and packing constraints frequently does not allow one to distinguish between competing models. In such cases the use of energetic criteria proves to be absolutely necessary.

The calculation of the lattice energy of a polymeric crystal exhibits some peculiarities associated with the requirements imposed by the crystal symmetry on the conformation of the polymer chain. An obvious symmetry requirement is that a polymer chain in a crystal should necessarily possess a translational symmetry with the period equal to the lattice period along the axis of the polymer chain. A higher symmetry may be observed when

[6] We shall not discuss the known biological substances, such as globular proteins, which do yield good single crystals [4.93].

the axis of the chain coincides with a screw axis of the crystal or passes through closed symmetry elements. Very often the polymer chains are assumed to possess a screw axis which is absent in the crystal itself. Thus, if a polymer chain has n monomer units per turn, it is usually assumed that the chain represents a regular helix of symmetry n_s $(s = 1, \ldots, n)$. Such an assumption is based on the chemical identity of the monomer residues, which should indeed lead to their conformational identity provided that the intermolecular forces operating in the crystal cannot significantly influence the macromolecular conformation. If the interchain interactions are comparable with the intrachain ones, the above assumption obviously becomes inadmissible and the chain symmetry should be chosen in complete accordance with the observed crystal symmetry.

In any case, in constructing model structures for polymeric crystals one should be able to generate chain conformations which possess helical or, at least, translational symmetry. The generation of such conformations is surprisingly difficult. To illustrate this point, let us consider a fragment of the helical chain of cellulose shown in Fig. 4.6a (the observed symmetry of the chain is 2_1). In the stereochemically-constrained model the bond lengths and angles are assumed rigid, so that the conformation of a residue is uniquely specified by the torsional angles τ_i $(i = 1, \ldots, 4)$ shown in Fig. 4.6a. The junction of two successive residues is described by the bond angle β and the two torsional angles τ' and τ''. If the angle β is kept fixed, only the

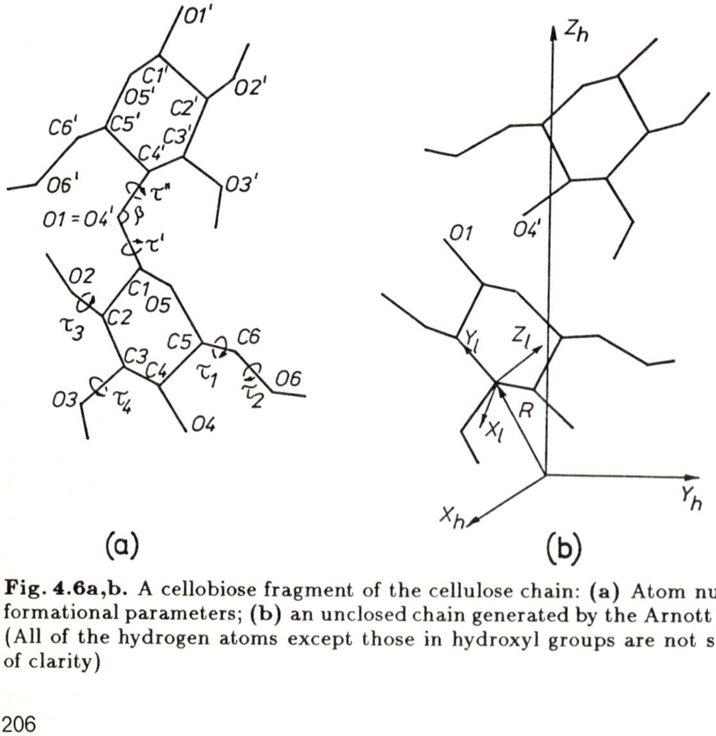

Fig. 4.6a,b. A cellobiose fragment of the cellulose chain: **(a)** Atom numbering and conformational parameters; **(b)** an unclosed chain generated by the Arnott algorithm [4.102]. (All of the hydrogen atoms except those in hydroxyl groups are not shown for the sake of clarity)

two parameters τ' and τ'' completely specify the junction. It is clear that τ' and τ'' cannot be chosen arbitrarily because the resulting chain will not necessarily have the desired helical symmetry. Hence, the conformational parameters at a junction cannot be treated as independent variables but should be derived from the parameters of the helix.

Explicit relations between the conformational parameters of a polymer chain and the respective helical parameters were first derived by *Sugeta* and *Miyazawa* [4.100]. These relations can be written in the general form

$$h = f_h(\tau, \beta, l), \tag{4.161}$$

$$\alpha = f_\alpha(\tau, \beta, l), \tag{4.162}$$

where α is the angle of rotation about the axis of the helix from an atom of one residue to the same atom of the following residue, h is the corresponding translation along the helix axis, f_h and f_α are rather complicated functions of the conformational parameters of a representative residue and the junction, and l, β and τ represent the bond lengths, bond angles and torsional angles in the backbone of the chain.

Using the Sugeta-Miyazawa equations, one can easily find the parameters h and α for a given set of conformational parameters τ, β and l. Unfortunately, except in a few specific cases, these equations cannot be converted into a suitable form that can be used to solve the reverse problem, i.e., the determination of some of the angles τ and β from the other conformational parameters and the parameters of the helix. It is this inverse problem which has to be solved in order to generate helical chains with a specified symmetry, period and chain conformation.

An attempt to solve the inverse problem has been undertaken by *McGuire* et al. [4.54] for the crystal packing of two poly-aminoacids. They used the Sugeta-Miyazawa equations in the direct form given in (4.161–162) and determined the unknown conformational parameters by successive iterations until the best fit between the observed values of h and α and those computed from (4.161,162) was obtained. However, such an approach suffers from two shortcomings. First of all, the necessity of an iterative procedure is displeasing from the computational point of view because the explicit expressions for f_h and f_α are very cumbersome and have to be solved repeatedly for any trial set of independent conformational parameters. The second shortcoming is that the suggested iterative procedure is not guaranteed to converge if the initial parameters are chosen too far from the final values.

Another procedure for generating helical chains was suggested by *Arnott* et al. [4.101,102]. Following this procedure one first calculates the atomic coordinates of an isolated residue using the internal conformational parameters τ, β and l. The atomic coordinates are computed in a local coordinate system $\{\boldsymbol{x}_l, \boldsymbol{y}_l, \boldsymbol{z}_l\}$ whose origin and axes are defined by any three successive atoms in the residue (Fig. 4.6b). A crystallographic coordinate system

$\{x_h, y_h, z_h\}$ is also introduced in which the z_h axis is coincident with the axis of the future helix. The local coordinate system is related to the crystallographic one by three Eulerian angles $\chi = \{\chi_1, \chi_2, \chi_3\}$ and a vector $\boldsymbol{R} = \{R_1, R_2, R_3\}$ joining the origins of the two systems. For any given values of χ and $\boldsymbol{R}$, the local atomic coordinates of the initial residue are transformed to the crystallographic coordinate system and then rotated by an angle α about the axis z_h of the helix and translated by a distance h along the axis to generate the following residue. Repeated applications of the rotation and translation generate the desired chain with specified values of α and h.

The procedure suggested by *Arnott* et al. [4.101,102] has an obvious deficiency: it does not ensure the closure of the chain. If the chain unit is defined as shown in Fig. 4.6b, with the junction atom belonging to two successive residues simultaneously (O1 $\equiv$ O4$'$), the application of the above algorithm does not ensure that the atoms O1 and O4$'$ are coincident. This is an inherent deficiency of the Arnott algorithm, suggesting that the parameters used as variables have not been appropriately constrained.

Unfortunately, the constraints necessary to ensure a closure of the chain are very difficult to derive in an explicit form. For this reason, in following the Arnott algorithm, one has to be content with implicit relationships between the variable parameters. The implicit constraint ensuring a closure of the chain is obvious:

$$|\boldsymbol{r}_{\mathrm{O1}} - \boldsymbol{r}_{\mathrm{O4'}}|^2 = 0. \tag{4.163}$$

In practical computations, this constraint can be readily incorporated using the penalty function

$$G = W|\boldsymbol{r}_{\mathrm{O1}} - \boldsymbol{r}_{\mathrm{O4'}}|^2, \tag{4.164}$$

which is added to the R-factor and then subjected to a minimization along with the latter. Additional constraints may also be used to yield the correct stereochemistry at the junction.

The necessity of dealing with penalty functions is, of course, a disadvantage of Arnott's algorithm, which considerably complicates the calculations. The main difficulty is associated with the choice of the weight factor W. If W is chosen too small, the polymer chain may remain open in the resulting optimum structure. If W is too large, the convergence of the search for the optimum structure will be too slow because the objective function of the search, $R + G$, is mainly sensitive to $|r_{\mathrm{O1}} - r_{\mathrm{O4'}}|^2$. Our experience has shown that it is sometimes very difficult to select a weight factor which simutaneously ensures the convergence of the search and the closure of the chain.

4.6.2 Variable Virtual Bond Method

The most convenient way of generating helical chains is provided by the variable virtual-bond method [4.103,104]. The details of the method are illustrated in Fig. 4.7 for the cellulose chain. The chain is represented as a succession of chemically and conformationally identical units defined so that the atom at the junction of two successive units belongs to both units simultaneously. The number of monomer residues in the chain unit is determined by the chain symmetry. For the cellulose chain (2_1 symmetry), the chain unit consists of a monomer residue and an additional glycosidic oxygen atom at the junction with the following residue.

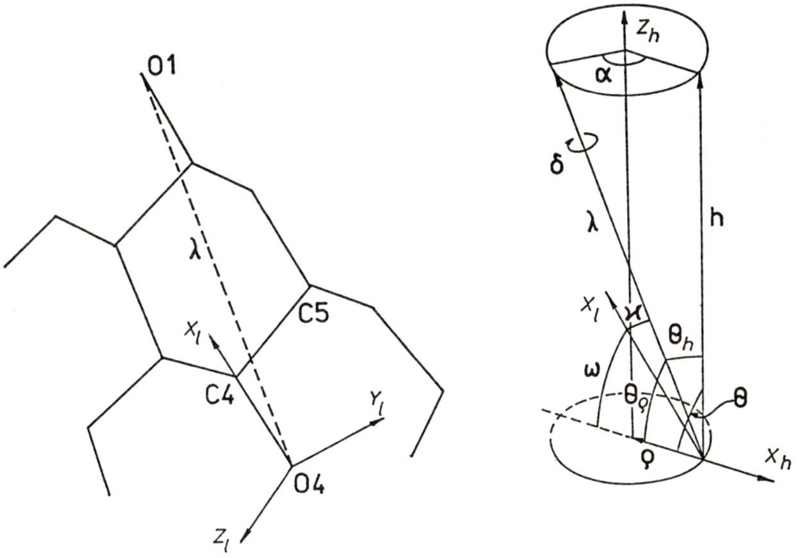

Fig. 4.7. An illustration of the variable virtual-bond method for generating helical chains

The construction of a chain starts with a computation of the atomic Cartesian coordinates of an isolated residue (Fig. 4.7). The local coordinate system, $\{x_l, y_l, z_l\}$, is chosen so that the origin is at the junction atom, O4, the x_l axis passes through C4, y_l lies in the plane of O4, C4 and C5 within the obtuse angle O4–C4–C5, and the z_l axis completes a right-handed orthogonal set. The atomic coordinates are determined from a specified set of bond angles β, bond lenghts l and torsional angles τ. Some of these conformational parameters may be treated as variables, while the others may be kept fixed at appropriate standard values. The algorithm needed to determine the atomic coordinates from the conformational parameters β, l and τ is widely known and may be found in [4.105].

Having computed the atomic coordinates, one can immediately find the length λ of the virtual bond (O4–O1) joining two successive junctions. In general, λ is not necessarily a constant but may depend on the variable conformational parameters of the chain unit. Given the helical parameters α and h, one can then calculate the helix radius ϱ:

$$\varrho = \left(\frac{\lambda^2 - h^2}{2(1 - \cos\alpha)}\right)^{1/2}. \tag{4.165}$$

The precise position of the helix axis with respect to the local coordinate system $\{\boldsymbol{x}_l, \boldsymbol{y}_l, \boldsymbol{z}_l\}$ is for the moment unknown.

Furthermore, one can determine the angles θ_h and θ_ϱ formed by the vector λ with the helix axis and the helix radius $\boldsymbol{\varrho}$, respectively

$$\cos\theta_\varrho = \varrho(1 - \cos\alpha)/\lambda, \quad 0<\theta_\varrho<90°, \tag{4.166}$$

$$\cos\theta_h = h/\lambda, \quad 0<\theta_h<90°. \tag{4.167}$$

At this step an additional parameter should be introduced to complete the set of parameters specifying the geometry of the chain. A convenient parameter is the angle δ which describes the rotation of the chain unit about the virtual bond. This angle may be chosen to be the dihedral angle between the plane defined by $\boldsymbol{\lambda}$ and $\boldsymbol{x}_l$ and the plane defined by $\boldsymbol{\lambda}$ and $\boldsymbol{\varrho}$; δ is equal to zero when $\boldsymbol{\varrho}$, $\boldsymbol{\lambda}$ and $\boldsymbol{x}_l$ are coplanar and is positive when $\boldsymbol{\varrho}$, $\boldsymbol{\lambda}$ and $\boldsymbol{x}_l$ form a right-handed set.

For any given δ one can find the angle ω formed by the vectors $\boldsymbol{x}_l$ and $\boldsymbol{\varrho}$:

$$\cos\omega = \cos\delta\,\sin\theta_\varrho\,\sin\kappa + \cos\theta_\varrho\,\cos\kappa, \tag{4.168}$$

where κ is the known angle between $\boldsymbol{\lambda}$ and $\boldsymbol{x}_l$ ($\cos\kappa = \lambda_x/\lambda$). A knowledge of ω completes the information necessary to find the position and orientation of the helix axis with respect to the local coordinate system $\{\boldsymbol{x}_l, \boldsymbol{y}_l, \boldsymbol{z}_l\}$. Indeed, from the vectors $\boldsymbol{\lambda}$, $\boldsymbol{x}_l$ and the angles θ_ϱ, ω one can immediately find the radius vector $\boldsymbol{\varrho}$ which joins the origin of the local coordinate system to the origin of the axial set of the helix $\{\boldsymbol{x}_h, \boldsymbol{y}_h, \boldsymbol{z}_h\}$. The direction of the z_h axis is uniquely specified by the vector $\boldsymbol{h}$, which can be readily found from the vectors $\boldsymbol{\lambda}$, $\boldsymbol{\varrho}$ and the angles θ_h, θ ($= 90°$). The choice of the $\boldsymbol{y}_h$ and $\boldsymbol{x}_h$ axes is, in principle, irrelevant. It is, however, convenient to choose an $\boldsymbol{x}_h$ that passes through the origin O4 of the chain and a $\boldsymbol{y}_h$ that completes a right-handed orthogonal set.

It is to be noted that the determination of the radius vector $\boldsymbol{\varrho}$ from $\boldsymbol{\lambda}$, $\boldsymbol{x}_l$, θ_ϱ and ω generally yields two distinct solutions corresponding to either a right- or left-handed set $\{\boldsymbol{\varrho}, \boldsymbol{\lambda}, \boldsymbol{x}_l\}$. The choice between the two solutions depends on the sign of δ: for $\delta>0$ we shall adopt the solution which yields a right-handed set $\{\boldsymbol{\varrho}, \boldsymbol{\lambda}, \boldsymbol{x}_l\}$, while for $\delta<0$ we shall adopt the solution corresponding to a left-handed set. Similarly, two distinct solutions are obtained for the vector $\boldsymbol{h}$, derived from $\boldsymbol{\lambda}$, $\boldsymbol{\varrho}$, θ_h and θ. However, only one of these solutions is consistent with the adopted symmetry of the chain. For $\alpha<180°$ one should choose the solution which yields a right-handed set $\{\boldsymbol{\varrho}, \boldsymbol{\lambda}, \boldsymbol{h}\}$, while for $\alpha>180°$ the solution should lead to a left-handed set (Fig. 4.7).

Thus, for any selected set of conformational parameters τ, β and l, helical parameters h and α, and angle δ, one can find the position and orientation of the helix axis with respect to the local coordinate system $\{\boldsymbol{x}_l, \boldsymbol{y}_l, \boldsymbol{z}_l\}$ by following the above algorithm. The following operations are then applied to the coordinates of the initial residue: (1) transformation to the Cartesian coordinate system $\{\boldsymbol{x}_h, \boldsymbol{y}_h, \boldsymbol{z}_h\}$; (2) transformation to cylindrical coordinates; (3) successive rotations about the $\boldsymbol{z}_h$ axis by an angle α and translations along the axis by distance h to generate a chain of a given length; and (4) inverse transformation of the cylindrical coordinates of the chain to the Cartesian coordinate system $\{\boldsymbol{x}_h, \boldsymbol{y}_h, \boldsymbol{z}_h\}$.

The application of the above algorithm results in a closed chain with predetermined symmetry, period, conformation of the repeating unit and rotation of the unit about the virtual bond. The only disadvantage of the algorithm is that the bond angle at the junction, C1–O1–C4′, cannot be treated as an independent variable and, hence, cannot be maintained at a desired value during the refinement of the crystal configuration. This disadvantage is not, however, very significant because the deviation of the angle from its normal value can readily be taken into account by introducing an appropriate deformation term in the intramolecular energy.

It is to be noted that in the variable virtual-bond method the torsional angles at the junction are also not independent variables of the chain, but are implicit functions of the helical parameters and of the conformational parameters of the monomer unit.

Generation of regular chains with a prescribed symmetry and period is the only methodological difference between the treatment of polymeric crystals and crystals of low-molecular-weight compounds. Having generated the chain, one computes the intrachain energy, then positions the chains in the unit cell and computes the interchain energy.

The intrachain energy per chain unit may be decomposed into the following terms,

$$E = E_{\text{unit}} + E_{\text{junct}} + E_{1-2} + E_{1-3} + \ldots, \tag{4.169}$$

where E_{unit} is the conformational energy of a representative unit, E_{junct} is the energy of all bond and torsional angles at the junction and E_{1-i} is the non-bonded interaction energy between the first and ith unit. If the unit cell contains crystallographically distinct chains, the intrachain energy should be calculated for each chain individually and then averaged according to (4.29).

4.6.3 Application to Cellulose

To demonstrate the utility of lattice-energy calculations in determining the structure of polymeric crystals, we shall now discuss the solution of the structural problem for cellulose II. For other examples the reader is referred to the work by *Scheraga* et al. on poly-L-alanine, poly(β-benzyl-L-aspartate) [4.54] and poly[β-(p-chlorobenzyl)-L-aspartate] [4.106], to the work by *Giglio* et al. on polyethylene, polyfluoroethylene [4.107], poly-isobutylene [4.108], poly-(L-azetidinecarboxylic acid) [4.109], poly-(vinyl-chloride) [4.110], kapton H [4.111], polymeric sulfur trioxide [4.112] and ethylene-butadiene copolymers [4.113] and to the work by *Sidorovich* et al. [4.114] on aromatic polyamide-polyimides.

Cellulose II is believed to be the most stable polymorph of cellulose. It can be obtained from native cellulose (cellulose I) either by a solution regeneration process or by mercerization which involves a swelling treatment with sodium hydroxide. Cellulose II has a monoclinic unit cell with dimensions $a = 7.93$, $b = 9.18$, $c = 10.34$ Å and $\gamma = 117.3°$ (average values over various preparations [4.115]). Due to the systematic absence of $00l$ reflections of odd order in the X-ray diffraction pattern, the space group of cellulose II is commonly assumed to be $P2_1$. The unit cell contains two crystallographically distinct 2_1 chains passing through the ab plane; one at the origin and the other at the center of the unit cell.

The model chosen by us [4.116] consists of an array of partially deformable chains packed together in accordance with the observed crystal symmetry and the unit-cell dimensions. In generating the chains the conformational parameters of the glucose rings were kept fixed at their average values [4.102]. All hydrogen atoms attached to the rings were generated at the expected positions using standard bond lengths and angles. Standard values were also used for the bond lengths and angles in the side groups. The only parameters that were allowed to vary in the residues were the torsional angles τ_i ($i = 1,\ldots,4$) describing the rotations of the methoxyl ($i = 1$) and hydroxyl ($i = 2,3,4$) groups. The particular bond sequences used in the definition of τ_i are: τ_i = C4-C5-C6-O6, τ_2 = C5-C6-O6-H, τ_3 = C1-C2-O2-H, τ_4 = C2-C3-O3-H (Fig. 4.6a). Each angle τ_i is equal to zero when the respective bond sequence, A-B-C-D, is cis. A counterclockwise rotation of the C-D bond when looking down the B-C bond represents a positive rotation. The orientation of the methoxyl group is also described in terms of the orientation of the C6-O6 bond relative to the C4-C5 and C5-O5 bonds: gg, gauche to C5-O5 and gauche to C4-C5 ($\tau_1 = -60°$); gt, gauche to C5-O5 and trans to C4-C5 ($\tau_1 = 180°$); and tg, trans to C5-O5 and gauche to C4-O5 ($\tau_1 = 60°$).

The monomer units were linked into chains of 2_1 symmetry using the variable virtual-bond method. The helical parameter h was kept fixed at $c/2$, while δ was allowed to vary.

The arrangement of the two symmetrically distinct chains in the unit cell is described by five parameters (φ^1, φ^2, p^1, p^2 and s) specifying the orientation, direction and relative shift of the chains. The angles φ^1 and φ^2 describe the rotations of the chains at the origin and center about their axes and are defined by the helix axes x_h and the crystallographic axis a' [$a' = a \cdot \cos(\gamma - 90°)$]. The position of a chain with O4 at $(0, -y, 0)$ for the origin chain and at $(1/2, 1/2 - y, z)$ for the center chain corresponds to $\varphi^1 = \varphi^2 = -90°$. A rotation from a to b represents a positive rotation.

The direction of a chain was defined to be positive ($p = 1$) when $z_{C1} > z_{C4}$ and negative ($p = -1$) otherwise. The O4 atom of the origin chain was always kept at $z = 0$, while that of the center chain was kept at $z = s$.

Altogether, there were 15 variable geometrical parameters: τ_i^k, δ^k, φ^k, p^k, and s, with $i = 1, \ldots, 4$ and k referring to either the origin or center chain.

The quality of a trial set of model parameters was assessed by computing the objective function

$$f = \phi + WR'', \tag{4.170}$$

where ϕ is the total potential energy of the system, R'' is the crystallographic reliability factor and W is a weighting factor.

The intermolecular contribution to ϕ was evaluated using the MKB74 parameter set of Table 3.1, supplemented by the KMN69 parameters of Table 3.3 for interactions involving oxygen atoms.

The intrachain energy, as given by (4.169), involves torsional potentials (3.72), bond-angle bending contributions (3.71), non-bonded interactions (3.2) and Morse potentials

$$\begin{aligned} f_{\text{H-bond}} = D\{ &\exp[-2n(r_{\text{O}\ldots\text{H}} - r^0_{\text{O}\ldots\text{H}})] \\ &- 2\exp[-n(r_{\text{O}\ldots\text{H}} - r^0_{\text{O}\ldots\text{H}})]\}, \end{aligned} \tag{4.171}$$

describing O–H...O interactions. The parametrization that was used to compute the terms of the intrachain energy was that of *Dashevskii* [4.117]. Only those terms which were affected by a variation of the conformational parameters τ_i and δ were included in the calculation.

The R''-factor in (4.170) was evaluated from

$$R'' = \left(\sum_{m=1}^{M} w_m |F_m^{\text{obs}} - F_m^{\text{calc}}|^2 / \sum_{m=1}^{M} w_m |F_m^{\text{obs}}|^2 \right)^{1/2}, \tag{4.172}$$

where F_m^{obs} and F_m^{calc} are the observed and calculated amplitudes of the structure factor, w_m is the weight applied to the mth observation and M is the number of observed structure amplitudes. Each F_m^{calc} is a function of the 15 geometrical parameters of the model, and a non-geometrical parameter – the average isotropic temperature factor B_T –

$$F_m^{\text{calc}} = K\left\{\sum_{hkl}\left[F_{hkl}^{\text{calc}} \exp\left(-B_T\varrho_m^2/4\right)\right]^2\right\}^{1/2} . \qquad (4.173)$$

The summation is carried out over all of the planes hkl that contribute to the mth reflection; ϱ_m is the reciprocal d spacing; and K is a scale factor.

The numerical values of F_m^{obs} and w_m for 29 observed and 11 non-observed reflections were taken from an X-ray diffraction study by *Kolpak* et al. [4.118] on mercerized cellulose. The scale factor was determined for each running set of F_m^{calc} so as to minimize (4.172).

The weighting factor W in (4.170) was chosen so as to make small changes in ϕ and R'' equally significant for the objective function. The considerations used in selecting W are as follows. Let us consider two alternative models, A and B, with the R''-factors R''_A and R''_B ($R''_A < R''_B$). Let us evaluate what differences in R'' are statistically significant. Each model is described by a total of 16 parameters (including B_T), six of which τ_2^k, τ_3^k, and τ_4^k ($k = 1, 2$), have a negligible effect on R''. Also, as will be shown below, two more parameters, p^1 and p^2, may be formally reduced to the single parameter p assuming three distinct values ($p = 0, \pm 1$) for one antiparallel and two parallel variants of the chain packing. Thus, the number of parameters important for the R'' factor is $16 - 7 = 9$. The number of observations (i.e., reflections) is 29. The hypothesis to be tested is: the 9 parameters corresponding to the poorer model B correctly describe the crystal structure. The dimension of the hypothesis is 9. The number of degrees of freedom is $29 - 9 = 20$. For the usual significance level of 1% we find, from *Hamilton*'s tables [4.119], that the hypothesis cannot be rejected if $R''_B/R''_A < 1.6$. Assuming the better model A to have an R''-factor of about 0.12 (a typical value for the best models of cellulose II), we can immediately see that a difference in R'' less than about 0.07 is insignificant.

On the other hand, we know that the accuracy of the atom-atom potential method in potential-energy calculations is typically of the order of 1 kcal/mol. It may, therefore, be roughly assumed that any two models are energetically equivalent if their energies differ be less than 1 kcal/mol. Thus, to make the objective function equally sensitive to small changes in ϕ and R'', the R''-factor in (4.170) should be given a weight of $1/0.07 \approx 15$ kcal/mol.

Formally, the search for the optimum model represents a search for the global minimum of the objective function f in 16-dimensional parameter space. Generally, this is a highly non-trivial problem. Fortunately, in our particular case the problem may be substantially simplified by using simple packing and symmetry considerations. Thus, based on the shape of the chain in the projection along c it is easy to see that a reasonable (non-overlapping) packing can only result when the glucose rings are nearly parallel to the ac plane. This corresponds to a variation of φ^k within rather narrow intervals about $\varphi^k \approx \pm 90°$ for $p^k = 1$ and $\varphi^k \approx 0$ and $180°$ for $p^k = -1$.

Furthermore, the parameter space to be scanned in the search for the global minimum may be reduced by resorting to the obvious symmetry relations

$$\begin{aligned} & f(\boldsymbol{\eta}^1, \varphi^1, p^1; \boldsymbol{\eta}^2, \varphi^2, p^2, s) \\ & \quad = f(\boldsymbol{\eta}^2, \varphi^2, p^2; \boldsymbol{\eta}^1, \varphi^1, p^1, -s) \\ & \quad = f(\boldsymbol{\eta}^1, \varphi^1 + 180°, p^1; \boldsymbol{\eta}^2, \varphi^2 + 180°, p^2, s) \\ & \quad = f(\boldsymbol{\eta}^1, \varphi^1, p^1; \boldsymbol{\eta}^2, \varphi^2 + 180°, p^2, s \pm \tfrac{1}{2}) \\ & \quad = f(\boldsymbol{\eta}^1, \varphi^1 \pm 180°, p^1; \boldsymbol{\eta}^2, \varphi^2, p^2, s \pm \tfrac{1}{2}), \end{aligned} \tag{4.174}$$

where $\boldsymbol{\eta}^k$ represents all conformational parameters of the kth chain, i.e., $\boldsymbol{\eta}^k = \{\tau_i^k, \delta^k\}$. Based on the symmetry relations, the subspace of the parameters φ^k, p^k and s may be subdivided into three crystallographically independent regions chosen as follows

$$\begin{aligned} & (1)\ p^1 = 1, \quad p^2 = -1, \quad -\tfrac{1}{2} < s \leq \tfrac{1}{2}, \quad \varphi^1 \approx -90°, \quad \varphi^2 \approx 0; \\ & (2)\ p^1 = p^2 = 1, \quad -\tfrac{1}{2} < s \leq \tfrac{1}{2}, \quad \varphi^1 \approx \varphi^2 \approx -90°; \\ & (3)\ p^1 = p^2 = -1, \quad -\tfrac{1}{2} < s \leq \tfrac{1}{2}, \quad \varphi^1 \approx \varphi^2 \approx 0. \end{aligned} \tag{4.175}$$

The direction parameters p^1 and p^2 may be formally reduced to a single polarity parameter $p = (p^1 + p^2)/2$; $p = 0$ corresponds to an antiparallel arrangement of the chains, while $p = \pm 1$ corresponds to parallel arrangements.

A further reduction of the computational work associated with the global search was possible due to the fact that reasonable values of δ lie within a rather narrow interval, $\delta \approx 45 \pm 10°$. Variation of δ outside of this interval led to either shortened O2...C6′ contacts or to unrealistic glycosidic bond angles [4.103].

Finally, the optimum value for the average isotropic temperature factor turned out to be close to 32 $Å^2$ and was only slightly dependent on the other parameters of the model.

Thus, the adopted model involved only nine "significant" parameters (τ_i^k and s) that had to be scanned throughout the entire range of their variation. These nine parameters were sampled from the corresponding parameter subspace using uniform random grids.

Each trial set of model parameters was refined by a local minimization of the objective function using *Powell*'s quasi-Newton algorithm [4.13]. The most favorable regions of parameter space were also explored by varying, in turn, one of the nine significant parameters, while allowing the rest of the 15 parameters to adjust to the minimum of f.

Before proceding with a calculation of the optimum crystal structure, it appeared reasonable to determine the most favorable conformations for an isolated chain of cellulose of 2_1 symmetry. In seeking the low-energy conformations all intrachain conformational parameters τ_i and δ, as well as the chain half-period h were allowed to vary. The six most favorable conformers that were found are depicted in Figs. 4.8–10. Their conformational and energetic characteristics are given in Table 4.4. All of the conformers contain a O5...HO3′ hydrogen bond with an energy of about 3.9 kcal/mol. The lowest-energy conformer, labeled *A*1, also contains a three-center (bifurcated)

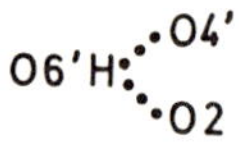

hydrogen bond which involves one donor and two acceptor oxygens. Bonds of this type will, hereafter, be referred to as two-acceptor bonds to distinguish them from two-donor

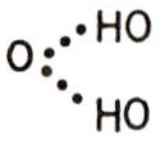

bonds.

The energetic preference of two-acceptor

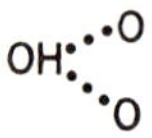

bonds over linear, OH...O, hydrogen bonds has recently been demonstrated by *Newton* et al. [4.120] in their ab initio $4-31G$ calculations on the water trimer. It was found that a second proton acceptor makes the system 1 to 2 kcal/mol more stable. In our case we found the energy difference to be 2.6 kcal/mol, which is comparable to the ab initio result.

A further argument in favor of the formation of two-acceptor hydrogen bonds in cellulose is the common occurrence of such bonds in crystals of carbohydrates. Accurate neutron-diffraction analysis of the structures of several carbohydrates [4.121] have shown that 25 of the 100 hydrogen bonds occurring in the crystals are of the two-acceptor type.

It can be seen in Table 4.4 that the methoxyl group of the *A*1 conformer is close to the tg position. A rotation of about 80° brings the group closer to the gg position and yields a new stable conformer, *A*2. In this conformer the

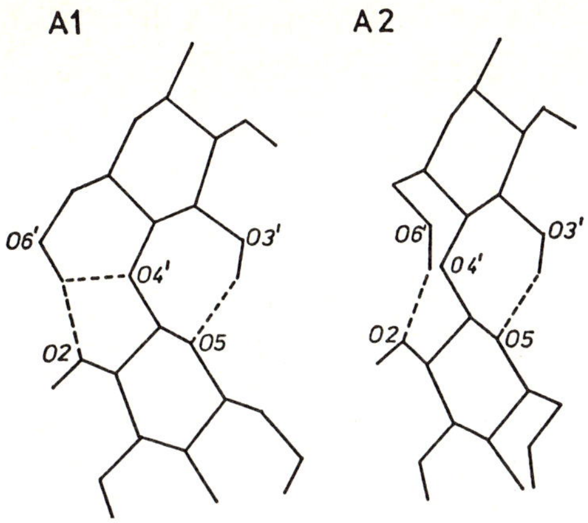

Fig. 4.8. Conformers A1 and A2 of an isolated cellulose chain. (Hydrogen bonds are indicated by broken lines)

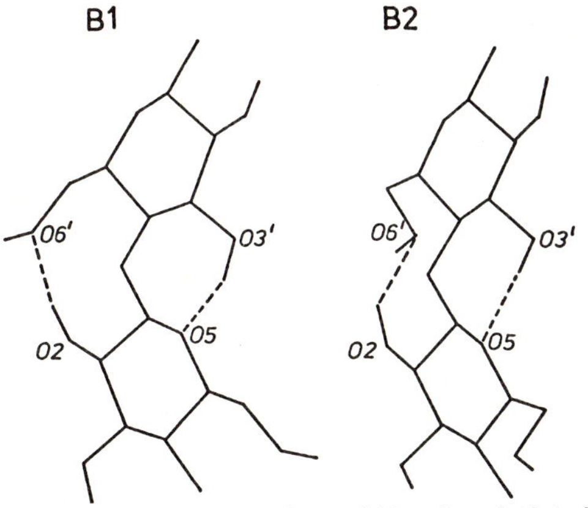

Fig. 4.9. Conformers B1 and B2 of an isolated cellulose chain

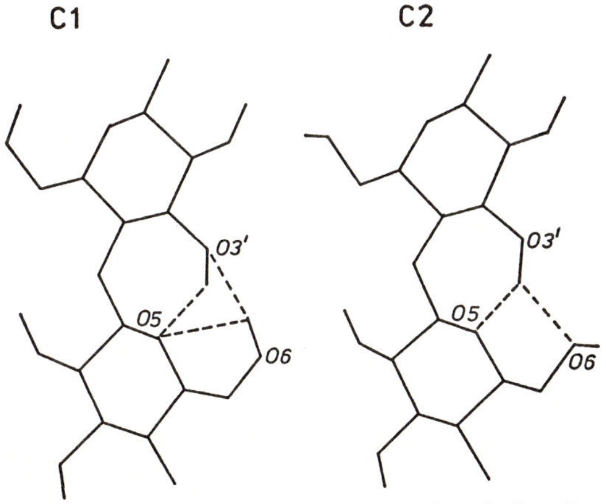

Fig. 4.10. Conformers C1 and C2 of an isolated cellulose chain

Table 4.4. Conformational parameters and energy for the six stable conformers of an isolated cellulose chain [deg, Å, kcal/mol] [4.103]

Conformer	τ_1	τ_2	τ_3	τ_4	δ	h	β	E
A1	64	-52	174	170	44	5.16	117	-4.8
A2	- 20	79	179	180	36	5.17	116	3.1
B1	68	167	-53	171	44	5.16	118	0.3
B2	- 17	184	38	180	36	5.17	116	-0.5
C1	-174	47	-56	-122	53	5.11	117	-0.1
C2	-168	179	-56	156	51	5.13	118	2.6

O6′H...O4′ bond breaks down, so that the two-acceptor bond transforms to a usual linear, O6′H...O2, bond with an energy of 3.8 kcal/mol.

The hydrogen bond between O6′ and O2 can also be formed through the hydrogen at O2, with O2 serving as a proton donor and O6′ as a proton acceptor. This is characteristic of the conformers of type *B* (Fig. 4.9). Again, two distinct orientations of the methoxyl group are possible. In the *B*1 conformer the methoxyl group is close to the tg-position, while in *B*2 it is closer to gg. The energy of the O2H...O6′ hydrogen bond is found to be 3.9 and 3.6 kcal/mol for *B*1 and *B*2, respectively.

There are also two stable conformers, labeled *C*1 and *C*2, whose methoxyl groups are within 13° of the gt-position (Table 4.4 and Fig. 4.10). The *C*1 conformer contains a two-acceptor,

$$\mathrm{O6H}\!:\!\begin{matrix}\cdot\cdot\,\mathrm{O3'}\\ \cdot\cdot\,\mathrm{O5}\end{matrix}\;,$$

and a two-donor,

$$\mathrm{O5H}\!:\!\begin{matrix}\cdot\cdot\,\mathrm{HO3'}\\ \cdot\cdot\,\mathrm{HO6}\end{matrix}\;;$$

bond, with energies of 4.7 and 4.9 kcal/mol, respectively. The *C*2 conformer contains a two-acceptor,

$$\mathrm{O3'H}\,\begin{matrix}\cdot\cdot\,\mathrm{O6}\\ \cdot\cdot\,\mathrm{O5}\end{matrix}\;,$$

bond with an energy of about 6 kcal/mol.

It is essential that the magnitude of the equilibrium period of the six stable conformers of the isolated chain is close to the experimentally observed value for cellulose. Also, free sites capable of participating in intermolecular hydrogen bonds exist in all of the conformers. For this reason, each of the six conformers may, in principle, exist in a crystal of cellulose II.

Table 4.5. Structural and energy parameters for different models of mercerized cellulose [deg, kcal/mol] [4.116]

Model	τ_1^1	τ_2^1	τ_3^1	τ_4^1	δ^1	φ^1	τ_1^2	τ_2^2	τ_3^2	τ_4^2	δ^2	φ^2	s	R"	ϕ	f
a_1	66	51	-151	174	43	-82	69	162	-53	163	46	- 3	0.106	0.179	-21.4	-18.7
a_2	68	-52	-152	176	42	-83	71	103	-57	163	45	- 2	0.102	0.183	-21.2	-18.5
a_3	65	57	- 79	177	42	-83	71	166	-52	175	46	- 2	0.122	0.178	-21.2	-18.5
a_4	- 63	173	-161	182	37	-90	69	163	-52	174	45	- 2	0.125	0.164	-20.1	-17.6
a_5	67	-50	-152	-55	42	-83	68	102	-56	161	47	- 6	0.100	0.186	-20.4	-17.6
a_6	178	164	- 97	174	44	-81	72	161	-56	167	42	1	0.226	0.124	-19.4	-17.5
a_7	67	-52	-153	176	42	-86	70	34	-54	72	45	1	0.116	0.208	-20.5	-17.4
p_1	64	64	- 70	172	45	-74	170	175	-77	169	45	-79	0.136	0.168	-19.4	-16.9
a_8	174	166	-123	177	42	-82	64	-65	177	166	43	- 1	0.217	0.126	-18.3	-16.4
a_9	177	162	-109	176	43	-82	76	172	183	167	43	0	0.226	0.121	-17.6	-15.8
a_{10}	-178	168	88	174	41	-83	-164	-62	-35	167	44	0	0.245	0.133	-17.1	-15.1
a_{11}	77	-48	-156	173	57	-61	102	162	-52	160	44	- 7	0.107	0.111	- 9.7	- 8.0
a_{12}	154	164	- 81	165	43	-80	157	80	-49	159	61	-18	0.234	0.108	- 9.4	- 7.8

In the search for the optimum crystal structure of cellulose II a large number of local minima for the objective function were found. The three deepest minima, with $f = -18.7$ to -18.5 kcal/mol, correspond to an antiparallel arrangement of the chains. The next four minima also correspond to an antiparallel packing, with f ranging from -17.6 to -17.4 kcal/mol. The deepest minimum for a parallel packing had an objective function equal to -16.9 kcal/mol. The parameters of these minima are given in Table 4.5 (minima a_1 to a_7 and p_1). Also included in the table are six additional minima which, according to the objective function, were not so good but had a low R''-factor (minima a_8 to a_{12}).

From Table 4.5 one can see that the best parallel model is markedly inferior to the best antiparallel models. The difference in the objective functions of the best parallel and antiparallel models allows one to reject the hypothesis that cellulose II may contain parallel chains [4.122].

It can also be concluded from Table 4.5 that it is indeed hardly possible to establish the structure of the polymer on the basis of diffraction data alone. The ratio R''_i/R''_j ($R''_i > R''_j$) does not exceed 1.6 for any of the models in Table 4.5 except a_7. This means that only the a_7 model can be rejected at a significance level greater than 1%, while the other models are statistically indistinguishable. A good illustration of the fact that low values of the R''-factor cannot be given much weight in interpreting the X-ray diffraction patterns of polymers is provided by the models a_{11} and a_{12}. These possess very low R''-factors but are unsatisfactory from the energetic point of view.

Of the best antiparallel models, we shall first discuss the a_6 model. With respect to its parameters, this model is similar to the models recently derived by *Kolpak* and *Blackwell* [4.115] (KB) and *Stipanovic* and *Sarko* [4.98] (SS) in a stereochemically-constrained refinement of the X-ray fiber diffraction data. Transformation of the parameters reported in [4.98,115] to our conventions yields $\tau_1^1 = 176°$, $\tau_1^2 = 80°$, $\varphi^1 = -91°$, $\varphi^2 = 2°$ and $s = 0.211$ for the SS model and $\tau_1^1 = 186°$, $\tau_1^2 = 70°$, $\varphi^1 = -90°$, $\varphi^2 = 6°$ and $s = 0.216$ for the KB model. These values are not too far from our values for the a_6 model.

Using the notation of *Stipanovic* and *Sarko* [4.98], the a_6 model may be classified into gt+tg models. The symbols gt and tg specify the positions of the methoxyl groups in the origin and center chain, respectively. The projections of the structure along the c and b axes according to the a_6 model are depicted in Fig. 4.11. The structure represents an array of alternating sheets parallel to the ac plane; each made up of translationally equivalent chains. The center chains, which form the (020) sheet, exhibit a conformation similar to that of $B1$ of the isolated chain. The sheet contains two intramolecular, $O5_4 \ldots HO3_3$ (3.66) and $O2_4H \ldots O6_3$ (3.98) hydrogen bonds and an intermolecular, $O6_3H \ldots O3_3^a$ (3.98), one[7].

The conformation of the chain in the (010) sheet is not typical of an isolated chain. It involves only one intramolecular, $O5_1 \ldots HO3_2$(3.96), hydro-

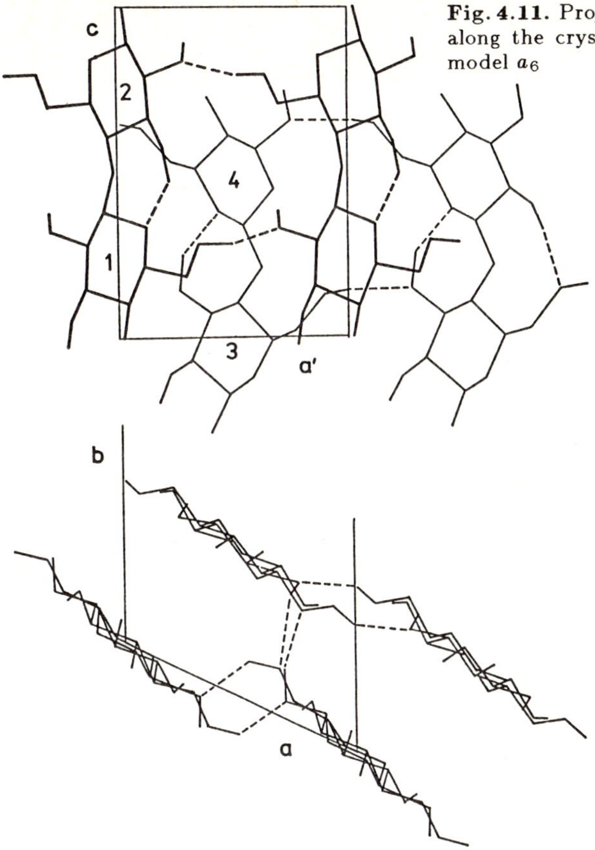

Fig. 4.11. Projections of the cellulose chains along the crystallographic axes c and b for model a_6

gen bond, while the methoxyl group is involved in a strong intermolecular, $O6_1H...O2_1^a(3.81)$ intrasheet bond.

Neighboring sheets are linked together through a strong $O2_4...HO2_1^a$ (3.26) bond and a weak $O3_3H...O6_1^b(1.65)$ bond. Thus, the proton at $O3_3$ is involved in a two-acceptor,

$$O3_3H \begin{array}{l} \cdot\cdot\, O6_1^b \\ \cdot\cdot\, O5_4 \end{array} ,$$

hydrogen bond.

In comparison to the KB model, the a_6 model possesses one extra intersheet bond, $O3_3H...O6_1^b$. This bond also appears in the SS model. The

[7] The symbols used to label the hydrogen bonds are the following: an atom subscript refers to one of the four basis residues shown in Fig. 4.11; an atom superscript indicates a translation to be applied to the basis residue and the number in parentheses represents the energy of the bond in kcal/mol. For each model we only present crystallographically distinct bonds. In selecting such bonds obvious symmetry relations such as $O_1H...O_1^{-a} = O_1^aH...O_1$, $O_3...HO_2^{b-c} = O_4^{-a}...HO_1$, etc., were used.

latter, however, exhibits an additional intersheet bond, $O6_3H\ldots O3_1^a$, which is involved in a two-acceptor,

$$O6_3H \begin{array}{l} \cdot\cdot O3_1^a \\ \cdot\cdot O3_3^a \end{array}$$

bond but is absent in the a_6 model. We attempted to find a structure in which such a bond should occur along with the other hydrogen bonds. The attempt, however, was unsuccessful. The trial structures with the hydrogen-bonding network of the SS model were all unstable and transformed to the a_6 model after a refinement.

Contrary to the findings of *Stipanovic* and *Sarko* [4.98] who reported that the gt+tg models were significantly favored over any of the possible O6 rotationally mixed models (i.e., those models whose origin and center chains had their methoxyl groups in different positions), we have found a model of the gg+tg type, that possesses a low potential energy and quite an acceptable R''-factor. This model is designated in Table 4.5 as a_4.

Quite an unexpected result was obtained with the models a_5 and a_7 (both are of the tg+tg type). A characteristic feature of these models is the absence of any intramolecular $O5\ldots HO3'$ bonds in one of the sheets. Usually, such structures are a priori excluded from consideration, with the argument that a breakdown of the $O5\ldots HO3'$ bond necessarily leads to an irreparable loss in energy. However, a calculation shows that this breakdown may be compensated for by new intermolecular bonds. In the a_5 model the liberated proton at $O3_1$ is involved in a $O3_1^aH\ldots O6_3$ (2.39) bond, while in the a_7 model the proton at $O3_3$ forms a two-acceptor,

$$O3_3^aH \begin{array}{l} \cdot\cdot O6_1^{a+b} \\ \cdot\cdot O6_3 \end{array},$$

hydrogen bond.

We now turn to the a_1 model corresponding to the global minimum of the objective function. According to the positions of the methoxyl groups, this model may be classified as a tg+tg model. Two projections of this model are shown in Fig. 4.12. The (020) sheet of the a_1 model is similar to the (020) sheet of the a_6 model. The chains in the (010) sheet exhibit a conformation similar to the one in $A1$, with characteristic hydrogen bonds: $O5_1\ldots HO3_2$ (3.94) and

$$O6_2H \begin{array}{l} \cdot\cdot O4_2\,(2.65) \\ \cdot\cdot O2_1\,(3.89) \end{array}.$$

There are no intermolecular intrasheet bonds in this sheet.

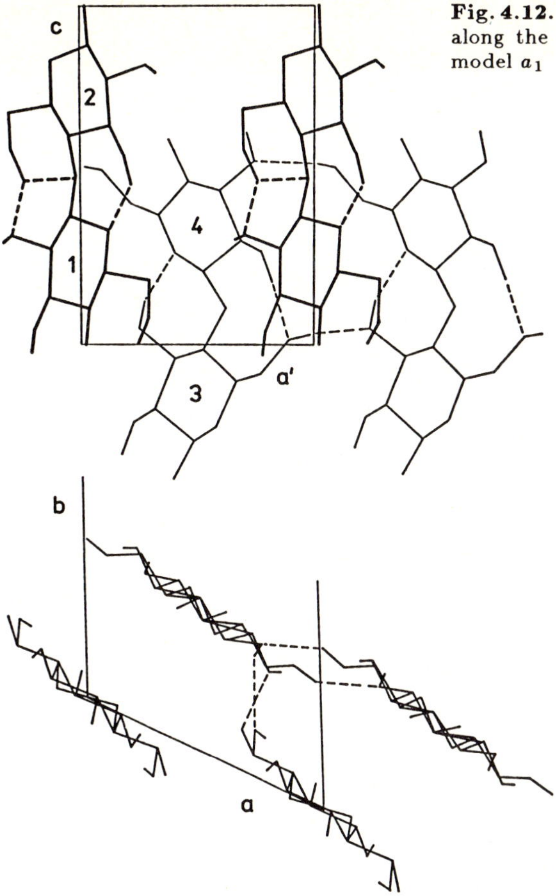

Fig. 4.12. Projections of the cellulose chains along the crystallographic axes *c* and *b* for model a_1

The intersheet variety of hydrogen bonds in model a_1 is repesented by $O2_1^aH...O2_4$ (3.29) and $O3_3H...O6_1^b$ (1.57). The proton at $O3_3$ is involved in a two-acceptor,

$$O3_2H \begin{matrix} \cdot\cdot O5_4 \\ \cdot\cdot O6_1^b \end{matrix} ,$$

bond.

The a_1 model differs from the a_2 and a_3 models mainly in the orientation of the HO6 and HO2 hydroxyl groups. Thus, the a_2 model may be derived from the a_1 model by a rotation of $HO6_2$ through $\sim 60°$ (see τ_2^2), and from model a_3 by a rotation of $HO6_1$ and $HO2_1$ through ~ 108 and $\sim 70°$, respectively (see τ_2^1 and τ_3^1). In the former case (a_2) the topography of the hydrogen-bonding network remains unaltered. The only change is a displacement of the proton of the $O6_2H...O3_2^a$ bond to a new position relative to the line joining the acceptor and donor oxygen atoms. In the lat-

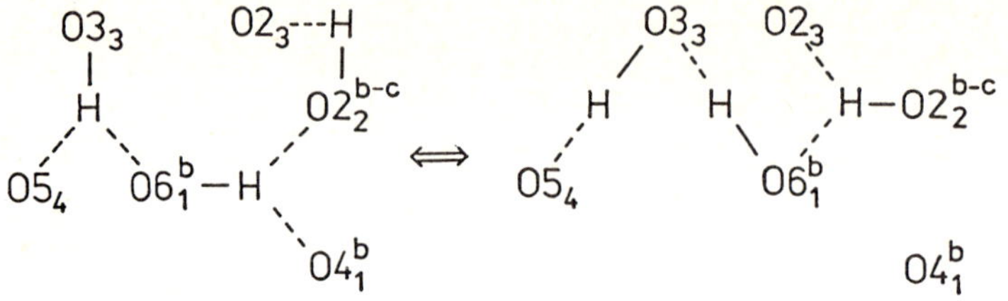

Fig. 4.13. Changes in the hydrogen-bonding network of cellulose II in going from the a_1 to the a_3 model

ter case (a_3) the hydrogen-bonding network undergoes substantial changes. These changes are shown schematically in Fig. 4.13.

It is essential that the three models, a_1, a_2 and a_3, are nearly equivalent in energy. A calculation of the barriers to the transitions $a_1 \rightarrow a_2$ and $a_1 \rightarrow a_3$, performed by successively varying the transition parameters τ_2^2, τ_2^1, and τ_3^1, while optimizing the other parameters, yields values as small as 0.2 and 1.0 kcal/mol, respectively (Fig. 4.14). It appears that all of the three structures may exist in the crystal at the same time and transform into one another by thermal migration of the protons. In other words, the models a_1, a_2 and a_3 may be treated as a single model (a_0) with a mobile hydrogen-bonding network.

The occurrence of soft degrees of freedom and, as a result, of several distinct conformational states makes the model a_0 not only advantageous from the energetic but also from the entropic point of view.

A comparison of the a_0 model with the a_6 model, an analogue of the KB and SS models, shows that a_0 possesses an acceptable R''-factor and is about 2 kcal/mol more stable. This energy difference seems to be quite significant since it refers to different modifications of the same substance (Sect. 4.3).

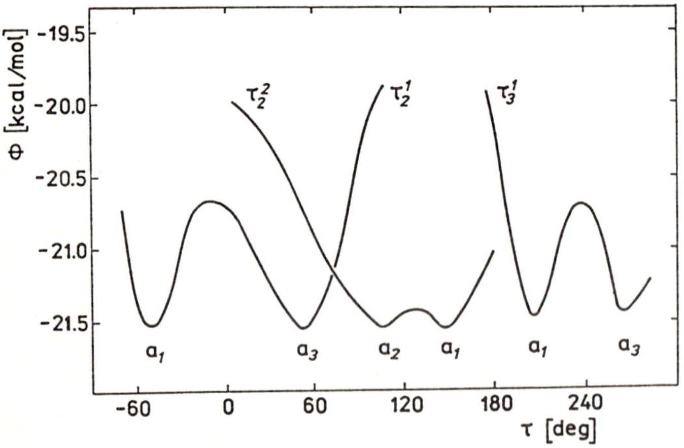

Fig. 4.14. Energy barriers of the $a_1 \leftrightarrow a_2$ and $a_1 \leftrightarrow a_3$ transitions in cellulose II

A further argument in favor of a_0 over a_6 is provided by a comparison of the structural amplitudes of the 002 and 004 meridional reflections. Due to the difficulties encountered in applying the Lorentz and polarization corrections to the meridional reflections, their intensities have not been estimated quantitatively [4.118] and have not been included in the R''-factor. These reflections have been found to be moderately weak and strong, respectively, which qualitatively agrees with model a_0 ($|F_{002}|/|F_{004}| = 1.5/26.3$) but disagrees with a_6 ($|F_{002}|/|F_{004}| = 28/22$).

To summarize, a scanning of the configurational space of a 16-parameter model of cellulose II reveals a variety of structures consistent with the available X-ray diffraction data. The use of an energy criterion based on the atom-atom approximation to the potential energy enables us to choose one antiparallel model (a_0) as the most probable. The remarkable feature of this model is the ability of its hydrogen-bonding network to easily undergo a rearrangement through a thermal migration of protons. Although the available experimental data are too meager to substantiate the optimum model in full detail, the success of the atom-atom potential method in describing the structure and energetics of organic crystals lends much weight to the predicted structure.

5. Lattice Dynamics

If a lattice in mechanical equilibrium is disturbed, every particle in the lattice undergoes certain vibrations about its equilibrium position. The frequencies, amplitudes and directions of movement of all particles can be found using the Born-von Karman formalism, provided that the interacting forces operating in the system are known.

In molecular crystals the forces between individual atoms of a molecule are much stronger than the forces between molecules. As a result, the vibrational spectrum of a crystal exhibits two more or less distinct bands. The lower-frequency band involves intermolecular or external vibrations, in which the molecules move essentially as rigid bodies. The higher-frequency vibrations represent internal vibrations, very similar to those occurring in an isolated molecule.

A knowledge of the intermolecular interaction energy as a function of the atomic coordinates allows one to predict the characteristics of the external modes and to evaluate the effect of crystal field on the intramolecular vibrations.

The dynamics of a lattice manifests itself in a variety of physical phenomena, such as inelastic neutron scattering, X-ray diffraction, infrared absorption and Raman scattering. By comparing the observed dynamical properties with those predicted from lattice-dynamical model calculations, one may assess the reliability of a particular model potential. On the other hand, the model calculations are of great value in interpreting the results of experimental measurements, even if the model potential that is used is not quite correct.

5.1 General Theory

A general theory of lattice dynamics has been formulated by *Born* and *von Karman* as early as 1912 [5.1]. In applying this theory to an organic molecular crystal one treats the crystal as a complex lattice of point masses associated with individual atoms. The position of an atom a in a unit cell l is specified by the vector $\boldsymbol{r}^{al}$ with the coordinates r_{α}^{al} ($\alpha = 1, 2, 3$) in a convenient crystal-fixed Cartesian coordinate frame. Given the position vector of the unit cell l, $\boldsymbol{r}^{l}$, and the position vector of the atom a with respect to the cell, $\boldsymbol{r}^{a}$, we have

$$\boldsymbol{r}^{al} = \boldsymbol{r}^{l} + \boldsymbol{r}^{a}. \tag{5.1}$$

The displacement of the atom from its equilibrium position, $\bar{\boldsymbol{r}}^{al}$, is described by the vector

$$\boldsymbol{u}^{al} = \boldsymbol{r}^{al} - \bar{\boldsymbol{r}}^{al} \quad . \tag{5.2}$$

In the neighborhood of the crystal configurational equilibrium the potential energy may be expanded in a Taylor series in u_α^{al}, i.e.,

$$\phi = \phi_0 + \sum_{\alpha al} \phi_\alpha^{al} u_\alpha^{al} + \frac{1}{2} \sum_{\alpha al} \sum_{\alpha' a' l'} \phi_{\alpha\alpha'}^{ala'l'} u_\alpha^{al} u_{\alpha'}^{a'l'} + \dots , \tag{5.3}$$

where

$$\phi_\alpha^{al} = \left(\frac{\partial \phi}{\partial u_\alpha^{al}} \right)_0 , \quad \text{and} \tag{5.4}$$

$$\phi_{\alpha\alpha'}^{ala'l'} = \left(\frac{\partial^2 \phi}{\partial u_\alpha^{al} \partial u_{\alpha'}^{a'l'}} \right)_0 . \tag{5.5}$$

The coefficient ϕ_α^{al} represents the αth component of the force on atom al. Since the expansion is about the equilibrium configuration, which is indicated by a subscript 0 in (5.3–5), all of the forces ϕ_α^{al} should vanish. In the harmonic approximation all terms in (5.3) of order greater than two are neglected. Choosing ϕ_0 as the zero of energy, we obtain

$$\phi = \frac{1}{2} \sum_{\alpha al} \sum_{\alpha' a' l'} \phi_{\alpha\alpha'}^{ala'l'} u_\alpha^{al} u_{\alpha'}^{a'l'} \quad . \tag{5.6}$$

The kinetic energy of the system, T, can also be expressed in terms of the same displacement coordinates. In the harmonic approximation, T can be readily shown to be given by

$$T = \frac{1}{2} \sum_{\alpha al} m_a (\dot{u}_\alpha^{al})^2 , \tag{5.7}$$

where $\dot{u}_\alpha^{al}$ denotes a differentiation of u_α^{al} with respect to time.

For an infinite crystal, both (5.6 and 7) represent divergent expressions. To overcome this difficulty, the lattice is subdivided into identical blocks, each containing $N = L^3$ unit cells. It is then assumed that the motion of the atoms in individual blocks is identical in each block. That is,

$$\boldsymbol{u}^{al} = \boldsymbol{u}^{al'} \quad , \tag{5.8}$$

for any two cells l and l' which occupy equivalent positions in different blocks. The condition (5.8) is usually called the cyclic boundary condition.

Under such condition, the summation over l in (5.6,7) may be restricted to the N unit cells of a representative block. There are now only $n_c N$ distinct dicplacement vectors $\boldsymbol{u}^{al}$ to be found (n_c being the number of atoms per unit cell).

The equations of motion for the atoms in a crystal can be readily derived from (5.6,7) using Hamilton's equations

$$m_a \ddot{u}_\alpha^{al} = - \sum_{\alpha' a' l'} \phi_{\alpha\alpha'}^{ala'l'} u_{\alpha'}^{a'l'} \quad . \tag{5.9}$$

Taking advantage of the periodicity of a crystal, the solution of (5.9) in the form of a traveling wave is

$$u_\alpha^{al} = u_\alpha^a(\boldsymbol{q}) \exp\{\mathrm{i}[\boldsymbol{q}\cdot\boldsymbol{r}^l - \omega(\boldsymbol{q})t]\}. \tag{5.10}$$

Such a wave is formally treated as a quasi-particle, a phonon, whose wave vector is $\boldsymbol{q}$. The N distinct values of $\boldsymbol{q}$ are to be found in the first Brillouin zone of the reciprocal lattice.

Substituting (5.10) into (5.9) yields

$$m_a \omega^2(\boldsymbol{q}) u_\alpha^a(\boldsymbol{q}) = \sum_{\alpha' a' l'} \phi_{\alpha\alpha'}^{ala'l'} u_{\alpha'}^{a'}(\boldsymbol{q}) \exp[\mathrm{i}\boldsymbol{q}(\boldsymbol{r}^{l'} - \boldsymbol{r}^l)]. \tag{5.11}$$

This equation can be rewritten as

$$\omega^2(q) w_\alpha^a(\boldsymbol{q}) = \sum_{\alpha' a'} D_{\alpha\alpha'}^{aa'}(\boldsymbol{q}) w_{\alpha'}^{a'}(\boldsymbol{q}), \quad \text{where} \tag{5.12}$$

$$w_\alpha^a(\boldsymbol{q}) = (m_a)^{1/2} u_\alpha^a(\boldsymbol{q}) \tag{5.13}$$

are the components of the mass-weighted displacement vectors and $D_{\alpha\alpha'}^{aa'}$ is the modified dynamical matrix defined by

$$\begin{aligned} D_{\alpha\alpha'}^{aa'}(\boldsymbol{q}) &= (m_a m_{a'})^{-1/2} \sum_{l'} \Phi_{\alpha\alpha'}^{ala'l'} \exp[\mathrm{i}\boldsymbol{q}\cdot(\boldsymbol{r}^{l'} - \boldsymbol{r}^l)] \\ &= (m_a m_{a'})^{-1/2} \sum_{l'} \phi_{\alpha\alpha'}^{aa'l'} \exp(\mathrm{i}\boldsymbol{q}\cdot\boldsymbol{r}^{l'}). \end{aligned} \tag{5.14}$$

In (5.14) we have used the invariance of the force constants with respect to lattice translations,

$$\phi_{\alpha\alpha'}^{ala'l'} = \phi_{\alpha\alpha'}^{a0a'(l'-l)} \quad , \tag{5.15}$$

as well as the fact that a summation over $l' - l$ is equivalent to a summation over l'. Following the convention of the previous chapter, we have dropped the superscript $l = 0$ labeling the unit cell at the origin.

Equation (5.12) represents a set of linear algebraic equations in $w_\alpha^a(\boldsymbol{q})$. The condition that this set has a non-trivial solution is

$$|D_{\alpha\alpha'}^{aa'}(\boldsymbol{q}) - \omega^2(\boldsymbol{q})\delta_{\alpha\alpha'}\delta_{aa'}| = 0. \tag{5.16}$$

This determinantal equation is of the order $3n_c$ and yields $3n_c$ eigenvalues $\omega^2(\boldsymbol{q}j)$ ($j = 1,\ldots,3n_c$). As functions of $\boldsymbol{q}$ the frequencies $\omega(\boldsymbol{q}j)$ form $3n_c$ dispersion hypersurfaces in the reciprocal space. Substituting each $\omega^2(\boldsymbol{q}j)$ back into (5.12) and solving the resulting set of equations with respect to $w_\alpha^a(\boldsymbol{q})$ yields $3n_c$ eigenvectors, $\boldsymbol{w}(\boldsymbol{q}j) = \{w_\alpha^a(\boldsymbol{q}j)\}$.

It is clear from (5.12) that the magnitude of $w_\alpha^a(\boldsymbol{g}j)$ is indeterminate to the extent of an arbitrary constant factor. It is convenient to use the normalized eigenvectors, $\boldsymbol{e}(\boldsymbol{q}j)$, such that

$$\sum_{\alpha a} e_\alpha^{a*}(\boldsymbol{q}j')e_\alpha^a(\boldsymbol{q}j) = \delta_{jj'} \quad , \tag{5.17}$$

$$\sum_{j} e_{\alpha'}^{a'*}(\boldsymbol{q}j)e_\alpha^a(\boldsymbol{q}j) = \delta_{aa'}\delta_{\alpha\alpha'} \quad , \tag{5.18}$$

and to represent the corresponding mass-weighted amplitudes of the atomic displacements as

$$w_\alpha^a(\boldsymbol{q}j) = a(\boldsymbol{q}j)e_\alpha^a(\boldsymbol{q}j). \tag{5.19}$$

In (5.19) $a(\boldsymbol{q}j)$ is a constant describing the amplitude of the excitation of mode $(\boldsymbol{q}j)$. The magnitude of $a(\boldsymbol{q}j)$ can be found from energetic considerations. Substituting (5.10), in which the wave amplitudes are related to the orthonormalized eigenvectors $\boldsymbol{e}(\boldsymbol{q}j)$ through (5.13,19), into the expressions for the kinetic and potential energies, one easily obtains the following equation for the total energy of the system

$$E = T + \phi = N\sum_{\boldsymbol{q}j}\omega^2(\boldsymbol{q}j)a^2(\boldsymbol{q}j). \tag{5.20}$$

It can be seen that there are no cross terms involving different modes in (5.20), which implies that the modes can be considered to be independent. The energy associated with a particular mode $\boldsymbol{q}j$ is

$$E(\boldsymbol{q}j) = N\omega^2(\boldsymbol{q}j)a^2(\boldsymbol{q}j). \tag{5.21}$$

By equating this energy to the average energy of a harmonic oscillator vibrating with frequency $\omega(\boldsymbol{q}j)$,

$$E(\boldsymbol{q}j) = \hbar\omega(\boldsymbol{q}j)\left(\frac{1}{2} + \frac{1}{\exp\left[\hbar\omega(\boldsymbol{q}j)/kT\right] - 1}\right), \tag{5.22}$$

one can find a closed-form equation for the unknown $\boldsymbol{a}(\boldsymbol{q}j)$.

We can now write an expression for the amplitude of the displacement of atom a in mode $\boldsymbol{q}j$:

$$u_{\alpha}^{a}(\boldsymbol{q}j) = (Nm_a)^{-1/2} e_{\alpha}^{a}(\boldsymbol{q}j)[E(\boldsymbol{q}j)]^{1/2}/\omega(\boldsymbol{q}j), \tag{5.23}$$

where $E(\boldsymbol{q}j)$ is defined by (5.22).

The most general solution for the atomic displacements is given by the superposition of all displacements from the travelling waves:

$$u_{\alpha}^{al} = (Nm_a)^{-1/2} \sum_{\boldsymbol{q}j} \frac{[E(\boldsymbol{q}j)]^{1/2}}{\omega(\boldsymbol{q}j)} e_{\alpha}^{a}(\boldsymbol{q}j) \exp\{\mathrm{i}[\boldsymbol{q}\cdot\boldsymbol{r}^{al} - \omega(\boldsymbol{q}j)t]\} \quad . \tag{5.24}$$

The above formalism has been used [5.2–4] to calculate the complete vibrational spectrum of an organic crystal. An obvious requirement for a practical application of this formalism is a knowledge of the total lattice energy as a function of the crystal configuration. Given a rule for evaluating $\phi(\ldots, \boldsymbol{r}^{al}, \ldots)$, one can easily compute the second derivatives in (5.5) (for example, numerically), substitute them into (5.14), and then solve the determinantal equation (5.16) for ω.

5.2 The Taddei-Califano Formalism and the Rigid-Molecule Approximation

The theory outlined in the previous section is quite rigorous and does not involve any assumptions other than the harmonicity of ϕ and T. In dealing with organic molecular crystals, some simplifications may be introduced into the theory, based on the fact that the forces between the molecules of a crystal are much weaker than the forces between the individual atoms of a molecule. In the vibrational spectra of organic crystals, this fact manifests itself through a separation of the normal vibrational modes into two more or less distinct bands. A high-frequency band is made up of modes mainly associated with the intramolecular motions. The frequencies of these internal modes are usually very similar to those corresponding to the free-molecule state. A low-frequency band consists of modes describing the intermolecular or external vibrations. If the unit cell of a crystal contains N_c molecules and each molecule contains n_m atoms, there will be $6N_c$ external modes and $3n_mN_c - 6N_c$ internal modes ($3n_m - 6$ for each molecule of the unit cell).

The fact that the intramolecular forces are much stronger than the intermolecular ones can be incorporated into the theory. First of all it can be assumed that the equilibrium molecular conformation in a crystal is the same as in the free-molecule state. Let $\{\boldsymbol{\xi}^{kl}, \boldsymbol{\eta}^{kl}, \boldsymbol{\varsigma}^{kl}\}$ be a Cartesian coordinate system whose axes coincide with the principal inertial axes of a molecule k in a unit cell l ($k = 1, \ldots, N_c$). The system $\{\boldsymbol{\xi}^{kl}, \boldsymbol{\eta}^{kl}, \boldsymbol{\varsigma}^{kl}\}$ is a moving coordinate system whose axes and origin follow the displacements of the molecular axes of inertia and the center of mass. The Cartesian coordinates of the atoms in this system will hereafter be denoted as ϱ_α^{akl} and their displacements as κ_α^{akl}. Of the $3n_m$ displacements κ_α^{akl}, only $3n_m - 6$ are independent.

The instantaneous position of the inertial coordinate system may be specified by three translational and three rotational parameters describing the displacement of the system $\{\xi^{kl}, \eta^{kl}, \varsigma^{kl}\}$ from its equilibrium position $\{\bar{\xi}^{kl}, \bar{\eta}^{kl}, \bar{\varsigma}^{kl}\}$. Just as *Taddei* et al. [5.5] we chose the displacement coordinates of the entire system to be:

Q_σ^{kl} – the internal free-molecule normal coordinates of a molecule k in the unit cell l ($\sigma = 1, \ldots, 3n_m - 6$); they are related to the atomic displacements κ^{akl} [5.4],

$$\kappa_\alpha^{akl} = \sum_\sigma L_{\alpha\sigma}^a Q_\sigma^{kl}; \tag{5.25}$$

t_α^{kl} – the mass-weighted translational displacement coordinates defined by

$$t_\alpha^{kl} = M^{1/2} v_\alpha^{kl}, \tag{5.26}$$

where M is the mass of a molecule and the v_α^{kl} are the instantaneous center-of-mass coordinates of a molecule kl in the equilibrium coordinate system $\{\bar{\xi}^{kl}, \bar{\eta}^{kl}, \bar{\varsigma}^{kl}\}$;

ψ_α^{kl} – the inertia-weighted rotational displacement coordinates defined by

$$\psi_\alpha^{kl} = I_\alpha^{1/2} \theta_\alpha^{kl}, \tag{5.27}$$

where the θ_α^{kl} are the infinitesimal rotations of a molecule kl about the axes $\{\bar{\xi}^{kl}, \bar{\eta}^{kl}, \bar{\varsigma}^{kl}\}$ and the I_α are the corresponding moments of inertia.

The coefficients $L_{\alpha\sigma}^a$ in (5.25) are independent of the particular molecule kl under consideration. This follows from the assumption that the molecular conformation in a crystal is the same as in the free state, and hence is the same for all molecules kl.

To simplify the notation, all of the above displacement coordinates will be denoted by W_τ^{kl}, where $\tau = 1, \ldots, 3n_m - 6$ refers to Q_σ^{kl}, $\tau = 3n_m - 5, \ldots, 3n_m - 3$ to t_α^{kl} and $\tau = 3n_m - 2, \ldots, 3n_m$ to ψ_α^{kl}.

The kinetic and potential energies of a crystal can now be written as

$$T = \frac{1}{2} \sum_{\tau kl} (\dot{W}_\tau^{kl})^2, \tag{5.28}$$

$$\phi = \frac{1}{2}\sum_{\tau kl}\sum_{\tau' k' l'}\left(\frac{\partial^2(\phi_{\rm intra}+\phi_{\rm inter})}{\partial W_\tau^{kl}\partial W_{\tau'}^{k'l'}}\right)_0 W_\tau^{kl} W_{\tau'}^{k'l'}$$

$$= \frac{1}{2}\sum_{\tau kl}\sum_{\tau' k' l'}\left(\frac{\partial^2\phi_{\rm intra}}{(\partial W_\tau^{kl})^2}\delta_{\tau\tau'}\delta_{kk'}\delta_{ll'} + \frac{\partial^2\phi_{\rm inter}}{\partial W_\tau^{kl}\partial W_{\tau'}^{k'l'}}\right)_0 W_\tau^{kl} W_{\tau'}^{k'l'}$$

$$= \frac{1}{2}\sum_{\tau kl}\sum_{\tau' k' l'}\left((\omega_\tau^0)^2\delta_{\tau\tau'}\delta_{kk'}\delta_{ll'} + \frac{\partial^2\phi_{\rm inter}}{\partial W_\tau^{kl}\partial W_{\tau'}^{k'l'}}\right)_0 W_\tau^{kl} W_{\tau'}^{k'l'} \quad , \qquad (5.29)$$

where ω_τ^0 denotes the frequency of a free-molecule normal mode. It is implied that $\omega_\tau^0 = 0$ for $\tau > 3n_m - 6$.

The advantage of using the displacement coordinates given in (5.25–27) now becomes apparent. Since Q_τ^{kl} are the free-molecule normal coordinates, the equations for the intramolecular components of T and ϕ only contain diagonal terms. We recall that this is only true if the effect of the crystal field on the molecular conformation is negligible.

Further transformations of (5.28,29) can be carried out in complete analogy to the Born-von Karman treatment. The resulting equation for the vibrational frequencies of the system is

$$|D_{\tau\tau'}^{kk'}(\boldsymbol{q}) - [\omega_\tau^2 - (\omega_\tau^0)^2]\delta_{\tau\tau'}\delta_{kk'}| = 0, \qquad (5.30)$$

where the elements of the dynamical matrix are now defined by

$$D_{\tau\tau'}^{kk'}(\boldsymbol{q}) = \sum_{l'} D_{\tau\tau'}^{klk'l'} \exp[-\mathrm{i}\boldsymbol{q}\cdot(\boldsymbol{r}^{l'} - \boldsymbol{r}^{l})]$$

$$= \sum_{l'} D_{\tau\tau'}^{kk'l'} \exp(-\mathrm{i}\boldsymbol{q}\cdot\boldsymbol{r}^{l'}), \quad \text{and} \qquad (5.31)$$

$$D_{\tau\tau'}^{klk'l'} = \left(\frac{\partial^2\phi_{\rm inter}}{\partial W_\tau^{kl}\partial W_\tau^{k'l'}}\right)_0. \qquad (5.32)$$

Assuming that $\phi_{\rm inter}$ can be expressed in terms of interactions as

$$\phi_{\rm inter} = \frac{1}{2}\sum_{klk'l'} \varepsilon^{kl,k'l'} \quad , \qquad (5.33)$$

we immediately obtain

$$D_{\tau\tau'}^{kk'}(\boldsymbol{q}) = \sum_{l'}{}'\left(\frac{\partial^2\varepsilon^{kl,k'l'}}{\partial W_\tau^{kl}\partial W_{\tau'}^{k'l'}}\right)_0 \exp[-\mathrm{i}\boldsymbol{q}(\boldsymbol{r}^{l'} - \boldsymbol{r}^{l})]$$

$$+ \delta_{kk'}\sum_{l'}\sum_{k''}\left(\frac{\partial^2\varepsilon^{kl,k''l'}}{\partial W_\tau^{kl}\partial W_{\tau'}^{kl}}\right)_0, \qquad (5.34)$$

where the prime indicates that $l' \neq l$ if $k = k'$ or $k = k''$. As in the previous chapter, $\varepsilon^{kl,k'l'}$ denotes the interaction energy of a molecule k and a molecule k' in the lth and l'th unit cell, respectively.

Unlike (5.16), the secular equation (5.30) only involves the intermolecular force constants. It yields the frequencies of the external modes, and the shifts in the internal frequencies produced by the crystal field. If necessary, the absolute values of the internal frequencies may be obtained by separately computing the free-molecule normal frequencies ω_τ^0.

The existence of two different scales of forces can also be incorporated into the theory by completely neglecting the interplay between the intra- and intermolecular vibrations. In this approximation the internal modes are assumed to be the same as in the free-molecule state, while the external modes can be evaluated by treating the crystal as a lattice of rigid molecules.

The dynamics of such a lattice can be discussed in complete analogy to the theory of *Born* and *von Karman* [5.1]. The only modification that is needed is to introduce three additional coordinates for each particle to specify their orientations. Another possibility is to take advantage of the above formalism and neglect all of the elements of the dynamical matrix (5.31) that couple the internal and external modes. In the latter case the secular determinant breaks up into two blocks, one corresponding to the internal and the other to the external vibrational modes. The latter block leads to the following secular equation for the external-mode frequencies

$$|D_{\tau\tau'}^{kk'}(\boldsymbol{q}) - \omega^2(\boldsymbol{q})\delta_{\tau\tau'}\delta_{kk'}| = 0, \tag{5.35}$$

where τ and τ' now assume values from 1 to 6 and refer to the three translations and three librations of the molecule as a whole.

The dynamical matrix elements of this equation are given by (5.31) with

$$D_{\tau\tau'}^{klk'l'} = (M_\tau M_{\tau'})^{-1/2}\phi_{\tau\tau'}^{klk'l'} \quad , \tag{5.36}$$

and

$$M_\tau = \begin{cases} M, & \tau = 1,2,3, \\ I_{\tau-3}, & \tau = 4,5,6, \end{cases} \tag{5.37}$$

$$\phi_{\tau\tau'}^{klk'l'} = \left(\frac{\partial^2\phi_{\text{inter}}}{\partial U_\tau^{kl}\partial U_{\tau'}^{k'l'}}\right)_0 . \tag{5.38}$$

The U_τ^{kl} are the components of the translational-rotational displacement vector defined by

$$U_\tau^{kl} = \begin{cases} v_\tau^{kl}, & \tau = 1,2,3, \\ \theta_{\tau-3}^{kl} \ , & \tau = 4,5,6. \end{cases} \tag{5.39}$$

The order of (5.35) is $6N_c$. The diagonalization of $D^{kkl'}_{\tau\tau'}(\boldsymbol{q})$ yields $6N_c$ eigenvalues $\omega^2(\boldsymbol{q}j)$ and $6N_c$ associated eigenvectors $\boldsymbol{e}(\boldsymbol{q}j)$ which can again be chosen to be orthonormal according to (5.17,18).

The actual amplitudes of the translational and rotational displacements can be found using exactly the same considerations as discussed in Sect. 5.1. The result is

$$U^k_\tau(\boldsymbol{q}j) = (NM_\tau)^{-1/2} e^k_\tau(\boldsymbol{q}j)[E(\boldsymbol{q}j)]^{1/2}/\omega(\boldsymbol{q}j), \tag{5.40}$$

or, in a more detailed form,

$$v^k_\alpha(\boldsymbol{q}j) = (NM)^{-1/2} e^k_\alpha(\boldsymbol{q}j)[E(\boldsymbol{q}j)]^{1/2}/\omega(\boldsymbol{q}j), \tag{5.41}$$

$$\theta^k_\alpha(\boldsymbol{q}j) = (NI_\alpha)^{-1/2} e^k_{\alpha+3}(\boldsymbol{q}j)[E(\boldsymbol{q}j)]^{1/2}/\omega(\boldsymbol{q}j), \quad \alpha = 1,2,3. \tag{5.42}$$

The most general solution for the molecular translational-rotational displacements can be found in complete analogy to (5.24).

5.3 Calculation of Force Constants Using the Atom-Atom Potential Method

In this section we shall derive explicit expressions for [5.6,7]

$$\left(\frac{\partial^2 \varepsilon^{kl,k'l'}}{\partial W^{kl}_\tau \partial W^{k'l'}_{\tau'}}\right)_0 \quad \text{and} \quad \left(\frac{\partial^2 \varepsilon^{kl,k'l'}}{\partial W^{kl}_\tau \partial W^{kl}_{\tau'}}\right)_0 \tag{5.43}$$

needed to calculate the force constants given in (5.34). We shall first consider the derivatives with respect to the coordinates of different molecules. In the atom-atom approximation these may be represented as

$$\left(\frac{\partial^2 \varepsilon^{kl,k'l'}}{\partial W^{kl}_\tau \partial W^{k'l'}_{\tau'}}\right)_0 = \sum_{aa'} \left(\frac{\partial^2 \varphi_{aa'}(r^{a'k'l',akl})}{\partial W^{kl}_\tau \partial W^{k'l'}_{\tau'}}\right)_0, \tag{5.44}$$

where $\varphi_{aa'}$ is the atom-atom potential and $r^{a'k'l',akl}$ is the length of the vector joining atom a of molecule k in the lth unit cell to atom a' of molecule k' in the l'th unit cell (interatomic distance);

$$r^{a'k'l',akl} \equiv |\boldsymbol{r}^{a'k'l'} - \boldsymbol{r}^{akl}|. \tag{5.45}$$

Using the notations of the previous chapter, a representative term in (5.44) may be written as

$$\left(\frac{\partial^2 \varphi_{aa'}}{\partial W^{kl}_\tau \partial W^{k'l'}_{\tau'}}\right)_0 = \varphi''_{aa'} \left(\frac{\partial r^{a'k'l',akl}}{\partial W^{kl}_\tau}\right)_0 \left(\frac{\partial r^{a'k'l',akl}}{\partial W^{k'l'}_{\tau'}}\right)_0$$

$$+\varphi'_{aa'}\left(\frac{\partial^2 r^{a'k'l',akl}}{\partial W_\tau^{kl}\partial W_{\tau'}^{k'l'}}\right)_0. \tag{5.46}$$

The derivatives of the interatomic distance are given by

$$\left(\frac{\partial r^{a'k'l',akl}}{\partial W_\tau^{kl}}\right)_0 = -\sum_\alpha A_\alpha^{aa'}\left(\frac{\partial r_\alpha^{akl}}{\partial W_\tau^{kl}}\right)_0, \tag{5.47}$$

$$\left(\frac{\partial^2 r^{a'k'l',akl}}{\partial W_\tau^{kl}\partial W_{\tau'}^{k'l'}}\right)_0 = \sum_{\alpha\alpha'} B_{\alpha\alpha'}^{aa'}\left(\frac{\partial r_\alpha^{akl}}{\partial W_\tau^{kl}}\right)_0\left(\frac{\partial r_{\alpha'}^{a'k'l'}}{\partial W_{\tau'}^{kl'}}\right)_0$$

$$-\delta_{kk'}\delta_{ll'}\sum_\alpha A_\alpha^{aa'}\left(\frac{\partial^2 r_\alpha^{akl}}{\partial W_\tau^{kl}\partial W_{\tau'}^{k'l'}}\right)_0, \tag{5.48}$$

where

$$A_\alpha^{aa'} = \left(\frac{\partial r^{a'k'l',akl}}{\partial r_\alpha^{a'k'l'}}\right)_0 = -\left(\frac{\partial r^{a'k'l',akl}}{\partial r_\alpha^{akl}}\right)_0$$

$$= r_\alpha^{a'k'l',akl}/r^{a'k'l',akl} \quad , \tag{5.49}$$

$$B_{\alpha\alpha'}^{aa'} = \left(\frac{\partial^2 r^{a'k'l',akl}}{\partial r_\alpha^{akl}\partial r_{\alpha'}^{a'k'l'}}\right)_0 = \frac{\delta_{\alpha\alpha'} - A_\alpha^{aa'}A_{\alpha'}^{aa'}}{r^{a'k'l',akl}}. \tag{5.50}$$

Using the notation

$$C_{\alpha\alpha'}^{aa'} = \varphi''_{aa'}A_\alpha^{aa'}A_{\alpha'}^{aa'} + \varphi'_{aa'}B_{\alpha\alpha'}^{aa'} \quad , \tag{5.51}$$

we may rewrite (5.46) in a compact form

$$\left(\frac{\partial^2\varphi_{aa'}}{\partial W_\tau^{kl}W_{\tau'}^{k'l'}}\right)_0 = -\sum_{\alpha\alpha'} C_{\alpha\alpha'}^{aa'}\left(\frac{\partial r_\alpha^{akl}}{\partial W_\tau^{kl}}\right)_0\left(\frac{\partial r_{\alpha'}^{a'k'l'}}{\partial W_{\tau'}^{k'l'}}\right)_0. \tag{5.52}$$

Now the problem is to relate the atomic Cartesian coordinates r_α^{akl} in the crystal-fixed frame to the displacement coordinates W_τ^{kl}. Let the orientations of the molecule-fixed coordinate systems with respect to the crystal-fixed one be specified by the direction cosines $\Lambda_{\alpha\beta}^{kl}$, where α denotes one of the crystal-fixed and β one of the molecule-fixed axes. If the molecule kl is displaced from its equilibrium orientation, the infinitesimal change in $\Lambda_{\alpha\beta}^{kl}$ may be expressed in terms of the rotational displacement coordinates θ_α^{kl} :

$$\lambda_{\alpha\beta}^{kl} \equiv \Lambda_{\alpha\beta}^{kl} - \overline{\Lambda}_{\alpha\beta}^{kl}$$

$$= \sum_{\gamma\lambda}\overline{\Lambda}_{\alpha\lambda}^{kl}\delta_{\lambda\beta\gamma}\theta_\gamma^{kl} - \frac{1}{2}\sum_{\gamma\lambda}\sum_{\mu\nu}\overline{\Lambda}_{\alpha\lambda}^{kl}\delta_{\lambda\nu\gamma}\delta_{\beta\nu\mu}\theta_\gamma^{kl}\theta_\mu^{kl}, \tag{5.53}$$

where $\delta_{\alpha\beta\gamma}$ is a permutation symbol equal to 0 unless α, β and γ are all different, +1 for the cyclic sequence xyz and −1 for the sequence zyx. The Greek indices in (5.53) assume the values 1, 2 and 3.

A superposition of the rotational ($\lambda_{\alpha\beta}^{kl}$), translational ($v_{\alpha}^{kl}$) and conformational displacements (κ_{α}^{akl}) shifts the Cartesian coordinates r_{α}^{akl} from their equilibrium positions to

$$r_{\alpha}^{akl} = \bar{r}_{\alpha}^{akl} + \sum_{\beta} \bar{\Lambda}_{\alpha\beta}^{kl}(v_{\beta}^{kl} + \kappa_{\beta}^{akl}) + \lambda_{\alpha\beta}^{kl}\bar{r}_{\beta}^{akl}. \tag{5.54}$$

Substituting (5.25–27,53) into (5.54) yields the desired expression for r_{α}^{akl} in terms of the components of $\boldsymbol{W}^{kl}$,

$$\begin{aligned} r_{\alpha}^{akl} =& \bar{r}_{\alpha}^{akl} + M^{-1/2}\sum_{\beta}\bar{\Lambda}_{\alpha\beta}^{kl}t_{\beta}^{kl} + \sum_{\mu}\Big(\sum_{\beta}\bar{\Lambda}_{\alpha\beta}^{kl}L_{\beta\mu}^{a}\Big)Q_{\mu}^{kl} \\ &+ \sum_{\beta}\Big(I_{\beta}^{-1/2}\sum_{\gamma\lambda}\bar{\Lambda}_{\alpha\lambda}^{kl}\bar{r}_{\gamma}^{akl}\delta_{\lambda\gamma\beta}\Big)\psi_{\beta}^{kl} \\ &- \frac{1}{2}\sum_{\beta\gamma}\Big[(I_{\beta}I_{\gamma})^{-1/2}\sum_{\lambda\mu\nu}\bar{\Lambda}_{\alpha\nu}^{kl}\delta_{\nu\mu\beta}\delta_{\lambda\mu\gamma}\bar{r}_{\lambda}^{akl}\Big]\psi_{\beta}^{kl}\psi_{\gamma}^{kl}. \end{aligned} \tag{5.55}$$

A differentiation of this expression with respect to the various displacement coordinates and a subsequent substitution of the results in (5.52) yields [5.6]

$$\left(\frac{\partial^2\varphi_{aa'}}{\partial Q_{\sigma}^{kl}\partial Q_{\sigma'}^{k'l'}}\right)_0 = -\sum_{\alpha\beta\lambda\mu} C_{\alpha\beta}^{aa'}\bar{\Lambda}_{\alpha\lambda}^{kl}\bar{\Lambda}_{\beta\mu}^{k'l'}L_{\lambda\sigma}^{a}L_{\mu\sigma'}^{a'}. \tag{5.56}$$

$$\left(\frac{\partial^2\varphi_{aa'}}{\partial t_{\alpha}^{kl}\partial t_{\alpha'}^{k'l'}}\right)_0 = -\sum_{\beta\gamma} M^{-1}C_{\beta\gamma}^{aa'}\bar{\Lambda}_{\beta\alpha}^{kl}\bar{\Lambda}_{\gamma\alpha'}^{k'l'} \quad , \tag{5.57}$$

$$\begin{aligned}\left(\frac{\partial^2\varphi_{aa'}}{\partial \psi_{\alpha}^{kl}\partial \psi_{\alpha'}^{k'l'}}\right)_0 =& -\sum_{\beta\gamma\beta'\gamma'\lambda\mu} (I_{\alpha}I_{\alpha'})^{-1/2}C_{\lambda\mu}^{aa'}\bar{\Lambda}_{\lambda\beta}^{kl}\bar{\Lambda}_{\mu\beta'}^{k'l'} \\ &\times\delta_{\beta\gamma\alpha}\delta_{\beta'\gamma'\alpha'}\bar{r}_{\gamma}^{akl}\bar{r}_{\gamma'}^{a'k'l'} \quad ,\end{aligned} \tag{5.58}$$

$$\left(\frac{\partial^2\varphi_{aa'}}{\partial Q_{\sigma}^{kl}\partial t_{\alpha}^{k'l'}}\right)_0 = -\sum_{\beta\lambda\mu} M^{-1/2}C_{\lambda\mu}^{aa'}\bar{\Lambda}_{\lambda\beta}^{kl}\bar{\Lambda}_{\mu\alpha}^{k'l'}L_{\beta\sigma}^{a} \quad , \tag{5.59}$$

$$\begin{aligned}\left(\frac{\partial^2\varphi_{aa'}}{\partial Q_{\sigma}^{kl}\partial \psi_{\alpha}^{k'l'}}\right)_0 =& -\sum_{\nu\lambda\mu\beta\gamma} I_{\alpha}^{-1/2}C_{\beta\gamma}^{aa'}\bar{\Lambda}_{\beta\gamma}^{kl}\bar{\Lambda}_{\gamma\lambda}^{k'l'}\bar{r}_{\mu}^{a'k'l'} \\ &\times\delta_{\lambda\mu\alpha}L_{\nu\sigma}^{a} \quad ,\end{aligned} \tag{5.60}$$

$$\begin{aligned}\left(\frac{\partial^2\varphi_{aa'}}{\partial t_{\alpha}^{kl}\partial \psi_{\alpha'}^{k'l'}}\right)_0 =& -\sum_{\beta\gamma\lambda\mu} (MI_{\alpha'})^{-1/2}C_{\lambda\mu}^{aa'}\bar{\Lambda}_{\lambda\alpha}^{kl}\bar{\Lambda}_{\mu\beta}^{k'l'} \\ &\times\bar{r}_{\gamma}^{a'k'l'}\delta_{\beta\gamma\alpha'} \quad .\end{aligned} \tag{5.61}$$

The derivatives with respect to two displacements of the same molecule can be derived similarly. The only modification that is needed is to replace (5.52) by [5.6]

$$\left(\frac{\partial^2\varphi_{aa'}}{\partial W_\tau^{kl}\partial W_{\tau'}^{kl}}\right)_0=\sum_{\alpha\alpha'}C_{\alpha\alpha'}^{aa'}\left(\frac{\partial r_\alpha^{akl}}{\partial W_\tau^{kl}}\right)_0\left(\frac{\partial r_{\alpha'}^{akl}}{\partial W_{\tau'}^{kl}}\right)_0$$
$$-\sum_\alpha\varphi'_{aa'}A_\alpha^{aa'}\left(\frac{\partial^2 r_\alpha^{akl}}{\partial W_\tau^{kl}\partial W_{\tau'}^{kl}}\right)_0. \tag{5.62}$$

The final expressions for the derivatives can be obtained from (5.56,57, 59–61) by changing the sign and replacing $k'l'$ by kl. An exception is the expression for two rotational coordinates:

$$\left(\frac{\partial^2\varphi_{aa'}}{\partial\psi_\alpha^{kl}\partial\psi_{\alpha'}^{kl}}\right)_0=\sum_{\lambda\mu\lambda'\mu'\beta\gamma}(I_\alpha I_{\alpha'})^{-1/2}C_{\beta\gamma}^{aa'}\bar{A}_{\beta\lambda}^{kl}\bar{A}_{\gamma\lambda'}^{kl}$$
$$\times\delta_{\lambda\mu\alpha}\delta_{\lambda'\mu'\alpha'}\bar{r}_\mu^{akl}\bar{r}_{\mu'}^{akl}+\sum_{\lambda\mu\nu\beta}(I_\alpha I_{\alpha'})^{-1/2}\varphi'_{aa'}$$
$$\times A_\beta^{aa'}\bar{A}_{\beta\lambda}^{kl}\delta_{\lambda\mu\alpha}\delta_{\nu\mu\alpha'}\bar{r}_\alpha^{akl}. \tag{5.63}$$

5.4 Symmetry Properties of Force Constants

The number of distinct force constants to be evaluated in solving the lattice-dynamical problem can be substantially reduced by taking into account the crystal symmetry. For the sake of simplicity, we shall consider crystals composed of rigid molecules and containing a single crystallographically distinct molecule in the unit cell. The symmetry properties of the force-constant matrix for such crystals have been comprehensively studied by *Pawley* [5.8].

We shall first consider the force constants $\phi_{\tau\tau'}^{klk'l'}$ which involve a differentiation with respect to the coordinates of different molecules ($kl\neq k'l'$). Let us choose one molecule in the origin unit cell as the basis molecule and denote it as molecule O. All other molecules in the crystal are generated from molecule O by applying all possible symmetry operations S. With this notation, a representative force constant which involves a displacement of the basis molecule is

$$\phi_{\tau\tau'}^{os}=\frac{\partial^2\phi}{\partial U_\tau^o\partial U_{\tau'}^s}, \tag{5.64}$$

where U_τ^o and $U_{\tau'}^s$ are the translational-rotational displacements of molecules O and s, respectively. We remind the reader that the displacement vectors $\boldsymbol{U}^o$ and $\boldsymbol{U}^s$ are defined in their own coordinate systems, $\{\bar{\boldsymbol{\xi}}^o,\bar{\boldsymbol{\eta}}^o,\bar{\boldsymbol{\zeta}}^o\}$ and $\{\bar{\boldsymbol{\xi}}^s,\bar{\boldsymbol{\eta}}^s,\bar{\boldsymbol{\zeta}}^s\}$.

The physical meaning of the force constant $\phi^{os}_{TT'}$ is quite clear – it relates the force and torque on molecule O to the translational-rotational displacement of molecule s. Denoting the six-dimensional force-torque vector as $\boldsymbol{\phi}^o = \{\phi^o_T\}$, we may write

$$\phi^o_T = -\sum_{T'} \phi^{os}_{TT'} U^s_{T'} \quad , \tag{5.65}$$

or, in matrix notation,

$$\boldsymbol{\phi}^o = -\underline{\underline{\phi}}^{os} \boldsymbol{U}^s. \tag{5.66}$$

Consider now the general symmetry operation $\underline{\underline{S}}$, whose rotational component consists of a 3×3 matrix $\underline{\underline{B}}$. The action of $\underline{\underline{S}}$ on a translational-rotational displacement vector $\boldsymbol{U}$ may be shown to be

$$\underline{\underline{S}}\boldsymbol{U} = \begin{pmatrix} \underline{\underline{B}} & \vdots & 0 \\ \cdots & \vdots & \cdots \\ 0 & \vdots & \sigma\underline{\underline{B}} \end{pmatrix} \boldsymbol{U}, \tag{5.67}$$

where $\sigma = \det|\underline{\underline{B}}|$; $\sigma = 1$ or -1 if $\underline{\underline{B}}$ is a proper or improper rotation, respectively.

Equation (5.66) may be now rewritten as

$$\underline{\underline{S}}\boldsymbol{\phi}^o = (\underline{\underline{S}}\,\underline{\underline{\phi}}^{os}\underline{\underline{S}}^{-1})\underline{\underline{S}}\boldsymbol{U}^s. \tag{5.68}$$

This equation is a good starting point for a discussion of the transformational properties of $\underline{\underline{\phi}}^{os}$.

Let $\underline{\underline{S}}_1$ be a symmetry element which leaves molecules O and s unchanged:

$$\boldsymbol{\phi}^o = -(\underline{\underline{S}}_1\underline{\underline{\phi}}^{os}\underline{\underline{S}}_1^{-1})\boldsymbol{U}^s. \tag{5.69}$$

Comparing this equation with (5.66), one immediately sees that

$$\underline{\underline{\phi}}^{os} = -(\underline{\underline{S}}_1\underline{\underline{\phi}}^{os}\underline{\underline{S}}_1^{-1}). \tag{5.70}$$

A second type of symmetry operation S_2 interchanges molecules O and s. For such an operation (5.68) transforms to

$$\boldsymbol{\phi}^s = -(\underline{\underline{S}}_2\underline{\underline{\phi}}^{os}\underline{\underline{S}}_2^{-1})\boldsymbol{U}^o, \tag{5.71}$$

which yields

$$\underline{\underline{\phi}}^{so} = \underline{\underline{S}}_2\underline{\underline{\phi}}^{os}\underline{\underline{S}}_2^{-1}. \tag{5.72}$$

A third type of symmetry operation transforms molecule s to s', but does not affect the basis molecule. If $\underline{\underline{S}}_3$ is such an operation, then

$$\underline{\underline{\phi}}^{os'} = \underline{\underline{S}}_3 \underline{\underline{\phi}}^{os} \underline{\underline{S}}_3^{-1}. \tag{5.73}$$

Finally, a fourth type of symmetry transformation, $\underline{\underline{S}}_4$, alters the positions of both molecules involved, transforming O to s' and s to s''. This type of symmetry operation leads to

$$\underline{\underline{\phi}}^{s's''} = \underline{\underline{S}}_4 \underline{\underline{\phi}}^{os} \underline{\underline{S}}_4^{-1}. \tag{5.74}$$

Equations (5.70,72–74) may be complemented by an obvious relation

$$\underline{\underline{\phi}}^{so} = \underline{\underline{\tilde{\phi}}}^{os} \quad , \tag{5.75}$$

where the tilde denotes a transposition. The latter relation immediately follows from (5.64)

$$\frac{\partial^2 \phi}{\partial U_\tau^o \partial U_{\tau'}^s} = \frac{\partial^2 \phi}{\partial U_{\tau'}^s \partial U_\tau^o}. \tag{5.76}$$

We now turn to a consideration of the force constants which only involve displacements of the basis molecule. These force constants, usually referred to as "self-terms", relate the force-torque on molecule O to its own displacement

$$\boldsymbol{\phi}^o = -\underline{\underline{\phi}}^{oo} \boldsymbol{U}^o. \tag{5.77}$$

Let us first consider a pure translational displacement $v_\alpha^o = v_\alpha$, $\boldsymbol{\theta}_\alpha^o = 0$ ($\alpha = 1,2,3$) [5.8]. The crystal configuration corresponding to this displacement is equivalent to the one obtained after applying a shift $-\boldsymbol{v}$ to all molecules s ($s \neq 0$), while maintaining the position of molecule O unaltered. The displacement $-\boldsymbol{v}$ of molecule s in the coordinate system of the basis molecule is, in turn, equivalent to the displacement $\underline{\underline{B}}_s^{-1}\boldsymbol{v}$ in the coordinate system of the molecule s. Considering this circumstance, we may write

$$-\underline{\underline{\phi}}^{oo} \begin{pmatrix} \boldsymbol{v} \\ \cdots \\ 0 \\ 0 \\ 0 \end{pmatrix} = \sum_s \underline{\underline{\phi}}^{os} \begin{pmatrix} \underline{\underline{B}}_s^{-1} \boldsymbol{v} \\ \cdots\cdots\cdots \\ 0 \\ 0 \\ 0 \end{pmatrix} = \sum_s \underline{\underline{\phi}}^{os} \left(\begin{array}{c:c} \underline{\underline{B}}_s^{-1} & 0 \\ \hdashline 0 & 0 \end{array}\right) \begin{pmatrix} \boldsymbol{v} \\ \cdots \\ 0 \\ 0 \\ 0 \end{pmatrix}, \tag{5.78}$$

that is,

$$-\underline{\underline{\phi}}^{oo} = \sum_s \underline{\underline{\phi}}^{os} \left(\begin{array}{c:c} \underline{\underline{B}}_s^{-1} & 0 \\ \hdashline 0 & 0 \end{array}\right). \tag{5.79}$$

To fill in the empty blocks of matrix (5.79), let us consider a pure rotational displacement $v_\alpha^o = 0$, $\theta_\alpha^o = \theta_\alpha$. As in the previous case, the application of a displacement $\boldsymbol{\theta}$ to the molecule O is equivalent to applying certain displacements to the surrounding molecules, while keeping molecule O fixed. Simple geometrical considerations have shown [5.8] that the required displacements are given by the vector

$$\begin{pmatrix} -\underline{\underline{R}}^s\,\boldsymbol{\theta} \\ \cdots\cdots\cdots \\ -\boldsymbol{\theta} \end{pmatrix}, \tag{5.80}$$

where $\underline{\underline{R}}^s$ is a 3×3 matrix defined by

$$\underline{\underline{R}}^s = \begin{pmatrix} 0 & r_3^s & -r_2^s \\ -r_3^s & 0 & r_1^s \\ r_2^s & -r_1^s & 0 \end{pmatrix}. \tag{5.81}$$

The center-of-mass coordinates r_α^s in (5.81) as well as the displacements (5.80) are defined in the coordinate system of the molecule O. In the coordinate system of molecule s, the displacement (5.80) is [5.8]

$$\begin{pmatrix} -\underline{\underline{B}}_s^{-1}\underline{\underline{R}}^s\boldsymbol{\theta} \\ \cdots\cdots\cdots \\ -\sigma_s\underline{\underline{B}}_s^{-1}\boldsymbol{\theta} \end{pmatrix} = \left(\begin{array}{c:c} & \underline{\underline{B}}_s^{-1}\underline{\underline{R}}^s \\ \cdots\cdots & \cdots\cdots \\ & \sigma_s\underline{\underline{B}}_s^{-1} \end{array}\right) \begin{pmatrix} 0 \\ 0 \\ 0 \\ \cdots \\ \boldsymbol{\theta} \end{pmatrix}. \tag{5.82}$$

Thus, the final expression for $\underline{\underline{\phi}}^{oo}$ has the following form,

$$-\underline{\underline{\phi}}^{oo} = \sum_s \underline{\underline{\phi}}^{os} \left(\begin{array}{c:c} \underline{\underline{B}}_s^{-1} & \underline{\underline{B}}_s^{-1}\,\underline{\underline{R}}^s \\ \cdots\cdots & \cdots\cdots \\ 0 & \sigma_s\,\underline{\underline{B}}_s^{-1} \end{array}\right). \tag{5.83}$$

This equation completes the set of symmetry relations involving the force constant matrices $\underline{\underline{\phi}}^{os}$ and $\underline{\underline{\phi}}^{oo}$. A knowledge of these force constants can readily be shown to be sufficient to construct the dynamical matrix. Let us consider a crystal whose unit cell contains two molecules, g^o and h^o, in addition to the molecule O. $\underline{\underline{G}}_o$ and $\underline{\underline{H}}_o$ are the symmetry operations which generate the molecules g^o and h^o from the molecule O. All of the molecules in the crystal can now be classified as being of type E, G or H, which are translationally equivalent to the molecules O, g^o and h^o, respectively. Let the group table be

Second operation \ First operation	$\underline{\underline{E}}$	$\underline{\underline{G}}$	$\underline{\underline{H}}$
$\underline{\underline{E}}$	$\underline{\underline{E}}$	$\underline{\underline{E}}$	$\underline{\underline{H}}$
$\underline{\underline{G}}$	$\underline{\underline{G}}$	$\underline{\underline{H}}$	$\underline{\underline{E}}$
$\underline{\underline{H}}$	$\underline{\underline{H}}$	$\underline{\underline{E}}$	$\underline{\underline{G}}$

, (5.84)

so that $\underline{\underline{G}}\,\underline{\underline{E}} = \underline{\underline{G}}$, $\underline{\underline{H}}\,\underline{\underline{G}} = \underline{\underline{E}}$, etc. The dynamical matrix in (5.31) may be represented as

$$\underline{\underline{D}}^{kk'}(\boldsymbol{q}) = \begin{pmatrix} \underline{\underline{D}}^{OE}(\boldsymbol{q}) & \underline{\underline{D}}^{OG}(\boldsymbol{q}) & \underline{\underline{D}}^{OH}(\boldsymbol{q}) \\ \underline{\underline{D}}^{G_oE}(\boldsymbol{q}) & \underline{\underline{D}}^{G_oG}(\boldsymbol{q}) & \underline{\underline{D}}^{G_oH}(\boldsymbol{q}) \\ \underline{\underline{D}}^{H_oE}(\boldsymbol{q}) & \underline{\underline{D}}^{H_oG}(\boldsymbol{q}) & \underline{\underline{D}}^{H_oH}(\boldsymbol{q}) \end{pmatrix}, \tag{5.85}$$

where $\underline{\underline{D}}^{OE}(\boldsymbol{q})$, $\underline{\underline{D}}^{OG}(\boldsymbol{q})$,... are 6×6 blocks whose elements are defined by

$$D^{OE}_{\tau\tau'}(\boldsymbol{q}) = (M_\tau M_{\tau'})^{-1/2} \sum_E \phi^{OE}_{\tau\tau'} \exp\left(-\mathrm{i}\boldsymbol{q}\boldsymbol{r}^E\right), \tag{5.86}$$

$$D^{OG}_{\tau\tau'}(\boldsymbol{q}) = (M_\tau M_{\tau'})^{-1/2} \sum_G \phi^{OG}_{\tau\tau'} \exp\left(-\mathrm{i}\boldsymbol{q}\boldsymbol{r}^G\right), \tag{5.87}$$

and so on.

The blocks in the first row of $\underline{\underline{D}}^{kk'}(\boldsymbol{q})$ determine the equation of motion for molecule O and only involve the force constants $\phi^{os}_{\tau\tau'}$ and $\phi^{oo}_{\tau\tau'}$. The remaining rows may be derived from the first row since the wave of the wavevector $\boldsymbol{q}$, as experienced by molecule O, is exactly the same as the wave of the wavevector $\underline{\underline{G}}^{-1}\boldsymbol{q}$, as experienced by g^o [5.8]. The use of this fact leads to

$$\underline{\underline{D}}^{G_oE}(\boldsymbol{q}) = \underline{\underline{D}}^{OG^{-1}}(\underline{\underline{G}}^{-1}\boldsymbol{q}). \tag{5.88}$$

It can be seen from the group table (5.84) that $\underline{\underline{G}}\,\underline{\underline{H}} = \underline{\underline{E}}$ and, hence, $\underline{\underline{G}}^{-1} = \underline{\underline{H}}$. Therefore, (5.88) may be rewritten as

$$\underline{\underline{D}}^{G_oE}(\boldsymbol{q}) = \underline{\underline{D}}^{OH}(\underline{\underline{H}}\boldsymbol{q}). \tag{5.89}$$

The other blocks in the second and third rows of $\underline{\underline{D}}^{kk'}(\boldsymbol{q})$ can be transformed in a similar fashion. The final result is

$$\underline{\underline{D}}^{kk'}(\boldsymbol{q}) = \begin{pmatrix} \underline{\underline{D}}^{OE}(\boldsymbol{q}) & \underline{\underline{D}}^{OG}(\boldsymbol{q}) & \underline{\underline{D}}^{OH}(\boldsymbol{q}) \\ \underline{\underline{D}}^{OH}(\underline{\underline{H}}\boldsymbol{q}) & \underline{\underline{D}}^{OE}(\underline{\underline{H}}\boldsymbol{q}) & \underline{\underline{D}}^{OG}(\underline{\underline{H}}\boldsymbol{q}) \\ \underline{\underline{D}}^{OG}(\underline{\underline{G}}\boldsymbol{q}) & \underline{\underline{D}}^{OH}(\underline{\underline{G}}\boldsymbol{q}) & \underline{\underline{D}}^{OE}(\underline{\underline{G}}\boldsymbol{q}) \end{pmatrix}. \tag{5.90}$$

It is now seen that the computation of the entire dynamical matrix $\underline{\underline{D}}^{kk'}(\boldsymbol{q})$ is reduced to a computation of the blocks $\underline{\underline{D}}^{OE}$, $\underline{\underline{D}}^{OG}$ and $\underline{\underline{D}}^{OH}$ at the three different wavevectors $\boldsymbol{q}$, $\underline{\underline{H}}\boldsymbol{q}$ and $\underline{\underline{G}}\boldsymbol{q}$. Hence, force constants other than $\phi^{os}_{\tau\tau'}$ and $\phi^{oo}_{\tau\tau'}$ need not be computed to set up the dynamical matrix.

Thus, we have shown that the use of crystal symmetry substantially reduces the number of matrix elements which need be evaluated in order to solve the lattice-dynamical problem. A further simplication can be attained using the methods of group theory and by considering the symmetry properties of the dynamical matrix and its eigenvectors in reciprocal space. The methods of group theory allow one to bring the dynamical matrix to a block-diagonal form, thereby reducing the initial dynamical problem to a number of problems of smaller dimensions.

Besides reducing the computational work associated with a diagonalization of the dynamical matrix, group theory is also of value in classifying and labeling the various vibrational modes according to their symmetry. For reasons of space, we shall not discuss these points here but refer the reader to the papers by *Maradudin* and *Vosko* [5.9], and by *Venkataraman* and *Sahni* [5.10]. The first paper illustrates the application of group theory to the standard Born-von Karman formalism, while the latter presents a comprehensive discussion of the group theory of external vibrational modes within the framework of the rigid-molecule approximation. More recently *Pawley* [5.11] introduced a diagrammatic description of phonon symmetries.

5.5 Experimental Tests

In this section we shall briefly discuss ways in which the results of lattice-dynamical calculations may be compared to experiment.

The most valuable information about the lattice dynamics of a crystal is obtained from the measurements of the coherent inelastic scattering of thermal neutrons. Figure 5.1 shows a typical energy distribution of scattered neutrons, as measured by *Pawley* et al. [5.12] for fully deuterated naphtha-

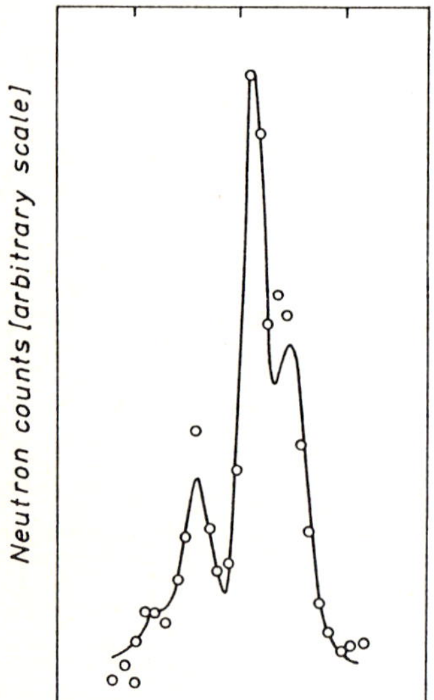

Fig. 5.1. Typical energy distribution of coherent inelastic scattered neutrons [5.12]

lene. The distribution corresponds to a specified wave transfer vector $\boldsymbol{Q}$. For the situation in which energy is lost, $\boldsymbol{Q}$ is related to the phonon wave vector $\boldsymbol{q}$ through $\boldsymbol{Q}+\boldsymbol{q}=\boldsymbol{B}$, where $\boldsymbol{B}$ is a vector of the reciprocal lattice. Each peak in the distribution curve corresponds to a scattering from a definite phonon mode. The abscissa of a peak yields the frequency of the mode. By measuring the scattered neutron distributions for various values of $\boldsymbol{Q}$, one obtains the dispersion relations, which can be compared in a straightforward way with the calculated dependences $\omega(\boldsymbol{q}j)$.

The scattered neutron spectra such as the one shown in Fig. 5.1 also provide a test for the calculated eigenvectors. This test is made by comparing the observed intensities of the peaks to those resulting from the calculated eigenvectors. The intensity of the scattering from a particular vibrational mode is related to the eigenvector of this mode through the coherent one-phonon structure factor defined by

$$G_j(\boldsymbol{Q}) = \sum_a b_a \boldsymbol{Q}\cdot\boldsymbol{u}^a(\boldsymbol{q}j)\exp(\mathrm{i}\boldsymbol{B}\cdot\boldsymbol{r}^a)\exp(-D_a), \tag{5.91}$$

where b_a, D_a and $\boldsymbol{r}^a$ are the scattering length, the Debye-Waller factor and the position vector of the ath atom, respectively. The displacement amplitudes $u^a_\alpha(\boldsymbol{q}j)$ are related to the eigenvectors $\boldsymbol{e}(\boldsymbol{q}j)$ by (5.23).

For a lattice of rigid molecules the eigenvector $\boldsymbol{e}(\boldsymbol{q}j)$ is made up of both translational and rotational components, which are related to the translational and rotational displacement amplitudes $\boldsymbol{v}^k(\boldsymbol{q}j)$ and $\boldsymbol{\theta}^k(\boldsymbol{q}j)$ through (5.41,42). In this case the vector amplitudes of the atomic displacements in mode $\boldsymbol{q}j$ are given by

$$\boldsymbol{u}^a(\boldsymbol{q}j) = \boldsymbol{v}^k(\boldsymbol{q}j) + \boldsymbol{\theta}^k(\boldsymbol{q}j)\times\boldsymbol{\varrho}^{ak} \tag{5.92}$$

and (5.91) transforms to [5.13]

$$G_j(\boldsymbol{Q}) = \sum_{ak} b_a\boldsymbol{Q}\cdot[\boldsymbol{v}^k(\boldsymbol{q}j) + \boldsymbol{\theta}^k(\boldsymbol{q}j)\times\boldsymbol{\varrho}^{ak}] \times \exp(\mathrm{i}\boldsymbol{B}\boldsymbol{r}^k)\exp(\mathrm{i}\boldsymbol{Q}\boldsymbol{\varrho}^{ak})\exp(-D_a). \tag{5.93}$$

Having solved the lattice-dynamical problem, one can compare the theoretical and observed spectra of neutron scattering and assess the reliability of the calculated eigenfrequencies and eigenvectors.

An expression similar to (5.93) can also be written for the diffuse scattering of X-rays [5.14]. In this case, however, a comparison with experiment is not so straightforward since the observed intensities involve appreciable contributions from Compton and many-phonon scattering.

A good test for the correctness of a lattice-dynamical model is afforded by experiments on the incoherent inelastic scattering of neutrons. The energy of the scattered neutrons is usually determined by the time-of-flight method [5.15] and the observed spectrum is described by

$$N(t_0) = \Delta t_0 \int dE_0\, dE\, dt\, F(E_0, E, t_0, t)\sigma^{\mathrm{inc}}(E_0, E, \varphi, T), \tag{5.94}$$

where $N(t_0)$ is the number of neutrons which arrive at the analyzer at the time t_0 for an interval Δt_0; $F(E_0, E, t_0, t)$ is the apparatus function, $\sigma^{\mathrm{inc}}(E_0, E, \varphi, T)$ is the differential cross section for incoherent inelastic scattering, E and E_0 are the energies of the incoming and outgoing neutrons, respectively, and φ is the scattering angle.

For hydrogeneous substances the dominant contribution to σ^{inc} is due to the one-phonon scattering from hydrogen atoms, which is given by [5.16]

$$\sigma_H^{\mathrm{inc},1} = \left(\frac{E}{E_0}\right)^{1/2} \frac{\hbar|Q|^2}{4M_c\omega} \sum_H (b_H^{\mathrm{inc}})^2 \times \frac{\exp(-2D_a)}{\exp\left(\frac{\bar{\omega}}{kT}\right) - 1} G_H^{\mathrm{inc}}(\omega), \tag{5.95}$$

where M_c is the mass of the unit cell and $G_H^{\mathrm{inc}}(\omega)$ is the hydrogen amplitude-weighted density of phonon states defined by

$$G_H^{\mathrm{inc}}(\omega) = \frac{M_c}{m_H} \frac{V_c}{8\pi^3} \frac{1}{3} \sum_j \int d\boldsymbol{q} |\boldsymbol{u}^H(\boldsymbol{q}j)|^2 \delta(\omega - \omega(\boldsymbol{q}j)). \tag{5.96}$$

The summation in (5.95) is carried out over all hydrogen atoms in the molecule.

It is worth noting that, in the case of crystals with one atom per unit cell, σ^{inc} is related to the phonon spectrum through the usual (non-weighted) density of states,

$$g(\omega) = \frac{V_c}{8\pi^3} \frac{1}{3} \sum_j \int d\boldsymbol{q}\, \delta(\omega - \omega(\boldsymbol{q}j)). \tag{5.97}$$

With molecular crystals, we need to know the weighted density of phonon states, which depends on the amplitudes $\boldsymbol{u}^H(\boldsymbol{q}j)$. For this reason, not only the eigenfrequencies but also the eigenvectors are put to a test on comparing the calculated and observed spectra.

Unfortunately, measurements of inelastic neutron scattering represent a very difficult experimental problem. In most cases one has to be content with information provided by infrared absorption and Raman scattering.

The interaction of infrared photons with an organic crystal is essentially a one-phonon process in which the absorption of a photon results in the creation of a phonon. The law of conservation of momentum gives

$$\boldsymbol{q}_{\mathrm{photon}} = \boldsymbol{q}_{\mathrm{phonon}} \quad . \tag{5.98}$$

For infrared photons, $|\boldsymbol{q}_{\mathrm{photon}}|$ is of the order of 10^1 to $10^2\mathrm{cm}^{-1}$, while

$|\boldsymbol{q}_{\text{phonon}}|$ ranges from 0 to 10^8cm^{-1}. That is, only those phonons whose wave vector is practically zero may be excited by infrared photons.

A similar situation is encountered in the case of Raman scattering, which represents a two-phonon process described by

$$\boldsymbol{q}_{\text{photon}}^{\text{incident}} = \boldsymbol{q}_{\text{photon}}^{\text{scattered}} + \boldsymbol{q}_{\text{phonon}} \quad . \tag{5.99}$$

Again, only phonons with $|\boldsymbol{q}|\approx 0$ may be excited.

Thus, measurements of the infrared absorption and Raman scattering give us the limiting frequencies, which can be directly compared to the calculated $\boldsymbol{q} = 0$ eigenfrequencies.

In principle, spectroscopic measurements can also be used to check the calculated eigenvectors $\boldsymbol{e}(0j)$. This can be done by comparing the observed band intensities to those resulting from the calculated eigenvectors. However, a rigorous calculation of the intensities requires a knowledge of the dipole moment and polarizability of the unit cell as a function of the crystal configuration. The difficulties involved in such calculations are quite obvious.

Most calculations of the intensity performed thus far have used the oriented-gas approximation which neglects the interactions of the molecular charge distributions in a crystal. However, the reliability of this approximation is too poor [5.17] for it to be used to test lattice-dynamical models.

Obviously measurments of the infrared absorption and Raman scattering cannot provide any information about the acoustic vibrations of a crystal. This missing information is partly determined from the measurements of sound velocities, which yield the slope of the acoustic dispersion curves in the relevant directions of the wave vector.

Given the velocities of sound for selected wave vectors, one can also evaluate the elastic constants. The basic relationships are corollaries of the equations of motion for elastic waves, which can be written (neglecting gravitational forces) as [5.18]

$$\frac{\partial \sigma_{\alpha\beta}}{\partial r_\beta} = \varrho \frac{\partial^2 u_\alpha}{\partial t^2} , \tag{5.100}$$

where the $\sigma'_{\alpha\beta}$s are the components of the stress tensor, the u'_αs are the components of the shift, and ϱ is the density of the crystal.

If the elastic properties of the crystal obey Hooke's law, (5.100) has solutions of the following form

$$u_\alpha = A_\alpha \exp\left[\frac{2\pi\mathrm{i}}{\lambda}(k_1 r_1 + k_2 r_2 + k_3 r_3 - v_k t)\right] \quad , \tag{5.101}$$

where v_k is the velocity of the wave in the direction $\boldsymbol{k} = \{k_1, k_2, k_3\}$ and λ is the wavelength. After simple transformations, (5.100) can be reduced to

$$\sum_{\beta}(M_{\alpha\beta} - \delta_{\alpha\beta}\varrho v_k^2)A_\beta = 0, \tag{5.102}$$

where $M_{\alpha\beta}(= M_{\beta\alpha})$ is a function of the elastic constants and the direction of the wave vector; viz.,

$$\begin{aligned} M_{11} =& k_1^2 C_{11} + k_2^2 C_{66} + k_3^2 C_{55} + 2k_2k_3C_{56} \\ &+ 2k_1k_3C_{15} + 2k_1k_2C_{16} \quad , \end{aligned} \tag{5.103}$$

$$\begin{aligned} M_{22} =& k_1^2 C_{66} + k_2^2 C_{22} + k_3^2 C_{44} 2k_2k_3C_{24} \\ &+ 2k_1k_3C_{46} + 2k_1k_2C_{26} \quad , \end{aligned} \tag{5.104}$$

$$\begin{aligned} M_{33} =& k_1^2 C_{55} + k_2^2 C_{44} + k_3^2 C_{33} + 2k_2k_3C_{34} \\ &+ 2k_1k_3C_{35} + 2k_1k_2C_{46} \quad , \end{aligned} \tag{5.105}$$

$$\begin{aligned} M_{12} =& k_1^2 C_{56} + k_2^2 C_{24} k_3^2 C_{34} + k_2k_3(C_{23} + C_{44}) \\ &+ k_1k_3(C_{36} + C_{45}) + k_1k_2(C_{25} + C_{46}), \end{aligned} \tag{5.106}$$

$$\begin{aligned} M_{13} =& k_1^2 C_{15} + k_2^2 C_{46} + k_3^2 C_{35} + k_2k_3(C_{35} + C_{46}) \\ &+ k_1k_3(C_{13} + C_{55}) + k_1k_2(C_{14} + C_{56}), \end{aligned} \tag{5.107}$$

$$\begin{aligned} M_{23} =& k_1^2 C_{16} + k_2^2 C_{26} + k_3^2 C_{45} + k_2k_3(C_{25} + C_{46}) \\ &+ k_1k_3(C_{14} + C_{56}) + k_1k_2(C_{12} + C_{66}). \end{aligned} \tag{5.108}$$

The condition that (5.102) has non-trivial solutions is

$$|M_{\alpha\beta} - \delta_{\alpha\beta}\varrho v_k^2| = 0. \tag{5.109}$$

In elasticity theory this determinantal equation, known as Christoffel's equation, is usually used to find, given the elastic constants, the phase velocities of the plane waves and the polarization directions in each of these waves for a selected direction in the crystal. In our case, however, we are interested in the inverse problem, i.e., to derive the elastic constants from the elastic wave velocities obtained experimentally or obtained from the slopes of the calculated acoustic dispersion curves at $\boldsymbol{q} = 0$. This inverse problem can be solved iteratively, starting from some trial elasticity tensor and adjusting it so as to obtain the best least-squares fit for the velocities. Such a procedure, however, can introduce additional uncertainties in the elastic constants besides those arising from the limited accuracy of the experiment. Moreover, in applying this procedure to crystals of low symmetry, one may encounter computational instabilities which may preclude an unambiguous determination of some elastic constants. For these reasons, the behavior of the calculated acoustic branches at $\boldsymbol{q}\rightarrow 0$ can be investigated best by comparing the elastic wave velocities themselves rather than the elastic constants.

An important, although indirect test for the reliability of the calculated eigenfrequencies and eigenvectors is provided by the anisotropic temperature factors obtained in standard crystallographic treatments of X-ray diffraction data. Let us consider, following *Pawley* [5.19], a particular molecule taking part in a vibrational mode with eigenvector $\boldsymbol{e}$ and frequency ω. Let us also write the complex displacement amplitudes $\boldsymbol{v}$ and $\boldsymbol{\theta}$, which are related to the components of $\boldsymbol{e}$ through (5.40–42), in the form

$$v_\alpha = x_\alpha + \mathrm{i} y_\alpha, \tag{5.110}$$

$$\theta_\alpha = \varphi_\alpha + \mathrm{i}\psi_\alpha. \tag{5.111}$$

If $\boldsymbol{\varrho}^a$ denotes the position vector of atom a, the vector displacement of this atom in the given mode is

$$\boldsymbol{u}^a = (\boldsymbol{x} + \boldsymbol{\varphi}\times\boldsymbol{\varrho}^a)\cos\omega t + (\boldsymbol{y} + \boldsymbol{\psi}\times\boldsymbol{\varrho}^a)\sin\omega t. \tag{5.112}$$

In (5.112) we have dropped a factor responsible for the correct phase relationships between different molecules in the unit cell. This factor is not important for our discussion.

The mean-square displacement of atom a in the direction of a unit vector $\boldsymbol{l}$ is

$$\frac{1}{2}\sum_{\alpha\beta} u_\alpha^a u_\beta^a l_\alpha l_\beta. \tag{5.113}$$

Substituting (5.112) into (5.113) and summing over all of the modes, we obtain the total mean-square displacement in the direction $\boldsymbol{l}$,

$$\langle u^2\rangle = \sum_{\alpha\beta}\Big\{\sum_{\text{modes}}[(\boldsymbol{x}+\boldsymbol{\varphi}\times\boldsymbol{\varrho}^a)_\alpha(\boldsymbol{x}+\boldsymbol{\varphi}\times\boldsymbol{\varrho}^a)_\beta$$
$$+ (\boldsymbol{y}+\boldsymbol{\psi}\times\boldsymbol{\varrho}^a)_\alpha(\boldsymbol{y}+\boldsymbol{\psi}\times\boldsymbol{\varrho}^a)_\beta]\Big\}. \tag{5.114}$$

We now note that the expression in the curly brackets is nothing but the $\alpha\beta$ component of the symmetric tensor $\underline{\underline{B}}^a$ used in crystallography to describe the anisotropic thermal motion of individual atoms in a crystal. We remind the reader that in crystallographic computations the thermal motion of an atom is taken into account by the smearing function,

$$\sigma(\boldsymbol{r}^a) = (2\pi)^{-3/2}\det(\underline{\underline{B}}^a)^{-1/2}\exp\left[-\frac{1}{2}\sum_{\alpha\beta}(\underline{\underline{B}}^a)^{-1}_{\alpha\beta} r_\alpha^a r_\beta^a\right] , \tag{5.115}$$

whose transform, when multiplied by the scattering factor of an atom at rest, yields the scattering factor of an atom in thermal motion. The transform of (5.115) is usually written in the form

$$g^a(h_1, h_2, h_3) = \exp\Big[-(b_{11}^a h_1^2 + b_{22}^a h_2^2 + b_{33}^a h_3^2 + 2b_{12}^a h_1 h_2 + 2b_{13}^a h_1 h_3 + 2b_{23}^a h_2 h_3)\Big] \quad , \tag{5.116}$$

where h_1, h_2 and h_3 are the Miller indices and

$$b_{11}^a = 2\pi^2 (a^*)^2 B_{11}^a \quad , \tag{5.117}$$

$$b_{12}^a = 2\pi^2 a^* b^* B_{12}^a \quad , \tag{5.118}$$

etc., and a^*, b^* and c^* are the reciprocal axis lengths.

In crystallographic applications the thermal parameters $b_{\alpha\beta}$ are treated as adjustable parameters and are obtained from a least-squares structure refinement. Thus, having solved the lattice-dynamical problem, one has an opportunity of assessing the reliability of the calculated eigenvectors by computing $b_{\alpha\beta}$ from (5.114–118) and comparing them to experiment. The comparison is more straightforward if the thermal motion of the atoms is interpreted in terms of the rigid-body translation, libration and screw correlation tensors $\underline{\underline{T}}, \underline{\underline{L}}$ and $\underline{\underline{S}}$, as introduced by *Cruickshank* [5.20], and by *Schomaker* and *Trueblood* [5.21]. The relation between $\underline{\underline{B}}^a$ and the three tensors is

$$\underline{\underline{B}}^a = \underline{\underline{T}} + \underline{\underline{R}}^a \underline{\underline{L}}\, \underline{\underline{\tilde{R}}}^a - \underline{\underline{R}}^a \underline{\underline{S}} - \underline{\underline{\tilde{S}}}\, \underline{\underline{\tilde{R}}}^a, \tag{5.119}$$

where $\underline{\underline{R}}^a$ is defined by

$$\underline{\underline{R}}^a = \begin{pmatrix} 0 & -\varrho_3^a & \varrho_2^a \\ \varrho_3^a & 0 & -\varrho_1^a \\ -\varrho_2^a & \varrho_1^a & 0 \end{pmatrix} . \tag{5.120}$$

In practice, the tensors $\underline{\underline{T}}$, $\underline{\underline{L}}$ and $\underline{\underline{S}}$ are usually adjusted by a least-squares method to give the best fit between the $\underline{\underline{B}}^a$'s of (5.119) and those derived in an X-ray analysis. It is also possible to make the rigid-body assumption at an early stage of a crystal structure refinement and determine $\underline{\underline{T}}$, $\underline{\underline{L}}$ and $\underline{\underline{S}}$ directly, without first computing the individual atomic temperature factors [5.22].

By equating the components of $\underline{\underline{B}}^a$ from (5.119) to those given in the curly brackets of (5.114), we immediately obtain expressions for $\underline{\underline{T}}$, $\underline{\underline{L}}$ and $\underline{\underline{S}}$ in terms of the real and imaginary coefficients of $\boldsymbol{v}$ and $\boldsymbol{\theta}$:

$$T_{\alpha\beta} = \frac{1}{2} \sum_{\text{modes}} (x_\alpha x_\beta + y_\alpha y_\beta), \tag{5.121}$$

$$L_{\alpha\beta} = \frac{1}{2} \sum_{\text{modes}} (\varphi_\alpha \varphi_\beta + \psi_\alpha \psi_\beta), \tag{5.122}$$

$$S_{\alpha\beta} = \frac{1}{2} \sum_{\text{modes}} (\varphi_\alpha x_\beta + \psi_\alpha y_\beta). \tag{5.123}$$

It is worth noting that if the site symmetry of a molecule is $\bar{1}$, the tensor $\underline{\underline{S}}$ vanishes so that the rigid-body motion of the molecule can be described in terms of $\underline{\underline{T}}$ and $\underline{\underline{L}}$ only. This is consistent with the fact that the translational and librational components of an eigenvector are $\pi/2$ out of phase for molecules centered on inversion centers [5.22]. In such a case, y_α and φ_α in (5.110,111) vanish and all of the terms in (5.123) are identically equal to zero.

5.6 Numerical Results

Although the theory of lattice dynamics was formulated over 70 years ago and its extension to molecular crystals did not present any problems, numerical applications to organic crystals did not appear until the 1960's. The earliest numerical studies were reported by *Cochran* and *Pawley* [5.14] who calculated the frequencies of the external vibrational modes of hexamethylenetetramine within the framework of the rigid-molecule approximation. However, the force constants were not evaluated from an intermolecular model potential but were treated as adjustable parameters. This was possible due to the high symmetry of the crystal, which substantially reduced the number of independent elements in the force constant matrix.

The first lattice-dynamical application of the atom-atom potential method was reported by *Kitaigorodsky* et al. [5.23] who evaluated the mean-square frequency of the external-mode of adamantane:

$$\overline{\omega^2} = \frac{1}{6} \sum_{\tau} \phi_{\tau}^{oo} / M_{\tau}. \tag{5.124}$$

The force constants in (5.124) were computed by numerical differentiations of the lattice energy with respect to molecular translations and rotations. The calculations were carried out at various temperatures, using the experimentally observed temperature dependence of the lattice period. The resulting $\overline{\omega^2}$ was related to the Debye characteristic temperature [5.18]

$$\theta_D = \sqrt{\frac{5}{3}\frac{\hbar}{k}} \sqrt{\overline{\omega^2}}, \tag{5.125}$$

and then used to evaluate the Grüneisen constant

$$\gamma = -\frac{V}{\theta_D}\frac{\partial \theta_D}{\partial V}. \tag{5.126}$$

A year later, *Kitaigorodsky* [5.24] outlined the various possible applications of the atom-atom potential method to organic solids. In particular, the possibility of solving the lattice-dynamical problem using force constants deduced from atom-atom potential calculations was emphasized.

Detailed calculations of the frequencies of external vibrational modes within the framework of the atom-atom potential method were first performed by *Oliver* [5.25,26], *Pawley* [5.27] and *Weulersse* [5.28] for the "classical" organic molecular crystals of benzene, naphthalene and anthracene. Similar calculations were carried out somewhat later by *Kitaigorodsky* and *Mukhtarov* [5.29,30] for biphenyl, naphthalene and anthracene. By that time, quite reliable parameters for the potentials of many atomic species were available, and the only remaining obstacle was actually a shortage of experimental data on external vibrational modes. This situation changed drastically in the 1970's due to the considerable progress in low-frequency infrared absorption and Raman scattering techniques. Also, the first applications of coherent, one-phonon scattering of slow neutrons to organic solids were reported to yield direct information on the dispersion relations.

Although a considerable shortage of experimental data on the dynamical properties of organic solids still exists, the measurements and model calculations reported thus far do allow some conclusions to be made with respect to the possibilities of the atom-atom potential method. In the following subsections we shall discuss the results for some selected organic crystals, concentrating when possible on the most recent work and omitting some earlier results.

5.6.1 Benzene

At normal pressures benzene crystallizes in an orthorhombic structure (space group *Pbca*) with the four molecules in the unit cell lying on inversion centers. The lattice vibrations of benzene have been studied extensively, both theoretically and experimentally. The optically active normal modes of benzene are available from far-infrared absorption [5.31–33] and Raman scattering [5.34–38] measurements. Similar measurements have been reported for fully deuterated benzene [5.35,39]. Neutron scattering experiments have been carried out for both C_6H_6 [5.40–42] and C_6D_6 [5.43].

A detailed study of the lattice vibrations of benzene was performed by *Taddei* et al. [5.5]. Using the formalism discussed in Sect. 5.2, they thoroughly analyzed both the internal and external vibrations and examined the intermode mixing. The results [5.5] are summarized in Tables 5.1,2 and refer to a temperature of 138 K.

Since the unit cell of benzene contains four molecules, there are $6\times4 = 24$ distinct external modes for a general wave vector in the Brillouin zone.

Table 5.1. Observed and calculated $q = 0$ lattice frequencies of benzene [cm^{-1}] [5.5, 5.42,44]

Symmetry	a	b	c	d	e	f	g
A_g	57	57	55	55	64	57	46
	79	84	75	76	82	68	73
	92	108	95	96	106	90	92
B_{1g}	57	63	60	61	71	53	59
	100	106	96	97	109	91	109
	128	145	128	131	142	128	132
B_{2g}	79	91	83	84	90	79	81
	90	105	93	94	106	87	90
	...	113	101	102	113	98	117
B_{3g}	61	79	66	67	72	56	64
	92	101	89	90	103	81	87
	128	143	128	129	142	127	128
A_u	ia	59	56	57	61	58	61
	ia	72	65	66	75	65	67
	ia	100	95	98	108	101	101
B_{1u}	70	76	72	72	79	75	75
	85	93	86	87	95	89	89
B_{2u}	53	59	57	59	66	59	61
	94	104	99	102	111	106	107
B_{3u}	53	53	53	53	59	55	55
	94	101	98	98	110	103	102
Standard deviation		10.57	3.31	4.10	12.43	6.54	6.31

[a]Observed at 140 K [5.31,37] except the lower value for B_{2u} which was measured at 173 K (the shorthand ia stands for inactive); [b]Calculated using the Taddei-Califano method and the W66 potentials; [c]Calculated as in b but with the W67 potentials; [d]Calculated using the rigid-molecule approximation and the W67 potentials; [e]Calculated as in d but with the "universal" potentials; [f]Calculated as in d but with the atom-atom potentials of Table 5.3 (without the quadrupole terms); [g]Calculated as in f but including the quadrupole terms

At the center (point Γ, $q = 0$) of the zone, the number of distinct modes is not reduced by symmetry. The symmetry species of the 24 external modes are given in Table 5.1, which compares the experimental and calculated values of the $q = 0$ frequencies.

To examine the effect of a potential on the frequencies, two different model potentials corresponding to the sets W66 and W67 in Table 3.1 were used. The results given in columns b (set W66) and c (set W67) of Table 5.1 are to be compared with the experimental frequencies in column a. It can be seen that a change in potentials markedly influences the frequencies. This is particularly true of the highest frequencies (B_g), which are shifted by as much as 15 cm^{-1}. The W67 set is on the whole superior to the W66 set in predicting the frequencies; the standard deviations being 3.3 and 10.6 cm^{-1}, respectively.

Column d of Table 5.1 lists the frequencies of the external modes calculated using the rigid-molecule approximation and the W67 parameter set. A comparison with the frequencies in column c shows that the effect of the internal vibrational modes on the external ones is to lower the latter

Table 5.2. Observed and calculated splittings for the infrared active internal models of benzene [cm^{-1}] [5.5]

Molecular mode	Symmetry D_{6h}	D_{2h}	Observed	Calculated
ν_{11}	a_{2u}	B_{1u}	0.0	0.0
		B_{2u}	25.3	16.0
		B_{3u}	-1.0	0.6
ν_{12}	b_{1u}	B_{1u}	0.0	0.0
		B_{2u}	1.5	0.4
		B_{2u}	1.0	0.0
ν_{14}	b_{2u}	B_{1u}	0.0	0.0
		B_{2u}	3.5	2.2
		B_{3u}	1.5	1.7
ν_{15}	b_{2u}	B_{1u}	0.0	0.0
		B_{2u}	7.8	6.8
		B_{3u}	5.5	4.6
ν_{16}	e_{2u}	B_{1u}	0.0	0.0
		B_{2u}	...	-0.3
		B_{3u}	0.8	1.4
		B_{1u}	...	14.3
		B_{2u}	...	3.0
		B_{3u}	14.5	11.8
ν_{17}	e_{2u}	B_{1u}	0.0	0.0
		B_{2u}	...	-2.4
		B_{3u}	4.0	0.0
		B_{1u}	...	8.7
		B_{2u}	2.0	3.2
		B_{3u}	12.8	8.2
ν_{18}	e_{1u}	B_{1u}	0.0	0.0
		B_{2u}	-2.0	0.2
		B_{3u}	-0.9	-0.8
		B_{1u}	4.5	4.9
		B_{2u}	5.0	2.7
		B_{3u}	4.0	3.9
ν_{19}	e_{2u}	B_{1u}	0.0	0.0
		B_{2u}	-0.9	-1.6
		B_{3u}	...	-0.8
		B_{1u}	4.5	2.9
		B_{2u}	1.7	0.4
		B_{3u}	2.2	0.7

by $1-3\,cm^{-1}$. This is within the accuracy of predicting the observed frequencies, so that the rigid-molecule approximation is quite adequate for a treatment of the external vibrations of benzene.

For a comparison, we have also included the rigid-molecule results of *Bokhenkov* et al. [5.42] (column e) and *Califano* et al. [5.44] (columns f, g) in Table 5.1. The numerical values in column e were obtained using the usual atom-atom model potential, whose parameters correspond to the universal set of Table 3.1. A somewhat more complicated model potential was used by *Califano* et al. [5.44] who combined a sum of (6–exp) atom-atom potentials with an electrostatic term describing the interaction of molecular quadrupoles. The parameters of this model potential, listed in Table 5.3, were determined by fitting to the lattice frequencies, equilibrium structures and heats of sublimation of benzene and naphthalene.

Table 5.3. Parameters of the model potential of *Califano* et al. [5.44] for benzene and naphthalene [kcal/mol, Å, esu·10^{26}]

Interaction	A	B	C
H...H	39	2939	3.746
H...C	75	8037	3.703
C...C	524	265834	3.909

Quadrupole moments

Benzene	$Q = -8.1$
Naphthalene[a]	$Q_{xx} = -14.4,\ Q_{yy} = 6.4,\ Q_{zz} = 8.0$

[a]The x axis is chosen perpendicular to the molecular plane, y is the long and z the short in plane symmetry axes

It can be seen from Table 5.1 that the universal potentials provide the worst fit to the experimental frequencies; the standard deviation being 12.4 and the largest deviation being 17 cm^{-1}. The model potential of *Califano* et al. [5.44] is much more successful, although somewhat worse than *Williams'* W67 model.

We now turn the reader's attention to Table 5.2, which compares the observed splittings of the IR-active internal modes with those calculated by *Taddei* et al. [5.5] using the W67 potentials. Let us recall that in the Taddei-Califano formalism the splittings are derived in a straightforward manner, i.e., as solutions of (5.30) for $\tau \leq 3n_m - 6$. It can be seen that the overall agreement with experiment is quite satisfactory.

An important calculation performed by *Taddei* et al. [5.5] involved the vibrational frequencies of benzene as a function of the truncation radius, r^{tr}. The results for some external and internal modes are shown in Fig. 5.2. It can be seen that the external modes are strongly dependent on r^{tr}, and that a convergence is only achieved if $r^{tr} > 5.5$ Å. The effect of the truncation radius on the internal modes is much less important: a summation within 3.5–4 Å is already sufficient to carry the frequencies to convergence.

Taddei et al. [5.5] have attempted to refine their parameters by fitting the calculated frequencies of the external modes, heats of sublimation and torques of benzene to the corresponding observed values. However, only a slight improvement in the fit was achieved, compared to that afforded by the W67 parameter set (the standard deviation between the calculated and observed frequencies was only reduced by 0.05 cm^{-1}). The best parameter sets, derived from the W66 and W67 sets, have already been listed in Table 3.1 (TW66 and TW67).

An extension of the lattice-dynamical calculations to $\boldsymbol{q} \neq 0$ vibrational modes has been reported by *Bonadeo* and *Taddei* [5.45]. The dispersion relations were determined for a number of directions in the Brillouin zone using the W67 potentials. Figure 5.3 shows the dispersion curves for some high-symmetry points and directions specified in Fig. 5.4. At the points Z, A

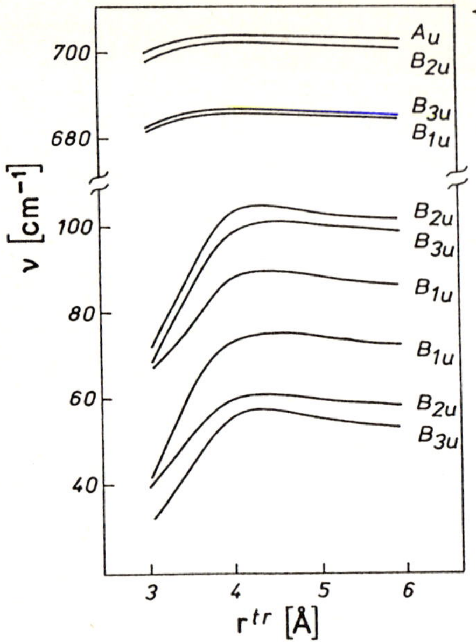

Fig. 5.2. Dependence of the calculated frequencies of the translational modes and ν_{11} components of C_6H_6 on the truncation radius [5.5]

Fig. 5.3. Predicted phonon dispersion relations for crystalline benzene at 138 K in four directions in the Brillouin zone [5.45] (see Fig. 5.4 for the specification of the directions)

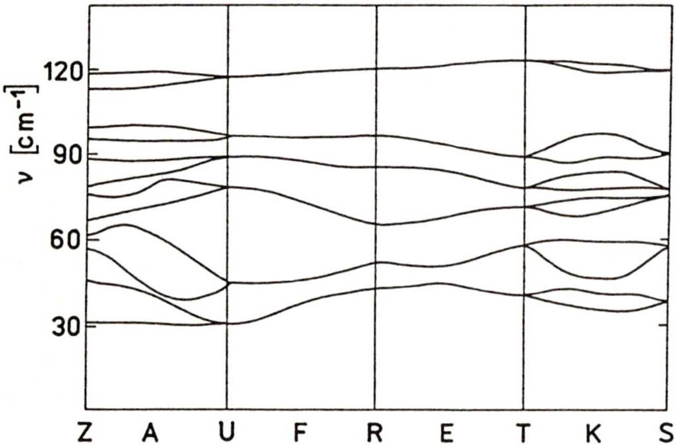

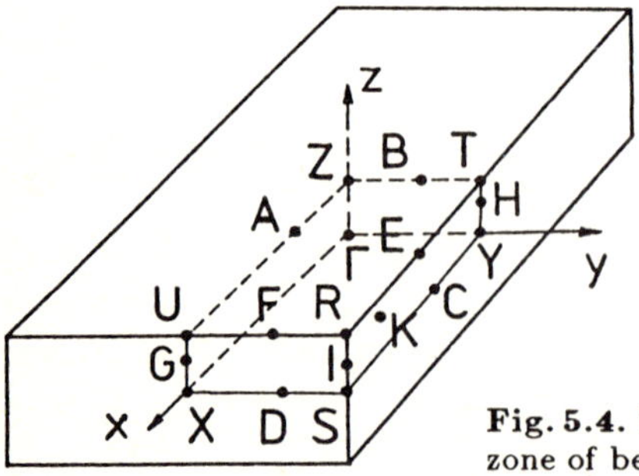

Fig. 5.4. High-symmetry points and lines in the Brillouin zone of benzene

and K the vibrational modes are two-fold degenerate and, hence, there are 12 distinct external modes. At the points U, F, R, E, T and S the external modes are four-fold degenerate and only six can be observed. These latter points are particularly suitable for coherent inelastic neutron scattering experiments because (1) the four-fold degeneracy should considerably enhance the scattering intensity, and (2) the appearance of only 6 distinct modes, instead of 24 for a general wave vector, imposes less stringent demands on the experimental resolution.

The frequencies of the external modes at the points U, S, T and R have been measured by *Powell* et al. [5.43] five years after the frequencies had been predicted by lattice-dynamical calculations. The measurements were made on fully deuterated benzene in order to avoid the strong incoherent scattering of neutrons from the hydrogen atoms. All six modes were resolved at U, S and R, while one of the modes at T was not observed. The frequencies measured by *Powell* et al. [5.43] are compared with the theoretical results of *Bonadeo* and *Taddei* [5.45] in Table 5.4. It is seen that the agreement is exceptionally good; the largest deviation being 5 cm^{-1}.

The observed intensities of the neutron scattering at the four points examined are shown in Fig. 5.5 [5.43]. The vertical lines represent the intensities computed from (5.93) using the calculated eigenvectors. One can

Table 5.4. Observed and calculated frequencies of the external modes of benzene-d_6 at four points in the Billouin zone [cm^{-1}]

Point in Brillouin zone	Symmetry	Observed [5.43]	Calculated [5.45]
R (π/a, π/b, π/c)	u	91	94
		54	50
		41	41
	g	106	109
		77	78
		60	60
S (π/a, π/b, 0)		114	111
		85	83
		75	74
		71	71
		56	56
		44	38
T (0, π/b, π/c)		112	113
		89	85
		71	73
		...	67
		56	56
		43	40
U (π/a, 0, π/c)		102	107
		87	91
		81	83
		67	72
		47	43
		32	29

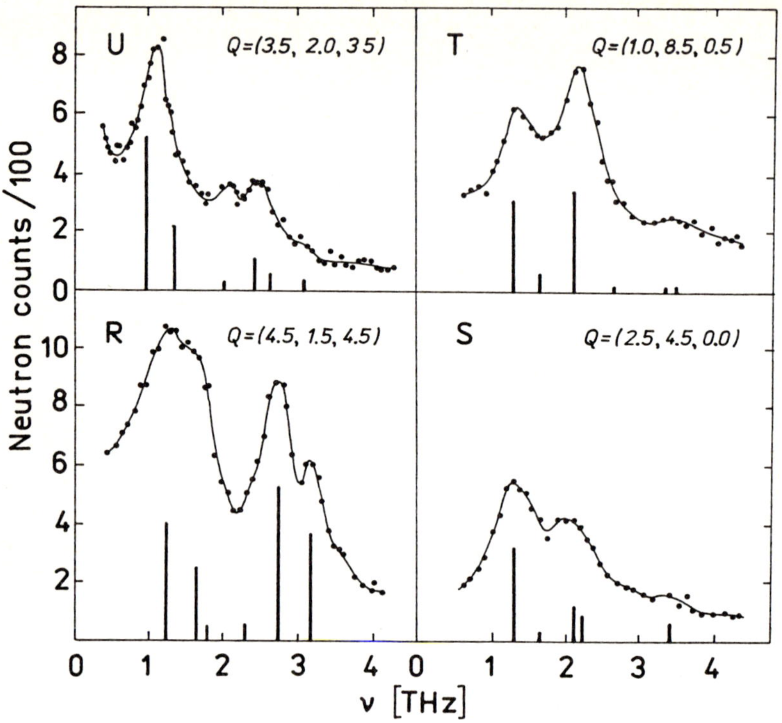

Fig. 5.5. Typical distributions of scattered neutrons for d-benzene, corresponding to four momentum-transfer vectors $\boldsymbol{Q}$. The calculated intensities are represented by the vertical lines [5.43]

see that the theoretical intensities show some discrepancies when compared with the observed ones. Most significant are the discrepancies at *R*. In this case the two highest vibrational modes are predicted to be more intense than the lowest vibrational modes. The experiment shows the opposite picture. In spite of these discrepancies, the overall agreement between experiment and theory is satisfactory, which suggests that the calculated eigenvectors are fairly good.

Another opportunity to test the lattice-dynamical model of *Bonadeo* and *Taddei* [5.45] arose fairly recently, due to *Bokhenkov* et al. [5.42], who reported a well-resolved spectrum of the incoherent inelastic scattering of neutrons from crystalline benzene. The external-mode part of the spectrum is shown in Fig. 5.6. Also shown in the same figure are two theoretical spectra. The labels "W67" and "universal" refer to the potential sets used in the calculations.

It can be seen in Fig. 5.6 that the calculated spectra show a fairly good qualitative agreement with experiment. There are, however, some quantitative discrepancies in the positions and relative intensities of the peaks. These discrepancies may arise from at least three sources. The first source

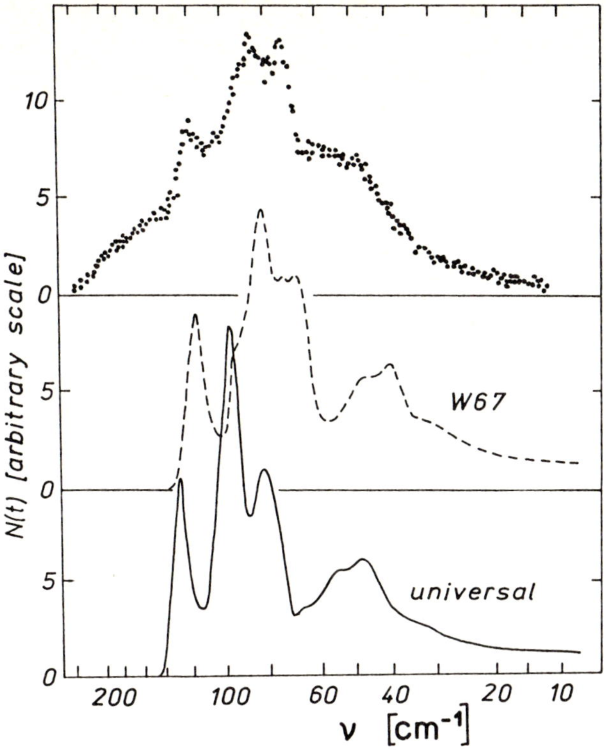

Fig. 5.6. Experimental and calculated spectra of the incoherent inelastic scattering of neutrons for benzene at 80 K in the external-mode region [5.42]

is the model potential used to determine the density of phonon states. The "universal" potential systematically leads to overestimated $\boldsymbol{q} = 0$ frequencies (Table 5.1). A similar trend is observed in Fig. 5.6 where the spectrum corresponding to the "universal" potential exhibits a marked shift to higher frequencies. The W67 potential yields better results for the positions of the peaks, although a slight shift to lower frequencies can clearly be observed. A discouraging feature of the W67 curve is the absence of a peak at $\sim 90\,\mathrm{cm}^{-1}$, which is well-resolved in the observed spectrum. The peak at $\sim 90\,\mathrm{cm}^{-1}$ can be observed in the universal spectrum, although its relative intensity is substantially underestimated.

A second possible source of discrepancies between the calculated and observed spectra is the neglect of many-phonon and multiple scattering processes in (5.95,96). However, the contribution of these two processes has been shown [5.42] to amount to 2 to 5% of the integral intensity, which cannot pronouncedly affect the calculated spectrum.

Finally, discrepancies may partly be due to the anharmonicity of the lattice vibrations. The fact is that the neutron scattering experiments and the lattice-dynamical calculations refer to different temperatures, 80 and 138 K, respectively. Had the experiments been carried out at 138 K, the spectrum would have been somewhat shifted towards lower frequencies (by about 5–6 cm^{-1}, judging from the observed shift of the optically active modes on going from 77 to 140 K [5.31]). Such a shift should improve the agreement between the experimental and the W67 spectra, while the fit provided by the universal potential should worsen.

5.6.2 Naphthalene

Naphthalene forms monoclinic crystals (space group $P2_1/a$) with two molecules in the unit cell. The dynamical properties of naphthalene are known particularly well. The optically active lattice vibrations of hydrogenous and deuterated samples have been studied at 4.7, 77 and 296 K [5.33,46–50]. The dispersion relations have been determined from the coherent inelastic scattering of neutrons in several directions in the Brillouin zone at 6, 77 and 98 K [5.12,13,51–53]. Well-resolved spectra of the incoherent inelastic neutron scattering have been obtained at 4.7, 80 and 296 K [5.52,54–57]. The anharmonicity of lattice vibrations have been separately studied [5.58–60]. The elastic constants of naphthalene are also known for a wide temperature range [5.61]. All this information provides a good basis for the testing of the various lattice-dynamical models.

Most calculations of the lattice dynamics of naphthalene have been performed using the rigid-molecule approximation. Thus, before discussing the results of these calculations it is worthwhile to determine the magnitude of the error introduced by treating the naphthalene molecule as being absolutely rigid. Compared to benzene, naphthalene is definitely more deformable, which is evidenced by a much smaller gap between the external and internal vibrational bands ($\sim$50 cm^{-1} compared with $\sim$280 cm^{-1} for benzene). A quantitative analysis of the interplay between the external and internal vibrations of naphthalene has been reported by *Krivenko* and *Sheka* [5.63], and by *Pawley* and *Cyvin* [5.62]. The method used by the latter authors to solve the lattice-dynamical problem has much in common with the method suggested by *Taddei* et al. [5.5] (Sect. 5.2). As in the *Taddei-Califano* method, the molecular conformation and intramolecular force field in the crystal are assumed to be the same as in the free-molecule state. The only difference lies in the choice of the displacement coordinates. Instead of using the free-molecule normal coordinates and the six displacement coordinates of the molecule, *Pawley* and *Cyvin* [5.62] preferred to use Cartesian atomic displacements. Each force constant given in (5.5) was represented as a sum of intra- and intermolecular contributions. The intermolecular contribution was evaluated using the universal atom-atom potentials. The intramolecular contribution to the force constants was taken from a phenomenological force field derived by *Cyvin* et al. [5.64] in a standard normal-coordinate analysis of the internal vibrational modes. The relation-

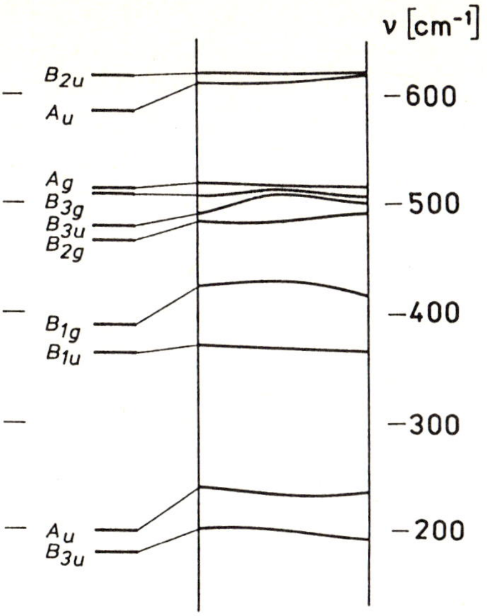

Fig. 5.7. Predicted dispersion of the ten lowest-frequency internal modes and their free-state values for naphthalene [5.62]

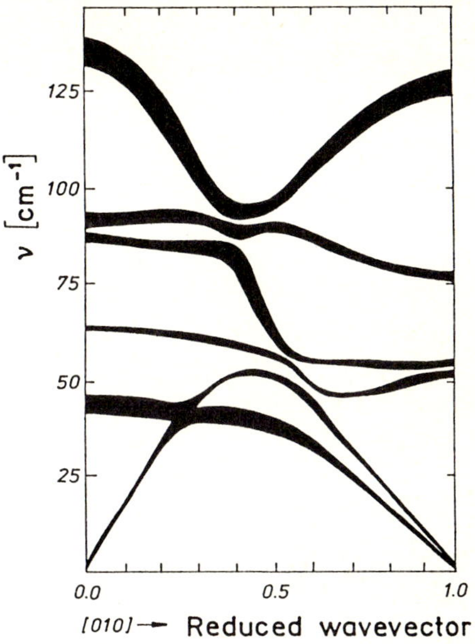

Fig. 5.8. Predicted external-mode dispersion for naphthalene [5.62]

ship between the force constants given in (5.5) and the normal-coordinate force constants is given by

$$\underline{\underline{\phi}} = \underline{\underline{\tilde{L}}}\,\underline{\underline{\phi}}_{\text{norm}}\,\underline{\underline{L}}, \tag{5.127}$$

where $\underline{\underline{L}}$ is the transformation matrix from the normal to the Cartesian atomic displacements.

The results of the calculations by *Pawley* and *Cyvin* [5.62] are presented in Figs. 5.7,8. Figure 5.7 shows the dispersion of the ten lowest-frequency internal modes in the symmetry direction [010]. For the sake of clarity, the symmetric and antisymmetric branches are illustrated separately; the former are shown in the 0 to 0.5 (the boundary of the Brillouin zone) range and the latter in the 0.5 to 1.0 range. The corresponding free-molecule modes are represented by the horizontal lines on the left. It is seen that the internal modes are appreciably raised in energy compared to the corresponding free-molecule modes. For the lowest modes A_u, B_{1g} and B_{3u} the shift is as much as 25–40 cm^{-1}. The dispersion of the internal modes is also rather pronounced. Thus, the range of frequency of the B_{3u} (next to the lowest) mode is 16 cm^{-1}.

The dispersion of the external vibrational modes is shown in Fig. 5.8. Again the point $q = 0.5$ separates the symmetric branches on the left from the antisymmetric branches on the right. The lower boundary of the bands in Fig. 5.8 corresponds to the deformable-molecule calculation. The upper boundary represents the rigid-molecule result. The largest band-width is about 10 cm^{-1}. A 5 cm^{-1} width is quite common. These values provide the reader with an idea of the accuracy that can be expected from a rigid-molecule treatment of naphthalene.

We shall now compare the results of the rigid-molecule calculations with the available experimental data. We shall start our discussion with the $q = 0$ external modes. For the crystal symmetry $P2_1/a$, $Z = 2(\bar{1})$, there are nine optical modes at the zone center; six modes, $3\Gamma_1(A_g) + 3\Gamma_3(B_g)$, represent pure librations while the other three, $2\Gamma_2(A_u) + \Gamma_4(B_u)$, represent pure translations. Eight of the nine optical modes have been observed spectroscopically at room temperature [5.46]. The observed frequencies are compared in Table 5.5 with the lattice-dynamical rigid-molecule predictions of *Filippini* et al. [5.65] and *Califano* et al. [5.44]. Four of the model potentials that were used are of the (6-exp) type. The model of *Califano* et al. [5.44], in which the sum of (6-exp) atom-atom potentials is complemented by a quadrupole-quadrupole term (Table 5.3), was also used.

Table 5.5. Observed and calculated $q = 0$ frequencies of the external modes of naphthalene at 296 K [cm^{-1}]

Symmetry	a	b	c	d	e	f	g
Librational modes							
B_g	46	45	47	48	42	42	44
A_g	51	52	58	61	52	48	49
B_g	71	69	79	76	64	70	75
A_g	74	81	92	91	76	73	73
A_g	109	114	122	138	118	101	106
B_g	125	114	122	128	106	101	124
Translational modes							
A_u	...	42	45	46	38	45	46
B_u	66	53	57	56	47	62	70
A_u	99	89	91	88	67	101	97
Standard deviation		8.2	10.4	14.6	16.7	9.9	2.8

[a]Observed [5.46,49]; [b–e]Calculated using W67 (b), W66 (c), "universal" (d) and TW67 (e) parameter sets [5.65]; [f,g]Calculated using model potential of *Califano* et al. [5.44] with (g) and without (f) quadrupole term

Inspection of Table 5.5 shows that Williams' W67 model potential is definitely the best of the simple models that were tried. The W66 and universal models systematically yield overestimated frequencies (compared to the W67 values), while the TW67 potential shifts the spectrum towards lower frequencies. Unlike the situation with benzene, the addition of a quadrupole term to the intermolecular potential leads to a substantial improvement in the fit. In particular, the splitting between the highest torsions is reproduced very well, which is not the case with the simple atom-atom models.

Table 5.6. Observed and calculated $q = 0$ frequencies of the external modes of naphthalene at 6 K cm^{-1}

Symmetry	a	b	c	d	e	f
		"harmonic model"		"quasiharmonic model"		
Librational modes						
B_g	54	54	69	51	58	51
A_g	64	59	75	58	71	55
B_g	79	80	93	79	92	84
A_g	84	94	121	97	115	84
A_g	112	130	154	127	158	113
B_g	130	138	162	133	153	129
Translational modes						
A_u	58	48	52	51	58	57
B_u	78	62	63	64	71	75
A_u	107	119	126	114	115	112
Standard deviation		11.3	25.8	9.6	22.2	4.4

[a]Observed [5.53]; [b,c]Calculated using the harmonic approximation and the W67 (b) and "universal" (c) parameters sets; [d,e]Calculated using the quasiharmonic approximation and the W67 (d) and "universal" (e) parameter sets; [f]Calculated using model potential of *Righini* et al. [5.66] (see Table 5.7)

The latter result seems to indicate a deficiency of the (6-exp) atom-atom potential in modeling the electrostatic interaction energy in naphthalene. In this respect, it would be interesting to see whether the predictions of the simple models may be improved by adding Coulombic atom-atom terms.

The fit observed at room temperature does not change markedly at lower temperatures. This is illustrated in Table 5.6 which compares the experimental and calculated $q = 0$ frequencies at 6 K. The calculated frequencies presented in the table were obtained as follows. The values under the heading "harmonic model" were calculated by *Natkaniec* et al. [5.53] using the minimum-energy unit-cell dimensions and molecular orientations. The frequencies under the heading "quasi-harmonic model" were evaluated using the experimental unit-cell dimensions measured at 4.7 K [5.56] and the molecular orientations ensuring zero torque on the molecules. The model

Table 5.7. Parameters of the model potential of *Righini* et al. [5.66] for naphthalene [kcal/mol, Å, esu·10^{26}]

Interaction	A	B	C
H...H	30.7	4252	3.928
C...H	83.9	5733	3.567
C...C	546.0	45544	3.916
Quadrupole moment[a]			
$Q_{xx} = -14.4$, $Q_{yy} = 4.4$, $Q_{zz} = 10.0$			

[a]The molecular axes are chosen as indicated in the footnote to Table 5.3

potentials W67 and universal were used. Also included in Table 5.6 are the quasiharmonic results reported by *Righini* et al. [5.66]. The model potential used by them was similar to the one used by *Califano* et al. [5.44]. It combined (6-exp) atom-atom potentials with an intermolecular quadrupole term. The parameters of this model are listed in Table 5.7. Unlike the parameters of *Califano* et al. [5.44] they were derived from the low-temperature properties of naphthalene only.

Comparing the standard deviations presented in Tables 5.5 and 5.6, one sees that the overall fit between the calculated and experimental low-temperature frequencies is somewhat worse than the fit observed at room temperature. This is rather surprising because the low-temperature data are free from errors arising from a harmonic treatment of the anharmonic vibrations. It is also seen, from Table 5.6, that the use of the minimum-energy unit-cell dimensions instead of the experimental ones worsens the fit. The latter result is not, however, surprising since any deficiency of the model potential manifests itself both explicitly (through errors in the force field) and implicitly (through errors in reproducing the equilibrium crystal structure).

The lattice-dynamical model of naphthalene, based on an atom-atom representation of the intermolecular energy, has been put to a more critical test by *Mackenzie* et al. [5.13]. They used the model to interpret the results of their measurements of the coherent inelastic scattering of neutrons at 77 K. The calculations were carried out in two steps. In the first step, the lattice-dynamical problem was solved using the universal atom-atom potentials. The resulting eigenvectors were substituted into (5.93) to yield an expected pattern of the scattering intensities. This pattern was of great use in the planning of the experiment, identifying the instrumental configuration which favored the scattering from a particular phonon mode. Some typical distributions measured by *Mackenzie* et al. [5.13] together with the calculated intensities are shown in Fig. 5.9. By comparing the observed distributions with the preliminary model calculation it was possible to assign the observed peaks to the appropriate phonon branches. In such a way the frequencies of most of the phonon branches have been determined. Only three branches, S_2, A_3 and A_4, have not been assigned. The S_2 branch was difficult to observe because its scattering intensity proved to be very low in the examined range of the wave-vector transfers. The A_3 and A_4 branches were difficult to resolve due to their extreme proximity and the resulting potential-dependent nature of their frequencies, eigenvectors and scattering intensities.

The assigned phonon frequencies are shown as points in Fig. 5.10. In the second step of the calculation, these frequencies were used to refine the atom-atom-potential parameters. To this end, the frequencies were first modified so that they corresponded to the values which would be observed if the naphthalene molecule were rigid. This was done by increasing the ob-

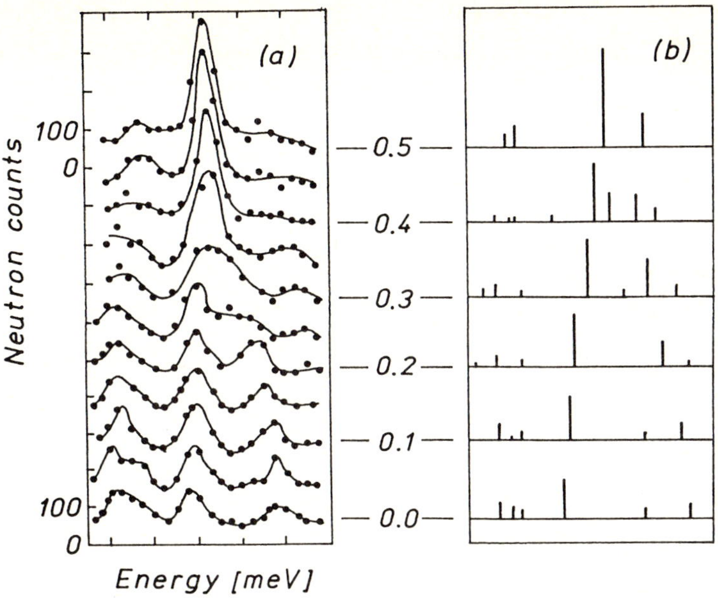

Fig. 5.9a,b. Distributions of scattered neutrons for d-naphthalene, measured (**a**) and calculated (**b**) at regular intervals along the [010] direction between the center and the boundary of the Brillouin zone centered at (3,2,0) [5.13]

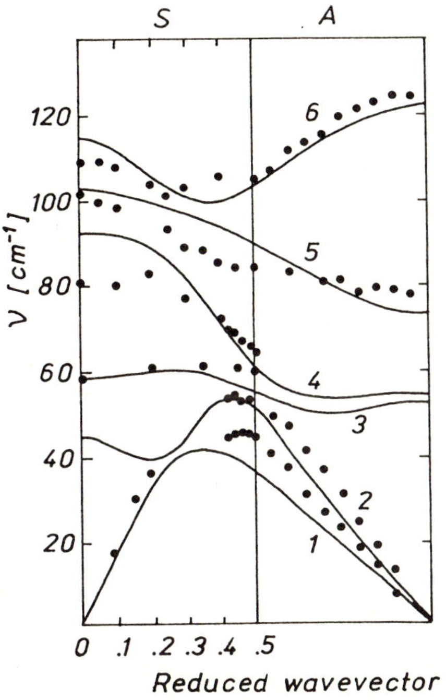

Fig. 5.10. Observed external phonon frequencies (points) and calculated dispersion curves along the [010] direction in the Brillouin zone of d-naphthalene at 77 K [5.13]. The six symmetric modes (S) are on the left side of the diagram and the six antisymmetric modes (A) on the right

served frequencies in proportion to the theoretical differences between the rigid-molecule and the deformable-molecule values, as calculated by *Pawley* and *Cyvin* [5.62] (see the discussion at the beginning of this section). The resulting frequencies were used in the subsequent least-squares fitting. At this stage the rigid-molecule approximation was introduced. Some details of the fitting have already been discussed in Chap. 3. The resulting parameter set was presented in Table 3.1 (set MPD77). The new parameter set was then used in a deformable-molecule calculation to yield the dispersion curves shown in Fig. 5.10. It is interesting to note that on going from the rigid-molecule to the deformable-molecule calculation two antisymmetric modes showed shifts opposite to those predicted by *Pawley* and *Cyvin* [5.62] using the universal potential. This suggests that the intuitive idea that the coupling between the internal and external modes should necessarily lower the latter is not always valid. Another result is that the approximate procedure of correcting the observed phonon frequencies for the expected difference between the rigid-molecule and deformable-molecule values is hardly justifiable when dealing with a refinement of parameters.

More detailed measurements of the coherent inelastic scattering of neutrons from deuterated naphthalene have been made by *Bokhenkov* et al. [5.51] at 98 K and by *Natkaniec* et al. [5.53] at 6 K. Figure 5.11 [5.53] shows the experimental phonon dispersion curves measured at 6 K along five directions in the Brillouin zone. The directions of the measurements are specified in Fig. 5.12.

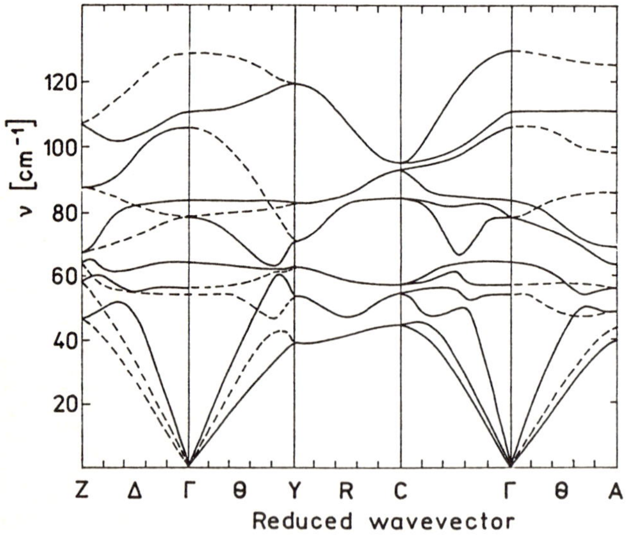

Fig. 5.11. Experimental phonon dispersion curves for d-naphthalene at 6 K [5.53]. The solid and dashed lines indicate the symmetric and antisymmetric modes, respectively. The directions of the measurements are specified in Fig. 5.12

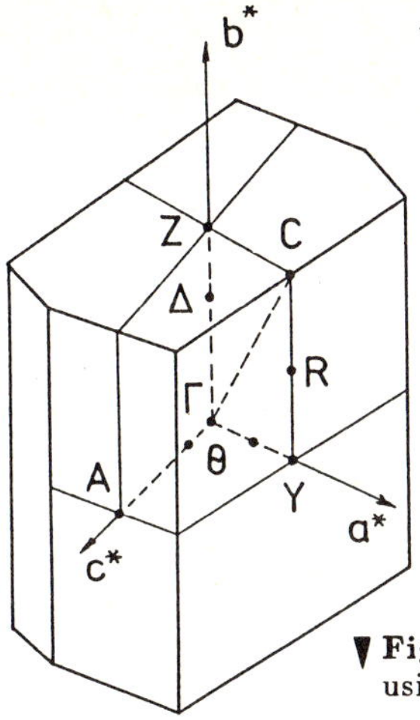

◄ **Fig. 5.12.** Brillouin zone for naphthalene, specifying the directions of the measurements of the coherent inelastic scattering of neutrons [5.53]

▼ **Fig. 5.13.** Predicted phonon dispersion curves for d-naphthalene using the W67 model potential [5.53]

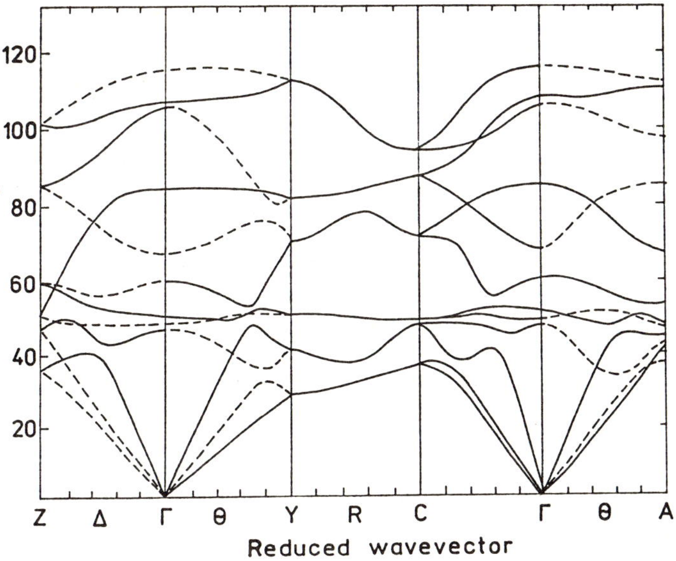

In interpreting the observed neutron distributions *Natkaniec* et al. [5.53] were again forced to resort to model calculations. The pattern of the phonon modes of naphthalene shown in Fig. 5.11 is very complicated. Thus, it is clear that the planning of the experimental scans to isolate the individual modes and the subsequent assignment of the peaks would have been impossible without preliminary model calculations.

The model calculations were carried out in the quasiharmonic approximation, i.e., the unit-cell dimensions were taken from experiment [5.56], while the molecular orientations were determined by minimizing the lattice energy. The theoretical dispersion curves calculated by *Natkaniec* et al. [5.53] using the W67 model potential are shown in Fig. 5.13. Comparing

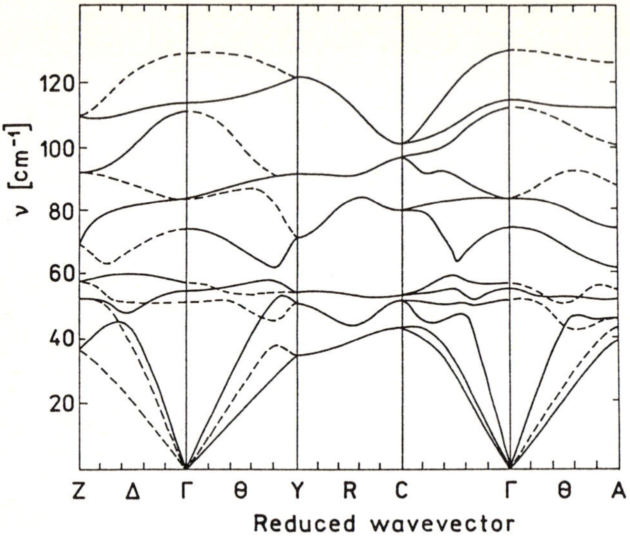

Fig. 5.14. Predicted phonon dispersion curves for d-naphthalene using the model potential of *Righini* et al. [5.66] (see Table 5.7)

Figs. 5.13 and 5.11, one can see that the overall qualitative behavior of the dispersion branches has been predicted fairly well, although there are some serious quantitative discrepancies.

A substantial improvement in the fit between the experimental and calculated dispersion relations was achieved by *Righini* et al. [5.66] by adding a quadrupole term to the sum of (6-exp) potentials. The dispersion curves calculated by *Righini* et al. [5.66] with the parameters of Table 5.7 are shown in Fig. 5.14. The deviations from the experimental data of *Natkaniec* et al. [5.53] do not exceed 10 cm^{-1}, which is surely an excellent result.

In analyzing the work by *Mackenzie* et al. [5.13], *Bokhenkov* et al. [5.51] and *Natkaniec* et al. [5.53], a critical reader may question the validity of the assignments of the neutron scattering results based on model calculations. The answer to this question can be found in [5.12]; *Pawley* et al. reported model-independent assignments for the five lowest-frequency phonon modes at point Y (Fig. 5.12). The assignments were based on extensive measurements of the scattering intensity at various values of the wave-vector transfer. A total of 24 different values of $\boldsymbol{Q}$ were taken to yield $5 \times 24 = 120$ measured intensities. These intensities were fitted to the theoretical ones computed from (5.93). The components of the eigenvectors were treated as adjustable parameters.

Thus, the solution for the eigenvectors was independent of any model assumption. It was only based on experimental information. The quality of fit was assessed by a least-squares method. As a starting set *Pawley* et al. [5.12] first used the model eigenvectors derived by *Mackenzie* et al. [5.13].

A refinement of the eigenvectors improved the fit between observed and theoretical intensities only slightly. The initial and final sets of eigenvectors were essentially similar.

To ensure that the final result is not determined by the starting conditions, four quite different starting sets of eigenvectors were tried in similar refinements. The result was that each refinement converged to the same point as obtained starting from the model eigenvectors. All of this supported the reliability of the model eigenvectors and hence of the assignments based on them.

Thus, we have discussed the potentialities of the atom-atom potential method in predicting the phonon eigenfrequencies and eigenvectors for particular directions of the wave vector. It is now of interest to see whether the method is also capable of predicting the entire phonon spectrum of naphthalene. The best way of doing this is to compare the theoretical and experimental spectra of the incoherent inelastic scattering of neutrons. The experimental data necessary for such a comparison are available [5.54–57]. Figure 5.15 shows the external-mode part of the scattering spectrum, as measured by *Belushkin* et al. [5.57] at 5 K. Also shown in the same figure are two theoretical spectra based on the hydrogen amplitude-weighted density of phonon states according to (5.95). The spectrum labeled "RCW80" was obtained using the model potential of *Righini* et al. [5.66] in which the (6-exp) potentials were complemented by a quadrupole term (Table 5.7). The spectrum labeled "W67" was computed using an ordinary atom-atom model and the Williams' W67 parametrization. Both calculations were carried out in the quasiharmonic rigid-molecule approximation.

As seen in Fig. 5.15, the model potential of *Righini* et al. [5.66] definitely provides a better fit to the observed spectrum than the W67 model. The discrepancies observed with the W67 model are most obvious in the high-frequency region. It is not unlikely that these discrepancies originate partly from defects in the rigid-molecule approximation. Such defects should not be so important for the model potential of *Righini* et al. [5.66] because it was specifically adjusted to the lattice frequencies of naphthalene and, therefore, had to absorb the effects of the molecular non-rigidity to some extent.

The velocities of the acoustic modes of naphthalene for six directions of the wave vector are presented in Table 5.8. The experimental values listed in the table were obtained by *Afanas'eva* [5.61] using an ultrasonic pulse method. The theoretical velocities were calculated by *Pawley* [5.27] using the universal model potential. The agreement between calculated and observed values is satisfactory. The discrepancies are normally 15–20% of the observed velocity. It is worth noting that the agreement for the elastic constants is somewhat worse, which reflects the uncertainties that arise in solving (5.102–109).

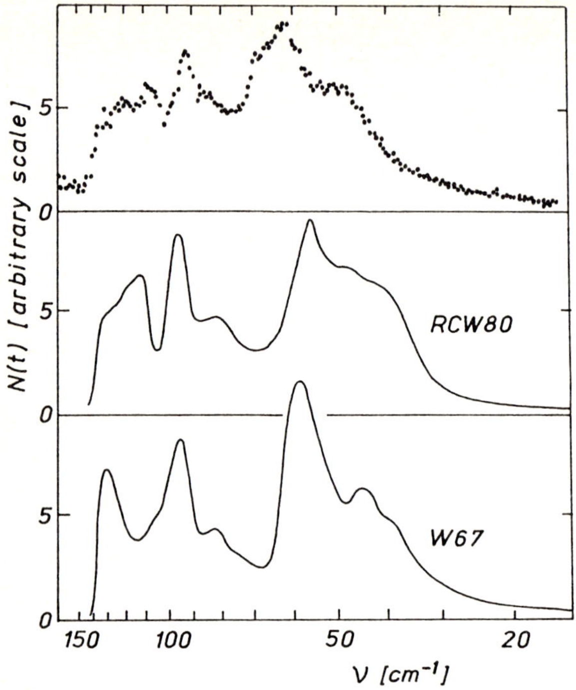

Fig. 5.15. Experimental and calculated spectra of the incoherent inelastic scattering of neutrons for naphthalene at 5 K [5.57]

Table 5.8. Observed and calculated velocities of the acoustic modes of naphthalene [10^5 cm/s]

Wave-vector direction	Observed [5.61]	Calculated [5.27]
001	3.36	4.05
	1.71	1.61
	1.16	1.27
110	3.18	3.00
	1.45	1.80
	1.31	1.11
010	2.93	3.10
	...	2.00
	...	1.57
101	2.36	2.49
	1.73	1.85
	1.98	1.75
100	2.62	2.42
	...	1.59
	1.94	1.57
011	3.03	3.58
	2.12	2.22
	1.24	1.21

Very recently *Della Valle* and *Pawley* [5.67] have attempted to simulate the lattice dynamics of naphthalene in a direct way, with no assumption of the harmonicity of the lattice vibrations. The thermal motion of the molecules was simulated with the method of molecular dynamics [5.68] for a system of 4096 rigid molecules with periodic boundary conditions. For some of the crystal properties (e.g., for the V-T dependence) the model predictions have shown a fairly good agreement with experiment. Despite this fact, the calculated results do not seem to be reliable in view of some doubtful assumptions made in the calculations.

In particular, in calculating the equilibrium crystal structure at zero pressure *Della Valle* and *Pawley* [5.67] neglected the contribution to pressure from the strain derivatives of the entropy. The importance of this contribution can be readily estimated using the experimental elastic constants and thermal expansions (Sect. 6.4.3). At room temperature the strain derivatives of the entropy of naphthalene are equal to about 50 e.u., which corresponds to a pressure of about 60 kbar. Clearly, this is not a negligible quantity.

Another doubtful point of the calculations is the use of familiar virial relations in which the virial of forces is represented by the average

$$\frac{1}{3}\left\langle \sum_{\alpha=1}^{3} \sum_{a,a'} r_\alpha^{aa'} \partial\varphi_{aa'}/\partial r_\alpha^{aa'} \right\rangle$$

($r_\alpha^{aa'}$ denotes the αth component of the interatomic distance) [5.69]. The above expression is only valid for particles interacting through an isotropic potential $\varphi_{aa'}$. That is to say, this expression would have been valid if the naphthalene crystal were composed of atoms not bound into the molecules. In fact, we are dealing with a system of rigid molecules whose interaction is described by an anisotropic intermolecular potential (a sum of atom-atom isotropic potentials).

A similar error has been made with respect to the strain derivatives of the potential energy of the crystal. The latter were represented in the form

$$\partial\phi/\partial u_{\alpha\beta} = \frac{1}{2} \sum_{a,a'} r_\beta^{aa'} \partial\varphi_{aa'}/\partial r_\alpha^{aa'}$$

which, again, is only true for a system of atoms not bound into the molecules. For a correct equation for the strain derivatives of the potential energy, the reader is referred to Sect. 4.1.

5.6.3 Anthracene

Unlike benzene and naphthalene, anthracene does not exhibit a frequency gap between the external and internal vibrational modes. This has been shown by *Dorner* et al. [5.70] in the coherent inelastic scattering of neutrons on fully deuterated anthracene. They observed sixteen low-frequency dispersion branches in the [ς00], [0ς0] and [00ς] directions (Fig. 5.16b) instead of only twelve branches expected for a rigid molecule. The four additional branches possibly originate from two low-frequency internal modes.

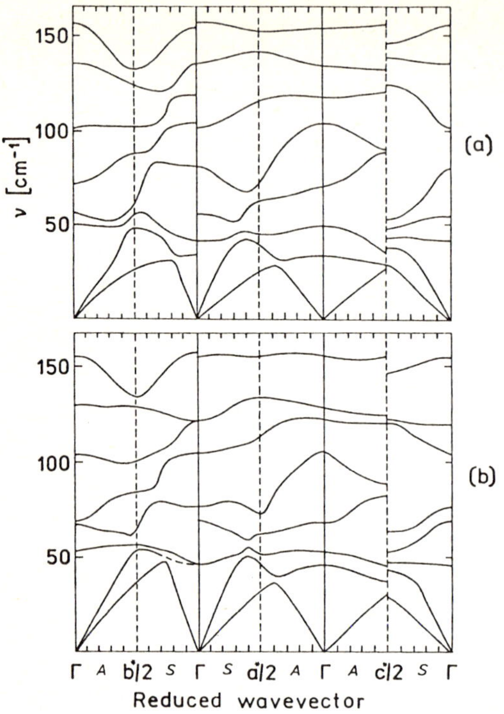

Fig. 5.16a,b. Calculated (**a**) and measured (**b**) dispersion relations for d-anthracene at 12 K [5.70,73]. The branches marked S and A are symmetric and antisymmetric with respect to the screw diad symmetry operation for the b^* direction and the glide plane operation for the a^* and c^* directions, respectively

According to a model calculation by *Evans* and *Scully* [5.71] based on a molecular force field, the two lowest-frequency modes of the free anthracene molecule are a butterfly vibration of B_{3u} symmetry and a twisting vibration of A_u symmetry (Fig. 5.17). In the butterfly motion the two outer rings of the molecule vibrate above and below the plane of the central ring so that each ring retains its π-bonded planarity. The twisting motion is described by an oscillating twist angle about the long molecular axis which results in atomic displacements proportional to the atomic coordinates along the axis. The existence of two kinds of molecular motion in the low-frequency lattice vibrations of anthracene was established by *Chaplot* et al. [5.72] via a least-squares eigenvector fitting similar to the one described in Sect. 5.6.2 for naphthalene.

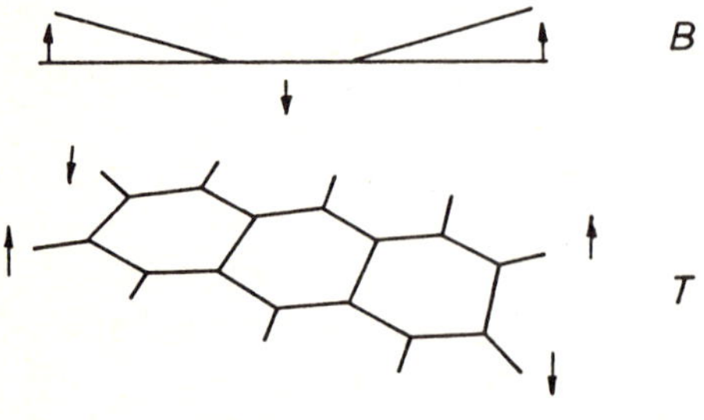

Fig. 5.17. Butterfly (B) and twisting (T) vibrational modes of anthracene

Detailed calculations of the lattice dynamics of anthracene, including the intermode mixing, were carried out by *Dorner* et al. [5.70], by *Chaplot* et al. [5.73] and by *Krivenko* et al. [5.74]. The former followed the Taddei-Califano formalism, allowing for the mixing between the external modes and the six lowest internal modes. Since the free-molecule frequencies of anthracene were not available, they were treated as parameters to be adjusted so as to reproduce the observed crystal-state frequencies.

Chaplot et al. [5.73] used a more general formalism [5.10] which can be applied when the internal displacement coordinates are not necessarily equal to the free-molecule normal coordinates. Each molecule was assumed to possess six external degrees of freedom and two internal ones describing the butterfly and twisting motions. The resulting equations for the eigenfrequencies were essentially identical to those of Taddei and Califano.

The dispersion relations calculated by *Chaplot* et al. [5.73] using the W67 potentials are compared with the experimental relations in Fig. 5.16. The agreement is generally good; a typical deviation being $10\,\mathrm{cm}^{-1}$.

The energy distribution of the butterfly and twisting motions in the upper three and all of the remaining branches of Fig. 5.16 is shown in Table 5.9. It is evident that the upper branches cannot be classified as being purely internal or purely external. Moreover, an appreciable fraction of the internal motion is present in the lower-frequency branches.

Despite the importance of the intermode mixing for a correct prediction of the dispersion relations in anthracene, many dynamical properties of the crystal may be accurately reproduced using the usual rigid-molecule approximation. This is obviously true of the sound velocities in the crystal, determined from the slopes of the acoustic branches at $q \to 0$. In Table 5.10 the experimental velocities of the acoustic modes, as found by *Afanas'eva* et al. [5.75], are compared with the rigid-molecule results reported by *Pawley* [5.27]. The agreement is satisfactory, at least not poorer than that achieved for naphthalene.

The rigid-molecule approximation can also be used to predict the six Raman-active lattice vibrations of anthracene. These represent purely librational motions and do not mix, for symmetry reasons, with the butterfly and twisting internal vibrations. The quality of the prediction of the librational frequencies is illustrated in Table 5.11, in which the experimental frequencies from [5.46] are compared with the theoretical rigid-molecule estimates reported by *Filippini* et al. [5.65].

Finally, the rigid-molecule approximation should lead to reliable estimates of the rigid-body tensors $\underline{\underline{T}}$ and $\underline{\underline{L}}$ (but not of the individual atomic temperature factors). The tensors derived from a Schomaker-Trueblood treatment (Sect. 5.5) absorb mostly the rigid-body component of the lattice vibrations and, therefore, should correlate well with the rigid-molecule lattice-dynamical estimates. A comparison of the experimental and theoretical $\underline{\underline{T}}$ and $\underline{\underline{L}}$ tensors of anthracene [5.65] is given in Table 5.12. The agreement is quite satisfactory.

Table 5.9. The distribution of the energy of the butterfly (B) and twisting (T) motions in the upper three branches of Fig. 5.16a [%] [5.73]

Reduced wavevector			B1	T1	B2	T2	B3	T3	B'	T'
0	0	0	6	-92	0	0	91	7	2	0
0	0.1	0	7	-89	1	3	88	6	2	1
0	0.2	0	9	-79	6	13	81	4	2	3
0	0.3	0	17	-58	11	34	68	2	3	5
0	0.4	0	36	-22	7	71	51	0	4	5
0	0.5	0	56	1	- 1	92	36	0	4	5
0	0.4	0	39	33	-30	57	25	1	4	7
0	0.3	0	22	56	-53	28	16	0	7	14
0	0.2	0	14	68	-24	4	48	-16	12	10
0	0.1	0	11	74	0	0	75	-18	12	7
0	0	0	10	76	0	0	76	-17	12	5
0	0	0	-6	92	0	0	91	7	2	0
0.1	0	0	-6	92	1	0	83	6	8	0
0.2	0	0	-5	93	5	0	71	5	17	0
0.3	0	0	-4	94	10	0	64	4	21	0
0.4	0	0	-1	95	16	0	60	3	22	0
0.5	0	0	0	87	19	- 9	59	1	20	1
0.4	0	0	3	75	15	-22	62	0	18	2
0.3	0	0	6	70	9	-24	68	0	14	4
0.2	0	0	8	71	4	-16	74	- 5	12	5
0.1	0	0	10	75	0	- 5	77	-13	11	6
0	0	0	10	76	0	0	76	-17	12	5
0	0	0.1	9	77	0	0	74	-17	15	5
0	0	0.2	8	78	0	0	68	-17	22	3
0	0	0.3	7	80	0	0	63	-16	28	2
0	0	0.4	7	81	0	0	60	-16	31	2
0	0	0.5	7	82	0	0	59	-15	33	1
0	0	0.5	25	-71	0	0	67	26	6	1
0	0	0.4	20	-74	2	- 1	69	22	7	1
0	0	0.3	12	-80	3	- 2	74	15	9	1
0	0	0.2	8	-86	1	- 1	77	9	11	1
0	0	0.1	7	-91	0	0	82	7	9	0
0	0	0	6	-92	0	0	91	7	2	0

B′ and T′ give the percentage of energy distributed over the remaining branches. The signs (except in B′ and T′) denote the phase relationships between the B and T motions within a particular mode

Table 5.10. Observed and calculated velocities of the acoustic modes of anthracene [10^5cm/s]

Wave vector direction	Observed [5.70]	Calculated [5.27]
001	3.51	4.06
	1.40	1.51
	1.47	1.61
110	3.15	3.12
	1.40	1.02
	1.00	1.72
010	3.05	3.28
	1.22	1.72
	1.94	1.19
101	2.60	2.66
	1.40	1.64
	1.94	1.65
100	2.69	2.55
	1.78	1.44
	1.33	1.92
011	3.01	3.41
	2.11	2.31
	1.40	1.18

Table 5.11. Observed and calculated Raman-active frequencies of the external modes of anthracene at room temperature [cm^{-1}]

Symmetry	a	b	c	d
B_g	125	122.5	131.4	141.7
A_g	121	128.2	140.0	160.0
A_g	70	72.2	78.4	80.2
B_u	65	60.6	67.0	62.7
B_g	45	46.4	50.8	48.2
A_g	39	39.5	45.2	42.5
Standard deviation		3.8	9.6	17.9

[a]Observed [5.46]; [b–d]Calculated using rigid-molecule approximation and the W67 (b), W66 (c) and "universal" (d) parameter sets [5.65]

Table 5.12. Rigid-body tensors $\underline{\underline{T}}$ and $\underline{\underline{L}}$ of anthracene [5.65]

	T [$A^2\ 10^4$]			L [$rad^2 10^4$]			Eigenvalues of L [deg^2]
	391	27	- 93	23	0	1	10.0
a		423	- 25		24	0	8.0
			500			18	3.3
	444	15	-101	33	0	- 5	19.2
b		442	- 15		26	-12	10.6
			514			53	6.9

[a]Schomaker-Trueblood treatment; [b]lattice-dynamical calculations

5.6.4 Hexamethylenetetramine

Hexamethylenetetramine (HMT) is a very convenient object for lattice-dynamical calculations. Due to the high symmetry of the crystal (space group $I43m$, $Z = 2$), the lattice dynamical problem can be solved in a purely phenomenological way – without resorting to an intermolecular model potential but treating the force constants as adjustable parameters. It is also possible to represent the intermolecular force constants in terms of phenomenological atom-atom force constants treated as adjustable parameters. By comparing the intermolecular and interatomic phenomenological models, one can examine the representability of the intermolecular interaction potential in terms of pairwise atom-atom interactions, without assuming a particular analytical expression for the atom-atom potential. Finally, a common analytical atom-atom model can also be used to gain an idea of the suitability of particular analytical functions to represent the potential.

An interesting analysis of the various lattice-dynamical models of HMT has been made by *Dolling* et al. [5.76,77]. The analysis was based on the

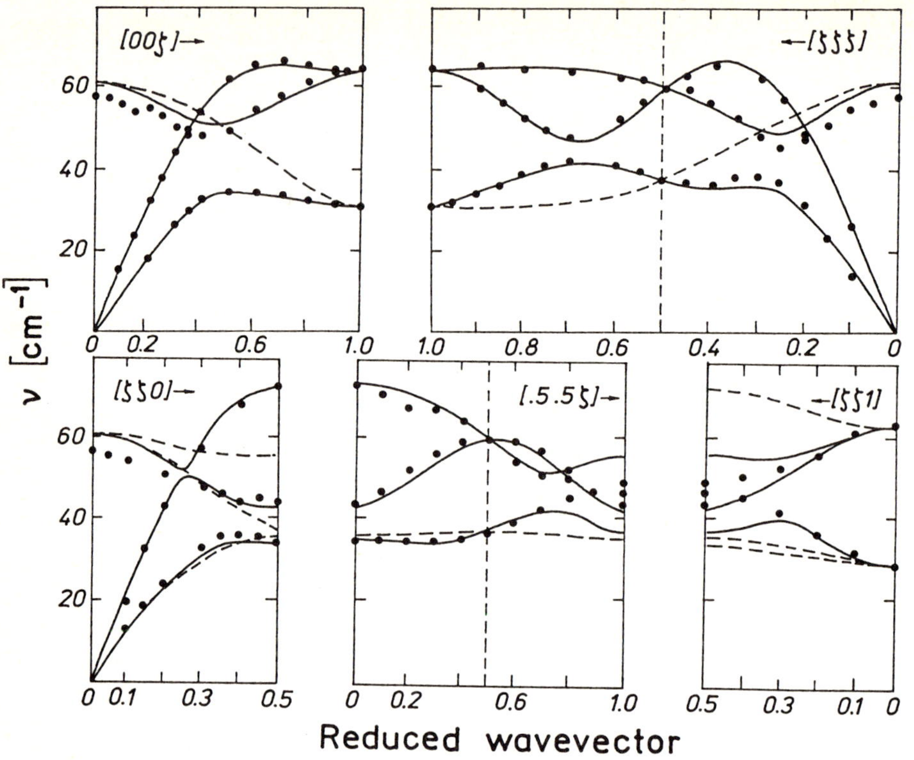

Fig. 5.18. Experimental (*points*) and calculated (*solid lines*, model AI1) phonon frequencies of d-hexamethylenetetramine at 298 K [5.77]. The dashed lines represent unobserved modes predicted by the molecular model M3

good-quality data of the coherent inelastic scattering of neutrons in five high-symmetry directions. To avoid high incoherent neutron scattering due to hydrogen atoms, fully deuterated HMT was used (DHMT). The experimental results are shown in Fig. 5.18 by points. The dashed lines represent unobserved modes which were computed from one of the force-field models to be discussed below.

The measured dispersion relations were first fitted with four molecular models, hereafter labeled M1 to M4. In these models the elements of the force constant matrix were expressed in terms of the contributions from individual molecular pairs, see (5.34). These contributions were treated phenomenologically so as to achieve the best fit between the calculated and experimental dispersion curves. In all of the models only the interactions between the molecule at the origin and its first coordination sphere were taken into account. These are the interactions with the eight nearest and the six next-nearest neighbor molecules. Due to the high symmetry of the crystal, only 12 independent parameters had to be determined. In model M1 all of the 12 parameters were allowed to vary. In model M2 some additional equilibrium constraints were applied and the DHMT molecule was assumed to possess a center of inversion. This reduced the number of adjustable parameters to eight. After the fitting, three of the eight parameters proved to be rather small and, hence, were neglected in the next model, M3. The dashed curves in Fig. 5.18 were computed using this model.

The last model, M4, is another modification of model M2, in which the interactions of the molecule at the origin with its next-nearest neighbors were neglected. In model M4 only four independent parameters had to be found.

The parameters involved in each particular model were determined by minimizing the quantity

$$\chi^2 = \Big(\sum_{i=1}^{N} \frac{[\omega_i^2(\text{fit}) - \omega_i^2(\text{obs})]^2}{[2\omega_i(\text{obs})\Delta\omega_i]^2}\Big)/(N-M), \qquad (5.128)$$

where N is the number of fitted frequencies, M is the number of adjustable parameters and $\Delta\omega_i$ is the experimental error in the frequency ω_i.

The results of the fitting are presented in Table 5.13. As might have been expected the best fit was observed with the twelve-parameter model M1, although the eight- and five-parameter models M2 and M3 also provided a very good fit.

Table 5.13. Quality-of-fit parameter χ^2 for the various lattice-dynamical models of deuterated hexamethylenetetramine [5.76,77]

Model	M1	M2	M3	M4	AI1	AI2	AI3	AI4	AI5	AI6	AA
Number of adjustable parameters	12	8	5	4	4	6	6	6	4	6	7
χ^2	0.46	0.71	0.71	3.05	1.27	1.05	1.11	1.15	1.13	1.06	0.69

The model M4 proved to be unsuccessful, although it differed from M3 by only one adjustable parameter. This result suggests that the interactions extending to the next-nearest neighbor molecules must be included in order to correctly describe the intermolecular force field.

In the second half of their work *Dolling* et al. [5.77] have interpreted the dispersion relations in terms of atom-atom interactions. The formalism that was used employed a crystal-fixed coordinate frame for molecular translations and a principal axes frame for molecular rotations. *Pawley* [5.8] presented this formalism in a more general form, using the principal-axes frame for both the translational and rotational displacements. To make the formulas consistent with the notations of the previous sections, we shall follow *Pawley*'s exposition [5.8].

In the atom-atom approximation, a representative force constant $\underline{\underline{\phi}}^{os}$ may be written as

$$\underline{\underline{\phi}}^{os} = \sum_{aa'} \underline{\underline{\phi}}^{aoa's} \quad , \tag{5.129}$$

where a and a' run over the atoms in the molecule at the origin and the molecule s, respectively. The 6×6 tensors $\underline{\underline{\phi}}^{aoa's}$ may be expressed in terms of the 3×3 tensors $\underline{\underline{\varphi}}^{aoa's}$ defined by

$$\varphi_{\alpha\alpha'}^{aoa's} = \frac{\partial^2 \phi}{\partial u_\alpha^{ao} \partial u_{\alpha'}^{a's}}, \tag{5.130}$$

where u_α^{ao} and $u_{\alpha'}^{a's}$ are the Cartesian atomic displacements for atoms a and a' in the principal axes frames of molecules O and s, respectively. These displacements are related to the translational-rotational rigid-body displacements, $\boldsymbol{v}$ and $\boldsymbol{\theta}$, through (5.92).

If ϕ is considered as an implicit function of $\boldsymbol{v}$ and $\boldsymbol{\theta}$, and is differentiated with respect to their components, one readily obtains an expression for $\underline{\underline{\phi}}^{aoa's}$, which can be written in a compact form [5.8]:

$$\underline{\underline{\phi}}^{aoa's} = \left(\begin{array}{c|c} \underline{\underline{\varphi}}^{aoa's} & \underline{\underline{\varphi}}^{aoa's} \underline{\underline{R}}^{a's} \\ \hline -\underline{\underline{R}}^{ao} \underline{\underline{\varphi}}^{aoa's} & -\underline{\underline{R}}^{ao} \underline{\underline{\varphi}}^{aoa's} \underline{\underline{R}}^{a's} \end{array} \right), \tag{5.131}$$

where $\underline{\underline{R}}^{ao}$ and $\underline{\underline{R}}^{a's}$ are the 3×3 matrices defined by (5.120). The procedure for obtaining $\underline{\underline{\phi}}^{os}$ by summing up the results of (5.131) is now straightforward, provided that the components of the atom-atom interaction tensors $\underline{\underline{\varphi}}^{aoa's}$ are available.

Dolling et al. [5.77] have tried several alternative models for the atom-atom interaction tensors. These models may be classified into three types. In the models of the first type the interaction energies were assumed to only depend on the distance between the interacting atoms. These models will, hereafter, be referred to as isotropic models and labeled AI. Each tensor $\underline{\underline{\varphi}}^{aoa's}$ contains only two independent parameters, i.e., the first and second derivatives of the interaction potential at a particular separation, $r^{a's,ao}$. In the AI models the derivatives $\varphi'_{aa'}$ and $\varphi''_{aa'}$ were treated as adjustable parameters. Six models of this type were considered. Only the shortest D...D and D...N non-bonded contacts 2.72 and 2.28 Å, respectively, were taken into account in model AI1, while the interactions between atoms further apart were neglected. Since each contact is characterized by two independent parameters, φ' and φ'', model AI1 involved a total of four adjustable parameters. As with the molecular models M1 to M4, the parameters were determined in a least-squares fitting of the calculated and experimental dispersion curves.

The dispersion curves based on model AI1 are shown in Fig. 5.18 as solid lines. It is seen that even this simple atom-atom model provides quite an acceptable description of the dispersion relations. The major discrepancy is the fit to the optical branch near the center of the Brillouin zone.

As simple as model AI1 is model AI5 in which the shortest D...N contacts were retained, while the D...D contacts were replaced by the shortest D...C contacts (3.17 Å). The other four AI models were constructed from the models AI1 and AI5 by including a third set of interatomic interactions. Models AI2, AI3 and AI4 are similar to AI1 but also include the contacts D...D, 2.97 Å, D...C, 3.17 Å, and D...N, 3.87 Å. Model AI6 is a modification of model AI5 that also includes the D...N contacts, 3.87 Å.

To assess the importance of isotropic potentials on the quality of fit, an anisotropic model was tried. In this model, labeled AA, the components of the interaction tensors $\underline{\underline{\varphi}}^{aoa's}$ were treated as adjustable parameters. The only constraint imposed on $\underline{\underline{\varphi}}^{aoa's}$ was that the resulting force field had to satisfy the crystal symmetry. As in the model AI1 only the shortest D...D and D...N contacts were taken into consideration, i.e., only two distinct tensors $\underline{\underline{\varphi}}^{aoa's}$ had to be found. Upon imposing the symmetry constraints, it was found that the D...D and the D...N contacts were specified by six and five independent force constants, respectively. These 11 constants could not, however, be determined independently since only particular linear combinations appear in the dynamical matrix. All in all seven such combinations were treated as adjustable parameters.

The quality-of-fit parameters χ^2 for all of the atomic models tried are listed in Table 5.13. It can be seen that the isotropic AI models all work fairly well, affording a fit not much poorer than that achieved with the molecular models M1–M3. This suggests that the representation of the intermolecular energy in terms of isotropic atom-atom interactions does not lead to serious errors in the description of the force field and is sufficiently flexible for lattice-dynamical applications.

The removal of the assumption of isotropic potentials is seen to markedly improve the quality of fit. The fit achieved with the seven-parameter model AA is even better than that provided by the eight-parameter molecular model M2.

The last model investigated by *Dolling* et al. [5.77] was a (6-exp) analytical one. The starting parameters were taken from the "universal" set. The nitrogen atoms were assumed to have the same parameters as the carbon atoms. A simultaneous refinement of the parameters proved to be unsuccessful because of the computational instabilities that arose in the fitting process.

The dynamical properties of DHMT were found to be mainly sensitive to the parameters $A_{\mathrm{H}\ldots\mathrm{H}}$ and $C_{\mathrm{H}\ldots\mathrm{H}}$. The best analytical model was derived in a refinement of these two parameters, while the other parameters were maintained at their "universal" values. The best numerical values for $A_{\mathrm{H}\ldots\mathrm{H}}$ and $C_{\mathrm{H}\ldots\mathrm{H}}$ were 162 Å^6 kcal/mol and 4.74 Å^{-1}, respectively. Comparing these values with the corresponding starting values in Table

3.1, we see that $C_{\text{H...H}}$ has remained practically unchanged, while $A_{\text{H...H}}$ increased by a factor of about three. The fit achieved with the analytical model was not as good as that achieved with the phenomenological models. Nevertheless, the calculated dispersion curves are in fairly good qualitative agreement with experiment; the discrepancies normally being within 5 to $10\,\text{cm}^{-1}$.

Even with such a rigid molecule as DHMT, the coupling between the external and internal vibrations is rather important. This was demonstrated by *Dolling* et al. [5.77] using the procedure suggested by *Pawley* and *Cyvin* [5.62] (Sect. 5.6.2). The effect of the molecular non-rigidity on the external vibrational modes of DHMT amounts to about 15% of the frequency and is always the largest for the modes of purely or largely librational character.

5.6.5 β-p-Dichlorobenzene

Of the three known polymorphs of p-dichlorobenzene, the high-temperature β polymorph is the most suitable for lattice-dynamical studies. The β polymorph is triclinic (space group $P\bar{1}$) with only one molecule in the unit cell, so that the crystal has a relatively large Brillouin zone and few phonon modes. This considerably simplifies the experiments and a subsequent assignment of the phonon branches.

A detailed study of the lattice dynamics of β-p-dichlorobenzene was made by *Reynolds* et al. [5.78] at 90 and 295 K using the coherent inelastic scattering of neutrons. The experimental dispersion curves are shown in Fig. 5.19. The assignment of the phonon groups to a particular dispersion branch was made by comparing the observed scattering intensities with those predicted from lattice-dynamical calculations.

In calculating the dispersion relations *Reynolds* et al. [5.78] tried a usual (6-exp) atom-atom model with parameters derived from a fitting of the crystal structures, heats of sublimation and optically active external vibrational frequencies of α-p-dichlorobenzene, 1,2,4,5-tetrachlorobenzene and hexachlorobenzene. This model potential has been briefly discussed in Chap. 3 where it was labeled RKW74. The fit provided by the RKW74 model potential to the experimental dispersion curves of β-p-dichlorobenzene was far from satisfactory. This is seen in Fig. 5.19 where the calculated results are compared with the experimental results. Both the calculated and experimental curves in this figure refer to 0 K. The former were evaluated in the harmonic approximation, using the crystal configuration corresponding to the minimum lattice energy. The latter were obtained in a linear extrapolation of the experimental phonon frequencies at 295 and 90 K to 0 K.

With the aim of improving the long-range part of their model potential, *Reynolds* et al. [5.78] ascribed an additional (anisotropic) polarizability to the molecule and dipole moments to the C–Cl and C–D bonds. However,

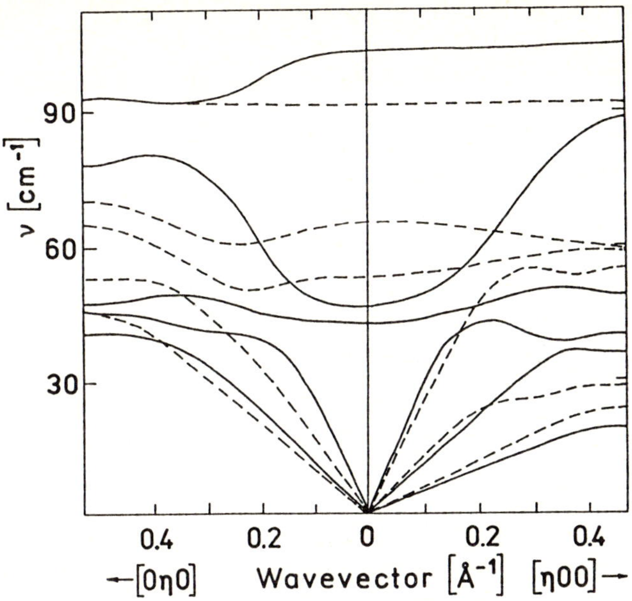

Fig. 5.19. Phonon dispersion curves for β-p-dichlorobenzene-d_4 at 0 K [5.78]. (—) represent calculated results obtained with the RKW74 model potential; (- - -) represent extrapolated experimental curves

an improvement in the fit was not achieved. Moreover, the corrected model potential predicted one of the phonon modes to be imaginary.

An attempt to calculate the dispersion in the quasiharmonic approximation, using the experimental unit-cell dimensions, also proved to be unsuccessful. Arguing that the dynamical properties are mainly sensitive to the short-range repulsions between the molecules, *Reynolds* et al. [5.78] attributed the failure of their calculation to the breakdown of the atom-atom approximation in the short-range region.

In our opinion, the large discrepancies between the calculated and observed dispersion curves in Fig. 5.19 do not stem from a deficiency in the atom-atom potential method itself but from a deficiency in the particular parameter set that was used. Such a conclusion is supported by the lattice-dynamical calculations of *Bonadeo* and *d'Alessio* [5.79] who used an ordinary (6-exp) model potential but another type of parametrization (parameter set BA73 in Table 3.5). The results of *Bonadeo* and *d'Alessio* [5.79] are compared with the experimental data of *Reynolds* et al. [5.78] in Fig. 5.20. It is seen that the fit provided by the BA73 parameter set is remarkably good. The discrepancies between the calculated and observed dispersion curves are normally within 5 cm^{-1}. Only one acoustic mode exhibits somewhat larger discrepancies when the wave vector approaches the boundary of the Brillouin zone.

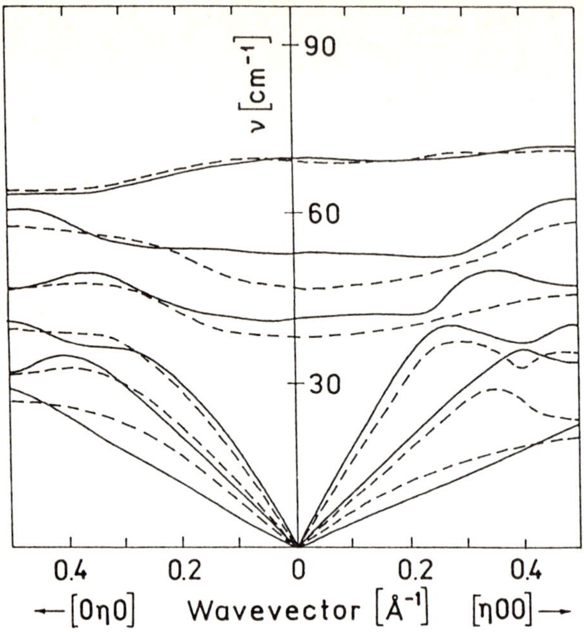

Fig. 5.20. Phonon dispersion curves for β-p-dichlorobenzene-d_4 at 295 K [5.79]. (—) represent calculated results obtained with the BA73 model potential; (- - -) represent experimental curves

5.6.6 Pyrazine

Pyrazine forms orthorhombic crystals (space group *Pnnm*) with two molecules in the unit cell. Both infrared [5.80] and Raman [5.81] data are available for the $q = 0$ phonon frequencies. The dispersion relations in the [$0\varsigma0$] direction are also known due to the coherent inelastic neutron-scattering measurements of *Reynolds* [5.82].

An attempt to reproduce the available dynamical data in a lattice-dynamical calculation has been undertaken by *Reynolds* [5.82]. The model potential that was used has already been discussed in Sect. 3.3.3. We remind the reader that the model was essentially of the atom-atom type, but also included an electrostatic term which describes the interaction between the CH bonds and the lone-pair dipoles of the nitrogen atoms. The lattice-dynamical calculations were carried out in the rigid-molecule approximation, which was justified by the large separation between the external and internal vibrational bands (236 cm^{-1} [5.80,81]).

The fit afforded by the model calculation is illustrated in Fig. 5.21. For the sake of clarity, the dispersion branches in Fig. 5.21 are divided into four symmetry species. The overall pattern of the dispersion branches was predicted reasonably well. In general, the quantitative agreement is also

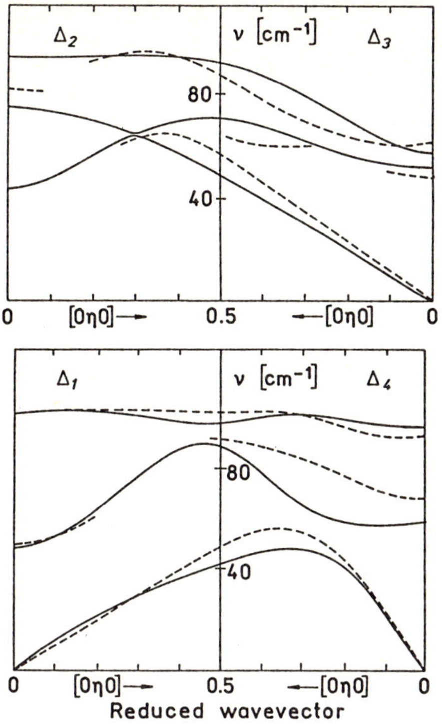

Fig. 5.21. Experimental (- - -) and calculated (—) atom-atom potential method) phonon frequencies of d-pyrazine [5.82]

acceptable (the average deviation being $7\,\mathrm{cm}^{-1}$), except for the lowest Δ_4 optical branch for which the discrepancies amount to $20\,\mathrm{cm}^{-1}$.

5.6.7 Concluding Remarks

In the previous subsections we have considered six organic molecular crystals whose dynamical properties can be treated, to a good approximation, using a rigid-molecule model. There are several other crystals whose lattice dynamics have been studied in a more or less extent. These are biphenyl [5.83–87], p-chloroaniline [5.88], phenanthrene, m-chloronitrobenzene [5.89], tetracene, pentacene [5.90], tetracyanoethylene [5.91] and acetylene [5.92]. The molecules of these compounds (except for acetylene) are essentially non-rigid. In calculating the lattice dynamics of these compounds, due allowance should therefore be made for the mixing between the internal and external phonons. This necessitates, in turn, a knowledge of the intramolecular force field (or the free-molecule vibration frequencies), which somewhat reduces the value of the calculation in testing intermolecular potentials.

A few words should be devoted to acetylene in which the phonon frequencies calculated with the available (6-exp) potentials showed extremely large deviations from the experimental values [5.92]. An attempt to derive a new atom-atom model potential appropriate specifically to acetylene was only successful with regards to the cubic phase of the crystal, while in the orthorhombic phase the calculations led to imaginary frequencies in some regions of the Brillouin zone.

Recently, *Gamba* and *Bonadeo* [5.93] have reported that the difficulties encountered with acetylene can be surmounted by explicitly including an electrostatic component in the model potential. It has been demonstrated that the crystal properties of both polymorphs of acetylene can only be adequately described if the electrostatic interactions up to the 2^6th pole are taken into account.

In general, the examples presented in the previous sections show that the atom-atom potential method is a very useful tool for investigating the lattice dynamics of organic molecular crystals. In most cases the method is capable of predicting phonon frequencies within $10\,\mathrm{cm}^{-1}$, provided that the parameter set is well calibrated. The eigenvectors of the modes seem to be less sensitive to the parameters. They can even be predicted reasonably well with a rough preliminary model. This is of great importance in the planning of neutron scattering experiments and the subsequent assignment of the observed groups to the appropriate phonon branches.

A fundamental result of the calculations is the demonstration of the fact that the atom-atom potentials derived from pure static crystal properties may be used to predict the lattice dynamics. This supports the view that the atom-atom model potentials are not merely phenomenological functions used to fit the experimental data, but indeed simulate the actual intermolecular potential (Sect. 3.2).

6. Thermodynamics

In this chapter we shall discuss two alternative approaches to an evaluation of the thermodynamic properties of an organic molecular crystal. The first is based on the formalism discussed in the previous chapter. A crystal is treated as a collection of independent harmonic oscillators whose frequencies can be determined by solving the lattice-dynamical problem. The anharmonicity of the vibrations is completely neglected, while the correlations in the motions of the individual molecules are described quite well. In the second approach, based on the so-called cell model, the vibrations are not assumed to be harmonic but, instead, the molecular motions are treated to be completely uncorrelated.

6.1 Quasi-Harmonic Approximation

In Chap. 5 we have shown that if the potential energy of a crystal is expanded as a Taylor series in the atomic displacements and truncated beyond the second-order terms, the resulting equations of motion describe a collection of $3Nn_m$ independent harmonic oscillators (N represents the number of molecules in the crystal and n_m the number of atoms in a molecule). The free energy of such a system is simply obtained by adding the contributions from each particular oscillator,

$$\frac{1}{2}\hbar\omega_i + kT \ \ln\left[1 - \ \exp\left(-\hbar\omega_i/kT\right]\right.,$$

to the static lattice energy ϕ_0

$$F = \phi_0 + \sum_{i=1}^{3Nn_m} \left\{\frac{1}{2}\hbar\omega_i + kT \ \ln\left[1 - \ \exp\left(-\hbar\omega_i/kT\right)\right]\right\}. \tag{6.1}$$

In the strict harmonic approximation F only depends on the lattice parameters through the static term ϕ_0. In this case purely anharmonic crystal properties such as thermal expansion cannot be described at all. In the quasi-harmonic approximation the effect of anharmonicity in the lattice energy is treated, to a first approximation, by allowing the frequencies of vibration to depend on the lattice parameters due to perturbation of the harmonic force constants. The quasi-harmonic equilibrium crystal structure corresponds to a minimum of the free (but not static) lattice energy, as given by (6.1). Since (6.1) is a function of temperature, the quasi-harmonic equilibrium crystal structure is temperature dependent.

The quasi-harmonic approximation leads, however, to underestimated thermal expansion [6.1] and, as a result, to thermodynamic properties really observed at higher pressures. To avoid this, experimental instead of calculated lattice parameters are frequently used (experimental version of the quasi-harmonic approximation).

Returning to (6.1) we note that the vibrational part of F does not contain any cross terms in ω_i and can be always represented as a sum of contributions from particular regions of the vibrational spectrum. The vibrational spectra of molecular crystals exhibit two well separated bands, corresponding to the external and internal vibrations. In this case the vibrational free energy can be separated into two contributions known as the internal (or intramolecular) and external (or intermolecular, or crystalline) vibrational free energies.

The static term in (6.1) can also be represented as a sum of internal and external contributions. Thus we can write

$$F = F_{\text{int}} + F_{\text{ext}} \quad \text{where} \tag{6.2}$$

$$F_{\text{int}} = \phi_{0,\text{int}} + \sum_{\text{int.modes}} \left\{ \frac{1}{2}\hbar\omega_i + kT \ \ln\left[1 - \exp\left(-\hbar\omega_i/kT\right)\right] \right\}, \tag{6.3}$$

$$F_{\text{ext}} = \phi_{0,\text{ext}} + \sum_{\text{ext.modes}} \left\{ \frac{1}{2}\hbar\omega_i + kT \ \ln\left[1 - \exp\left(-\hbar\omega_i/kT\right)\right] \right\}. \tag{6.4}$$

The calculation of F_{int} does not present any difficulties. Since the effect of a crystal field on the internal vibrations is small (compared to the magnitudes of the vibrational frequencies themselves), quite reliable estimates for F_{int} can be obtained using the free-molecule vibrational frequencies. Being interested in the external contribution to the thermodynamic properties, we shall make no attempt to predict F_{int} from dynamical calculations. Instead, F_{int} will be regarded as an experimentally known quantity that can be evaluated from observed internal frequencies. The static part of F_{int} will be assumed to be independent of volume and will therefore cancel in a calculation of the excess free energy, $F(T) - F(0)$. To simplify the notation, we shall drop the labels "int" and "ext", implying that we shall only be dealing with the external contribution.

In the expression for the external free energy we actually have a double summation, one over the phonon branches and the other over the wavevectors in the Brillouin zone

$$F = \phi_0 + \sum_j \sum_{\boldsymbol{q}} \left\{ \frac{1}{2}\hbar\omega(\boldsymbol{q}j) + kT \ \ln\left[1 - \exp\left(-\hbar\omega(\boldsymbol{q}j)/kT\right)\right] \right\}. \tag{6.5}$$

This equation can be rewritten using the normalized density-of-states function $g(\omega)$

$$F = \phi_0 + 6N \int_0^\infty g(\omega) \left\{ \frac{1}{2}\hbar\omega + kT \ \ln\left[1 - \ \exp\left(-\hbar\omega/kT\right)\right] \right\} d\omega. \tag{6.6}$$

It is also worthwhile to write down the expressions for the internal energy and specific heat

$$E = \phi_0 + 6NkT \int_0^\infty g(\omega) \left(\frac{1}{2} + \frac{1}{\exp\left(\hbar\omega/kT\right) - 1} \right) \frac{\hbar\omega}{kT} d\omega, \tag{6.7}$$

$$C_v = 6Nk \int_0^\infty g(\omega) \left(\frac{\hbar\omega}{kT} \right)^2 \frac{\exp\left(\hbar\omega/kT\right)}{\left[\exp\left(\hbar\omega/kT\right) - 1\right]^2} d\omega. \tag{6.8}$$

We shall not give the explicit expression for the entropy which is readily derived from (6.6,7) as

$$S = (E - F)/T. \tag{6.9}$$

Earlier calculations of the thermodynamic properties of organic molecular crystals followed the Debye approximation [6.2]. In this approximation the actual density of phonon states, $g(\omega)$, is replaced by the function,

$$g_D(\omega) = \begin{cases} \text{const } \omega^2, & \omega \leq \omega_D \\ 0, & \omega > \omega_D \end{cases}, \tag{6.10}$$

where ω_D is a limiting frequency which defines the Debye characteristic temperature via

$$\theta_D = \hbar\omega_D/k. \tag{6.11}$$

The Debye characteristic temperature is related to the root-mean-square vibration frequency by (5.125) and can be evaluated from the intermolecular potential using (5.124).

Figure 6.1 shows the temperature dependence of θ_D for four organic crystals, as calculated by *Kitaigorodsky* and *Mukhtarov* [6.3] using the "universal" atom-atom potentials. Also given in Fig. 6.1 are the experimental characteristic temperatures. These were evaluated by fitting Debye's equation for the entropy,

$$S_D = -R[6 \ \ln(\theta_D/T) + 8], \tag{6.12}$$

to the experimental external entropy derived by subtracting the internal vibrational entropy from the third-law calorimetric entropy. The calculated θ'_Ds show a substantial disagreement with experiment, the deviations rang-

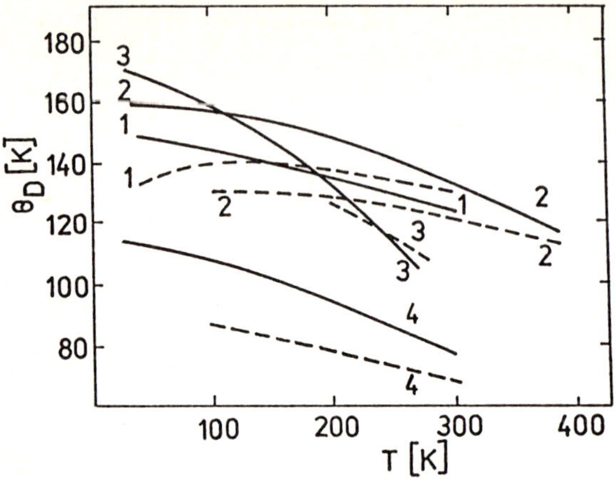

Fig. 6.1. Temperature dependence of the Debye characteristic temperature for naphthalene (*1*), anthracene (*2*), benzene (*3*), and diphenylmercury (*4*) [6.3]. The solid and dashed lines represent the calculated and experimental results, respectively

ing to about 20%. The magnitude of the error that would have been resulted had the same thermodynamic model been used to predict the entropy itself can be readily estimated. Thus, for $\Delta\theta_D = 20\%$

$$\Delta S = 6R\Delta\theta_D/\theta_D \simeq 2.4\,\text{e.u.} \tag{6.13}$$

A somewhat more sophisticated thermodynamic model was used by *Bonadeo* et al. [6.4] in their calculations of the thermodynamic properties of p-dichlorobenzene and 1,2,4,5-tetrachlorobenzene. *Bonadeo* et al. [6.4] only applied the Debye approximation to the three acoustic branches, assuming the Debye frequency to be equal to the lowest optical $q = 0$ frequency. Each optical branch was treated in the Einstein approximation; the dispersion of the optical phonons was neglected and the Einstein frequencies assumed to be equal to the $q = 0$ optical frequencies.

For a comparison, the thermodynamic properties were also evaluated in the quasi-harmonic approximation, using the density of phonon states computed lattice-dynamically. An important result was that the Debye-Einstein estimates of the thermodynamic functions were very close to the corresponding strict values, the deviations were less than 3%. This suggests that the assumptions involved in the Debye-Einstein model do not introduce a significant error in the thermodynamic quantities.

The above-mentioned work by *Bonadeo* et al. [6.4] will be discussed in more detail in Sect. 6.5.1. We shall now consider some of the computational aspects of the quasi-harmonic treatment of the thermodynamic functions.

In principle, the vibrational free energy can be directly evaluated from (6.5) by sampling a sequence of wavevectors, $\boldsymbol{q}^{(i)}$, in the Brillouin zone, solving the lattice-dynamical problem for each wavevector, and accumulating the contributions to F from each particular phonon branch and wavevector. An obvious way of sampling $\boldsymbol{q}^{(i)}$ is to use a uniform grid (regular or random). In this case the contributions from all sampled points should be given equal weight. However, such a procedure is not the most efficient one because the contributions of the various $\boldsymbol{q}^{(i)}$ are not equally important. Equation (6.5) shows that the contributions from the acoustical branches tend to infinity as $q \to 0$. It is desirable therefore to concentrate the sample points near $\boldsymbol{q} = 0$ using an appropriate distribution function for $\boldsymbol{q}^{(i)}$.

A similar situation is encountered in calculating the Cruickshank-Schomaker-Trueblood tensors $\underline{\underline{T}}$, $\underline{\underline{L}}$ and $\underline{\underline{S}}$ (Sect. 5.5). Here again, the contribution of the points close to $\boldsymbol{q} = 0$ is the most important and uniform samplings prove to be inefficient.

Several sampling procedures have been examined by *Filippini* et al. [6.5]. The following sampling algorithm was found to be the most efficient. Each reciprocal axis was divided into n unequal intervals $\Delta q_\alpha^{(i)}$ ($\alpha = 1, 2, 3$; $i = 1, 2, \ldots, n$) whose centers specified the coordinates of the sampled points. The total number of sampled points was n^3. The width of the intervals was chosen to be proportional to $[i^2/(3 + i^2)]^{1/3}$. In other words, the closer a point was to $\boldsymbol{q} = 0$, the "thicker" was the sampling. To take into account the non-uniformity of the grid, the contribution from each point $\boldsymbol{q}^{(jkl)} = \{q_1^{(j)}, q_2^{(k)}, q_3^{(l)}\}$ was given a weight proportional to the volume, $\Delta q_1^{(j)} \Delta q_2^{(k)} \Delta q_3^{(l)}$.

In calculating the $\underline{\underline{T}}$, $\underline{\underline{L}}$ and $\underline{\underline{S}}$ tensors the advantage of the non-uniform grid over the regular one ($\Delta q_\alpha^{(i)} = \text{const}$) is remarkable. In this case a non-uniform sampling of only $4^3 = 64$ points resulted in nearly the same accuracy as the uniform sampling of $32^3 = 32768$ points. For the thermodynamic functions, however, the non-uniform sampling gives practically no advantage.

The problem of sampling efficiency in quasiharmonic calculations of the thermodynamic properties has been studied in [6.6], too. To concentrate the sample points near $\boldsymbol{q} = 0$, use was made of the following distribution function

$$g(\boldsymbol{q}) = C \prod_{\alpha=1}^{3} (q_\alpha^2 + c^2)^{-1}, \tag{6.14}$$

where C is a normalizing factor, and c a parameter governing the "sharpness" of the distribution. The sample points were generated at random, using the formula

$$q_\alpha^{(i)} = c \tan\left[\arctan \frac{2\gamma^{(i)} - 1}{2c}\right], \tag{6.15}$$

where $\gamma^{(i)}$ are random numbers uniformly distributed in the interval (0,1). It can be readily shown that the points generated by (6.15) indeed occur with the probability density (6.14).

Several values for the "sharpness" parameter were tried but no perceptible improvement in sampling efficiency was observed as compared to the uniform random sampling. In the latter case it took as few as 100 to 200 sampled points to achieve an acceptable accuracy in the thermodynamic functions. The numerical results of the quasi-harmonic calculations will be presented in Sect. 6.4.3, together with the corresponding experimental data and cell-model results.

6.2 Cell Model

The cell model may be regarded as an alternative to the quasi-harmonic approximation in treating the statistical mechanics of the solid state. Unlike the quasi-harmonic approximation, the cell model does not assume that the potential energy can be described by a harmonic function. Instead, it assumes that the motions of individual molecules in a crystal are independent, so that the correlations between the motions can be neglected. As noted by *Barker* [6.7], this is justified at high temperatures where the dominant contribution to the thermodynamic functions is made by high vibration frequencies, close to that corresponding to independent motion of the molecules. (In a sense, the cell model is an analogue of Einstein's approximation, except that the latter additionally assumes that the crystal lattice vibrations are harmonic.)

At low temperatures, quite the reverse, low-frequency vibrations become important, and the molecules move almost in unison, so that the correlations between their motions can not be neglected. The quasi-harmonic approximation treats these correlations quite rigorously.

To summarize, the cell model and the quasi-harmonic approximation are a high- and a low-temperature model of the solid state, respectively. The cell model is expected to work well at high temperatures where the anharmonicity is important while the correlations are not. By contrast, the quasi-harmonic approximation should be adequate at low temperatures, where the anharmonicity can be neglected, while the correlations cannot.

The basic assumptions of the cell model suggested by *Lennard-Jones* and *Devonshire* [6.8] as early as 1937 can be formulated as follows:

1. The volume of a system may be divided into identical cells, one for each molecule;
2. only those configurations of the system need be considered in which each cell is occupied by just one molecule;
3. the cells may be chosen so that their centers form a regular lattice; and
4. the molecules move independently in their cells, so that the instantaneous position of a molecule is independent of the positions of its surroundings.

Although the cell model was originally intended to deal with fluids, assumptions (1)–(3) make it much more appropriate to crystalline solids. A comparison of the cell-model predictions with numerical Monte-Carlo experiments and the experimental properties of condensed inert gases has shown that the cell model is an excellent model for solids but not for fluids [6.7].

The assumption of independent molecular motions means that the potential energy of a system can be represented as a sum of terms, each depending on the coordinates of just one molecule

$$\phi = C + \sum_i \psi^i(\boldsymbol{u}^i) \tag{6.16}$$

where C is a constant, and $\boldsymbol{u}^i$ the vector displacement of a molecule i from the center of its cell. For the present we shall consider point molecules, whose positions are specified by $3N$ Cartesian coordinates and whose interaction energy is only dependent on the intermolecular separation distance.

Since all the cells are assumed to be identical, the functions ψ^i in (6.16) are also identical, so that the superscript may be dropped. In order to exclude the constant C from (6.16), we introduce the potential energy ψ_0 which refers to the situation in which all of the molecules are located at their cell centers ($\boldsymbol{u}^i = 0$). For this particular situation, we may write

$$\phi_0 = C + \sum_i \psi^i(0) \tag{6.17}$$

and (6.16) becomes

$$\phi = \phi_0 + \sum_i [\psi(\boldsymbol{u}^i) - \psi(0)]. \tag{6.18}$$

The quantity $\psi(\boldsymbol{u}^i) - \psi(0)$ describes the change in potential energy when the molecule i is displaced from the center of its cell to a position $\boldsymbol{u}^i$, while all other molecules remain at their cell centers. Thus $\psi(\boldsymbol{u}^i)$ may be ascribed the meaning of the interaction energy of molecule i at $\boldsymbol{u}^i$ with all surrounding molecules fixed at their cell centers. Note that $\phi_0 = (1/2)N\psi(0)$.

Inserting (6.18) into the expression for the classical configuration integral,

$$Q = \frac{1}{N!}\int \dots \int \exp(-\phi/kT)\, d\boldsymbol{u}^1 d\boldsymbol{u}^2 \dots d\boldsymbol{u}^N, \tag{6.19}$$

one gets

$$Q \cong \exp(-\phi_0/kT) V_f^N, \tag{6.20}$$

where V_f is a so-called free volume defined by

$$V_f = \int_{\Delta_i} \exp\{-[\psi(\boldsymbol{u}^i) - \psi(0)]/kT\} d\boldsymbol{u}^i. \tag{6.21}$$

The integral is taken over the interior of the cell. The index i labeling the cell may again be dropped since V_f is independent of what particular cell is dealt with.

A knowledge of Q as a function of temperature and unit-cell dimensions allows all thermodynamic properties of a system to be calculated. The relevant formulas will be given below.

In dealing with crystals, it is natural to choose the cell centers to coincide with the equilibrium positions of the molecular centers. With this choice, ϕ_0 becomes the usual static lattice energy. Since the displacements of the molecules from their equilibrium positions are small, compared to the distance between the molecular centers, the choice of the cell boundaries is, in principle, not important. Thus, the cell may be defined to be the Wigner-Seitz cell [6.9].

In the original theory of Lennard-Jones and Devonshire a further assumption was made, besides the four listed above. It was also assumed that the force field $\psi(\boldsymbol{u})$, in which a molecule moves can be adequately described with a spherically symmetric function dependent only on the distance $|\boldsymbol{u}|$ of the molecule from the center of its cell. For molecules interacting via a Lennard-Jones potential, an analytical expression for $\psi(|\boldsymbol{u}|)$ can be derived by "smearing" the surrounding molecules over the surfaces of concentric spheres of appropriate radius. With anisotropic intermolecular potentials, such as those used to describe the interactions of organic molecules, this additional assumption becomes inadequate. We shall not, therefore, use this assumption and shall calculate $\psi(\boldsymbol{u})$ numerically, by a direct summation of the interactions of a molecule at $\boldsymbol{u}$ with all its neighbors fixed at their equilibrium lattice sites.

The above derivation of the cell model can hardly satisfy a critical reader because it does not contain any means of improving the model and any criterion for assessing the quality of the assumptions involved.

An attempt to construct a theory, in which the cell model appears as one of successive approximations leading to the rigorous result, has been made by *Kirkwood* [6.10]. This theory is based on the classical variational principle according to which the free energy calculated with an approximate probability density function must be higher than the actual free energy, as given by the correct Boltzmann expression for the probability density function. Assuming a particular analytical form for the probability density function, one may thus find the optimum approximation of this particular form by requiring the calculated free energy to be a minimum.

Following *Kirkwood* [6.10], one can represent the Boltzmann probability density function,

$$P(\boldsymbol{r}^1, \ldots, \boldsymbol{r}^N) = \exp(-\phi/kT) Q^{-1}, \tag{6.22}$$

as a product of identical one-particle distribution functions,

$$P(\boldsymbol{r}^1,\ldots,\boldsymbol{r}^N) = \prod_{i=1}^{N} \varrho(\boldsymbol{r}^i - \overline{\boldsymbol{r}}^i) = \prod_{i=1}^{N} \varrho(\boldsymbol{u}^i), \tag{6.23}$$

where $\overline{\boldsymbol{r}}^i$ specifies the position of cell i. This approximation is analogous to Hartree's approximation of a wave function. The derivation of the optimum $\varrho(\boldsymbol{u})$ is based on the same principle as used to derive the optimum one-electron orbitals.

To find the optimum form for $\varrho(\boldsymbol{u})$, Kirkwood substituted (6.23) into the expression for the configurational free energy, i.e.,

$$\begin{aligned} F &= E - TS \\ &= \int\ldots\int P\phi\, d\boldsymbol{r}^1\ldots d\boldsymbol{r}^N + kT\int\ldots\int P\ \ln P\, d\boldsymbol{r}^1\ldots d\boldsymbol{r}^N, \end{aligned} \tag{6.24}$$

and then required that the change in F due to a small variation in ϱ is equal to zero. For the functions $\varrho(\boldsymbol{u})$ subject to the normalization condition

$$\int_{\Delta} \varrho(\boldsymbol{u})d\boldsymbol{u} = 1, \tag{6.25}$$

this results in a closed set of equations, which may be written as

$$\varrho(\boldsymbol{u}) = \exp\{-[\eta(\boldsymbol{u})/kT]\}\Big(\int_{\Delta} \exp\{-[\eta(\boldsymbol{u})/kT]\}d\boldsymbol{u}\Big)^{-1} \tag{6.26}$$

$$\eta(\boldsymbol{u}) = \int_{\Delta'} [\psi(\boldsymbol{u}-\boldsymbol{u}') - \overline{\psi}]\varrho(\boldsymbol{u}')d\boldsymbol{u}'. \tag{6.27}$$

Here $\psi(\boldsymbol{u})$ has the same meaning as in the theory of Lennard-Jones and Devonshire, i.e., it is the interaction energy of a molecule at $\boldsymbol{u}$ with all surrounding molecules, which are fixed at their cell centers. The average field $\overline{\psi}$ in (6.27) is defined by

$$\overline{\psi} = \int_{\Delta}\int_{\Delta'} \psi(\boldsymbol{u}-\boldsymbol{u}')\varrho(\boldsymbol{u})\varrho(\boldsymbol{u}')d\boldsymbol{u}\, d\boldsymbol{u}'. \tag{6.28}$$

If $\eta(\boldsymbol{u})$ is a solution of (6.26–28), substituting into (6.24) reduces the configurational free energy to the form

$$F = N\Big\{\overline{\psi}/2 + kT\ \ln\int_{\Delta} \exp[-\eta(\boldsymbol{u})/kT]d\boldsymbol{u}\Big\}, \tag{6.29}$$

which is equivalent to writing the configuration integral as

$$Q = \Big\{\exp(-\overline{\psi}/2kT)\int_{\Delta} \exp[-\eta(\boldsymbol{u})/kT]d\boldsymbol{u}\Big\}^N. \tag{6.30}$$

Equations (6.26–28) may be regarded as equations of self-consistency for $\varrho(\boldsymbol{u})$ and $\eta(\boldsymbol{u})$. The usual iterative procedure to solve these equations is to take some trial $\varrho(\boldsymbol{u})$, substitute it into (6.27,28), find $\eta(\boldsymbol{u})$, substitute the result into (6.26) and so on. Considering that $\varrho(\boldsymbol{u})$ has a sharp maximum at the cell center, it is reasonable to choose a zero approximation to $\varrho(\boldsymbol{u})$ to be a delta function

$$\varrho^{(0)}(\boldsymbol{u}) = \delta(\boldsymbol{u}). \tag{6.31}$$

Substituting (6.31) into (6.27,28) leads to the first approximation

$$\begin{aligned}\overline{\psi}^{(1)} &= \int_{\Delta}\int_{\Delta'} \psi(\boldsymbol{u}-\boldsymbol{u}')\delta(\boldsymbol{u})\delta(\boldsymbol{u}')d\boldsymbol{u}\,d\boldsymbol{u}'\\ &= \psi(0),\end{aligned} \tag{6.32}$$

$$\begin{aligned}\eta^{(1)}(\boldsymbol{u}) &= \int_{\Delta'} [\psi(\boldsymbol{u}-\boldsymbol{u}') - \psi(0)]\delta(\boldsymbol{u}')d\boldsymbol{u}'\\ &= \psi(\boldsymbol{u}) - \psi(0).\end{aligned} \tag{6.33}$$

Inserting these equations into (6.30) and considering that $N\psi^{(1)}/2 = \phi_0$, one gets exactly the same expression for Q as given by the cell model. Based on this result, one might regard the cell model as an approximation to Kirkwood's variational theory and expect further iterations to improve the result. Unfortunately, calculations on the rigid-sphere and Lennard-Jones systems [6.7] have shown that this is not the case. In fact, the results afforded by the Kirkwood theory are less accurate than those given by the cell model. In particular, the cell model yields values of the free energy lower than the optimum values of the variational theory. At first sight, this seems to be a contradiction of the assertion that the solution of the self-consistency equations (6.26–28) minimizes the free energy (6.24). Actually this is not a contradiction because (6.32,33) are not solutions of the self-consistency equations.

Thus, the variational theory of *Kirkwood* [6.10] can hardly be regarded as a justification for the cell model. Moreover, as noted by *Barker* [6.7], "it is much nearer to the truth to regard the variational theory as justified insofar it approximates to the cell model".

A consistent analysis of the basic assumptions of the cell model has been undertaken by *Barker* [6.7] based on entirely different considerations. Let us divide [6.7] the available volume of the system into a lattice of N cells. The exact configuration integral of the system may be written in the form

$$Q = (N!)^{-1} \sum_{m_1 \ldots m_N} Q^{(m_1 \ldots m_N)} N!/\Pi m_s! \quad (\Sigma m_s = N) \tag{6.34}$$

where $Q^{(m_1 \ldots m_N)}$ is a restricted configuration integral taken over that part of the total configurational space in which m_1 molecules occupy cell 1, m_2 molecules occupy cell 2, and so forth. For solids, one can neglect a possible multiple occupance of the cells and write

$$Q = Q^{(1 \ldots 1)}. \tag{6.35}$$

To derive a tractable expression for $Q^{(1 \ldots 1)}$, let us represent the interaction energy of molecules i and j in the form of the following identity

$$\begin{aligned} \varepsilon(\boldsymbol{r}^i, \boldsymbol{r}^j) = & \varepsilon(\bar{\boldsymbol{r}}^i, \bar{\boldsymbol{r}}^j) + \varepsilon(\boldsymbol{r}^i, \bar{\boldsymbol{r}}^j) - \varepsilon(\bar{\boldsymbol{r}}^i, \bar{\boldsymbol{r}}^j) \\ & + \varepsilon(\bar{\boldsymbol{r}}^i, \boldsymbol{r}^j) - \varepsilon(\bar{\boldsymbol{r}}^i, \bar{\boldsymbol{r}}^j) + \Delta_{ij} \end{aligned} \tag{6.36}$$

where

$$\Delta_{ij} = \varepsilon(\boldsymbol{r}^i, \boldsymbol{r}^j) - \varepsilon(\boldsymbol{r}^i, \bar{\boldsymbol{r}}^j) - \varepsilon(\bar{\boldsymbol{r}}^i, \boldsymbol{r}^j) + \varepsilon(\bar{\boldsymbol{r}}^i, \bar{\boldsymbol{r}}^j). \tag{6.37}$$

With this definition of Δ_{ij} the total potential energy is given by

$$\begin{aligned} \phi &= \frac{1}{2} \sum_{i \neq j} \varepsilon(\boldsymbol{r}^i, \boldsymbol{r}^j) \\ &= \frac{1}{2} \sum_{i \neq j} \varepsilon(\bar{\boldsymbol{r}}^i, \bar{\boldsymbol{r}}^j) + \sum_i \sum_{j \neq i} [\varepsilon(\boldsymbol{r}^i, \bar{\boldsymbol{r}}^j) - \varepsilon(\bar{\boldsymbol{r}}^i, \bar{\boldsymbol{r}}^j)] \\ &\quad + \frac{1}{2} \sum_{i \neq j} \Delta_{ij}. \end{aligned} \tag{6.38}$$

The first sum is equal to the static lattice energy ϕ_0 while the sum over j in the second term is equal to the cell field $\psi(\boldsymbol{u}^i) - \psi(0)$, as it appears in (6.18). Thus one can rewrite (6.38) as

$$\phi = \phi_0 + \sum_i [\psi(\boldsymbol{u}^i) - \psi(0)] + \frac{1}{2} \sum_{i \neq j} \Delta_{ij}. \tag{6.39}$$

Defining

$$f_{ij} = \exp(-\Delta_{ij}/kT) - 1, \tag{6.40}$$

we may write the restricted configuration integral in the form

$$\begin{aligned} Q^{(1 \ldots 1)} = & \exp(-\phi_0/kT) \int_{\Delta_1} \cdots \int_{\Delta_N} \prod_i \exp\{-[\psi(\boldsymbol{u}^i) - \psi(0)]/kT\} \\ & \times \prod_{j>k} (1 + f_{jk}) d\boldsymbol{u}^1 \ldots d\boldsymbol{u}^N. \end{aligned} \tag{6.41}$$

Expanding the product of the factors $(1 + f_{jk})$ and integrating over the displacements of molecules not involved in f_{jk} in a given term leads to the result,

$$Q^{(1\ldots1)} = \exp(-\phi_0/kT)V_f^N\left(1 + \sum\langle f_{ij}\rangle + \sum\langle f_{ij}f_{kl}\rangle + \ldots\right). \quad (6.42)$$

In this case V_f has the same meaning as in (6.21). The angular brackets denote an averaging according to

$$\langle f_{ij}\ldots f_{kl}\rangle = \int_{\Delta_i}\ldots\int_{\Delta_l}\prod_{p=i,\ldots,l}\exp\{-[\psi(\boldsymbol{u}^p) - \psi(0)]/kT\}$$
$$\times V_f^{-1} f_{ij}\ldots f_{kl}d\boldsymbol{u}^i\ldots d\boldsymbol{u}^l. \quad (6.43)$$

The leading term in (6.42) coincides with the expression given by the cell model for the configuration integral, while the other terms are corrections to $Q^{(1\ldots1)}$ due to correlation effects.

It is clear that a rigorous evaluation of the correlation effects is equivalent in complexity to the initial problem, i.e., to a direct evaluation of the configuration integral. An approximate estimate of the corrections can be obtained by replacing the average of the product $f_{ij}\ldots f_{kl}$ by the product of averages

$$\langle f_{ij}\ldots f_{kl}\rangle = \langle f_{ij}\rangle\ldots\langle f_{kl}\rangle. \quad (6.44)$$

This equation is exactly true if there is no repeated index among $i, j, \ldots, k, l$. If this is not so, (6.44) is an approximation. Thus, while (6.44) correctly takes into account the correlated motions involving two different molecules (binary correlations), it neglects the correlated motions involving three or more molecules (ternary and higher-order correlations).

Using (6.44), the configurational integral of a crystal is easily reduced to

$$Q^{(1\ldots1)} = Q_{\text{cell model}}\left[\prod_{j>1}^{N}(1 + f_{1j})^{1/2}\right]^N. \quad (6.45)$$

The factor in brackets defines the correction to the free energy due to binary correlations,

$$F_{\text{bin corr}} = -N\,kT\frac{1}{2}\sum_{j>1}^{N}\ln(1 + \langle f_{1j}\rangle). \quad (6.46)$$

To estimate the contribution of binary correlations to the total correlation correction, *Barker* [6.7] considered a system of harmonic oscillators, for which the configuration integral could be evaluated exactly using the Born-von Karman formalism. It turned out that the correction to Q due to binary correlations accounts for 75–85% of the total correlation correction (which itself was rather small).

Since the correlation effects are not explicitly related to harmonicity of vibrations, it is expected that (6.46) will also give a rough estimate of the correlation correction when the forces are not harmonic.

As seen from (6.43), an evaluation of the binary correlation corrections $\langle f_{1j} \rangle$ is reduced to the computation of a $2n$-fold integral, n being the number of degrees of freedom per molecule. This is a matter of considerable difficulty when the forces are not harmonic. An approximate method to evaluate $\langle f_{1j} \rangle$ based on an analytical approximation to the cell field has been suggested by *Barker* [6.7]. For a 6–12 Lennard-Jones system in a high-density region the averages $\langle f_{1j} \rangle$ were found to vary from 0.02 to 0.05 and were practically independent of the temperature.

For systems with three translational and three rotational degrees of freedom per molecule, the evaluation of $\langle f_{1j} \rangle$ involves the computation of 12-fold integrals. These integrals were estimated numerically using the "importance sampling" method to be discussed in Sect. 6.4.2 [6.11]. Table 6.1 lists the results for the benzene crystal at 270 K. To calculate the cell field $\psi(\boldsymbol{u})$ the MKB74 atom-atom potentials were used (Table 3.1). Under the heading "symmetry operation" Table 6.1 presents the symmetry element by which the jth molecule is generated from the central (first) molecule. The first six molecules ($j = 2$ to 7) are within the nearest coordination sphere about molecule 1. The other six molecules in the nearest coordination sphere are related to the first six by an inversion center and yield the same $\langle f_{1j} \rangle$'s. In the last five rows of Table 6.1 (molecules 8 to 12) the $\langle f_{1j} \rangle$'s of several more distant neighbors of molecule 1 are listed.

Table 6.1. Binary correlation corrections $\langle f_{ij} \rangle$ for the benzene crystal at 270 K [6.11]

Molecule j	Symmetry operation			$\langle f_{1j} \rangle$
Nearest neighbors				
2	$\frac{1}{2} + x$,	y,	$\frac{1}{2} - z$	-0.034
3	$-\frac{1}{2} + x$,	y,	$\frac{1}{2} - z$	-0.13
4	$\frac{1}{2} - x$,	$\frac{1}{2} + y$,	z	0.033
5	$\frac{1}{2} - x$,	$-\frac{1}{2} + y$,	z	0.14
6	x,	$\frac{1}{2} - y$,	$\frac{1}{2} + z$	-0.0003
7	x,	$\frac{1}{2} - y$,	$-\frac{1}{2} + z$	0.04
More remote neighbors				
8	$1 + x$,	y,	z	0.0004
9	x,	$1 + y$,	z	0.0001
10	x,	y,	$1 + z$	-0.004
11	$\frac{1}{2} - x$,	y,	$1 + z$	0.0002
12	x,	y,	$2 + z$	0.000002

It can be seen that the dominant contribution to the correlation correction is accounted for, as expected, by the nearest neighbors. The contribution of the more remote molecules is negligible. It is interesting to note that the contributions from different molecules are of different signs and greatly cancel out. For this reason, the total correlation correction to the free energy, as given by (6.46), proves to be as small as 0.02 kcal/mol.

Since the cancellation of individual $\langle f_{1j} \rangle$'s (e.g.,–0.13 and 0.14, –0.034 and 0.033) may be fortuitous, we recalculated $\Delta F_{\text{bin corr}}$ by assuming all $\langle f_{1j} \rangle$'s to be of the same sign. In this case the correlation correction was 0.2 kcal/mol, a value which is an order of magnitude greater than the previous result. However, it is still small compared to the total thermal contribution to the free energy ($F - \phi_0 = -4.1$ kcal/mol).

6.3 Comparison of the Cell Model and the Quasi-Harmonic Approximation with Computer Experiments

Any thermodynamic calculation based on a model interaction potential involves two kinds of errors: the errors produced by the statistical mechanical approximations and the errors arising from flaws in the model potential. A comparison of the theoretical results with experiment is, therefore, not a good criterion for assessing the quality of statistical mechanical assumptions because good data may be the result of a mutual cancellation of statistical and "potential" errors.

To assess the merits of a statistical-mechanical model, it is better to make a comparison with the results of computer experiments using the same model potential. In this case the errors due to flaws in the potential can be avoided because the computer experiments yield the thermodynamic properties that would really be observed for a system of particles interacting through the particular model potential.

There are two basic methods of carrying out computer experiments: the Monte-Carlo method and the method of molecular dynamics. Since the former will be used in the cell model calculations, it is worthwhile to devote some attention to it.

The Monte-Carlo method is used to calculate thermodynamic quantities that may be represented as an ensemble average

$$\langle \varphi \rangle = \left\{ \int_{\Omega} \exp\left[-\phi(\boldsymbol{R})/kT \right] d\boldsymbol{R} \right\}^{-1} \int_{\Omega} \varphi(\boldsymbol{R}) \exp\left[-\phi(\boldsymbol{R})/kT \right] d\boldsymbol{R}, \quad (6.47)$$

where $\boldsymbol{R}$ denotes the coordinates that specify the configuration of the system, Ω is the available volume of the configurational phase space, and $\varphi(\boldsymbol{R})$ is a function of the coordinates $\boldsymbol{R}$.

Using the notation of (6.47), we may write down expressions for most thermodynamic properties:

the internal energy,

$$E = \frac{n}{2}NkT + \langle\phi\rangle \tag{6.48}$$

where n is the number of degrees of freedom per particle;

the pressure,

$$P = [NkT - \langle(\partial\phi/\partial \ln V)_T\rangle]/V; \tag{6.49}$$

the heat capacity at constant volume,

$$C_v = \frac{n}{2}NkT + k(\langle\phi^2\rangle - \langle\phi\rangle^2)/(kT)^2; \tag{6.50}$$

the isothermal bulk modulus,

$$\begin{aligned} B_T =& P + \frac{1}{V}\langle(\partial^2\phi/\partial \ln V^2)_T\rangle - \frac{1}{VkT} \\ &\times[\langle(\partial\phi/\partial \ln V)_T^2\rangle - \langle(\partial\phi/\partial \ln V)_T\rangle^2]; \end{aligned} \tag{6.51}$$

and the Grüneisen function,

$$\begin{aligned} \gamma =& \{Nk - [\langle\phi(\partial\phi/\partial \ln V)_T\rangle \\ &- \langle\phi\rangle\langle(\partial\phi/\partial \ln V)_T\rangle]/(kT^2)\}/C_v. \end{aligned} \tag{6.52}$$

Other thermodynamic functions can be derived from the above functions using familiar thermodynamic relations. Thus, the adiabatic bulk modulus and the heat capacity at constant pressure are given by

$$B_S = B_T - TC_v\gamma^2/V, \tag{6.53}$$

$$C_p = C_vB_S/B_T. \tag{6.54}$$

The configuration integral itself, which may be formally expressed through the ensemble average $\langle \exp(\phi/kT)\rangle$ and the volume of the system, cannot be determined by the Monte-Carlo method due to poor convergence properties of this ensemble average. Hence, important thermodynamic functions such as the free energy and entropy are beyond the scope of the Monte-Carlo method.

The principal difficulty involved in an evaluation of the ensemble averages of (6.47) is due to the fact that the energies of the overwhelming majority of configurations in Ω are too large to contribute significantly to the multidimensional integrals in (6.47). Hence, any uniform sampling of Ω would be completely impracticable. The Monte-Carlo method generates a chain of random configurations whose energies are close to $\langle\phi\rangle$ and, hence, contribute most to the ensemble averages. The chain is generated so that each sampled configuration $\boldsymbol{R}^{(i)}$ appears with a probability proportional to the Boltzmann factor $\exp[\phi(\boldsymbol{R}^{(i)})/kT]$. (A procedure for generating such a chain will be described in Sect. 6.4.2 for a one-particle system.) The ensemble averages are then evaluated as direct averages over the chain

$$\langle\varphi\rangle = \frac{1}{M}\sum_{i=1}^{M}\varphi(\boldsymbol{R}^{(i)}). \tag{6.55}$$

The only serious error involved in a computer experiment, using either the Monte-Carlo method or the method of molecular dynamics, is associated with the finite number of particles in the model system (the number is severely restricted by computational facilities). To reduce this error, the whole space is usually supposed to be filled with periodic repetitions of a fundamental cell, each repetation containing N molecules in the same relative positions. The artificial extra periodicity imposed upon the system may markedly affect the results when dealing with liquids, particularly in the region of the solid-liquid transition [6.12]. With crystals, the extra periodicity does not seem to be of much importance. Nevertheless, it is always desirable to have as large a system as possible because a small system may fail to reproduce the crystal properties determined by high-order derivatives of the free energy. This particularly applies to low temperatures at which the long-wavelength motions become important. If the size of the fundamental cell is too small, the long-wave-length vibrations cannot be properly simulated.

An assessment of the quality of the cell model and the quasi-harmonic approximation relative to the results of computer experiments was reported by *Holt* et al. [6.13] for a system of single-point particles and by *Gibbons* and *Klein* [6.14] for a system of particles interacting via a "dumbbell" model potential. The dumbbell model represents the intermolecular interactions by pair atom-atom potentials between the ends of linear molecules. It was used to describe the interactions in solid carbon dioxide, although for our purposes it is not important what particular system was simulated nor how well the calculations reproduced the experimentally observed properties.

The system investigated by *Gibbons* and *Klein* [6.14] is of particular interest to us since it represents the simplest model of a molecular crystal with all its inherent features arising from the occurrence of librational motions. Table 6.2 lists the thermal contributions to the thermodynamic functions given in (6.48–53), as determined by the Monte-Carlo method (with $N = 108$ and $N = 32$), the cell model and the quasi-harmonic ap-

Table 6.2. Comparison of the Monte-Carlo, cell model and quasiharmonic results for the thermodynamic properties of CO_2 at 194 K and a volume of 28.21 cm^3/mol [6.14]

Thermodynamic function	Static contribution	Thermal contributions a	b	c	d
E/NkT	-16.50	4.66	4.57	4.59	5.00
PV/NkT	-10.95	16.02	15.48	16.33	20.04
C_V/Nk	0.0	4.49	4.17	4.30	5.00
γ	0.0	3.17	2.94	3.21	3.92
B_TV/NkT	81.5	12.0	20.6	15.3	-11.7
B_SV/NkT	81.5	57.2	56.7	59.7	63.7

[a]Monte-Carlo $N = 108$; [b]Monte-Carlo $N = 32$; [c]Cell model; [d]Quasi-harmonic approximation

proximation. The listed values refer to a volume of 28.21 cm^3/mol and a temperature of 194 K (just below the melting point of CO_2).

As seen from Table 6.2, the cell model offers a very good approximation to the exact results, the approximation being even better than the Monte-Carlo results for the 32 particle system. The agreement is particularly good for the energy and pressure. The corresponding thermal contributions are reproduced by the cell model to within 2%. The cell model results for the other quantities are only slightly worse, all being within 4% of the exact values.

The quasi-harmonic approximation is seen to give much worse results. The deviations from the Monte-Carlo $N = 108$ values are about 10% for E, 20% for PV, 25% for B_T, 5% for B_S, 10% for C_v and 20% for γ.

A similar situation occurred at a lower temperature (158 K). The cell model was markedly superior to the other two models (Monte-Carlo with $N = 32$ and quasi-harmonic). The quasi-harmonic approximation was only found to be reliable at low temperatures (below $\theta_D/2$) where both the cell model and the Monte-Carlo method began to fail.

6.4 Extension of the Cell Model to Organic Molecular Crystals

6.4.1 Basic Formulas

For particles with three translational and three rotational degrees of freedom the cell model expression for the configuration integral may be written in the form

$$Q = \exp(-\phi_0/kT)V_f^N, \quad \text{with} \tag{6.56}$$

$$V_f = \int_\Delta \exp\{-[\psi(\boldsymbol{R}) - \psi(\overline{\boldsymbol{R}})]/kT\}d\boldsymbol{R}. \tag{6.57}$$

In the last equation $\boldsymbol{R}$ is a six-dimensional translation-rotation vector specifying the position and orientation of a molecule in its cell; $\overline{\boldsymbol{R}}$ refers to the equilibrium position and orientation; and $\psi(\boldsymbol{R})$ is the interaction energy of a molecule at $\boldsymbol{R}$ and its neighbors at their equilibrium sites. Note that

$$\phi_0 = \frac{1}{2}N\psi(\overline{\boldsymbol{R}}). \tag{6.58}$$

To make the formulas consistent with the ones introduced in Chap. 4, the first three components of $\boldsymbol{R}$ will be defined to be the fractional unit-cell coordinates $\boldsymbol{X} = \{X_\alpha\}$ of the origin of the principal molecular axes, while the other three to be the Euler angles $\chi = \{\chi_\alpha\}$ describing the orientation of the latter with respect to the crystal-fixed coordinate system defined in Sect. 4.2.1. With these definitions, the equation for the free volume becomes

$$\begin{aligned} V_f &= L_{11}L_{22}L_{33}\int_\Delta \exp\{-[\psi(\boldsymbol{R}) - \psi(\overline{\boldsymbol{R}})]/kT\}d\boldsymbol{R} \\ &= L_{11}L_{22}L_{33}\int_\Delta \exp\{-[\psi(\boldsymbol{X},\chi) - \psi(\overline{\boldsymbol{X}},\overline{\chi})]/kT\}dX_1dX_2dX_3 \\ &\quad \times \sin\chi_3 d\chi_1 d\chi_2 d\chi_3. \end{aligned} \tag{6.59}$$

We remind the reader that

$$\begin{aligned} L_{11}L_{22}L_{33} &= \det|\underline{\underline{L}}| \\ &= a_x b_y c_z \\ &= V_c, \quad \text{the unit-cell volume.} \end{aligned} \tag{6.60}$$

The exponential in (6.59) is generally sharply peaked in the neighborhood of $\overline{\boldsymbol{R}}$. For this reason the choice of cell boundaries is not important. It will be implied that the translational part of Δ is defined to be the Wigner-Seitz cell [6.9]. With our particular choice of $\boldsymbol{R}$, the volume of the translational part of Δ is dimensionless and equal to $V/(NV_c) = 1/N_c$. The rotational part of Δ may be conveniently chosen to only include symmetrically distinct orientations about the equilibrium orientation $\overline{\chi}$ (the cell center). With these choices, the total volume of the six-dimensional cell is given by

$$\Omega_\Delta = 8\pi^2/(N_c\sigma), \tag{6.61}$$

where σ is the molecular index of symmetry [6.15,16].

Explicit expressions for the thermodynamic functions are derived from the partition function

$$Z = [\lambda(T)]^N Q, \tag{6.62}$$

where $\lambda(T)$ is responsible for the ideal contribution to the thermodynamic functions,

$$\lambda(T) = \left(\frac{\sqrt{2\pi MkT}}{h}\right)^3 \left(\frac{\sqrt{2\pi kT}}{h}\right)^3 \sqrt{I_1 I_2 I_3}. \tag{6.63}$$

It should be noted that (6.63) does not contain σ because the integration in (6.59) has already been assumed to have been taken over symmetrically distinct orientations only.

The Helmholtz free energy of a crystal may now be written as

$$\begin{aligned} F &= -kT \ \ln Z \\ &= -NkT \ \ln \lambda(T) + \phi_0 - NkT \ \ln V_f. \end{aligned} \tag{6.64}$$

The equation for the internal energy may be derived from (6.64) using the familiar relation,

$$E = F - T(\partial F/\partial T)_L. \tag{6.65}$$

The subscript L means that the six independent parameters of the matrix $L_{\alpha\beta}$, as given by (4.47–52), are kept fixed. This is analogous to keeping the volume constant in the case of isotropic systems.

By differentiating (6.64) with respect to temperature and substituting the result in (6.65), one gets

$$E = \frac{6}{2} NkT + \phi_0 + N\langle \Delta\psi \rangle \tag{6.66}$$

where $\Delta\psi \equiv \psi(\boldsymbol{R}) - \psi(\overline{\boldsymbol{R}})$ and the angular brackets denote an averaging over the states of the central (moving) molecule,

$$\langle \varphi \rangle = V_f^{-1} V_c \int_{\Delta} \varphi(\boldsymbol{R}) \exp\{ - [\psi(\boldsymbol{R}) - \psi(\overline{\boldsymbol{R}})]/kT\} d\boldsymbol{R} \tag{6.67}$$

for any function $\varphi(\boldsymbol{R})$.

Once F and E are known, one can immediately evaluate the entropy from

$$S = (E - F)/T. \tag{6.68}$$

Differentiation of (6.66) with respect to temperature yields the specific heat at constant volume (or, more precisely, at constant $L_{\alpha\beta}$)

$$C_v = Nk[(\langle \Delta\psi^2 \rangle - \langle \Delta\psi \rangle^2)/(kT)^2 + 3]. \tag{6.69}$$

To determine the equilibrium unit-cell parameters we shall also need the expression for the components of the stress tensor $\sigma_{\alpha\beta}$. These can be readily evaluated from the derivatives $\partial F/\partial L_{\alpha\beta}$ using (4.10,59) in which $\partial\phi_c/\partial L_{\alpha\beta}$ should be substituted by $N_c^{-1}\partial F/\partial L_{\alpha\beta}$.

A differentiation of F with respect to $L_{\alpha\beta}$ does not present problems. Considering that the limits of integration in (6.59) are independent of $L_{\alpha\beta}$, one easily find that

$$\frac{\partial F}{\partial L_{\alpha\beta}} = \delta_{\alpha\beta}\frac{NkT}{L_{\alpha\alpha}} + \frac{\partial \phi_0}{\partial L_{\alpha\beta}} + N\left\langle \frac{\partial \Delta\psi}{\partial L_{\alpha\beta}} \right\rangle. \tag{6.70}$$

The expression for the second term on the right-hand side is given by (4.57,91,95). An explicit expression for $\partial\psi(\boldsymbol{R})/\partial L_{\alpha\beta}$ needed to evaluate the average in (6.70) can be derived in a manner similar to that used to obtain the expression for $\partial\phi_c/\partial L_{\alpha\beta}$ (Sect. 4.2.2).

For the most common case of one rigid molecule in the asymmetric part of the unit cell, one readily gets

$$\begin{aligned}\partial\psi(\boldsymbol{R})/\partial L_{\alpha\beta} = {} & \sum_{aa's'} \varphi'_{aa'} r_{\alpha}^{a,a's'}/r^{a,a's'} \\ & \times \Big(\sum_{\gamma} B_{\beta\gamma}^{s'} \overline{X}_{\gamma} + T_{\beta}^{s'} - X_{\beta}\Big),\end{aligned} \tag{6.71}$$

where

$$\begin{aligned} r_{\alpha}^{a,a's'} = {} & \sum_{\mu\nu} B_{\alpha\mu}^{s'} G_{\mu\nu}(\overline{\chi}) \varrho_{\nu}^{a'} - \sum_{\mu} G_{\alpha\mu}(\chi)\varrho_{\mu}^{a} \\ & + \sum_{\mu} L_{\alpha\mu}\Big(\sum_{\nu} B_{\mu\nu}^{s'} \overline{X}_{\nu} + T_{\mu}^{s'} - X_{\mu}\Big). \end{aligned} \tag{6.72}$$

In these equations a runs over all atoms in the central molecule and a' runs over the atoms in the surrounding molecules; s' labels the symmetry operations by which the surrounding molecules are generated from the central molecule (when the latter is at its equilibrium position); X_α and χ_α are the components of $\boldsymbol{R}$ and $\overline{X}_\alpha$ and $\overline{\chi}_\alpha$ are the components of $\overline{\boldsymbol{R}}$.

The expression for the elasticity tensor c_{ij} can be derived by a double differentiation of (6.64) with respect to the components of the strain tensor ε_i. Before doing this we note that the free energy, when calculated from (6.64), is a function of both the external and internal strains. The former are specified by the unit cell parameters $L_{\alpha\beta}$ while the latter by the translation-rotation vector $\overline{\boldsymbol{R}}$. The relation between ε_i and $L_{\alpha\beta}$ is trivial. Thus, $\varepsilon_1 = \Delta L_{11}/L_{11}$, $\varepsilon_2 = \Delta L_{22}/L_{22}$, $\varepsilon_5 = \Delta L_{31}/L_{33}$ and so forth.

The internal strains $\overline{\boldsymbol{R}}$ are not independent variables but are determined by the external ones through the equation

$$\left(\frac{\partial F}{\partial \overline{R}_{\alpha}}\right)_{\varepsilon_i} = 0. \tag{6.73}$$

Considering that F depends on ε_i both explicitly and implicitly, one obtains

$$\frac{DF}{D\varepsilon_i} = \frac{\partial F}{\partial \varepsilon_i} + \sum_\alpha \frac{\partial F}{\partial \overline{R}_\alpha} \frac{\partial \overline{R}_\alpha}{\partial \varepsilon_i}. \tag{6.74}$$

Both $D/D\varepsilon_i$ and $\partial/\partial\varepsilon_i$ denote differentiation with respect to ε_i; the difference is that in the former case $\overline{R}_\alpha$ is assumed to follow a variation in ε_i according to (6.72), while in the latter case $\overline{R}_\alpha$ is kept fixed, as if representing an independent variable.

Applying the operator $D/D\varepsilon_j$ to (6.74) one finds

$$\begin{aligned} c_{ij} &= \frac{D^2 F}{D\varepsilon_i D\varepsilon_j} \\ &= \frac{\partial^2 F}{\partial\varepsilon_i \partial\varepsilon_j} + \sum_\alpha \frac{\partial^2 F}{\partial\varepsilon_i \partial\overline{R}_\alpha} \frac{\partial \overline{R}_\alpha}{\partial\varepsilon_j} \\ &\quad + \sum_\alpha \frac{\partial F}{\partial \overline{R}_\alpha} \frac{\partial^2 \overline{R}_\alpha}{\partial\varepsilon_i \partial\varepsilon_j} + \sum_\alpha \frac{D}{D\varepsilon_j}\left(\frac{\partial F}{\partial \overline{R}_\alpha}\right) \frac{\partial \overline{R}_\alpha}{\partial\varepsilon_i}. \end{aligned} \tag{6.75}$$

In this equation the last two sums over α are easily seen to vanish insofar as $\overline{R}_\alpha$ satisfy (6.73) and

$$\frac{D}{D\varepsilon_j}\left(\frac{\partial F}{\partial \overline{R}_\alpha}\right) = 0. \tag{6.76}$$

For the non-vanishing sum over α there are two equivalent representations:

$$\begin{aligned} \sum_\alpha \frac{\partial^2 F}{\partial\varepsilon_i \partial\overline{R}_\alpha} \frac{\partial \overline{R}_\alpha}{\partial\varepsilon_j} &= \sum_\alpha \frac{\partial^2 F}{\partial\varepsilon_j \partial\overline{R}_\alpha} \frac{\partial \overline{R}_\alpha}{\partial\varepsilon_i} \\ &= \sum_\alpha \sum_\beta \frac{\partial^2 F}{\partial\overline{R}_\alpha \partial\overline{R}_\beta} \frac{\partial \overline{R}_\alpha}{\partial\varepsilon_i} \frac{\partial \overline{R}_\beta}{\partial\varepsilon_j}. \end{aligned} \tag{6.77}$$

The validity of this can be readily demonstrated using (6.76).

Another useful relation may be obtained by applying the operator $D/D\varepsilon_i$ to the entropy S. Given the derivatives $DS/D\varepsilon_i$ and the elastic constants c_{ij}, one may find the thermal expansion tensor by solving

$$\frac{DS}{D\varepsilon_i} = V \sum_j c_{ij} \alpha_j \tag{6.78}$$

with respect to α_j.

The expression for $DS/D\varepsilon_i$ is similar to (6.74)

$$\frac{DS}{D\varepsilon_i} = \frac{\partial S}{\partial \varepsilon_i} + \sum_\alpha \frac{\partial S}{\partial \overline{R}_\alpha} \frac{\partial \overline{R}_\alpha}{\partial \varepsilon_i}. \tag{6.79}$$

As seen from (6.75,79), a calculation of the implicit contribution to c_{ij} and $DS/D\varepsilon_i$ requires a knowledge of the derivatives $\partial\overline{R}_\alpha/\partial\varepsilon_i$. These latter can be approximately evaluated using a Newton-Raphson estimate for $\overline{R}_\alpha$

$$\overline{R}_\alpha = \overline{R}_\alpha^0 - \sum_\beta (F_R^{-1})_{\alpha\beta} \frac{\partial F}{\partial \overline{R}_\beta}, \tag{6.80}$$

where $\overline{R}_\alpha^0$ denotes an approximation to $\overline{R}_\alpha$ and F_R^{-1} is the inverse of the matrix

$$(F_R)_{\alpha\beta} = \frac{\partial^2 F}{\partial \overline{R}_\alpha \partial \overline{R}_\beta}. \tag{6.81}$$

Note that both F_R and $\partial F/\partial \overline{R}_\beta$ in (6.80) are to be computed at $\overline{\boldsymbol{R}}^0$.

Differentiating (6.80) with respect to ε_i, one gets

$$\frac{\partial \overline{R}_\alpha}{\partial \varepsilon_i} = -\sum_\beta (F_R^{-1})_{\alpha\beta} \frac{\partial^2 F}{\partial \varepsilon_i \partial \overline{R}_\beta} - \sum_\beta \frac{\partial (F_R^{-1})_{\alpha\beta}}{\partial \varepsilon_i} \frac{\partial F}{\partial \overline{R}_\beta}, \tag{6.82}$$

where again, the terms on the right-hand side are all taken at $\overline{\boldsymbol{R}}^0$. If $\overline{R}_\alpha^0$ tend to the respective solutions of (6.73), the last sum over β in (6.82) vanishes.

The problem of calculating the derivatives $\partial\overline{R}_\alpha/\partial\varepsilon_i$ can be further simplified by assuming that the equilibrium translation-rotation $\overline{\boldsymbol{R}}$ of a molecule in a crystal is determined by the static (but not free) lattice energy. The reasonableness of this is evidenced by the wide experience gained in crystal structure calculations by minimization of the static lattice energy. Formally, the above assumption means the replacement of F by ϕ_0 throughout in (6.73,80–82).

Thus far our analysis did not utilize the particular form assumed by the free energy in the cell model. Turning to (6.64), it can be shown that all the derivatives needed to calculate c_{ij} are given by the general equation

$$\frac{\partial^2 F}{\partial p_m \partial p_n} = \frac{\partial^2 \phi_0}{\partial p_m \partial p_n} + N\left\langle \frac{\partial^2 \Delta\psi}{\partial p_m \partial p_n} \right\rangle + \frac{N}{kT}\left(\left\langle \frac{\partial \Delta\psi}{\partial p_m} \right\rangle \left\langle \frac{\partial \Delta\psi}{\partial p_n} \right\rangle - \left\langle \frac{\partial \Delta\psi}{\partial p_m} \frac{\partial \Delta\psi}{\partial p_n} \right\rangle \right), \tag{6.83}$$

where p_m and p_n denote either of the parameters ε_i and $\overline{R}_\alpha$.

Using the same notations, the derivatives of the entropy in (6.79) can be written as

$$\frac{\partial S}{\partial p_m} = \delta_m^{(id)} + \frac{Nk}{(kT)^2}\left(\langle \Delta\psi \rangle \left\langle \frac{\partial \Delta\psi}{\partial p_m} \right\rangle - \left\langle \Delta\psi \frac{\partial \Delta\psi}{\partial p_m} \right\rangle \right). \tag{6.84}$$

where

$$\delta_m^{(id)} = \begin{cases} Nk, & \text{if } p_m = \varepsilon_i, i \leq 3 \\ 0, & \text{otherwise.} \end{cases}$$

For the intermolecular interaction potential expressed as a sum of atom-atom potential functions, the derivatives in the angular brackets in (6.83,84) can be calculated analytically. For reasons of space, we shall not write down the relevant equations and only note that these are very similar to the equations for the derivatives of the static lattice energy ϕ_c (Chap. 4). The analogy between the equations for the derivatives of ψ and ϕ_c can be appreciated by comparing (4.57,91,95) with (6.71,72).

In deriving the above formulas it was implied that all molecules in the crystal are symmetrically equivalent, so that all cells, cell fields and one-particle probability density distributions are identical. The equations obtained can be readily extended to the case of two and more crystallographically distinct molecules in the unit cell. Suppose, for instance, that N_A of the N molecules in a crystal are crystallographically equivalent to the basis molecule A, and N_B molecules are equivalent to another basis molecule, B $(N_A + N_B = N)$. Then, (6.18) may be rewritten

$$\phi = \phi_0 + \sum_{i=1}^{N_A} [\psi^i(\boldsymbol{U}^i) - \psi^i(0)] + \sum_{j=1}^{N_B} [\psi^j(\boldsymbol{U}^j) - \psi^j(0)], \tag{6.85}$$

where $\psi^i(\boldsymbol{U}^i)$ is the interaction energy of molecule i displaced from its cell center by $\boldsymbol{U}^i$ and its surrounding molecules located at their cell centers (similarly for $\psi^j(\boldsymbol{U}^j)$). It is clear that the cell fields ψ^i are all equivalent to the field experienced by the basis molecule A, while the ψ^j's are equivalent to that experienced by the basis molecule B

$$\psi^i(\boldsymbol{U}^i) = \psi^A(\boldsymbol{U}^i), \quad i = 1, \ldots, N_A, \tag{6.86}$$

$$\psi^j(\boldsymbol{U}^j) = \psi^B(\boldsymbol{U}^j), \quad j = 1, \ldots, N_B. \tag{6.87}$$

Instead of (6.58) we now have

$$\phi_0 = \frac{N}{2}[n_A\psi^A(0) + n_B\psi^B(0)], \quad \text{where} \tag{6.88}$$

$$n_A = N_A/N, \quad n_B = N_B/N. \tag{6.89}$$

If one substitutes (6.85) into the equation for the configuration integral and uses the coordinates $\boldsymbol{R}^i$ instead of the displacements $\boldsymbol{U}^i$, one finds that

$$Q = \exp(-\phi_0/kT)(V_f)^{N_A}(V_f)^{N_B}, \quad \text{where} \tag{6.90}$$

$$V_f^A = V_c \int_{\Delta_A} \exp\{-[\psi^A(\boldsymbol{R}^A) - \psi^A(\overline{\boldsymbol{R}}^A)]/kT\} d\boldsymbol{R}^A, \tag{6.91}$$

and similarly for V_f^B.

The free energy now becomes

$$F = -NkT \ \ln\lambda(T) + \phi_0 - NkT(n_A \ \ln V_f^A + n_B \ \ln V_f^B). \tag{6.92}$$

It is implied that the difference in conformation between molecules A and B is negligible, so that the principal moments of inertia are the same for both molecues. If this is not so, the ideal term in (6.92) should be replaced by

$$-NkT[n_A \ \ln\lambda_A(T) + n_B \ \ln\lambda_B(T)] \tag{6.93}$$

where λ_A and λ_B can be calculated from (6.63) with their own moments of inertia.

The equations for the other thermodynamic properties transform similarly to (6.92). For instance,

$$E = 3NkT + \phi_0 + N(n_A\langle\Delta\psi^A\rangle_A + n_B\langle\Delta\psi^B\rangle_B), \tag{6.94}$$

$$C_v = Nk\Big\{n_A[\langle(\Delta\psi^A)^2\rangle_A - \langle\Delta\psi^A\rangle_A^2]/(kT)^2 + n_B[\langle(\Delta\psi^B)^2\rangle_B - \langle\Delta\psi^B\rangle_B^2]/(kT)^2 + 3\Big\}, \tag{6.95}$$

where $\langle\,\rangle_A$ and $\langle\,\rangle_B$ denote an averaging over the states of molecules A and B, respectively.

6.4.2 Evaluation of Six-Fold Integrals

From a mathematical point of view, a calculation of the thermodynamic properties of a crystal within the cell model is reduced to the evaluation of six-fold integrals of the general form

$$I = \int_{\Delta} \varphi(\boldsymbol{R}) \exp[-\Delta\psi(\boldsymbol{R})/kT] d\boldsymbol{R}, \tag{6.96}$$

where $\varphi(\boldsymbol{R})$ is either equal to unity, $\Delta\psi(\boldsymbol{R})$, $[\Delta\psi(\boldsymbol{R})]^2$, or a derivative of $\Delta\psi(\boldsymbol{R})$.

Let $h(\boldsymbol{R})$ denote the integrand of (6.96). The simplest way of evaluating I is to average $h(\boldsymbol{R})$ over a sequence of random points $\boldsymbol{R}^{(i)}$ uniformly distributed over Δ with the probability density $p(\boldsymbol{R}) = \Omega_{\Delta}^{-1}$:

$$\theta^{(n)} = \frac{\Omega_\Delta}{n} \sum_{i=1}^{n} h(\boldsymbol{R}^{(i)}) \underset{n\to\infty}{\longrightarrow} \int_\Delta h(\boldsymbol{R}) d\boldsymbol{R}. \tag{6.97}$$

However, for the particular integrands appearing in (6.96) such a procedure will not be effective since the overwhelming majority of points $\boldsymbol{R}^{(i)}$ will not contribute significantly to I.

A much more suitable procedure is the "importance sampling" method [6.17]. Let us choose a positively valued function $p(\boldsymbol{R})$ such that

$$\int_\Delta p(\boldsymbol{R}) d\boldsymbol{R} = 1, \tag{6.98}$$

and then consider a function $g(\boldsymbol{R})$ defined to be equal to the ratio $h(\boldsymbol{R})/p(\boldsymbol{R})$. If P is a random point defined in Δ with the probability density $p(\boldsymbol{R})$, then the mathematical expectation of the random quantity $g(P)$ is equal to the integral I, i.e.,

$$Mg(P) = \int_\Delta g(\boldsymbol{R}) p(\boldsymbol{R}) d\boldsymbol{R} = I. \tag{6.99}$$

Furthermore, if $\boldsymbol{R}^{(1)}, \ldots, \boldsymbol{R}^{(n)}$ are independent realizations of the random point P, then the quantity

$$\theta^{(n)} = \frac{1}{n} \sum_{i=1}^{n} g(\boldsymbol{R}^{(i)}) = \frac{1}{n} \sum_{i=1}^{n} h(\boldsymbol{R}^{(i)})/p(\boldsymbol{R}^{(i)}) \tag{6.100}$$

converges to I.

The aim of the "importance sampling" method is to concentrate the distribution of sampled points $\boldsymbol{R}^{(i)}$ in the parts of Δ that are of "most importance". This means that one should choose $p(\boldsymbol{R})$ large in those parts of Δ where $|h(\boldsymbol{R})|$ is large, and choose $p(\boldsymbol{R})$ small where $|h(\boldsymbol{R})|$ is small. The best choice is to take $p(\boldsymbol{R})$ proportional to $|h(\boldsymbol{R})|$. However, such a choice is possible if $h(\boldsymbol{R})$ is given by a simple analytical function.

To select $p(\boldsymbol{R})$ appropriate to our problem, we notice that the behavior of the integrand in (6.96) is mainly determined by the exponential $\exp[-\Delta\psi(\boldsymbol{R})/kT]$. The exponent $\Delta\psi(\boldsymbol{R})/kT$, in turn, can be expanded in powers of the displacements $U_\alpha = R_\alpha - \overline{R}_\alpha$ and truncated beyond the second-order terms

$$\frac{\Delta\psi(\boldsymbol{R})}{kT} = \frac{1}{kT} \sum_{\alpha,\alpha'=1}^{6} \frac{\partial^2\psi}{\partial R_\alpha \partial R_{\alpha'}} U_\alpha U_{\alpha'}. \tag{6.101}$$

Thus, a reasonable choice for $p(\boldsymbol{R})$ is

$$p(\boldsymbol{R}) = C \exp\left(-\frac{1}{kT} \sum_{\alpha,\alpha'=1}^{6} \frac{\partial^2\psi}{\partial R_\alpha \partial R_{\alpha'}} U_\alpha U_{\alpha'}\right), \tag{6.102}$$

where C is a normalizing factor defined by (6.98).

To generate a sequence of random points occurring in Δ with the probability density $p(\boldsymbol{R})$, it is convenient to diagonalize the matrix $\partial^2\psi/\partial R_\alpha \partial R_\alpha$, and reduce $p(\boldsymbol{R})$ to the product of six one-dimensional distributions

$$p(\boldsymbol{W}) = \prod_{\alpha=1}^{6} p_\alpha(W_\alpha) = \sum_{\alpha=1}^{6} C_\alpha \exp\left(-\frac{\varepsilon_\alpha}{kT} W_\alpha^2\right). \tag{6.103}$$

Here ε_α are the eigenvalues of $\partial^2\psi/\partial R_\alpha \partial R_{\alpha'}$, and $\boldsymbol{W} = \{W_\alpha\}$ denotes new displacement coordinates related to the old ones through

$$\boldsymbol{W} = \underline{\underline{M}}\boldsymbol{U}, \tag{6.104}$$

where $\underline{\underline{M}}$ is a 6×6 matrix whose columns are the eigenvectors of $\partial^2\psi/\partial R_\alpha \partial R_{\alpha'}$. Since the distributions p_α are sharply peaked at the cell center, the normalization factors C_α may well be taken equal to those corresponding to the infinite integration limits.

Computer simulation of random points $\boldsymbol{W}^{(i)}$ with the product distribution (6.103) presents no problem since each component $W_\alpha^{(i)}$ can be generated independently of the others, with its own variance $\sigma_\alpha^2 = kT/2\varepsilon_\alpha$. The inverse transformation of the $\boldsymbol{W}^{(i)}$ to $\boldsymbol{U}(i)$ and thereafter to the molecular translation-rotation vectors $\boldsymbol{R}^{(i)}$ is made using (6.104).

The success of the above-described procedure for numerical evaluation of the six-fold integrals (6.96) is illustrated in Fig. 6.2. Shown as solid curves are the importance sampling estimates for some selected properties of benzene as a function of the number of sampled points. A fairly good approximation to the thermodynamic properties is seen to result after sampling as few as 2×10^3 random points in Δ. This offers an accuracy of the order of 0.01 kcal/mol for F and E, 0.1 cal/mol K for S and C_v, 0.1×10^{10}dyn/cm^2 for c_{ij} and 1 cal/mol K for $DS/D\varepsilon_i$.

For the sake of comparison, the thermodynamic properties of benzene were also evaluated using the traditional Monte-Carlo method by treating the cell-model average in (6.67) as the ensemble average of (6.47) for a one-particle system. The chain of configurations $\boldsymbol{R}^{(i)}$ was generated as follows. The central molecule was initially given a random displacement and the initial change in energy, $\Delta\psi(\boldsymbol{R}^{(0)})$, was evaluated. The molecule was then given a random move, and if the new configuration $\boldsymbol{R}^{(0)}_{\text{new}}$ had a lower $\Delta\psi$, the move was accepted, i.e., the next configuration of the chain, $\boldsymbol{R}^{(1)}$, was adopted to be $\boldsymbol{R}^{(0)}_{\text{new}}$. If $\Delta\psi(\boldsymbol{R}^{(0)}_{\text{new}})$ was greater than $\Delta\psi(\boldsymbol{R}^{(0)})$, the quantity $\exp\{[\Delta\psi(\boldsymbol{R}^{(0)}) - \Delta\psi(\boldsymbol{R}^{(0)}_{\text{new}})]/kT\}$ was computed and compared with a number chosen randomly from the interval (0,1). If the exponential was greater, the move was accepted. If it was smaller, the move was rejected, i.e., the next configuration of the chain was taken to be $\boldsymbol{R}^{(0)}$.

It can be shown that if such a chain is made sufficiently long, the frequencies of occurrence of the different $\boldsymbol{R}^{(i)}$ in the chain become proportional

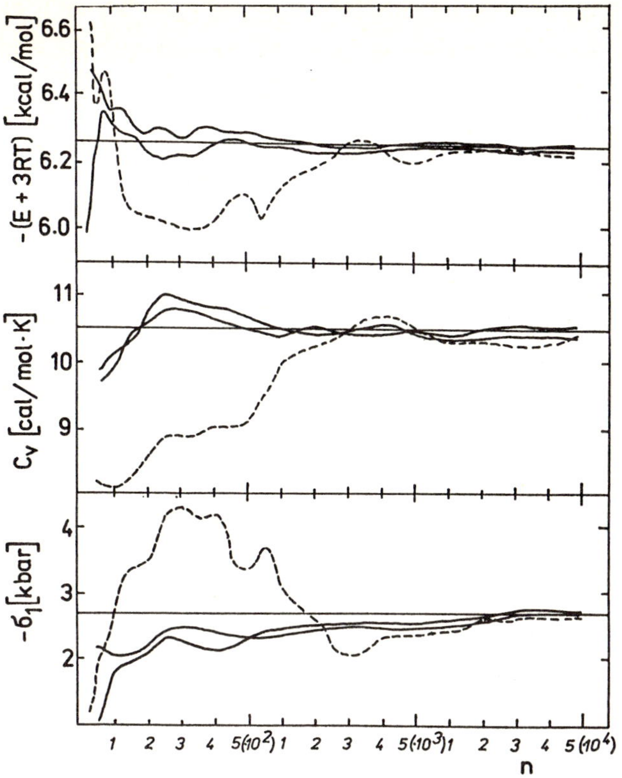

Fig. 6.2. Convergence curves for some selected thermodynamic properties of benzene at 270 K. The full and broken lines represent the importance sampling and Monte-Carlo results, respectively. The horizontal lines show the importance sampling results for $n = 5 \times 10^4$

to $\exp\left[-\Delta\psi(\boldsymbol{R}^{(i)})/kT\right]$. The averages in (6.67) are then given by the simple averages over the chain

$$\langle\varphi\rangle = \frac{1}{n}\sum_{i=1}^{n}\varphi(\boldsymbol{R}^{(i)}). \tag{6.105}$$

The results of the Monte-Carlo calculations are shown in Fig. 6.2 as dashed curves. It is seen that the final estimates of the thermodynamic functions are in good agreement with the "importance sampling" data. At the same time, one can observe that the Monte-Carlo results converge to their final values much more slowly.

The poor convergence of the Monte-Carlo estimates is not surprising. The matter is that a large portion (generally, a half) of sampled configurations is here expended as if in vain. We mean the rejected configurations which do not participate in the actual computation of the ensemble averages and are only needed to generate a chain with the desired (Boltzmann's) probability density distribution. By contrast, the configurations generated in an "importance sampling" run are all useful, in the sense that they all contribute to the ensemble averages.

The comparatively slower convergence of the Monte-Carlo estimates of the thermodynamic functions, as well as the difficulties associated with an evaluation of the free energy and entropy, definitely makes the Monte-Carlo method inferior to the "importance sampling" method in cell model calculations.

6.4.3 Numerical Results. Comparison with Experiment and Quasi-Harmonic Approximation

As with the quasi-harmonic approximation, a cell-model calculation of the thermodynamic properties of a crystal can be performed in two ways. In the first, more consistent way one proceeds from experimentally observed crystal symmetry, finds the structural parameters corresponding to the minimum free energy at a given temperature T, and then evaluates all required thermodynamic properties with these structural parameters.

Such an approach has been used to predict the pressure-induced phase transition in benzene [6.11] and the plastic-phase transition in adamantane [6.18]. These calculations will be described in detail in Sect. 6.5. Here we only note that the cell model was found to underestimate thermal expansion to yield denser equilibrium crystal structures than those observed experimentally.

The same shortcoming is inherent in the quasi-harmonic approximation, although for different reasons. While the quasi-harmonic approximation underestimates thermal expansion because it neglects the anharmonicity of vibrations, the cell model does so because it fails to correctly describe low-frequency motions. As a result, the low-frequency contribution to thermal expansion is lost and the whole expansion is underestimated.

The other way in which the cell model can be applied is to use not calculated but observed unit-cell parameters, as is done in the experimental version of the quasi-harmonic approximation. This second approach may be expected to give more reliable results for thermodynamic properties because it is free from errors associated with incorrect description of the equilibrium crystal structure.

Table 6.3 presents the cell-model results for the basic thermodynamic properties of naphthalene [6.6]. The calculated values are compared with the corresponding quasi-harmonic results and experimental external contribution to the thermodynamic properties. Both the cell model and quasi-harmonic calculations were carried out with the experimental unit-cell parameters, as measured by *Ryzhenkov* and *Kozhin* [6.19] in the temperature range 78 to 323 K. The molecular orientations were adjusted so as to min-

Table 6.3. The external contribution to the basic thermodynamic properties of naphthalene [kcal/mol, e.u.]. CM – cell model, QA – quasi-harmonic approximation. The observed heat capacities in column a are based on thermal expansions α obtained by graphical interpolation of the α-T data from [6.19] to the temperatures listed. Those in column b are based on thermal expanions α derived by least-squares fitting of the V-T data from [6.19] with a quadratic function

T[K]	ΔF			ΔE			S			C_v			
	obs	CM	QA	obs	CM	QA	obs	CM	QA	obs a	obs b	CM	QA
78	4.24	3.98	4.45	2.81	3.08	3.12	10.2	9.3	11.5	9.9	10.1	11.5	10.4
123	3.66	3.44	3.82	2.31	2.51	2.58	15.2	14.9	16.7	11.1	11.1	11.3	11.3
155	3.13	2.96	3.29	1.93	2.05	2.13	18.0	17.8	19.6	11.5	11.2	11.3	11.5
173	2.79	2.65	2.93	1.70	1.82	1.89	19.5	19.2	21.1	11.6	11.3	11.3	11.6
293	-	-	-	-	-	-	26.8	26.6	28.9	10.0	11.0	11.1	11.8
313	-0.76	-0.51	-0.57	-0.32	-0.35	-0.37	27.9	27.6	30.1	9.7	11.4	11.0	11.9
323	-0.83	-0.75	-0.85	-0.49	-0.57	-0.61	28.3	28.2	30.8	9.6	11.7	11.0	11.9

imize the static lattice energy. Atom-atom non-bonded interactions were treated using William's W67 parametrization and assuming the C-H bond length to be 1.027 Å. The summation radii used in calculating the harmonic force constants and all the terms involving the difference $\Delta\psi$ were 4, 5 and 6 Å for the H...H, H...C and C...C interactions, respectively. The static contribution to the thermodynamic properties was evaluated with a 20 Å summation radius.

As the experimental values, Table 6.3 presents the differences

$$f_{\rm ext} = f_{\rm cal} - f_{\rm int}, \tag{6.106}$$

where $f_{\rm cal}$ denotes the total (calorimetric) value of a thermodynamic function and $f_{\rm int}$ the respective internal contribution. The $f_{\rm cal}$'s were derived from the heat-capacity data of *McCullough* et al. [6.20]. In calculating $f_{\rm int}$ use was made of the fundamental frequencies from [6.21], except that the lowest a_u frequency of 195 cm^{-1} (unobserved) was replaced by a more commonly accepted value of 400 cm^{-1} [6.22–24].

The calculated F and E in Table 6.3 are compared with experiment with reference to their values at the room but not zero temperature [$\Delta F = F(T) - F(293)$, $\Delta E = E(T) - E(293)$]. We preferred such a comparison in order to avoid an error involved in evaluation of $F(0)$ and arising from extrapolation of the unit cell parameters to zero temperature.

The experimental heat capacities C_v in Table 6.3 were estimated using the familiar relations

$$C_v = C_p - \frac{\alpha^2 TV}{\beta_T} = C_p - \frac{1}{\frac{\beta_S}{\alpha^2 TV} + \frac{1}{C_p}}. \tag{6.107}$$

where α is the volume-expansion coefficient, and β_T and β_S are the isothermal and adiabatic compressibilities, respectively. The numerical values for β_S and α were obtained using the elastic constants from [6.25] and the unit-cell volumes from [6.19]. The resulting estimates for the experimental C_v

seem to be highly inaccurate in view of a strong dependence of the results on uncertainties in thermal expansion α (cf. columns a and b in Table 6.3).

As seen from Table 6.3, the cell model provides a definitely better fit to the experimental data than the quasi-harmonic approximation. This particularly applies to the entropy which can be compared with experiment in a straightforward way, without reference to such or another temperature. The cell model results for the entropy agree with experiment to within 0.3 e.u., except at the lowest temperature, close to $\theta_D/2$. (The Debye temperature of naphthalene is about 140 K; see Fig. 6.1).

For the elastic constants the cell-model predictions are not so successful (Table 6.4). Thus the calculation gives too low c_{22} and c_{13} and a too high c_{15}. In general, however, the agreement may be regarded as accetable considering a low accuracy of measuring the elastic constants, particularly the off-diagonal ones.

Table 6.4. Elastic constants in naphthalene at 293 K [10^{10}dyn/cm^2]. Only those constants are presented which are essential for calculating thermal expansions by (6.85)

	c_{11}	c_{22}	c_{33}	c_{55}	c_{12}	c_{13}	c_{15}	c_{23}	c_{25}	c_{35}
observed:	8.2	10.0	12.4	2.3	5.6	3.2	0.2	3.5	1.9	-2.9
calculated:										
total	6.5	6.6	12.2	2.7	4.4	0.7	2.2	1.4	1.5	-3.6
static contribution	6.1	6.0	9.8	1.9	4.6	1.6	2.0	2.6	1.2	-2.5
implicit contribution	0.7	0.3	0.3	1.3	-0.3	-1.1	-0.9	0.0	0.2	0.8

As seen from Table 6.4, the static contribution to the elasticity tensor is dominant although for some of the components (e.g., c_{33}, c_{13}, c_{23}) the thermal contribution is fairly significant. The implicit contribution to c_{ij}, as given by (6.77), is seen to be of minor importance.

An assessment of the cell-model results for thermal expansions α_i involves difficulties in view of large uncertainties in the experimental values for α_i. The latter fact can be appreciated from Table 6.5 which lists, aside from the calculation results, four sets of thermal expansions derived from different sources of experimental data. The most reliable data seem to be those in column b. Compared to these data, the calculated expansion is too small along c^* and too high along a and, particulary, b. As to the "shear" component of thermal expansion, the calculation failed to give a reliable estimate for it: the solution of (6.78) for α_5 was unstable in the sense that small variations in c_{ij} and $DS/D\varepsilon_i$ gave rise to drastic changes in α_5. This was not surprising since the elastic constants c_{i5} all were comparatively small.

It is interesting that despite pronounced anisotropy in thermal expansion, the tensor $DS/D\varepsilon_i$ is practically isotropic. This is seen from Table

Table 6.5. Thermal expansions in naphthalene at 293 K $[10^{-6}\mathrm{K}^{-1}]$. For the observed values, the following four sets are presented: (a) obtained by graphical interpolation of the $\alpha_i - T$ data from [6.19] to 293 K; (b) obtained by least-squares fitting of the unit-cell dimensions from [6.19] with quadratic functions; (c) taken from [6.26]; (d) determined from the unit-cell dimensions at 293 K [6.27] and 123 K [6.28]

		α_1	α_2	α_3	α_5
observed	a	168	45	174	160
	b	144	42	161	155
	c	112	40	105	104
	d	98	49	104	106
calculated		175	88	110	not converged

Table 6.6. Strain derivatives of entropy in naphthalene at 293 K [e.u.]. The observed values were derived from (6.78) using the experimental elastic constants from [6.25] and four sets of thermal expansions (a to d) in Table 6.5

		$\frac{DS}{D\varepsilon_1}$	$\frac{DS}{D\varepsilon_2}$	$\frac{DS}{D\varepsilon_3}$	$\frac{DS}{D\varepsilon_5}$
observed	a	57	59	62	-0.3
	b	51	54	56	0.1
	c	39	41	39	1
	d	37	41	38	2
calculated		41	39	42	2

6.6 which compares the calculated derivatives $DS/D\varepsilon_i$ with those obtained from (6.78) using the experimental elastic constants and thermal expansions. Although the experimental derivatives $DS/D\varepsilon_i$ show large variations depending on the thermal expansions used, in all the cases the "diagonal" derivatives are very close to each other and the "shear" one is insignificant. This characteristic feature of the tensor $DS/D\varepsilon_i$ is well reproduced by the calculation.

Aside from naphthalene, cell-model calculations were also performed for five other hydrocarbons composed of rigid molecules; viz., for benzene, anthracene, pyrene, acenaphthene and adamantane. The unit cell dimensions at appropriate temperatures were again taken from experiment [6.29–35] and the intermolecular interactions were treated with the W67 potential functions. The results for the excess free energy and entropy are compared in Table 6.7 with the corresponding experimental data and quasi-harmonic values. The internal energies have not been included in the table because the quality of predicting ΔE can be readily judged from that for ΔF and S. In evaluating the external contribution to the experimental free energies and entropies use was made of the calorimetric data from [6.36–40] and intramolecular fundamental frequencies from [6.41–46].

Inspection of Table 6.7 shows a remarkable agreement between the cell-model results and experiment. At high temperatures, the deviations for

Table 6.7. Excess free energy and entropy for five hydrocarbon crystals composed of rigid molecules [kcal/mol, e.u.]

T [K]	ΔF			S		
	obs	CM	QA	obs	CM	QA
Benzene[a]						
138	2.81	2.70	3.05	16.0	15.5	17.6
218	1.27	1.22	1.38	22.5	22.2	24.5
270	-	-	-	26.3	25.6	28.3
Anthracene[a]						
78	4.44	4.21	4.71	10.5	10.2	12.5
100	4.18	3.96	4.41	13.1	13.2	15.2
150	3.40	3.20	3.55	17.7	18.0	19.9
200	2.42	2.26	2.50	21.3	21.5	23.5
250	1.28	1.18	1.34	24.2	24.5	26.5
300	-	-	-	26.8	27.1	29.4
350	-1.40	-1.28	-1.45	29.2	29.4	32.1
400	-2.91	-2.66	-3.00	31.5	31.6	34.5
Pyrene						
133	4.18	3.35	3.73	18.9	17.6	19.6
155	3.73	2.97	3.30	20.8	19.5	21.6
191	2.93	2.29	2.55	23.2	22.3	24.3
213	2.41	1.85	2.04	24.5	23.6	25.8
244	1.63	1.15	1.26	26.3	25.6	27.8
293	-	-	-	28.8	28.3	30.6
Acenaphthene[b]						
133	3.58	3.23		16.9	16.6	
158	3.13	2.83		18.9	18.8	
201	2.25	2.05		21.8	23.9	
233	1.53	1.38		23.5	21.9	
262	0.82	0.75		25.0	25.6	
293	-	-		26.5	27.2	
Adamantane						
163	-	-	-	20.6	19.3	20.3

[a]For benzene and anthracene the excess free energies are given with reference to 270 and 300 K, respectively; [b]Quasi-harmonic calculations were not performed since the available computer program did not allow the treatment of crystals with symmetrically independent molecules

entropy are within 0.7 e.u. and they do not exceed 1.5 e.u. as the temperature is reduced. For pyrene and adamantane the quasi-harmonic predictions are quite satisfactory, while for benzene and anthracene they are 2–3 e.u. too high.

To check the dependence of the calculation results on the potential parameters used, the above discussed calculations were also performed using the MKB74 parameter set. The cell model results for the entropy were nearly as good as those for the W67 parametrization. The difference was that at low temperatures the MKB74 potentials provided a slightly better overall fit to experiment, while the high-temperature results were generally ~1.5 e.u. too high. Considering that the cell model is essentially a high-temperature model of the crystalline state, the W67 parameter set, which gives better results just at higher temperatures, may be concluded to be somewhat superior to the MKB74 set.

The cell-model and quasi-harmonic calculations have been extended to cover chlorine-, nitrogen- and fluorine-containing molecules; viz., chlorobenzene, 1,4-dichlorobenzene, 1,3,5-trichlorobenzene, 1,2,4,5-tetrachlorobenzene, hexachlorobenzene, hexamethylenetetramine and hexafluorobenzene [6.47,48]. The available experimental data on these compounds are complete enough to evaluate the external contribution to the basic thermodynamic properties and to apply the experimental version of the cell model. The unit cell parameters of the crystals are known at different temperatures from X-ray diffraction measurements [6.34,48–54]. Full calorimetric data are also available [6.48,55–57] as well as the complete sets of fundamental frequencies [6.58–61]. All the molecules are sufficiently rigid, so that the interplay between the internal and external vibrations is expected to be negligible.

To describe the atom-atom interactions in chlorinated benzenes the BB74a and W67 parameter sets were used. The calculated results for the excess free energy and entropy are compared with experiment in Table 6.8. It is seen that for all crystals with the exception of tetrachlorobenzene the cell model provides an excellent fit to experimental data. The quasi-harmonic results are seen to be systematically overestimated (by ~2 e.u. for entropy).

For tetrachlorobenzene, the cell model gives unusually low entropy values and even the quasi-harmonic approximation somewhat underestimates the entropy. Good results for this crystal were only obtained by replacing the BB74a and W67 potentials by "softer" MC78 and MKB74 ones. In

Table 6.8. Excess free energy and entropy for five chlorobenzene crystals [kcal/mol, e.u.]

T [K]	ΔF			S		
	obs	CM	QA	obs	CM	QA
Chlorobenzene						
120	-	-	-	16.1	15.7	
1,4-dichlorobenzene[a]						
100	4.30	4.38	4.87	13.1	12.8	14.8
300	-	-	-	28.8	28.7	31.2
1,3,5-trichlorobenzene						
90	4.57	5.01	5.13	14.0	13.8	15.1
293	-	-	-	28.7	28.4	30.7
1,2,4,5-tetrachlorobenzene						
211	2.28	2.27	2.47	25.1	23.0	24.8
239	1.56	1.58	1.60	26.9	24.6	26.8
266	0.81	0.82	0.91	28.4	26.2	28.0
294	-	-	-	30.0	27.6	29.7
Hexachlorobenzene						
124	4.00	4.41	4.76	17.6	17.5	19.3
143	3.64	4.02	4.38	19.2	19.3	20.8
190	2.65	2.92	3.15	22.7	23.0	24.9
216	2.04	2.25	2.49	24.4	24.7	26.3
263	0.84	0.94	1.12	27.2	27.3	28.9
270	0.65	0.72	0.86	27.4	27.8	29.5
293	-	-	-	28.5	29.0	31.1

[a] The high-temperature phase (293 K) was treated as positionally disordered [6.49], with the contribution to entropy due to disorder assumed to be R ln2 (Sect. 6.5.1)

this case the cell model reproduced the entropy almost exactly, while the quasi-harmonic approximation gave, as usually, an entropy about 2 e.u. too high.

Being appropriate for tetrachlorobenzene, the MC78 and MKB74 potentials were, however, inadequate for the four other chlorinated benzenes. When used in the framework of the cell model, these potentials overestimated the entropy by 1.5–4 e.u., and in the case of quasiharmonic approximation by 3–6 e.u.

In our view, the difficulties with tetrachlorobenzene arise not from defects of the BB74a and W67 potentials but, rather, from an error in the crystal structure data used in the calculations. The matter is that the crystal structure of this substance was determined very approximately, from Zeeman splitting of the ^{35}Cl n.q.r. spectrum and a two-dimensional X-ray diffraction analysis [6.54]. A calculation of the lattice energy at the experimental crystal configuration has revealed some shortened intermolecular contacts, and the subsequent energy minimization rotated the molecules by as great as 20° and lowered the energy by ~5 kcal/mol. It is not unlikely that the energy minimum found was a false minimum, not corresponding to the real crystal structure.

In calculating the intermolecular interactions in hexamethylenetetramine we combined the W67 potentials with Govers' potential for N...N interactions (Table 3.3). The C...C interactions in hexafluorobenzene were again simulated by the W67 potential, while the fluorine atoms were assumed to interact through the potential

$$\varphi(r) = -125.1\,r^{-6} + 105700\,\exp(-4.608\,r). \qquad (6.108)$$

This is the *Mason-Rice* [6.62] potential for Ne...Ne interactions, which afforded the best fit to the experimental entropy and heat of sublimation of hexafluorobenzene in the quasiharmonic calculations by *Filippini* et al. [6.63]. The experimental data needed for the cell-model calculations and the subsequent comparison of the results with experiment were taken from [6.48,64–70].

The cell-model results shown in Fig. 6.3 for hexamethylenetetramine fit the experimental data remarkably well, except at the lowest temperatures, close to $\theta_D/2 \approx 40$ K [6.1].

For hexafluorobenzene, the calculated entropy at 120 K was 18.7 e.u. which is only 0.4 e.u. higher than the observed value.

So far, we have only dealt with comparatively rigid molecules for which the interplay between the internal and external motions was of little importance. We shall now attempt to extend the cell-model calculations to molecules which contain rotating methyl groups. Such an extension is justified by the fact that a crystal field has a negligible effect (~0.1 kcal/mol) on the internal rotational barriers of the methyl groups [6.71]. Hence, the in-

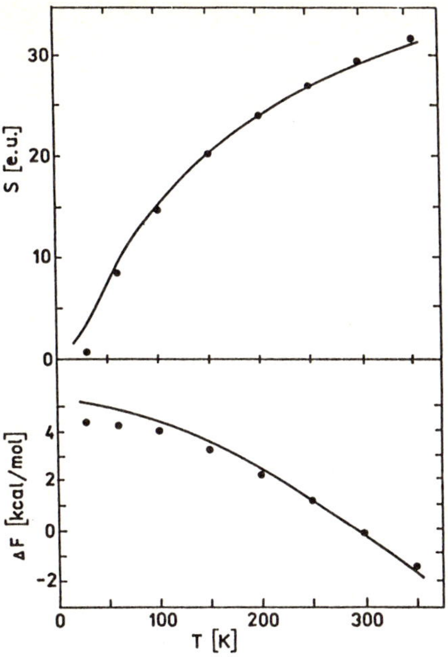

Fig. 6.3. Temperature dependence of the entropy and excess free energy in hexamethylenetetramine. The solid lines represent the experimental values and the circles represent the cell model results

ternal and external molecular motions can again be considered to be uncorrelated and the separation approximation written in (6.2) may be applied.

The best object to check this point is the crystal of toluene. The molecule of toluene possesses the lowest rotation barrier of the methyl-substituted benzenes, so that the correlation effects between the internal and external motions should be the greatest.

The crystal structure of toluene is only known at one temperature [6.72]. Thus, the only thermodynamic property that can be compared with experiment in a straightforward way is the entropy at this temperature (165 K). The cell model estimate of the external contribution to the entropy yields a value of 20.0 e.u. with a probable error of 0.2 e.u. [6.71] (the values corresponding to a W67 parametrization).

The internal contribution to the entropy of toluene may be evaluated as

$$S_{\text{int}} = S_{\text{int vib}} + S_{\text{int rot}}, \tag{6.109}$$

where $S_{\text{int vib}}$ is the vibrational contribution to S_{int} and $S_{\text{int rot}}$ is the contribution due to internal rotation. The numerical values for these two contributions are given in Table 6.9. The value of $S_{\text{int vib}}$ was calculated from the fundamental frequencies reported by *La Lau* and *Snyder* [6.73]. The two limiting values of $S_{\text{int rot}}$ corresponding to the lowest and highest rotation barriers reported in the literature, $\phi_{\text{int rot}} = 0$ (free rotation) and 0.75 kcal/mol (a slightly hindered rotation) are also given [6.74,75]. The numerical values for $S_{\text{int rot}}$ were taken from *Pitzer's* tables [6.16] where

Table 6.9. Calculated and observed entropy for solid toluene at 165 K [e.u.]

S_{ext}	20.0	
$S_{int\ vib}$	2.2	
$S_{int\ rot}$	3.0[a]	2.5[b]
S_{total}	25.2	24.7
$S_{total\ observed}$	24.7	

[a]Free rotation; [b]Hindered rotation with $\phi_{int\ rot} = 0.75$ kcal/mol

they are tabulated for selected values of $\phi_{int\ rot}$ and the free rotation partition function.

The sum $S_{ext} + S_{int\ vib} + S_{int\ rot}$ is compared in Table 6.9 with a third-law total entropy determined in a low-temperature calorimetric study by *Scott* et al. [6.74]. The agreement is excellent which justifies the assumptions made in the calculation.

The success with toluene allowed us to apply the same approach to refine the internal rotation barrier in hexamethylbenzene [6.71]. Two limiting estimates had already been reported in the literature for the rotation barrier, 2.985 and 7.870 kcal/mol. These values were obtained by *Frankosky* and *Aston* [6.76] by combining calorimetric, vapor pressure, heat-of-sublimation and spectroscopic data. Some intermediate results, the total third-law entropy of solid hexamethylbenzene and its internal contribution at 323.15 K, are presented in Table 6.10. The former was obtained in the usual way, as the integral

$$S_{solid} = \int_{0}^{323.15} C_p/T\, dT. \tag{6.110}$$

The latter was evaluated as the difference

$$S_{int} = S^0_{gas} - (S_{trans} + S_{rot}), \tag{6.111}$$

Table 6.10. Calculated and observed entropy for solid hexamethylbenzene at 323.15 K [e.u.]

S_{ext}	26.8		
S_{int}	40.4 - 45.5[a]	48.0[b]	49.7[c]
S_{total}	67.2 - 72.3	74.8	76.5
$S_{total\ observed}$	76.6		

[a]Calculated using experimental $\Delta H_{subl} = 17.8 - 19.6$ kcal/mol [6.71]; [b]Calculated using $\Delta H_{subl} = 20.2$ kcal/mol, as given by the cell model calculation; [c]Calculated as $S_{int\ vib} + S_{int\ rot}$ using experimental internal mode frequencies and calculated rotation barrier of 2.2 kcal/mol

where S^0_{gas} is the entropy of the hexamethylbenzene vapor at the standard pressure and 323.15 K, S_{trans} and S_{rot} are the translational and rotational contributions to S^0_{gas}, respectively. The value of S^0_{gas} was calculated from the total third-law entropy using the heat-of-sublimation and vapor pressure data [6.77]

$$S^0_{gas} = S_{solid} + \Delta H_{subl}/T - R \ln(P_{atm}/P_{eq}). \tag{6.112}$$

The last term represents the correction for the difference between the standard ($_{atm}$) and equilibrium ($_{eq}$) pressures.

Two estimates for S^0_{gas} were obtained in [6.76], corresponding to two different estimates for the heat of sublimation, ΔH_{subl} = 17.8 and 19.6 kcal/mol at 323.15 K. Accordingly, two different values were derived from (6.111,112) for S_{int}. These two values are presented in Table 6.10 as the limits of S_{int} (marked by footnote a).

To evaluate the internal rotation barrier, *Frankosky* and *Aston* used (6.109) in which $S_{int\,vib}$ was determined from the experimental internal frequencies [6.75].

To verify the consistency of the *Frankosky* and *Aston* estimates of S_{solid} and S_{int}, we separately evaluated the external contribution to the entropy using the cell model. In deriving the unit cell parameters of the crystal at 323.15 K we have used the room-temperature crystal structure data of *Brockway* and *Robertson* [6.78] and the thermal expansion parameters of *Woodward* [6.79]. The coordinates of the hydrogen atoms in hexamethylbenzene, not determined in [6.78], were found by minimizing the potential energy of an isolated molecule, ϕ_{int}, with respect to the six torsional angles describing the orientation of the six methyl groups. The energy ϕ_{int} only included the methyl-methyl non-bonded interactions, and did not include any terms related to the methyl group alone. Such a model for ϕ_{int} was justified in view of the low internal rotation barrier observed in toluene. The minimum energy conformation may be schematically depicted as

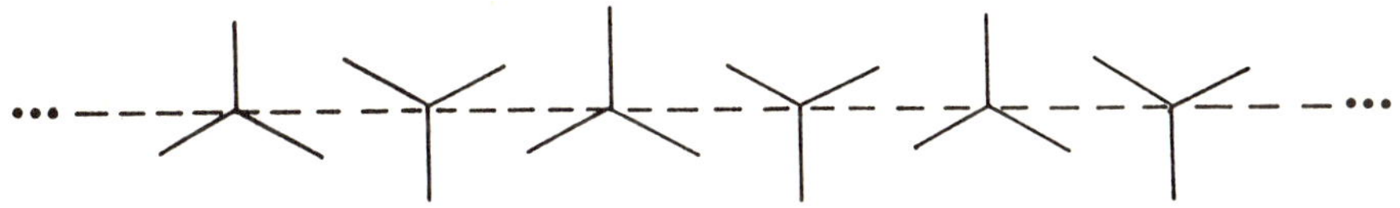

(for convenience, the benzene ring is shown as opened). Note that the molecule possesses S_6 symmetry, in agreement with spectroscopic evidence [6.80].

The cell model calculations were then carried out to give S_{ext} = 26.8 e.u. The addition of this quantity to S_{int} reported by *Frankosky* and *Aston* [6.76] gives a total entropy much lower than the measured third-law entropy. In our opinion, the discrepancy stems from an inaccurate determination of S_{int} from vapor-pressure and heat-of-sublimation data. Indeed, at 323.15 K even an error of 1–2 kcal/mol, not unusual for the heat of sublimation, produces an uncertainty in S_{int} as large as 3–6 e.u.

The heats of sublimation used by *Frankosky* and *Aston* [6.76] in their calculations seem to be too low. An evaluation of ΔH_{subl} using a cell-model estimate of the internal energy ($\Delta H_{\text{subl}} \cong 4RT - E$) yields a value of 20.2 kcal/mol which, when substituted to (6.112), markedly improves the result (Table 6.10, footnote b). Nevertheless, a perceptible discrepancy still remains which may arise from the last term in (6.122).

We have next attempted to evaluate S_{int} independently [6.71], without using vapor-pressure and heat-of-sublimation data. The calculations were based on the relation (6.109). The vibrational contribution to S_{int} was determined from the fundamental frequencies quoted by *Frankosky* and *Aston* [6.76], with the exception of those corresponding to the torsional vibrations of the methyl groups. The situation proved to be more complicated for $S_{\text{int rot}}$. If the displacements of the methyl groups from their equilibrium orientations are small, $S_{\text{int rot}}$ may be evaluated using the harmonic approximation. To find the torsional vibration frequenices, an approach similar to that used by *Born* and *von Karman* [6.81] to treat the vibrations of an infinite atomic chain can be applied. The following are two modifications that one should make: (i) the equation for forces should be replaced by that for torques, and (ii) a torque, depending on the displacements of a given methyl alone and not associated with methyl-methyl interactions, should be taken into account (the neglect of this torque would result in a zero vibrational frequency).

With these modifications, the expression for the torsional vibrational frequencies can be readily shown to be

$$\omega = [(4f_1 \sin^2 \pi s + f_2)/I]^{1/2}, \tag{6.113}$$

where f_1 and f_2 are the force constants due to methyl-methyl non-bonded interactions and to the own torsional potential, respectively; I is the moment of inertia of the methyl group and s is the wavenumber, $-1/2 < s \leq 1/2$.

Equation (6.113) is valid for an opened, infinite chain, such as that shown above for the equilibrium configuration of the hexamethylbenzene molecule. The closure of the chain is accomplished by imposing cyclic boundary conditions, according to which the displacements of the ith and $(i+6)$th methyls must be identical. This immediately gives rise to the following allowed values of the wavenumber: $s = 0, \pm 1/6, \pm 1/3, 1/2$.

The harmonic model described may, in principle, be used to evaluate S_{int} at low temperatures, since the force constants, at least f_1 and the crystal field contribution to f_2, can be readily estimated using the atom-atom potential method. A direct evaluation of the internal rotation barrier with the help of atom-atom potentials shows however that the harmonic model will be grossly incorrect at 323.15 K. For a methyl group rotating in the field of its neighbors fixed at their equilibrium orientations, the calculation yields $\phi_{\text{int rot}} = 2.2\,\text{kcal/mol}$. This means that at 323.15 K the probabilities of occupying the highest and lowest energy levels, 60° apart, are related as $\exp(-\phi_{\text{int rot}}/RT) = 0.03$. Clearly, with such a perceptible probability of a 60° rotation the displacements of a methyl from its equilibrium orientation cannot be considered to be small.

Instead of the harmonic model, a model of independent hindered rotation of a methyl in the field of its fixed neighbors may be tried as an alternative. The latter model may be justified just in the same terms as the cell model, which makes a similar assumption with respect to molecular motions.

Using *Pitzer's* tables [6.16] for the entropy of hindered rotation and assuming $\phi_{\text{int rot}} = 2.2\,\text{kcal/mol}$, we found the internal rotation contribution to the entropy from the six methyl groups to be 16.8 e.u. The resulting value for S_{int} is given in Table 6.10 (marked by footnote c). The addition of it to the cell model result for S_{int} yields a total entropy almost coincident with the measured third-law value.

Thus, the cell model and atom-atom potential approximation, combined with the assumption of independent hindered rotations of the methyls with $\phi_{\text{int rot}} = 2.2\,\text{kcal/mol}$, provide quite a consistent model for calculating the entropy of solid hexamethylbenzene.

The above calculation is a good example of how a cell model prediction for S_{ext}, combined with experimental third-law entropy data, may provide important information about intramolecular motions. It is clear that not only internal rotational barriers but also unobserved internal vibration modes can be elucidated in such a way.

Aside from compounds considered in this section, the cell-model and quasi-harmonic calculations have also been carried out for ethylene, phenanthrene, triethylenediamine, sulfur, trioxane and 1,3,5-trichloro-2,4,6-trifluorobenzene [6.82]. In all cases the results were much similar to those in Tables 6.3,7–10. In the high-temperature range the cell-model predictions for entropy were accurate to about 0.5 e.u., while the quasi-harmonic approximation gave entropy values 2–3 e.u. too high.

6.5 Calculations of Polymorphic Transitions

As noted in Chap. 4, polymorphic transitions provide a very critical test to check the reliability of a theoretical model. Since the energy changes involved in polymorphic transitions are generally very small, they can only be reproduced using very refined models. Some applications of the atom-atom potential method to polymorphic transitions have been discussed in Chap. 4. At that stage our possibilities were too limited, i.e., we could only determine the static potential energies of the polymorphs involved and then check whether the calculation puts the energies in the correct order. Based on the formalism of the preceding sections, we shall now attempt to carry out the calculations on a higher level by evaluating not only the static but also the thermal contribution to the thermodynamic quantities.

6.5.1 Temperature-Induced Transitions in p-Dichlorobenzene and 1,2,4,5-Tetrachlorobenzene

Some details of the polymorphic transitions in p-dichlorobenzene have already been described in Chap. 4. We remind the reader that three polymorphs of p-dichlorobenzene are known: γ, stable below 273.8 K, α, stable between 273.8 and 304.4 K, and β, stable between 304.4 K and the melting point at 326.2 K [6.57]. The γ and α polymorphs are monoclinic and belong to the space group $P2_1/c$ and $P2_1/a$, respectively [6.49]. The β polymorph is triclinic and belongs to the space group $P\bar{1}$.

An important detail concerning the structures of the p-dichlorobenzene polymorphs has been reported by *Wheeler* and *Coulson* [6.49] in their X-ray diffraction study at 100 K. It was found that the molecules of the low-temperature polymorph γ are not strictly planar. Instead, the Cl atoms are shifted 0.05 Å above and below the benzene ring according to the observed site symmetry $\bar{1}$. By contrast, the molecules of the high-temperature phases α and β are excessively planar, in the sense that the deviations of the chlorine atoms from the plane of the benzene ring are much smaller than the accuracy of the X-ray diffraction measurement [6.49]. This excessive planarity can be attributed to the existence of positional disorder, according to which the actual non-planar molecules (such as those of the γ phase) are distributed among two equally probable positions related by a 180° rotation. In an X-ray diffraction experiment one sees an average molecule, in which the deviations from planarity cancel out.

A further argument in favor of the existence of a positional disordering in the β polymorph is an unusually low temperature dependence of the anisotropic temperature factors [6.49]. This finding suggests the existence of a constant contribution to the temperature factors, which is independent of temperature and may well be due to positional disordering.

An attempt to predict one of the polymorphic transitions in p-dichlorobenzene, $\alpha \to \beta$, was undertaken by *Bonadeo* et al. [6.4]. The model they used combined the quasi-harmonic approximation and the BA73 model potential for the intermolecular interactions (Table 3.5). No allowance was taken for a possible positional disorder. Figures 6.4,5 show the temperature dependence of the density of phonon states, $g(\omega)$, and the free energy of both polymorphs. Despite a marked difference in $g(\omega)$, the free energy curves are very close to one another and do not intersect.

The failure of Bonadeo's calculation may be due to the neglect of thermal expansion in evaluating the free energies. The fact is that the calculation was carried out with the unit cell parameters fixed at the experimentally

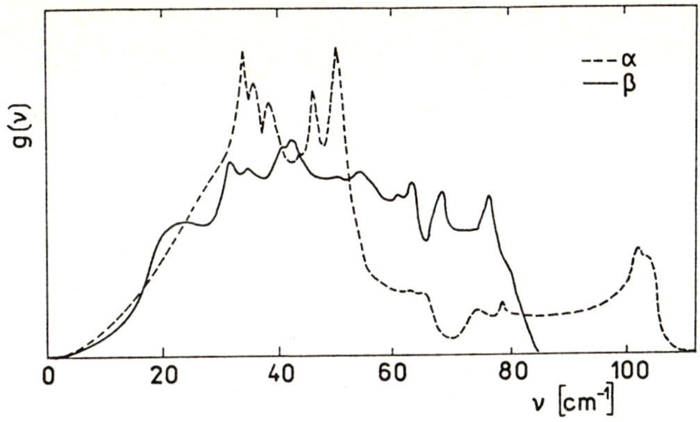

Fig. 6.4. Calculated densities of phonon states for α and β polymorphs of p-dichlorobenzene [6.4]

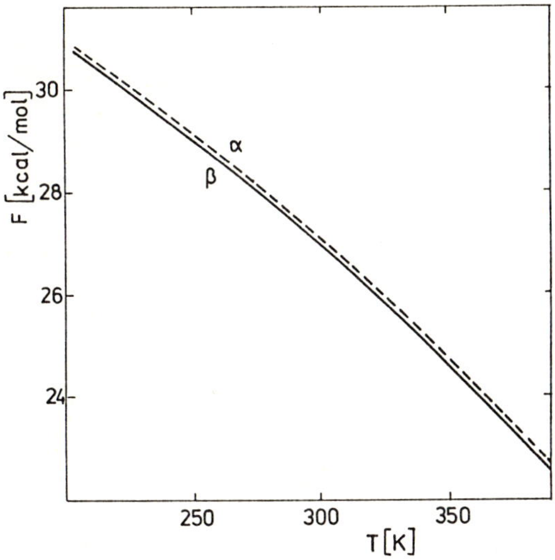

Fig. 6.5. Temperature dependence of the free energy of the α and β polymorphs of p-dichlorobenzene [6.4]

observed values at 295 K for α and 300 K for β polymorph. This is equivalent to assuming that the free energy depends on temperature only explicitly, while the implicit dependence, through the dependence of the force constants on the lattice parameters, is negligible. The validity of such an assumption seems questionable in view of the actually observed temperature dependence of the vibration frequencies. Although this dependence is rather weak, it may well influence the small energy changes involved in the transition. Positional disordering does not seem to be of importance for the $\alpha \rightarrow \beta$ transition because it takes place in both polymorphs.

Recently [6.47] we have attempted to reproduce the polymorphism in the p-dichlorobenzene crystal within the framework of the cell model. Considering very small energy changes involved in the transitions and a low accuracy of the "importance sampling" averages, it did not seem reasonable to hope for success in predicting the transition points. Thus, the aim of our work was more modest, i.e., to see whether we could predict the correct sequence for the free energies and entropies of the three polymorphs at a single temperature. The necessary structural data are available from *Wheeler* and *Coulson* [6.49]. At 100 K, the expected sequence for F and S is: $F_\gamma < F_\alpha < F_\beta$, $S_\gamma < S_\alpha < S_\beta$.

The "importance sampling" estimates were obtained using the then available computer program where the potential surface $\Delta\psi(\boldsymbol{R})/kT$ in (6.101) was approximated by a diagonal quadratic function $\sum(U_\alpha)^2/2\sigma_\alpha^2$. The parameters σ_α were determined by least-squares fitting of the quadratic function to $\Delta\psi(\boldsymbol{R})/kT$. Three potential models were tried in the calculations, viz. MC78, RKW74 and BB74a (Table 3.5). The correct sequence for both F and S was only observed when we used the BB74a potentials and treated the α and β phases as being disordered. The positional disordering was taken into account by simply adding the terms $R\ln 2$ and $-RT\ln 2$ to S and F, respectively. These additional terms corresponded to the above-discussed model for positional disordering, according to which each molecule might occupy, with equal probabilities, two distinct positions.

More recently, we have recalculated the γ to α transition in p-dichlorobenzene using an improved version of the computer program. In this version the potential surface $\Delta\psi(\boldsymbol{R})/kT$ was approximated by a quadratic function of the general form (6.101), with the derivatives $\partial^2\psi/\partial R_\alpha \partial R_{\alpha'}$ computed analytically. (Unfortunately, the triclinic β phase could not be treated with this program in view of symmetry restrictions in the algorithm used.) For comparison, the thermodynamic functions of the γ and α polymorphs were also calculated in the quasi-harmonic approximation.

Cell-model and quasi-harmonic results for the BB74a and MC78 parameter sets are summarized in Table 6.11. It is seen that the MC78 potentials arrange the free energies of the γ and α polymorphs in the reverse order, regardless of the statistical mechanical approximation used and no matter whether the α polymorph is treated as ordered or disordered. The BB74a potentials provide the correct order for both F and S, and the correction for possible disordering leaves the situation unchanged.

Table 6.11. Calculated free energies and entropies for the α and γ polymorphs of p-dichlorobenzene at 100 K [kcal/mol, e.u.]

Potential model	Polymorph	*Ordered model* F		S		*Disordered model* F		S	
		CM	QA	CM	QA	CM	QA	CM	QA
MC78	α	-15.30	-15.42	15.2	17.2	-15.44	-15.56	16.6	18.6
	γ	-15.25	-15.35	14.4	16.3	-15.25	-15.35	14.4	16.3
BB74a	α	-14.69	-14.79	13.0	15.1	-14.83	-14.93	14.4	16.5
	γ	-15.11	-15.20	12.8	14.8	-15.11	-15.20	12.8	14.8

Thus, the calculation of the relative stability of the p-dichlorobenzene polymorphs indicates an advantage of the BB74a potential model over the MC78 one, which is in agreement with the above discussed results for the thermodynamic properties of chlorinated benzenes.

We now turn to a consideration of 1,2,4,5-tetrachlorobenzene. Unlike p-dichlorobenzene, the tetrachlorobenzene crystal exhibits an extremely sharp polymorphic transition. No metastable phases have been observed. *Bonadeo* et al. [6.4] have shown that this transition can be accurately described by a simple Debye-Einstein approximation in which the dispersion of the optic phonons is neglected and the Debye frequency is assumed to be the lowest optic frequency. As with p-dichlorobenzene, they ignored thermal expansion and used unit cell parameters observed experimentally at a single temperature (300 K for the high-temperature phase β and 150 K for the low-temperature phase α). The BA73 model potential was used throughout the calculation. The temperature dependence of the free energy [6.4] is shown in Fig. 6.6. A well defined intersection point is evident at ~200 K, while the actually observed transition takes place at 188 K [6.52]. In view of the assumptions involved in the calculation the agreement is surprisingly good.

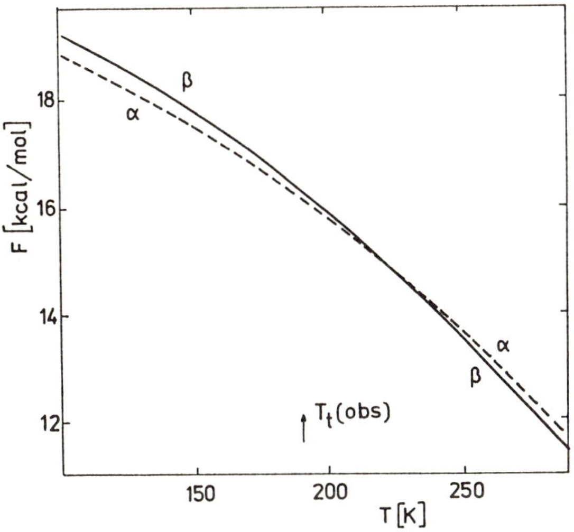

Fig. 6.6. Temperature dependence of the free energy of the α and β polymorphs of 1,2,4,5-tetrachlorobenzene [6.4]

6.5.2 Pressure-Induced Transition in Benzene

There are two known phases of crystalline benzene [6.83,84]. In the phase diagram shown in Fig. 6.7 these are labeled as O and M. At low pressures benzene crystallizes in an orthorhombic phase O (space group *Pbca*) with four molecules in the unit cell. The high-pressure phase M is monoclinic

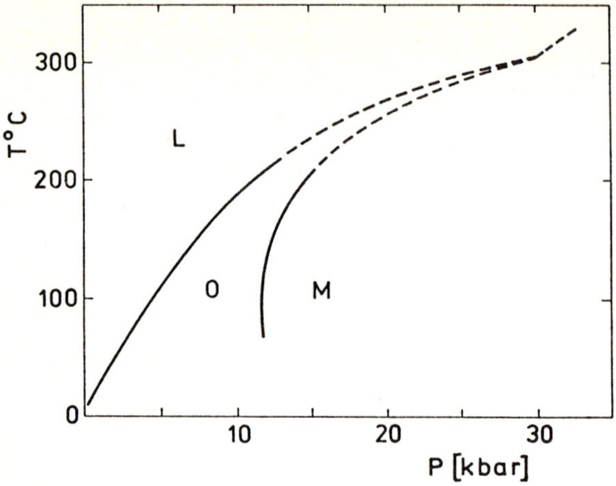

Fig. 6.7. Phase diagram for benzene (L: liquid, O: orthorhombic solid, M: monoclinic solid)

(space group $P2_1/c$) with two molecules in the unit cell. The detailed structure of both phases is available from X-ray and neutron diffraction studies [6.30].

An attempt to predict the pressure-induced transition in benzene was made by *Pertsin* [6.11]. The calculations were based on the cell model while the MKB74 parametrization was used to treat the intermolecular interactions.

Since detailed data on the pressure dependence of the unit cell parameters were not available, a rigorous version of the cell model was used in which the equilibrium crystal structure was determined by minimizing the free energy. The calculations were performed for $T = 270$ K. For each of the polymorphs, we chose some selected unit-cell volumes V_c and then, for each V_c, minimized the free energy using the constrained-volume algorithm described in Chap. 4. This yielded a set of equilibrium crystal structures corresponding to the hydrostatic pressure conditions given in (4.68). The corresponding pressure to a given structure was determined from the diagonal components of the stress tensor. Given the pressure, volume and free energy, we could immediately evaluate the Gibbs thermodynamic potential from the familiar relation,

$$G = F + PV. \tag{6.114}$$

By plotting G against P, we could then obtain two curves $G(P)$ whose intersection point is the transition pressure.

The calculation results are shown in Figs. 6.8–10. In the lower part of Fig. 6.8 we show the PV-isotherms for both polymorphs. The isotherms are

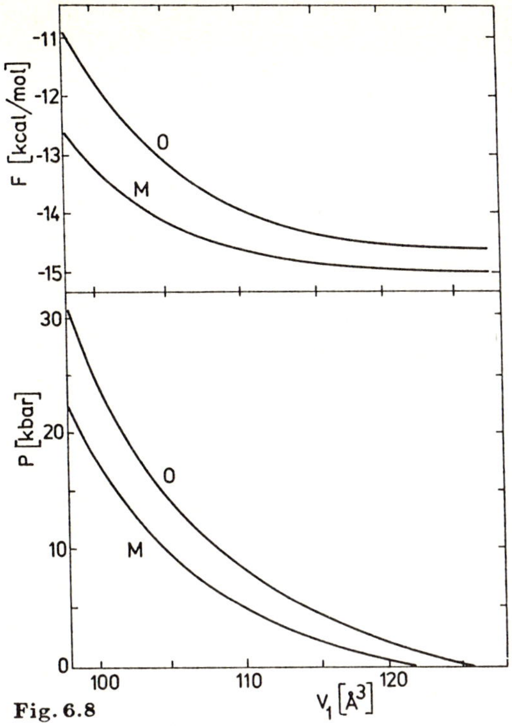

Fig. 6.8

Fig. 6.8. Free energy and pressure as a function of the volume per molecule for the orthorhombic (O) and monoclinic (M) polymorphs of benzene at 270 K [6.11]

Fig. 6.9. Pressure dependence of the Gibbs thermodynamic potential of the orthorhombic (O) and monoclinic (M) forms of solid benzene at 270 K [6.11]

Fig. 6.10. Pressure dependence of the internal energy and entropy of the orthorhombic (O) and monoclinic (M) forms of benzene at 270 K [6.11]

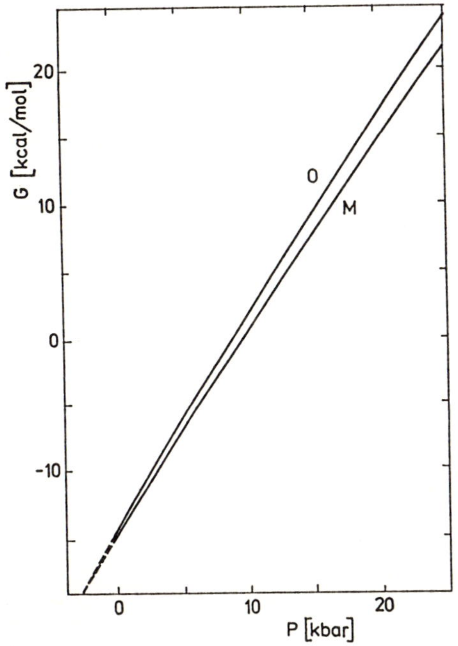

Fig. 6.9

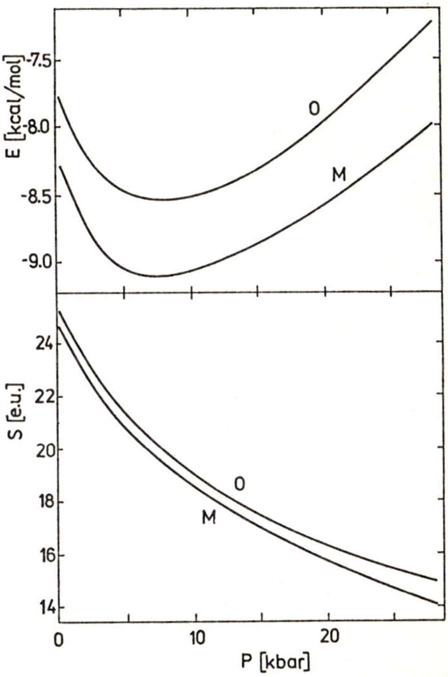

Fig. 6.10

Table 6.12. Experimental and calculated compressibility parameters γ^a for benzene [6.11]

	Pressure [kbar]					
	5	10	15	20	25	30
	orthorhombic phase		monoclinic phase			
$\gamma_{observed}$	0	0.041	0	0.020	0.037	0.051
$\gamma_{calculated}$	0	0.045	0	0.021	0.037	0.050

[a] Defined by (6.136) with $P' = 5$ and 15 kbar for the orthorhombic and monoclinic phases, respectively

in good quantitative agreement with the compressibility data of *Bridgman* [6.83,84]. This is illustrated in Table 6.12 where the calculated and observed *P-V* relations are compared in terms of the quantity

$$\gamma(P) = [V(P) - V(P')]/V(0). \tag{6.115}$$

The reference pressure P' in (6.115) is taken as 5 kbar for phase O and 15 kbar for phase M.

Although successful in predicting the relative changes in volume, the calculations lead however to absolute volumes that are too low. Thus, for phase M a volume of 97.5 Å^3 per molecule corresponds to a pressure of 23.1 kbar, while experimentally, a volume of 103.2 Å^3 was observed at a pressure of 25 kbar [6.30].

As seen from Fig. 6.8, the calculations correctly predict that the high-pressure phase M has a smaller volume than the low-pressure phase O at any given pressure. It is this difference in volume that gives rise to the increasing preferrence of phase M with increasing pressure.

We now turn our attention to the upper part of Fig. 6.8 which shows the volume dependence of the Helmholtz free energy for both polymorphs. The results are discouraging: the free energy of M is lower than that of O at any given pressure. Since $PV_M < PV_O$, the calculation predicts, in contradiction with experimental evidence, that the monoclinic phase M is stable throughout the examined pressure range. This is well seen in Fig. 6.9 which shows the pressure dependence of the thermodynamic potential for both polymorphs. Although the $G(P)$ curves do approach one another with decreasing pressure, they do not intersect until a negative pressure of ~2 kbar is reached.

In principle, the incorrect behavior of the free-energy curves in Fig. 6.8 may arise from two sources associated with the entropic and energetic contributions to F. The behavior of these two contributions with pressure is illustrated in Fig. 6.10. It can be seen that the difference in entropy between O and M is very small, ranging from 0.5 e.u. at $P = 0$ to ~1 e.u.

at $P = 30$ kbar. The near coincidence of the entropy curves correlates well with the experimental fact that the transition pressure in solid benzene is almost independent of temperature (Fig. 6.7).

Unlike the entropy, the internal energies of O and M differ substantially. The internal energy of the high-pressure polymorph is 0.5–0.8 kcal/mol lower than that of the low-pressure polymorph. Since the difference in entropy is negligible, the difference in internal energy is transferred practically unchanged to the difference in free energy.

Thus, of the three contributions appearing on the right-hand side of the equation

$$\Delta G = \Delta E - T\Delta S + P\Delta V, \tag{6.116}$$

the last two are predicted qualitatively correctly while the first one is predicted with a wrong sign.

It is essential that nearly all of ΔE is accounted for by the difference in static potential energy (the thermal contribution to ΔE is an order of magnitude smaller than ΔE itself). This suggests that quite a reliable analysis of the polymorphism of benzene can be made in terms of the static potential energy by replacing (6.116) by the equation,

$$\Delta G = \Delta\phi_0 + P\Delta V. \tag{6.117}$$

Such an analysis has been carried out by *Hall* et al. [6.85] who determined ΔG at 0 and 25 kbar. With the W67 (6-exp) model potential the result was similar to ours: the monoclinic polymorph was stable at both 0 and 25 kbar. Suspecting the failure of the calculations to stem from flaws in the model potential, *Hall* et al. [6.85] tried a more refined potential of a (6-exp-1) type, with parameters corresponding to set WS77 in Table 3.1. This new potential proved to be successful in that it yielded the proper signs for ΔG at 0 and 25 kbar.

6.5.3 Plastic-Phase Transition in Adamantane

The crystal of adamantane (tricyclo $[3.3.1.1.^{3,7}]$ decane) undergoes a typical transition to the plastic state at 208.62 K [6.40]. As seen from Fig. 6.11, the molecule of adamantane is almost spherical, which allows a rotational disordering at sufficiently high temperatures. The transition in adamantane is accompanied by an entropy change as large as 3.87 e.u. [6.40].

The low-temperature phase β of adamantane is tetragonal (space group $P\bar{4}2_1c$) with two molecules in the unit cell [6.35]. At temperatures above the transition point (α phase) adamantane forms a face-centered cubic lattice. The space groups $F\bar{4}3m$ and $Fm3m$ are equally consistent with X-ray diffraction data [6.35]. The $F\bar{4}3m$ symmetry corresponds to a perfectly

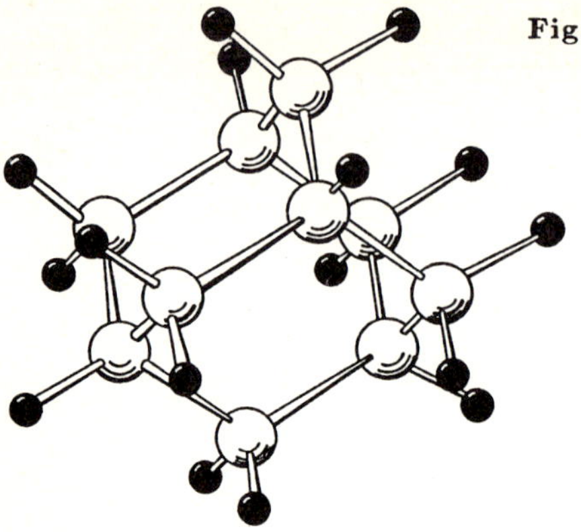

Fig. 6.11. Molecular structure of adamantane

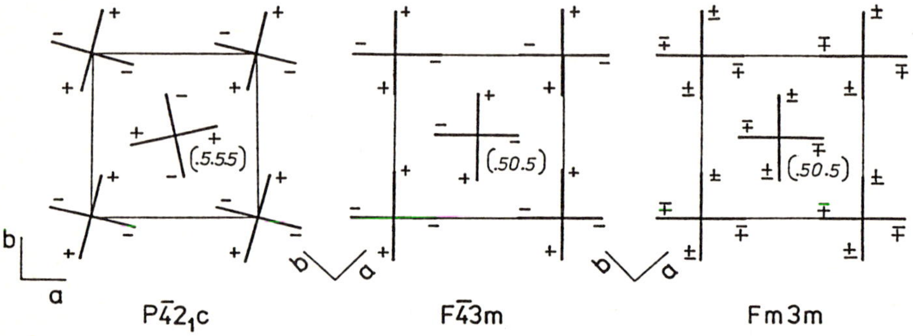

Fig. 6.12. A schematic representation of the $P\bar{4}2_1c$, $F\bar{4}3m$ and $Fm3m$ crystal structures of adamantane

ordered structure, while the $Fm3m$ symmetry corresponds to a structure in which the molecules are disordered among two positions related by a 90° rotation of the molecules about their $\bar{4}$ axes. All three structures are schematically shown in Fig. 6.12. Each adamantane molecule is represented by two non-crossing edges of a tetrahedron formed by the four methine carbon atoms.

A trivial calculation shows that the ordered model cannot be used to describe the high-temperature phase. Figure 6.13 shows the calculated dependence of the cell field $\psi(\boldsymbol{R})$ on the rotation angle of a molecule about its $\bar{4}$ axis in the $F\bar{4}3m$ structure. The angle $\varphi = 0$ corresponds to the perfectly ordered structure $F\bar{4}3m$, while the angle $\varphi = 90^\circ$ corresponds to a structure in which the central molecule is rotated by 90° with respect to the surrounding molecules. The orientation with $\varphi = 90^\circ$ is definitely favored over that with $\varphi = 0$. This means that each molecule in the high-temperature phase should tend to surround itself by molecules in the opposite orientation, which should make the $F\bar{4}3m$ structure unstable.

In general, plastic-phase transitions have much in common with the phenomenon of melting, the basic difference being that the latter involves

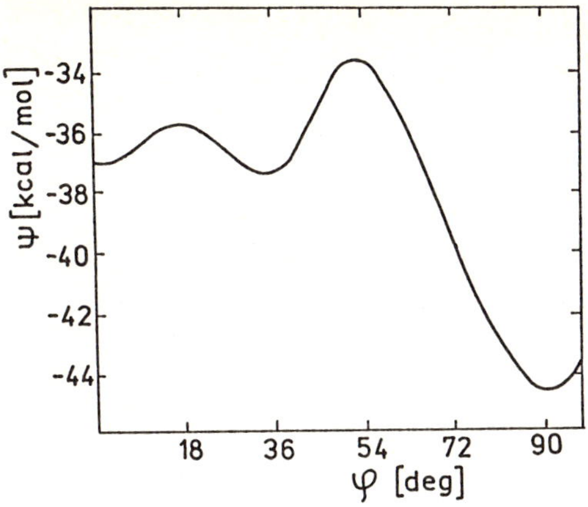

Fig. 6.13. The interaction energy of an adamantane molecule with its surroundings in the $F\bar{4}3m$ lattice as a function of the rotation about a $\bar{4}$ axis [6.11]

both rotational and translational disordering. Accordingly, the appearance of rotational disorders in a crystal can, in principle, be simulated by computer experiments, which have successfully been used to simulate the melting in simple systems.

First applications of computer experiments to the problem of plastic-phase transitions have already been published. Thus, *Neusy* et al. [6.86] have simulated a plastic-phase transition in bicyclo[2.2.2]octane using a constant-pressure modification of the molecular dynamics method [6.87], as adapted to molecular systems by *Nose* and *Klein* [6.88]. Such an approach, although promising, is however too time-consuming and suffers from general shortcomings inherent in a computer experiment as such (e.g., neither the molecular dynamics nor the Monte-Carlo method is capable of predicting the transitional entropy). For this reason, a need still exists for a simplified treatment of the problem.

The simplest approach to the problem of rotational disordering is based on the Ising formalism. In applying this formalism to adamantane, one represents the crystal by a three-dimensional network of spins, in which each spin is only allowed two distinct orientations [6.89,90]. The energy parameters describing the interaction of spins are easily computed with the atom-atom potential method. The Ising problem can then be solved using a more or less accurate approximation.

The concept of a finite number of rotational states is certainly oversimplified to provide satisfactory estimates for the transitional quantities. In the case of adamantane, for example, the approximate Ising model estimates of the entropy change at the transition are bounded above by a value of $R\ln 2 = 1.38$ e.u., which is to be compared with the observed entropy change of 3.87 e.u. This indicates that, besides the reorientations, the change in the vibrational motion of the molecules must be taken into account in a more realistic model.

An attempt to do this has been made by *Reynolds* [6.1]. The basic assumption used was that the rotational disordering only influenced the lattice vibrations in that it has an effect on the unit cell dimensions. The explicit dependence of the vibrations on disordering was completely neglected. To assess the reliability of this assumption, the density of phonon states was evaluated using: (i) the observed ordered-phase structure and (ii) an approximation to a disordered-phase structure in which each molecular site contained half a molecule in each of the two orientations. It was found that the difference between the density of phonon states in (i) and (ii) was small, provided that (i) and (ii) had the same volume.

Based on the above assumption, the free energy at a given volume and temperature can be represented as

$$F = F_{\text{vib}} + F_{\text{Ising}} + F_{\text{stretch}}, \tag{6.118}$$

where F_{vib} is the phonon contribution, F_{Ising} is the orientational order-disorder contribution and F_{stretch} is the energy required to stretch the lattice from its equilibrium, with phonon and no disorder, structure to the final disordered structure with volume V.

The phonon contribution, F_{vib}, was evaluated in the usual way by solving the lattice-dynamical problem for perfectly ordered structures [6.1]. The stretching energy was calculated for selected lattice periods and then fitted with a simple analytical function.

An approximate mean-field solution of the Ising model was at first tried for F_{Ising}. It was found that almost all the energy difference between the high- and low-temperature polymorphs was due to the eight nearest-neighbor interactions of the type (0,0,0)–(1/2,1/2,1/2). The appropriate Ising model was chosen to be body-centered cubic. In this model the energy and entropy contributions to F_{Ising} are given by

$$E_{\text{Ising}} = 4J(1-\eta^2), \tag{6.119}$$

$$S_{\text{Ising}} = -R\left(\frac{1+\eta}{2}\ln\frac{1+\eta}{2} + \frac{1-\eta}{2}\ln\frac{1-\eta}{2}\right), \tag{6.120}$$

where J denotes the energy parameter of the model, defined so that J and $-J$ are the interaction energies for parallel and antiparallel spins, respectively. The order parameter η is a solution of the equation

$$\frac{16J\eta}{kT} = \ln\frac{1+\eta}{1-\eta}. \tag{6.121}$$

The square of the order parameter defines the probability that a molecule's nearest neighbor has the same orientation. In the mean-field approximation η^2 is also the probability of any neighbor having the same orientation, and $(1+\eta)/2$ is the probability that a given molecule has a given orientation with respect to an external coordinate frame.

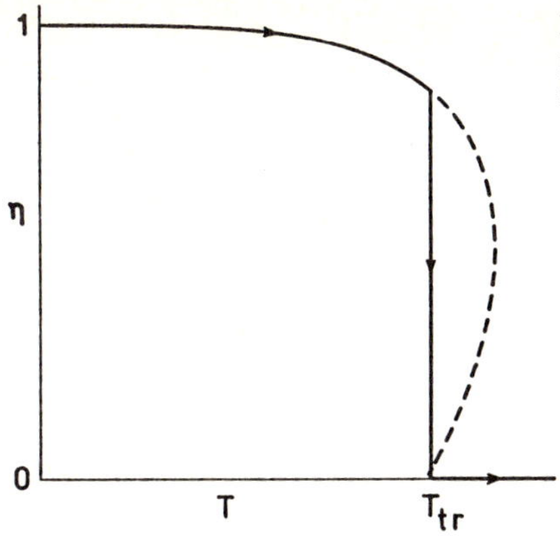

Fig. 6.14. Temperature dependence of the order parameter in adamantane, calculated using the mean-field approximation [6.1]

The change in the order parameter with temperature, calculated by *Reynolds* [6.1] using the mean-field model, is depicted in Fig. 6.14. At temperatures above the transition point the behavior of η is qualitatively incorrect as it completely neglects small amounts of ordering in the high-temperature phase. In spite of this fact, the simple mean-field approximation proves to be accurate enough to describe the properties of the individual phases, as well as most of the transitional quantities. This is illustrated in Table 6.13 which presents the basic results of Reynolds' calculation (marked by footnote a). The agreement with experiment is reasonable good (within 30%), except for the thermal expansions. The failure to predict these is to be expected since the quasiharmonic approximation was used to treat the lattice vibrations.

It is important to note that the high inaccuracy of evaluating the thermal expansions markedly influenced the other quantities, in particular the transition point. To avoid this, Reynolds was forced to use, when necessary, the experimental thermal expansions instead of the theoretical ones. In this sense the methodology followed by Reynolds was similar to that used in the experimental versions of the quasi-harmonic approximation and the cell model.

The success of the simple mean-field model in predicting most of the quantities listed in Table 6.13 is not surprising considering that these quantities are not markedly sensitive to the details of the rotational disordering and that the observed transition is extremely sharp (the order parameter rapidly decreases from 1 to 0 at the transition point).

A less favorable situation can be expected for those properties which depend on small disorders in the ordered phase or, conversely, on small amounts of ordering in the disordered structure. The specific heat is an example of such a property. *Reynolds* [6.1] has shown that the mean-field model is grossly incorrect in describing the specific heat and that only much more refined Ising-model solutions allow this thermodynamic property to be properly reproduced.

Table 6.13. Calculated and observed properties of the low- and high-temperature polymorphs of adamantane (all quantities correspond to atmospheric pressure unless otherwise indicated)

	Calculated a	b	c	Observed [6.35,40,91-94]
Transitional quantities				
T_{tr} [K]	198	260	325	208.6 315 (P=5)
$\partial T_{tr}/\partial P$ [K/kbar]	11.1	-	27.5	21.1
ΔS [e.u.]	5.4	4.6	8.5	3.9
ΔV [cm^3/mol]	3.5	-	9.3	2.8 - 3.4 1.8 (P=4.1)
Lattice periods [Å]				
a_β (T = 163)	-	-	6.47	6.60
c_β (T = 163)	-	-	8.60	8.81
a_α (T = 295)	-	-	9.30	9.43
Thermal expansion [$K^{-1}10^4$]				
β_β (T = 295)	1.4	-	2.7	5.0 1.0 (P=10)
β_α (T = 295)	1.6	-	3.2 2.3 (P=2)	4.7 2.7 (P=2)
Compressibilities [$kbar^{-1}$]				
α_β (T = 295)	0.012 (P=4.1)	-	0.008 (P=4.1)	0.011 (P=4.1) 0.009 (P=6.1)
α_α (T = 295)	0.020	-	0.018	0.021 0.011 (P=4.1)

[a]Quasi-harmonic approximation [6.1]; [b]Experimental version of the cell model; [c]Rigorous version of the cell model [6.11,18]

An alternative approach to deal with plastic-phase transitions can be developed on the basis of the cell model. In this case one does not have to assume that the free energy can be separated into vibrational and order-disorder contributions as the cell model involves the model of "discrete molecular states" in a natural way, as a zero approximation. We remind the reader that the cell model may be formally regarded as a first approximation to Kirkwood's variational theory, with a zero approximation chosen to be a delta function,

$$\varrho_i^{(0)}(\boldsymbol{R}^i) = \delta(\boldsymbol{R}^i - \overline{\boldsymbol{R}}^i). \tag{6.122}$$

If a molecule has two preferable orientational states, $\overline{\boldsymbol{R}}^{i,1}$ and $\overline{\boldsymbol{R}}^{i,2}$, then the zero approximation to $\varrho_i(\boldsymbol{R}^i)$ takes the form

$$\varrho_i^{(0)}(\boldsymbol{R}^i) = w_i\delta(\boldsymbol{R}^i - \overline{\boldsymbol{R}}^{i,1}) + (1 - w_i)\delta(\boldsymbol{R}^i - \overline{\boldsymbol{R}}^{i,2}), \tag{6.123}$$

where w_i is the probability that the ith molecule in the state $\overline{\boldsymbol{R}}^{i,1}$.

The use of (6.123) instead of (6.122) does not change any of the basic equations of the cell model, except that in evaluating the cell field $\psi(\boldsymbol{R})$ each surrounding molecule should be averaged according to

$$\psi(\boldsymbol{R}) = \sum_i [w_i \varepsilon(\boldsymbol{R}, \overline{\boldsymbol{R}}^{i,1}) + (1 - w_i)\varepsilon(\boldsymbol{R}, \overline{\boldsymbol{R}}^{i,2})],$$

where $\varepsilon(\boldsymbol{R}, \overline{\boldsymbol{R}}^{i,1})$ is the interaction energy of the central molecule at $\boldsymbol{R}$ with the ith molecule at $\overline{\boldsymbol{R}}^{i,1}$; similarly for $\varepsilon(\boldsymbol{R}, \overline{\boldsymbol{R}}^{i,2})$.

The evaluation of the probabilities w_i is equivalent to evaluating the distribution of the molecules over the discrete molecular states $\overline{\boldsymbol{R}}^{i,1}$ and $\overline{\boldsymbol{R}}^{i,2}$. In principle, the latter can be performed using the Ising formalism in complete analogy to Reynolds' treatment [6.1].

Considering the extreme sharpness of the phase transition in adamantane, it seems reasonable to try a simplified model for the rotational disordering, in which the low-temperature phase is perfectly ordered and the high-temperature phase is completely disordered. With such a model for the order-disorder transition, there is no longer any need to solve the Ising problem. Instead, two structures are to be separately considered: a perfectly ordered, tetragonal, $P\bar{4}2_1c$ and a completely disordered, cubic, $Fm3m$ one. For each of the structures, individually, one evaluates the thermodynamic functions for a wide range of temperatures and pressures and then locates the transition points as the points of intersection of the respective Gibbs energy curves.

Again, two different versions of the cell model may be tried: an experimental version, which uses the experimentally observed thermal expansions, and a rigorous version, in which the thermal expansions are calculated via a constrained-volume minimization of the free energy.

The results for the experimental version are presented in Fig. 6.15. The calculation employed the thermal expansions from [6.1,91] and the MKB74 parametrization of the interaction potential. Also given in Fig. 6.15 are the experimental external entropy curves derived from the calorimetric data of *Chang* and *Westrum* [6.40] and the fundamental frequencies from *Bailey* [6.46]. The calculated free-energy curves in Fig. 6.15 shows a well-defined point of intersection at $T \approx 260\,\mathrm{K}$, in good agreement with the observed transition point. The entropy change also compares well with the observed value.

In the "rigorous" version of the calculations we chose some selected values for temperature and volume, and then used the constrained-volume algorithm to find the equilibrium lattice periods. For the low-temperature tetragonal phase the procedure was exactly the same as the one described in the previous section for benzene. With the high-temperature, cubic phase, there was no need for a free-energy minimization since the equilibrium crystal structures were uniquely specified by the prescribed volumes.

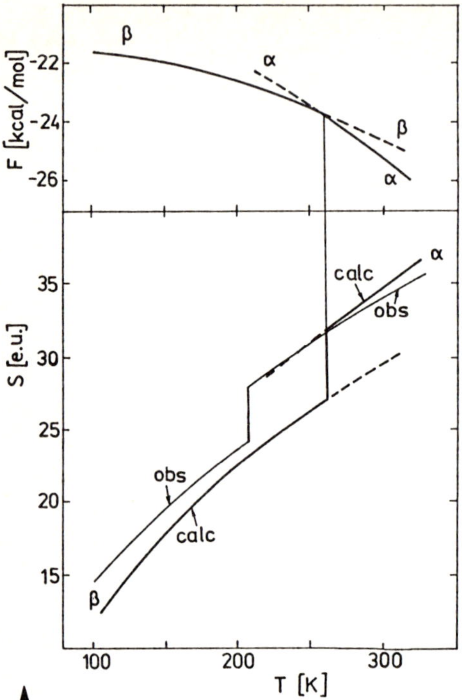

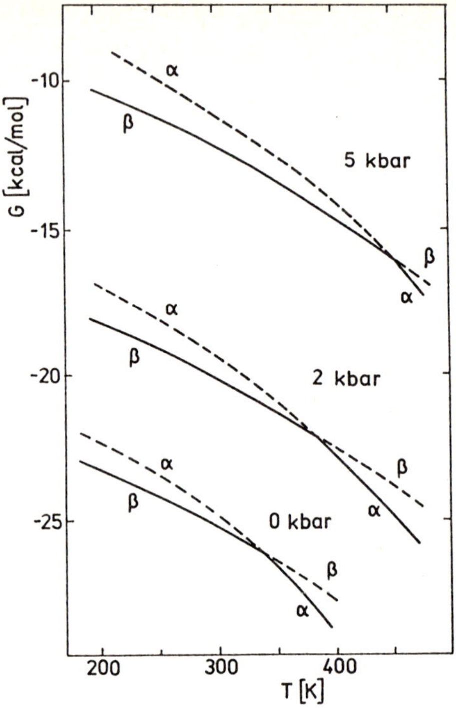

Fig. 6.15. Temperature dependence of the free energy and entropy of the low- (β) and high-temperature (α) phases of adamantane [6.11]

Fig. 6.16. The Gibbs thermodynamic potential of the α and β polymorphs of adamantane as a function of the temperature at 0, 2 and 5 kbar [6.11]

Figure 6.16 shows the predicted temperature dependence of the Gibbs thermodynamic potential for both polymorphs at three different pressures. The calculations correctly predict the intersection of the $G(T)$ curves.

The basic transitional quantities and the properties of the individual phases are given in Table 6.13. Inspection of the table shows that a rigorous cell model treatment leads, as with benzene, to somewhat denser equilibrium structures. As a result, the transitional quantities and the properties of the individual phases are systematically shifted to values which are really observed at higher pressures. (To illustrate this point, some experimental quantities in Table 6.13 are given at two different pressures.) The observed shift in pressure is not, however, very large so that the overall thermodynamic behavior of adamantane is predicted fairly well.

The only prediction that cannot be regarded to be satisfactory refers to the volume change, ΔV, at the transition. From Table 6.13 it can be seen that ΔV has been markedly overestimated instead of being underestimated, which might have been expected from the shift of the transition to higher pressures. It is worth noting that the error in ΔV mainly stems from too low equilibrium volumes predicted for the ordered phase. This, in turn, probably arises from the neglect of disorders in the low-temperature phase, which may become important near the transition point.

As seen from Table 6.13, *Reynolds*'s calculation [6.1] is superior to ours in predicting ΔV. However, this is not surprising considering that Reynolds treated the vibrational contribution to the thermal expansions using the experimental expansions. By contrast, no experimental data other than the molecular geometry and crystal symmetry were used in the cell model calculation of ΔV.

7. Imperfect Crystals

It has been recognized for a long time that many important properties of solids depend on the various imperfections in their structures, rather than on their chemical nature and the geometry of their lattices. Thus, the conductivity of semiconductors may be completely governed by minute amounts of impurities. Diffusion in solids strongly depends on defects. Color and luminescence of many crystals are also related to defects and impurities. The mechanical properties of solids are generally governed by dislocations. This list can be continued and a number of relevant examples can be found in standard textbooks on solid-state physics.

With organic crystals the most striking examples of defect-sensitive properties are the solid-state reactions. It is known that many reactions in organic solids are lattice-controlled, in the sense that the reactivity of a solid and the nature of the reaction products are governed by the crystal packing of the reactant. This is true, for example, of many dimerization and polymerization reactions in the solid state. A prerequisite of such reactions is that the distance between the potentially reactive sites in the monomer is below a certain threshold value (usually, 4–4.5 Å) and that the displacement of the monomer molecules to yield the product is minimal [7.1,2]. Reactions of this kind are usually known as topochemical reactions. The activation energy needed for such a reaction to take place may be supplied by irradiation, heating or mechanical stress.

The simple principles relating the crystal packing to the reactivity of organic solids were first formulated by *Hirshfeld* and *Schmidt* [7.3] in their topochemical preformation theory. The term preformation is used to stress the fact that the reaction product is as if preformed within the matrix of the parent solid. Since then a number of examples have been reported that demonstrate the validity of the topochemical preformation theory [7.1,2]. At the same time, there have been a number of instances where the theory, or at least its idealization which deals with a perfect lattice, has failed. Thus, there are situations where the topochemical preformation theory predicts photostability, yet a reaction does occur. In other cases the topochemical principles predict a certain product, but another is actually formed. Such anomalous behavior can only be rationalized by assuming that the reaction takes place in certain defective regions, whose structure may be quite different from that of the perfect crystal.

With photostimulated solid-state reactions, the interest in defective regions is two-fold. Besides enabling neighboring molecules to be appropriately oriented so as to facilitate a reaction, the defects also serve as exciton traps. The depths of the exciton traps can be estimated provided that the precise locations and orientations of the molecules in the defective regions are known [7.4,5].

Unfortunately, neither of the experimental techniques used to study crystal imperfections in organic solids can fully reveal the details of the local structure at various kinds of defects. However, these details can be readily simulated with the aid of a computer, provided that we are in a position to evaluate the energy of a local region for any given configuration. Thus, if we are interested in the local structure in the vicinity of an impurity molecule, we have to replace a host molecule by an impurity molecule and then allow the lattice to relax so as to minimize the potential energy of the system. Similarly, we may insert an extra half-plane in a lattice to simulate an edge dislocation, or remove a molecule from a lattice to simulate a vacancy. The computational approach not only yields the detailed structure of a defective region but also its energetics. This permits an evaluation of the equilibrium concentration of various defects and an assessment of their relative importance in a particular solid.

The simulation of various kinds of imperfections is also of value to organic gas-solid reactions. Again, a marked influence of the structural defects on reactivity can be observed [7.6]. Hence, a detailed knowledge of the structure of defective regions is needed to gain a deeper insight into the mechanisms of these reactions. Of particular importance is the surface, which is a kind of two-dimensional fault whose structure can be established by means of potential energy calculations.

In this chapter we shall consider several kinds of imperfections and give some examples that demonstrate the usefulness of the atom-atom potential method in predicting the local structure of defective regions.

7.1 Point Defects

7.1.1 Microscopic Model

The evaluation of the lattice energy of a defective crystal is much more complicated than the corresponding evaluation for a perfect crystal. This can readily be seen by writing the lattice energy in the form

$$\phi = \frac{1}{2} \sum_{i \neq j}^{N} \varepsilon_{ij} = \frac{1}{2} \sum_{i=1}^{N} \psi_i, \qquad (7.1)$$

where ε_{ij} is the interaction energy of molecules i and j, ψ_i is the interaction energy of molecule i and its surroundings and N is the number of molecules in the crystal. In a perfect crystal with one molecule in the asymmetric unit all of the ψ_i's are equal by symmetry, and (7.1) becomes

$$\phi = \frac{1}{2} N \psi_i, \tag{7.2}$$

where i refers to an arbitrary molecule in the crystal. In other words, an evaluation of ϕ requires the computation of the interaction energy ψ for only one molecule.

The occurrence of a point defect in a crystal is necessarily followed by the complete or partial loss of symmetry in the vicinity of the defect. As a result, the ψ_i's in (7.1) become different from one another and have to be individually computed to evaluate the lattice energy.

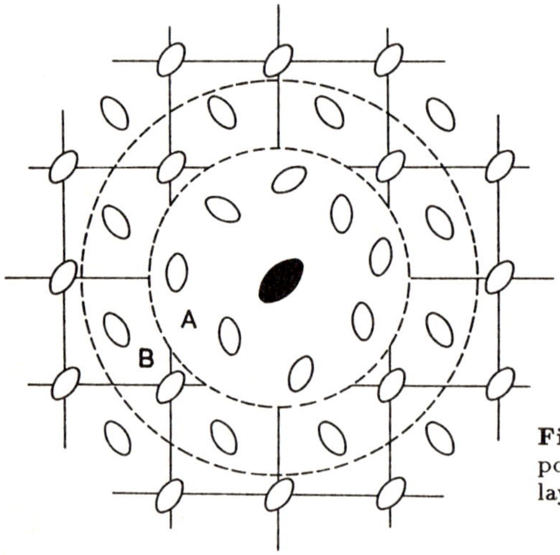

Fig. 7.1. Two-layer cluster model for a point defect. *A*: relaxable layer; *B*: rigid layer

A convenient microscopic model to deal with point defects is illustrated in Fig. 7.1. The model represents a cluster which contains the point defect at the center and two layers of surrounding molecules. The inner layer A contains molecules whose positions and orientations differ from those in the perfect lattice. The inner layer and the defect itself will hereafter be referred to as the core of the cluster. The thickness of the inner layer is chosen large enough to be able to neglect the lattice disturbance in the outer layer B. The thickness of the outer layer is chosen so that the interactions between the molecules in the cluster core and those outside the cluster can be neglected.

The potential energy of a crystal containing N_d point defects may be written as

$$\phi = (\phi_c + \phi_{c,s}) N_d + \phi_s. \tag{7.3}$$

ϕ_c denotes the potential energy of the cluster

$$\phi_c = \frac{1}{2}\sum_{i\neq j}^{Z} \varepsilon_{ij}, \tag{7.4}$$

where Z is the number of molecules in the cluster, $\phi_{c,s}$ is the interaction energy of the cluster and its surrounding molecules, and ϕ_s is the contribution to ϕ due to the interactions of the surrounding molecules with themselves.

We now imagine that the particles in the cluster core, as well as the defect itself, are replaced by molecules at their regular lattice sites. The lattice energy of the perfect crystal obtained in such a way may be formally represented as

$$\phi^0 = (\phi_c^0 + \phi_{c,s}^0)N_d + \phi_s^0, \tag{7.5}$$

where the superscript indicates that we are dealing with a perfect crystal. Since the interactions between the cluster core and the molecules outside the cluster are negligible, the following equation holds

$$\phi_{c,s}^0 = \phi_{c,s}. \tag{7.6}$$

A similar equation can be written for ϕ_s^0. Considering these equalities and subtracting (7.5) from (7.3) one obtains

$$\phi = \phi^0 + N_d(\phi_c - \phi_c^0) \tag{7.7}$$

The difference in parentheses may be taken as a definition for the defect energy.

Neglecting thermal effects, the equilibrium structure of a defective crystal is seen to correspond to a minimum of the cluster energy ϕ_c. Since the cluster core has been chosen large enough to neglect the lattice disturbance in the outer layer, ϕ_c may be minimized with respect to the positions and orientations of the core molecules. The molecules in the outer layer are fixed at their regular lattice sites.

To reduce the computational work associated with a minimization of ϕ_c, it is reasonable to exclude the terms in (7.4) which do not depend on variable parameters of the model. This can be easily done by rewriting (7.4) in the form

$$\phi_c = \frac{1}{2}\left(\sum_{i\neq j}^{Z_A} \varepsilon_{ij}^{(AA)} + \sum_{i=1}^{Z_A}\sum_{j=1}^{Z_B} \varepsilon_{ij}^{(AB)} + \sum_{i=1}^{Z_B}\sum_{j=1}^{Z_A} \varepsilon_{ij}^{(BA)} + \sum_{i\neq j}^{Z_B} \varepsilon_{ij}^{(BB)}\right), \tag{7.8}$$

where Z_A and Z_B are the number of molecules in the cluster core and the outer layer, respectively, $\varepsilon_{ij}^{(AB)}$ denotes the interaction energy of a molecule i in the core and molecule j in the outer layer (similarly for the other ε_{ij}'s).

The last sum in (7.8) is the same for both the defective and the perfect crystal and cancels out in calculating the defect energy $\phi_c - \phi_c^0$. Thus, we may rewrite the cluster energy as

$$\phi_c = \frac{1}{2}\left[\sum_{i=1}^{Z_A} \psi_i + \sum_{i=1}^{Z_A} \sum_{j=1}^{Z_B} \varepsilon_{ij}^{(AB)}\right], \tag{7.9}$$

where ψ_i is the interaction energy of a molecule i and its surroundings

$$\psi_i = \sum_{j=1;j\neq i}^{Z_A} \varepsilon_{ij}^{(AA)} + \sum_{j=1}^{Z_B} \varepsilon_{ij}^{(AB)}. \tag{7.10}$$

The cluster energy is not merely a sum of interaction energies ψ_i over the cluster core, but also involves a term which represents the interaction energy of the core and the outer layer [the double sum in (7.9)]. In computing the defect energy this latter term largely cancels out and is frequently neglected [7.7,8].

The minimization of the cluster energy with respect to the positions and orientations of the core molecules can be further simplified by using a cyclic process in which only one core molecule is displaced at a time. Let k denote a core molecule whose coordinates are varied in a given cycle. Equation (7.9) can then be represented in the form

$$\begin{aligned}\phi_c &= \frac{1}{2}\left[\sum_{i\neq k}^{Z_A}\left(\sum_{j\neq i;j\neq k}^{Z_A} \varepsilon_{ij}^{(AA)} + 2\sum_{j=1}^{Z_B} \varepsilon_{ij}^{(AB)}\right) + 2\sum_{i\neq k}^{Z_A} \varepsilon_{ik}^{(AA)} + 2\sum_{j=1}^{Z_B} \varepsilon_{kj}^{(AB)}\right] \\ &= \mathrm{const}(k) + \psi_k, \end{aligned} \tag{7.11}$$

where const(k) is a term independent of the position and orientation of the kth molecule.

Hence, the minimization of the cluster energy with respect to the variables of the kth molecule is equivalent to a minimization of the potential ψ_k. In other words, the advantage of the cyclic process is that only one term in (7.9) need be recalculated in each cycle.

The relation between ϕ_c before and after minimization is obvious:

$$\phi_c^{(\mathrm{after})} = \phi_c^{(\mathrm{before})} - \psi_k^{(\mathrm{before})} + \psi_k^{(\mathrm{after})}. \tag{7.12}$$

Before proceeding with a detailed discussion of point defects it is to be noted that the two-layer cluster model described above can be readily extended to linear and planar faults. Two modifications are needed: (1) to introduce a one- or two-dimensional periodicity along the fault core and (2) to change the shape of the cluster from spherical to cylindrical or slab-like, respectively [7.9].

7.1.2 Vacancies

The disturbance of a crystal structure in the vicinity of a vacancy was first calculated by *Craig* et al. [7.10] for naphthalene and anthracene. The model used consisted of a cluster of 21 molecules, containing an active (removable) molecule at the center and twenty nearest neighbors within a sphere of radius 10 Å. At the starting point the surrounding molecules were generated from the central one by applying crystal-symmetry operations. In seeking the minimum-energy configuration the molecules in the cluster were all allowed to move, i.e., an outer rigid layer was not used. The molecules in such a free-surface cluster were displaced independently of one another, except that the structure was assumed to retain a center of symmetry at the cluster center.

Craig et al. [7.10] obtained a more or less intact cavity when the active molecule was removed from the lattice and the surrounding molecules were allowed to relax to a minimum-energy structure. In both naphthalene and anthracene the effect of the structural changes was to make the cavity more spherical. The nearest molecules retreated from the vacancy while the more remote ones approached the vacancy. The same pattern was obtained for a divacancy, formed by removing two neighbor molecules along the *c* axis.

Table 7.1. Direction cosines of the principal axes of a molecule adjacent to a vacancy in the naphthalene crystal [7.10]

		Unit-cell axes a	b	c sinβ
Principal[a] molecular axes	L	-0.4249(-0.4413)[b]	0.1828(0.1953)	0.8866(0.8758)
	M	0.2854(0.2973)	-0.9024(-0.8892)	0.3229(0.3480)
	N	0.8591(0.8468)	0.3902(0.4137)	0.3313(0.3344)

[a] L and M lie in the long and short in-plane directions, while N is perpendicular to the molecular plane; [b]Given in parentheses are the results for the minimum-energy perfect crystal configuration

Some of the results of *Craig* et al. [7.10] are given in Table 7.1, which lists the direction cosines of the principal axes of the most affected molecule in the naphthalene cluster. A comparison with the corresponding regular lattice values (bracketed) shows that the orientational displacements are indeed very small. In addition, no molecule was found to be displaced more than 0.1 Å from its regular lattice site.

The gain in packing energy due to a relaxation of the lattice about the vacancy was estimated to be about 0.5 kcal/mol for naphthalene. This is a negligible quantity compared to the total lattice energy. Thus, it will be a good approximation to assume that the energy of formation of a vacancy in a crystal of naphthalene is equal to the enthalpy of sublimation.

In general, the single layer (free-surface) cluster model used by *Craig* et al. [7.10] suffers from an obvious shortcoming in that it introduces artificial surface effects and neglects the external pressure exerted upon the defective region by the surrounding, nearly unaffected lattice. This shortcoming should become particularly important in simulating a lattice relaxation about an impurity larger than a host molecule. The free-surface cluster model should exaggerate the swelling of the structure about the impurity, compared to the actual situation in which the disturbed region is embedded in a matrix of practically fixed molecules.

A more appropriate, two-layer cluster model similar to the one described in Sect. 7.1.1 was used by *Koehler* [7.11] to describe the lattice relaxation about point defects in a crystal of n-octane. The model represented a parallelepiped of 96 n-octane molecules initially occupying their regular lattice sites. Two kinds of vacancy defects were considered, i.e., two- and three-molecule vacancies formed by removing two or three consecutive molecules along the *a* axis. The shape of the microcrystal was chosen so that all crystal planes about the defects contained at least two sheets of molecules. The molecules closest to the defects were allowed translational and rotational displacements, while the other molecules were kept fixed.

The lattice relaxation about the defects was studied using the Monte-Carlo method briefly described in the previous chapter. However, ensemble averages were not computed and the method was solely used to scan the low-energy regions of configurational space. The temperature was taken as zero for a two-molecule vacancy and 60 K for a three-molecule vacancy. In the former case, as follows from the Monte-Carlo algorithm described in Sect. 6.2.3, a random move was only accepted if the energy of the new configuration was lower than that of the preceding configuration. In other words, the Monte-Carlo method was used as a stochastic method of locally minimizing the potential energy. At 60 K, there was a non-zero probability of accepting a higher-energy configuration. Thus, it was possible, using the Monte-Carlo method, to surmount the local minima and reach the global minimum, provided that the chain of sampled configurations was long enough.

The results obtained by *Koehler* [7.11] were similar to those discussed above for naphthalene and anthracene; i.e., the system did not relax very much about the defects.

Although the number of relevant examples is still very limited, it seems that the stability of a lattice about a vacant site against collapse is a phenomenon common to most organic molecular crystals. The origin of this phenomenon probably lies in the specific nature of the binding forces in the crystals. Since the binding forces rapidly fall off with intermolecular separation, it proves more advantageous for the molecules to retain a dense local packing than to assume a new packing with a higher average density.

7.1.3 Orientational Defects

The possibility of the formation of an orientational point defect, i.e., a misoriented molecule at a regular lattice site, has been explored by *Ramdas* et al. [7.7] to explain the unusual behavior of 1,5-dichloroanthracene upon photodimerization. The crystal of 1,5-dichloroanthracene has two known polymorophs and the nature of the photodimerization product strongly depends on the particular polymorph that is being irradiated. The monoclinic polymorph (space group $A2/a$, $Z = 4$) yields only the head-to-head dimer (Fig. 7.2), in full accordance with the topochemical preformation theory. Neighboring molecules in the monoclinic crystal are separated by a distance of 4.05 Å and are arranged so that only the head-to-head dimer should result. Upon UV irradiation the triclinic form (space group $P\bar{1}$, $Z = 1$) generates 80% head-to-head and 20% head-to-tail dimer.

Since the detailed crysal structure of the triclinic polymorph was unknown, *Ramdas* et al. [7.7] were forced to resort to the atom-atom potential method. Based on the known molecular geometry and the unit-cell dimensions, a minimum-energy crystal configuration was found with a packing energy about 2 kcal/mol lower than that of the monoclinic polymorph. However, the resulting crystal structure could not explain the appearance of the

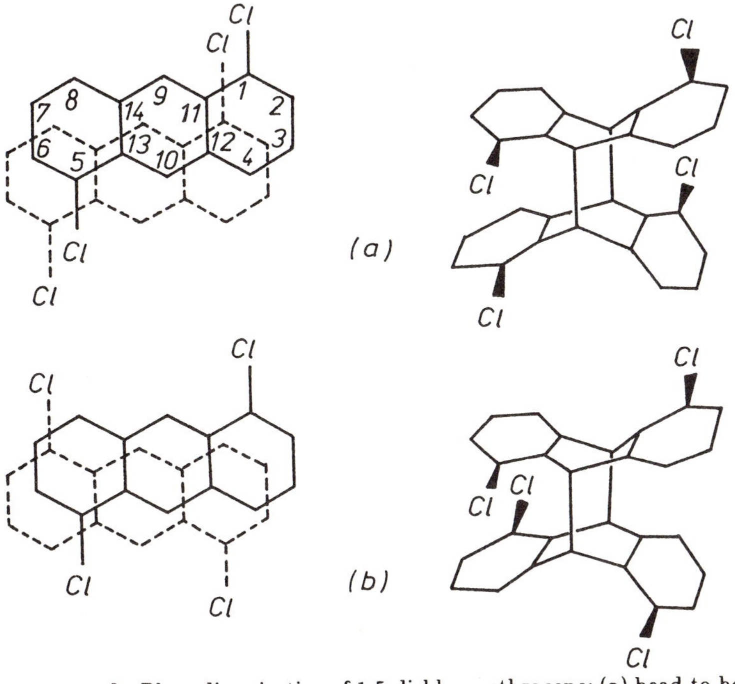

Fig. 7.2a,b. Photodimerization of 1,5-dichloroanthracene: (**a**) head-to-head and (**b**) head-to-tail dimer [7.7]

head-to-tail dimer because the orientations of the potentially dimerizable molecules are essentially the same as in the monoclinic form. An attempt to attribute the formation of the head-to-tail dimer to linear or planar faults was also unsuccessful: no slip system that brings neighboring molecules into a potential head-to-tail orientation was found.

It was then proposed that the required head-to-tail orientation is due to the presence of misoriented molecules rotated from the normal orientation by an angle 180° about the short or long molecular axis. Such misoriented molecules cannot, of course, be generated by thermal motion but may appear on the crystal surface during crystal growth and may be subsequently incorporated into the bulk of the crystal.

To test the above proposal *Ramdas* et al. [7.7] evaluated the extra energy imparted to a lattice by imposing an orientational defect into both the triclinic and monoclinic forms. The models that were used involved two-layer clusters with a misoriented molecule at the center, 8 to 10 molecules in the inner, relaxable layer and 16 to 22 molecules in the outer, rigid layer. The results of the energy minimization showed that the triclinic structure was much more adaptable to the orientational defect. Although the defect molecule itself underwent significant relaxation in both structures, the lattice disturbances about the defect were quite different: in the triclinic structure the neighboring molecules retained their normal orientations and positions, while in the monoclinic structure a considerable relaxation of the neighbors took place. The defect formation energy was found to be about 25 kcal/mol for the monoclinic structure and only 1 kcal/mol for the triclinic structure.

Thus, the calculations by *Ramdas* et al. [7.7] have at least provided a qualitative explanation of why the triclinic form of 1,5-dichloroanthracene can yield the head-to-tail product while the monoclinic form cannot. Unfortunately, the calculated defect formation energy of 1 kcal/mol, relating to the bulk of the crystal, cannot be used to evaluate the defect concentration or to predict the yield of head-to-tail dimer. Since it is hardly possible for the orientational defect to migrate across the entire lattice or to be annihilated by a 180° reorientation about the long molecular axis, almost all of the defects formed on the crystal surface during crystal growth will be incorporated into the crystal as a non-equilibrium defect concentration in the bulk. Thus, to predict the defect concentration it would be necessary to consider the crystal surface and evaluate the defect formation energy at the surface.

7.1.4 Conformational Defects

Conformational point defects are to be expected in crystals whose constituent molecules possess soft internal degrees of freedom, e.g., rotations

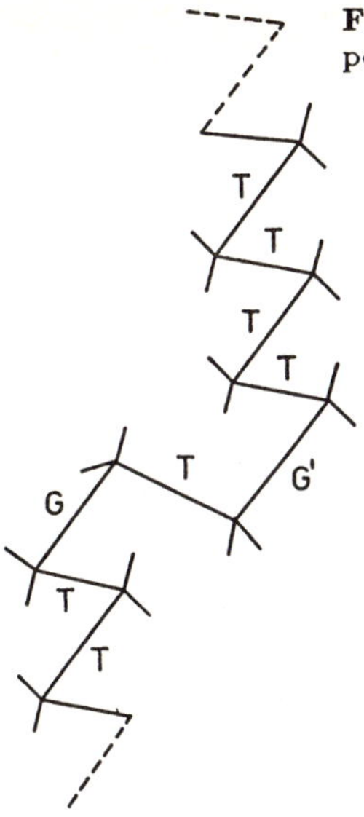

Fig. 7.3. A ...$TTGTG'TT$... conformational defect in crystalline polyethylene [7.8]

about single bonds. They may be generated thermally provided that the conformational change does not markedly alter the overall shape of the molecule. If the barriers separating the regular and defect conformations are too high, conformational defects may appear on the surface and can then be incorporated into the bulk of the crystal during crystal growth.

An important example of a conformational point defect has been studied by *Boyd* [7.8] in crystalline polyethylene. The defect, ...$TTGTG'TT$..., represents a kink which may be conveniently described in terms of the trans and gauche conformations of the skeletal C-C bonds (Fig. 7.3).

The defect formation energy can be divided into two components. The first involves the non-bonded energy associated with the swelling of the lattice about the kink and the crystallographic mismatch of the zig-zag stems along the chain axis as a result of the shortening of the chain. The second component takes into account the conformational changes that occur in the kinked chain and its surroundings.

The model chosen by *Boyd* [7.8] consisted of a two-layer cluster with a kinked chain at the center. The atoms in the cluster core were allowed to relax to a minimum energy. No constraints were imposed upon the valence angles and bond lengths nor upon the torsional angles and positions of the chains.

The total potential energy of the cluster was calculated as a sum of non-bonded atom-atom contributions and conformational terms as given by (3.70–72).

The energetics of the defect can be judged from Table 7.2. To demonstrate the effect of varying the number of deformable neighbors, several

Table 7.2. Effect of surroundings on the conformational (...$TTGTG'TT$...) defect energy of polyethylene [kcal/mol] [7.8]

	Number of deformable coordination shells			
	0	1	2	2 (no rigid outer shell)
Defect energy	7.6	6.6	6.7	7.9

situations were considered. The first refers to the case in which a kinked molecule is embedded in the rigid lattice (no coordination shells are free to deform). The others refer to situations in which the deformable layer of the cluster contains one and two nearest coordination shells, respectively, and the number of deformable molecules are six and twelve. The last column refers to two deformable coordination shells and no outer rigid shell.

It is seen from Table 7.2 that deformable surroundings play an important, but not crucial role in the evaluation of the defect energy. The defect energy converges rapidly with increasing number of deformable shells. An unexpected result is that the free-surface (single-layer) cluster possess a higher defect energy. This is probably an artefact arising from the inclusion of surface energy in the defect energy.

The lattice strains about the kink were found to be rather small and not very sensitive to the level of inclusion of neighboring deformable shells. For the unit cell containing the defect the relative change in volume is as low as 4%.

The conformational changes in the defect chain are listed in Table 7.3. The given values of the torsional and bond angles for the backbone of the defect chain are to be compared with the ideal torsional angles of $\pm 180°(T)$, $60°(G)$ and $-60°(G')$, and a skeletal bond angle of 111° (typical of n-paraffins). The conformational distortions are seen to be the largest at the kink and steadily decrease towards the ends of the chain.

7.1.5 Substitutional Impurities

A simple geometrical analysis of the structures of organic molecular crystals shows that the voids in the structures are indeed small compared to the size of the molecules themselves [7.12,13]. Hence, it is highly improbable that these voids can be filled by foreign molecules. In the following text we shall only deal with situations in which the foreign molecule behaves as a substitutional impurity, i.e., it is inserted into the matrix of host by replacing one or more host molecules at their regular lattice sites.

Early calculations on substitutional impurities were performed using the rigid-lattice approximation, with no allowance for the relaxation of the host lattice [7.14–16]. For this reason, it is hardly worthwile discussing the details and the numerical results. The only result that appears worth not-

Table 7.3. Torsional and bond angles of the defect polyethylene chain [deg] [7.8]

Bond[a]	Torsional angle	Skeletal bond angle
	-	111.8
1	178.2	-
	-	112.6
2	-176.5	-
	-	110.7
3	173.3	-
	-	114.4
4	75.3	-
	-	113.0
5	-178.1	-
	-	113.1
6	- 76.5	-
	-	114.8
7	-173.0	-
	-	110.1
8	174.7	-
	-	113.4
9	-176.9	-
	-	111.0
10	178.3	-
	-	112.3
11	-179.1	-
	-	112.0

[a] The C-C bonds are numbered sequentally from 1 to 11 when going upward along the fragment of the chain shown in Fig. 7.3; for a given bond the corresponding torsional angle specifies rotation about this bond, while the bond angles are defined by the adjacent bonds on both sides of this bond

ing refers to crystals containing symmetrically distinct molecules in the unit cell. It has been found that the energy of substitution may markedly depend on the particular molecule being substituted; e.g., an α-nitronaphthalene molecule in the lattice of acenaphthene [7.14] and a diphenylmercury molecule in the matrix of tolane or stilbene [7.15,16]. It is interesting to note that the impurity was always found to prefer replacing the host molecule that occupies a lower-energy position in the pure host crystal. Thus, the calculated energy of substitution of acenaphthene by α-nitronaphthalene was lower for the acenaphthene molecule parallel to the (001) plane, which was $\sim$0.6 kcal/mol lower in energy than the other symmetrically independent molecule.

The energetic preference of particular lattice sites in the above-mentioned systems was supported by X-ray diffraction measurements.

An analysis of the lattice relaxation in the vicinity of an impurity molecule was first undertaken by *Craig* et al. [7.10,17] for anthracene and naphthalene. The model that was used involved a single-layer free-surface 21-molecule cluster similar to the one described in Sect. 7.1.2.

When the host molecule was replaced by an impurity smaller in size, such as naphthalene in anthracene and benzene in naphthalene, the situation resembled that for a vacancy: a marked disturbance of the host lattice was not observed. In addition, the orientation of the impurity did not markedly differ from that of the replaced host. There was, however, a substantial center-of-mass displacement of the impurity molecule. Thus, the naphthalene molecule, placed initially at an inversion center in the host anthracene crystal, undergoes a 1 Å shift to minimize its energy and almost superimposes itself on the center and end rings of the replaced anthracene molecule. The energy of substitution of anthracene by naphthalene was calculated to be ~13 kcal/mol. This is quite a reasonable value considering that the substitution is nearly equivalent to the loss of a benzene ring.

If the host molecule is replaced by a larger impurity, such as anthracene in naphthalene and tetracene in anthracene, the relaxation of the host is, as might be expected, of great importance. For the anthracene/naphthalene and the tetracene/anthracene systems this relaxation reduces the substitutional energy from 28 to 0.24 kcal/mol and from 41 to 6 kcal/mol, respectively.

It is essential that the energy stresses in the vicinity of the oversized guest are not completely relieved by lattice relaxation. Thus, in the relaxed tetracene/anthracene system a pair of host molecules still experiences a repulsive interaction with the tetracene guest. In our opinion this is due to the small number of particles in the system and the use of unrealistic free-surface conditions. Since the layer about the guest molecule is very thin (monomolecular), the swelling of the structure, caused by the oversized guest, leads to too large an increase in the surface energy, which precludes a further reduction of the energy stresses.

The importance of including a sufficiently large number of particles in the relaxable layer was demonstrated by *Markey* [7.18]. When the number of anthracene molecules surrounding the tetracene guest was increased from 20 to 54, the substitutional energy dropped from 6 to 3.6 kcal/mol.

The structural changes occurring in the anthracene layer about the tetracene guest were surprisingly small in view of the fact that the difference in length between the anthracene and tetracene molecule is 2.4 Å. The largest angular displacement of the principal axis in the relaxed core was 7.8° and the largest center-of-mass displacement was 0.2 Å (for a 21-molecule cluster). The inclusion of successive neighbor shells to the core relieved the energy stresses, but affected the structure only marginally. Similar results were obtained for the anthracene/naphthalene system.

The orientation of the tetracene guest in the anthracene lattice, as well as of the anthracene guest in the naphthalene lattice, proved to be practically the same as that of the pure host, in agreement with experimental evidence [7.19].

Of the more recent studies on substitutional impurities, that by *McCool* et al. [7.20] deals with packing of 1,7- and 1,3-diazanaphthalenes in a lattice of durene. The model chosen was a two-layer cluster of 500 molecules, with 24 molecules allowed to relax. The calculations have shown that guest in-plane rotations are to be expected and can be several tens of degrees from the perfect orientation. This finding is in a qualitative agreement with the ESR spectra [7.20] which have revealed substantial deviations from the perfect guest substitution.

So far we have dealt with cases in which one guest molecule replaced just one host molecule. If the guest and host molecules differ significantly in size, it is easy to imagine situations in which one host is substituted by two or more guests or, vice versa, one guest replaces two or more hosts.

A good example of the former situation is revealed by the solid-state photocleavage of dianthracene which produces sandwich-like dimers of anthracene within the parent crystal [7.17]. There is some experimental information available about this system. *Fergusson* and *Mau* [7.21] observed excimer emission from the anthracene dimers and identified one type of emission as characteristic of isolated dimer sites in the ordered dianthracene lattice. The cleavage reaction was found to be photochemically reversible. The activation energy for the reverse reaction is about 1.7 kcal/mol. The excimer emission could only be observed from dimers produced in an excited state by photocleavage, but not by excitation of the dimer ground state. This suggests that after excimer emission the dimer relaxes to a configuration from which the excimer structure, and photodimerization, are only attainable through thermal activation. Additional information about the anthracene dimers was afforded by a polarization of the dimer absorption spectrum [7.22], which indicated that the orientation of the dimer was close to that of the parent dianthracene molecule.

A detailed configuration of the anthracene dimer encapsulated in the dianthracene lattice was derived by *Craig* and *Markey* [7.17] from packing-energy calculations. The calculations were carried out using a two-layer cluster containing 14 host molecules in the inner (relaxable) layer. The initial position of the guest anthracenes was an eclipsed one in which the principal axes are parallel to those of the replaced dianthracene host. After the energy minimization the anthracene rings of the dimer are still parallel but are displaced away from the eclipsed position so that the carbon atoms of one anthracene molecule lie over the centers of carbon-carbon bonds in the other anthracene molecule. This result is fully compatible with the experimental evidence discussed above. While the initially formed anthracene dimer had an eclipsed structure and was both the source of excimer emission and the starting point for the reverse reaction, the relaxed dimer was no longer eclipsed and had to be activated before it could be photodimerized.

The orientation of the cleaved dimer in the relaxed structure is practically the same as that of the parent host, in agreement with the polarization of the dimer absorption spectrum.

It is interesting to note that the relaxed cluster had a much lower potential energy (by 6.7 kcal/mol) than the corresponding cluster of the pure host. This was a result of the appearance of an additional attractive term in the potential energy, a term arising from the interactions between the anthracene molecules themselves. Actually, the structure experienced energy stresses which made the molecules in the inner layer 1–2 kcal/mol less stable than those in the pure crystal.

The situation in which a guest molecule replaces two and more host molecules is well exemplified by the porphyrin substitutional defects in a n-octane matrix. There is much information available about such defects. First of all, the system is known to exhibit the Shpolskii effect: i.e., the spectral lines of the guest molecule are very narrow and well defined, which indicates a weak interaction between porphyrin and its surroundings. Also, from ESR and fluorescence spectra [7.23–25], it is known that there are two physically distinct substitution sites in the host. The orientation of the porphyrin molecule in one site is available from ESR measurements. It is also known that the porphyrin plane in the other site is nearly parallel to the a axis of the host crystal. In addition, the lowest librational frequencies of the guest molecule are available [7.26]. All of this information is a good basis for checking the reliability of model calculations.

A microscopic picture of the porphyrin substitutional defects in a n-octane matrix was obtained by *Koehler* [7.11] using a two-layer 96-molecule cluster model similar to the one described in Sect. 7.1.2. A preliminary geometrical analysis showed that the most likely defect structures involve a porphyrin molecule, lying nearly parallel to the bc or ac plane, that has replaced two or three n-octane molecules which were neighbors along the crystallographic b or a axes, respectively (Figs. 7.4,5).

The search for the minimum-energy configuration yielded three most probable defect sites, one being a two-vacancy defect (site A) and the other two three-vacancy defects (sites B and B'). The structures of the A and B sites are shown in Figs. 7.4,5. The structure of the B' site resembled that of the B site, but exhibited a somewhat different orientation of the guest molecule.

The most stable site was the A site, which was about 28 kcal/mol lower in energy than the B site. The third site (B') was about 1.6 kcal/mol higher in energy than B. The other local minima found in the search for the optimum defect structure were considerably less stable.

The orientation of a guest molecule in the A site agreed well with the ESR data. This is demonstrated in Table 7.4 which compares the experimental and calculated values of the angle between the molecular and crystal axes. For the B and B' sites the angle that the porphyrin plane makes with the a axis was 5° and 10°, respectively. In other words, the orientations of the preferred B site agreed more closely with the little that was known experimentally about the second site.

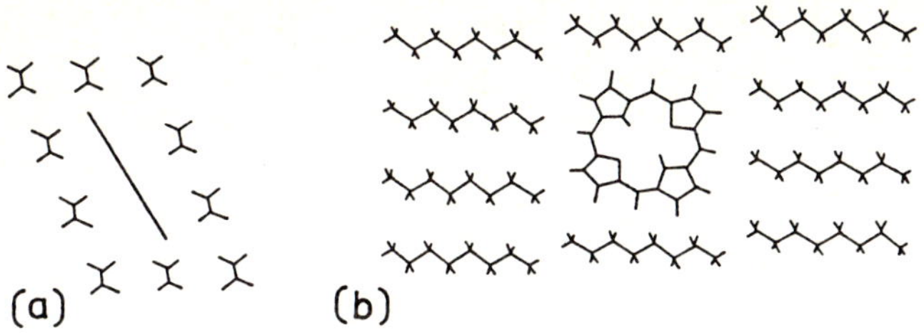

7.4a,b. Porphyrin substitutional defect in n-octane matrix [7.11]: (a) View of the *A* site along the long axis of the octane molecules. The projection of the *a* axis is horizontal; (b) View of the *A* site perpendicular to the *bc* plane. The *c* axis is horizontal

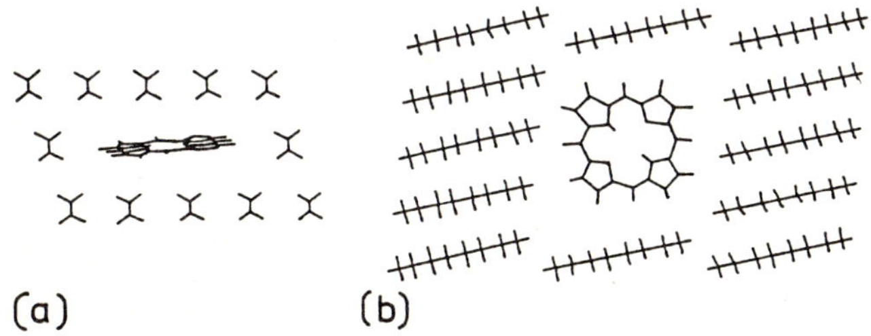

Fig. 7.5a,b. Porphyrin substitutional defect in n-octane matrix [7.11]: (a) View of the *B* site along the long axis of the octane molecules. The projection of the *a* axis is horizontal; (b) View of the *B* site perpendicular to the *ab* plane. The *c* axis is horizontal

Table 7.4. Angle between molecular and crystal axes for porphyrin in n-octane matrix (*A* site) [deg]

Axes[a]	Angles Observed [7.25]	Calculated [7.11]
z - a	27 ± 4	28
y - a	90 ± 5	86
y - c	60 ± 10	56
x - a	65 ± 5	62

[a] The molecular axes *x* and *y* were chosen to lie in the porphyrin plane, with *x* parallel to the NH bonds; the *z* axis is perpendicular to the porphyrin plane

Besides agreeing with the experimental orientation, the *A* and *B* defect models explained other experimental observations. For example, the two-vacancy *A* site was a tighter fit than the three-vacancy site and thus led to higher librational frequencies. A rough estimate of the librational frequencies, made by considering the guest molecule to oscillate in the field of its fixed neighbors, yielded $45\,\mathrm{cm}^{-1}$ for the *A* site and $25\,\mathrm{cm}^{-1}$ for the *B* site. The experimentally observed values were 30 and $15\,\mathrm{cm}^{-1}$, respectively [7.26].

As in the case of the two- and three-molecule vacancies discussed in Sect. 7.1.2, the n-octane lattice proved to be almost unaffected by the substitutional defect. The largest center-of-mass displacement of an octane molecule was only 0.2 Å and the change in energy due to lattice relaxation was less than 0.04%. This suggests a weak coupling of the guest molecule to the host lattice and explains the observed Shpolskii effect.

Another result which explained the experimental Shpolskii effect was that the porphyrin guest molecule interacted with the host lattice mainly through its hydrogen atoms. The porphyrin π-system, therefore, yielded narrow spectral lines typical of an isolated porphyrin molecule because it was as if shielded from the matrix effects.

Thus, the microscopic picture of the porphyrin substitutional defects in a n-octane matrix obtained from packing-energy calculations agreed in all of its predictions with what is known experimentally about this system.

Thus far we were mainly interested in the structural aspects of the substitutional impurities and did not attach much weight to the numerical values of the substitutional energies. We now turn to the energetic aspect of the problem to see whether or not we are capable of quantitatively predicting the equilibrium concentration of the impurities.

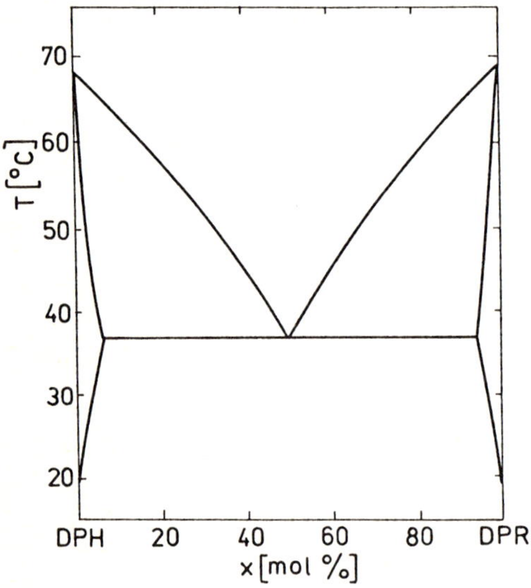

Fig. 7.6. Phase diagram of the diphenyl/α,α'-dipyridyl (DPH/DPR) system [7.27]

As an example, we shall consider the binary system, diphenyl/dipyridyl (DPH/DPR). The phase diagram of this system is shown in Fig. 7.6 [7.27]. It is seen that each of the components can form solid solutions in the other. At the eutectic point (37.5°C) the solid solution of DPR in DPH (α phase) may contain up to 8% DPR, while the solid solution of DPH in DPR (β phase) may contain up to 5% DPH. X-ray diffraction measurements [7.27] have shown that both the α and β phases retain the structure of their host crystals.

In constructing microscopic models for the α and β phases, due allowance should be made for the difference in symmetry between the DPH and DPR molecules. It is clear from general considerations that the orientation of a substitutional impurity in a host lattice will only be unique if the symmetry of the impurity molecule is the same as, or higher than that of the host molecule. If the symmetry group of the impurity molecule is a subgroup of the closed symmetry group of the host molecule, there will be more than one possible orientation for the impurity. The former situation is exemplified by the replacement of DPR by DPH (β phase), while the replacment of DPH by DPR (α phase) affords an example of the latter situation. Since the DPR molecule may be superimposed upon the DPH molecule in two different ways, related by a 180° rotation of DPR about its short or long molecular axis, there will be two distinct orientations of the DPR impurity in the α phase.

We shall first consider the simpler case in which the impurity is only allowed to adopt a single orientation in the host crystal. Let N be the total number of molecules in the solid solution crystal, N_h the number of host molecules and N_g the number of guest molecules ($N_h + N_g = N$). It will be assumed that $N_g \ll N_h$, so that each impurity molecule can be considered to be only surrounded by host molecules. With this assumption, each impurity molecule may be taken to be the center of a two-layer cluster such as the one described in Sect. 7.1.1.

Using (7.7), the potential energy of the system may be written as

$$\phi = \phi^0 + N_g(\phi_c - \phi_c^0), \tag{7.13}$$

where ϕ^0 is the potential energy of the pure host crystal, ϕ_c is the cluster energy minimized with respect to the parameters of the cluster core and ϕ_c^0 is the energy of the pure cluster (i.e., the same cluster used to compute ϕ_c except that the impurity has been replaced by a host molecule).

Neglecting the lattice vibrations the partition function for the system may be written as

$$Z = \frac{N!}{N_h! N_g!} \exp\left(-\frac{\phi}{kT}\right). \tag{7.14}$$

The free energy per particle is given by

$$F/N = kT[(1-x)\ln(1-x) + x\ln x] + \varphi^0 + x(\phi_c - \phi_c^0), \tag{7.15}$$

where $x = N_g/N$ and $\varphi^0 = \phi^0/N$.

We now turn to the case in which the guest molecule may assume two distinct orientations, g' and g'', in the host lattice. In this case there will be two distinct types of clusters in the crystal, with the energies ϕ_c' and ϕ_c''. The energy spectrum of the system will now be given by

$$\phi(g'g'g'\ldots g') = \phi^0 + N_g(\phi_c' - \phi_c^0), \tag{7.16}$$

$$\phi(g''g'g'\ldots g') = \phi^0 + (N_g - 1)\phi'_c + \phi''_c - N_g\phi^0_c$$
$$= \phi^0 + N_g(\phi'_c - \phi^0_c) + (\phi''_c - \phi'_c), \tag{7.17}$$
$$\phi(g''g''g'\ldots g') = \phi^0 + N_g(\phi'_c - \phi^0_c) + 2(\phi''_c - \phi'_c), \tag{7.18}$$
$$\vdots \qquad \qquad \vdots$$
$$\phi(g''g''g''\ldots g'') = \phi^0 + N_g(\phi'_c - \phi^0_c) + N_g(\phi''_c - \phi'_c), \tag{7.19}$$

where $\phi(g'g'g'\ldots g')$ is the potential energy of the system in which all impurities assume the orientation g', $\phi(g''g'g'\ldots g')$ is the potential energy of the system in which one impurity assumes the orientation g'' and the others the orientation g' and so on.

The partition function of the system is given by the sum

$$Z = \frac{N!}{N_h!N_g!}\exp\left(-\frac{\phi(g'g'\ldots g')}{kT}\right)$$
$$+ \frac{N!}{N_h!(N_g-1)!1!}\exp\left(-\frac{\phi(g''g'\ldots g')}{kT}\right)$$
$$+ \frac{N!}{N_h!(N_g-2)!2!}\exp\left(-\frac{\phi(g''g''g'\ldots g')}{kT}\right)$$
$$+ \ldots + \frac{N!}{N_h!N_g!}\exp\left(-\frac{\phi(g''\ldots g'')}{kT}\right)$$
$$= \frac{N!}{N_h!N_g!}\exp\left(-\frac{\phi^0 + N_g(\phi'_c - \phi^0_c)}{kT}\right)$$
$$\times \sum_{j=0}^{N_g} \frac{N_g!}{(N_g - j)!j!}\exp\left(-\frac{j(\phi''_c - \phi'_c)}{kT}\right). \tag{7.20}$$

Using the identity,

$$\sum_{j=0}^{n} \frac{n!}{(n-j)!j!} b^j \equiv (1+b)^n, \tag{7.21}$$

we may rewrite (7.20) in the form

$$Z = \frac{N!}{N_h!N_g!}\exp\left(-\frac{\phi^0 + N_g(\phi'_c - \phi^0_c)}{kT}\right)$$
$$\times \left[1 + \exp\left(-\frac{\phi''_c - \phi'_c}{kT}\right)\right]^{N_g}. \tag{7.22}$$

Taking the logarithm of (7.22), we immediately obtain the free energy

$$F/N = kT[(1-x)\ln(1-x) + x\ln x] + \varphi^0 + x(\phi'_c - \phi^0_c)$$
$$- kTx\ln\left[1 + \exp\left(-\frac{\phi''_c - \phi'_c}{kT}\right)\right]. \tag{7.23}$$

The resulting equation can be readily generalized to the case in which the impurity molecule assumes an arbitrary number n of possible orientations

$$F/N = kT[(1-x)\ln(1-x) + x\ln x] + \varphi^0 + x(\phi_c^1 - \phi_c^0) - kTx\,\ln\sum_{i=1}^{n}\exp\left(-\frac{\phi_c^i - \phi_c^1}{kT}\right). \quad (7.24)$$

Applying (7.15 and 23) to the β and α phases, respectively, one can easily find the limiting impurity concentration in both phases. This is best done graphically by plotting the free-energy curves $F_\alpha(x)$ and $F_\beta(x)$ and finding the common tangent (Fig. 7.7).

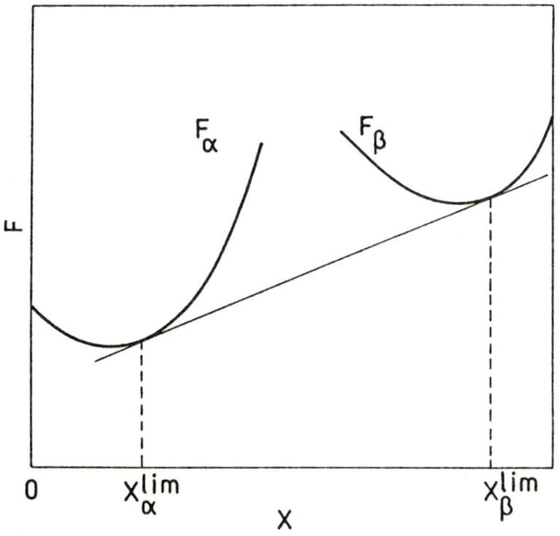

Fig. 7.7. Graphical determination of the limiting impurity concentrations in a binary system. (Arbitrary units)

Numerical calculations for the DPR/DPH system have been performed by *Nikitin* and *Pertsin* [7.28]. The defective regions in the α phase were simulated using a two-layer cluster containing 16 DPH molecules in the inner layer and 58 DPH molecules in the outer layer. The cluster model for the β phase had 14 and 50 DPR molecules in the inner and outer layers, respectively. The intermolecular interactions in the systems were treated using the W67 and G74 potentials of Tables 3.1,3.

Since the equilibrium cluster structure was defined to have the lowest potential energy, it seemed reasonable to use, as reference host structures, the structures corresponding to a minimum of the potential energy instead of the observed crystal structures.

The results of the energy minimization for the pure host crystals are summarized in Table 7.5. The calculated structures are seen to be appreciably denser than the observed structures. The difference in volume between the observed and calculated unit cells is 7.6% for DPH and 10.7% for DPR.

Table 7.5. Results of the lattice energy minimization of pure diphenyl and dipyridyl crystals [Å, deg, kcal/mol] [7.28]

Unit cell parameters					Euler angles[a]			Lattice energy
β	a	b	c	V_c	χ_1	χ_2	χ_3	
Diphenyl								
initial structure (observed)								
94.4	8.12	5.64	9.47	432.4	74.9	79.4	-55.4	-18.89
final structure								
93.0	7.60	5.49	9.59	399.6	76.8	81.4	-62.0	-21.06
Dipyridyl								
initial structure (observed)								
118.8	5.66	6.24	13.46	416.8	50.0	54.2	40.7	-19.54
final structure								
119.5	5.23	6.48	12.61	372.0	46.4	53.4	42.7	-21.08

[a] The molecular axes x and y were chosen to lie in the long and short in-plane directions, z is perpendicular to the molecular plane

Although the volume differences do not cause much trouble and may well be attributed to thermal expansion, the predictions of the individual lattice parameters do not appear to be satisfactory. This is particularly true of the parameters c and b of DPH and DPR, respectively. The calculated values are higher than the observed ones instead of being lower. There is also a marked discrepancy between the calculated and observed orientation of the DPH molecules, which is appreciably higher than the discrepancies typical of potential energy calculations on hydrocarbon crystals.

The introduction of impurities to host crystals did not produce large energy stresses. Thus, the starting (unrelaxed) cluster of the β phase was only about 11 kcal/(mole of cluster) less stable than the corresponding pure cluster. This is only 0.7 kcal/mol, when related to one molecule. Even smaller energy differences were observed for the two forms of the α phase [within 2 kcal/(mole of cluster)].

The reduction in the cluster energy due to lattice relaxation was found to be 6.3 and 0.4 kcal/(mole of cluster) for the α and β phases, respectively. Marked structural changes were not observed in the relaxed structures. Rotational displacements in the α phase were all within 2° while the center-of-mass displacements were within 0.002 of the corresponding lattice periods. The largest rotational displacement in the β phase was 7° for the impurity and 3° for the host molecules. The center-of-mass displacements were all within 0.01 Å.

The final results for the DPH/DPR system are summarized in Table 7.6. Given in the upper part of the table are the values obtained, as described above, through a relaxation of the molecules over both the rotational and translational degrees of freedom. The resulting limits of solubility are seen to differ significantly from the corresponding experimental values. This is particularly true of the β phase in which the calculated solubility is two orders of magnitude lower than the observed one. Considering that the DPH molecule is somewhat larger in size than the DPR molecule (due to the presence of two extra hydrogen atoms), it may be that the low solubility in the β phase results from the neglect of thermal expansion in the matrix lattice (Table 7.5). Indeed, had thermal expansion been taken into account, the matrix would be less tight. As a result the lattice would lose less energy in accomodating a guest molecule.

To verify this we recalculated all of the quantities in Table 7.6 using the experimental unit-cell dimensions for the pure host crystals and allowing only rotational displacements for the molecules. Accordingly, only the rotational parameters were allowed to vary in the α and β phase clusters, while the center-of-mass coordinates were kept fixed at the values observed in the corresponding host crystals. The results are given in the lower part of Table 7.6. It is seen that the agreement with experiment is now better, although still far from satisfactory.

Table 7.6. Energetic characteristics [kcal/mol] and limiting impurity concentrations [mole %] for the diphenyl/dipyridyl system [7.28]

Phase	ϕ^0	$\phi'_c - \phi^0_c$	$\phi''_c - \phi'_c$	x^{lim}
complete relaxation				
α	-21.06	1.18	0.10	20.0
β	-21.08	4.93	-	0.03
relaxation over orientational variables alone				
α	-19.24	2.09	0.12	5.7
β	-19.73	0.92	-	18.0

The failure of the calculations to yield good quantitative estimates of the limiting solubilities may arise from at least three sources. The first source of error lies in the model potential that was used. Some indication of this is provided by the results obtained for the pure host crystals (Table 7.5).

The second source lies the neglect of the vibrational contribution to the free energy, which was made by writing the free energy in the form of (7.24). The most serious consequence of this neglect is the inability of the calculation to take into account thermal expansion in both the pure and defective crystals. The importance of thermal expansion can be appreciated by comparing the values in the upper and lower parts of Table 7.6.

Finally, errors in the calculated limits of solubility may partly arise from the assumption that the impurity molecules are only surrounded by host molecules. Actually, at 5–10% concentrations the impurity molecules may be separated by only one or two host molecules so that the interactions between the impurities and the interpenetration of their surrounding layers may be of great importance.

Thus, the results discussed in this section indicate that packing-energy calculations provide a fairly reliable means of predicting the detailed structure of defective regions about a substitutional impurity. At the same time, the estimates of the defect energy must be treated cautiously because they appear to be very sensitive to the assumptions and approximations of the computational model that is used.

We conclude this section by noting that the microscopic cluster model used to treat point defects can also be used to evaluate the barriers to molecular reorientations. This can easily be done by considering the reorienting molecule to be a substitutional impurity and calculating the energy of such a defect as a function of the rotational angle. Calculations of this kind are widespread in the literature [7.29–39]. The difficulties involved in such calculations are similar to those discussed above for substitutional impurities, so that it is hardly reasonable to hope for success in predicting the energetics of reorientations.

7.2 Linear Faults

Thus far, the only kind of linear fault that has been studied using the atom-atom potential method is a b-stacking fault in a crystal of anthracene. The core of the fault consists of a chain of molecules parallel to the crystallographic axis b (a direction of strong intermolecular coupling). Each molecule in this chain has been rotated through 180° from the perfect crystal configuration and displaced a distance $b/2$. The defect chain is terminated at each end by half vacancies.

The presence of such faults in anthracene has been used by *Craig* et al. [7.40] to explain the unusual photoreactivity of the anthracene crystal, which is not consistent with the principles of topochemical preformation theory (the herringbone packing in the crystal is such that there is very little overlap between adjacent molecules and there is no preformation of the photodimers). *Craig* et al. [7.40] have put forward the following arguments in favor of the photodimerization taking place at the b-stacking faults. When a b-chain is misoriented as described above, the local environment of a molecule in the chain becomes $P\bar{1}$. The result is a compression within the $P2_1/a$ structure of anthracene. This compression should lead to exciton trapping since it increases the dispersive stabilization of the upper electronic state more than the ground state. The misorientation should facilitate the dimerization of the pairs of molecules that are $a/2 \approx 4.3$ Å apart and parallel to one another. The scheme proposed by *Craig* et al. [7.40] is consistent with the experimental observation that the photodimerization reaction propagates in the direction of the monoclinic b-axis.

The hypothesis that the b-stacking faults serve as active sites for the photodimerization of anthracene has been tested by *Markey* and *Walmsley*

[7.41]. The model chosen for the calculations consisted of a cluster with free-surface boundary conditions, similar to those described in Sect. 7.1.2. Due to the severe restrictions imposed upon the size of the cluster by the computational facilities, the defect chains that were examined were very short: (1) a two-molecule chain in which the molecules at $(0, \pm 1/2, 0)$ replace three regular-lattice molecules at (0,0,0) and $(0, \pm 1, 0)$, and (2) a four-molecule chain in which the molecules at $(0, \pm 1/2, 0)$ and $(0, \pm 3/2, 0)$ replace five regular-lattice molecules at (0,0,0), $(0, \pm 1, 0)$ and $(0, \pm 2, 0)$.

For both forms of the defect chain, two situations were examined, i.e., the defect was either on the *ab*-surface or in the bulk of the crystal. The energetic characteristics of the defects are presented in Table 7.7. The subheading "ideal" refers to the starting configuration in which the defect molecules are placed at the above indicated sites and then rotated 180° from the perfect-lattice orientation. All of the energy values are given relative to the energy of a monovacancy in the bulk of the crystal.

Table 7.7. Defect energies of a *b*-stacking fault in anthracene [7.41] [kcal/mol]

	Two-molecule stack	Four-molecule stack
Surface defect		
ideal	37.5	71.0
core relaxation	13.6	24.1
complete relaxation	12.0	20.8
Bulk defect		
ideal	62.6	121.4
core relaxation	20.3	38.0
complete relaxation	12.9	21.9

As seen from Table 7.7, the relaxation substantially reduces the energy stresses in the defect. Thus, for a two-molecule bulk defect the relaxation of the core molecules reduces the potential energy by 42.3 kcal/mol. A further 7.4 kcal/mol reduction results from the extended defect region. It is also seen that the energy of the stacking fault is almost the same at the surface or in the bulk (when complete relaxation is allowed).

Another trend observed in Table 7.7 is that the stability of the fault decreases with increasing length. Although the defect energy is seen to increase more slowly than the length of the stack, it is hardly reasonable to expect the formation of extended stacking faults in the crystal.

The calculated defect energies in Table 7.7 are by themselves very high. The resulting equilibrium concentrations of the defects are less than 10^{-3} mol/mol [7.41]. Thus, it seems very unlikely that the *b*-stacking faults play a significant role in the photodimerization of anthracene.

The stacking fault in the crystal was found to have a negligible effect on the surrounding molecules. However, the structural changes in the fault chain were significant. This is seen in Table 7.8 which compares the direction

Table 7.8. Direction cosines of the principal axes of a two-molecule stack in the bulk defect at equilibrium [7.41]

		Unit-cell axes		
		a	b	c sinβ
Principal molecular axes	L	0.500(0.511)[b]	0.174(-0.130)	-0.848(-0.850)
	M	-0.360(-0.306)	0.933(0.896)	-0.021(-0.321)
	N	0.788(0.804)	0.316(0.424)	0.529(0.418)

[a] L is the long, M the short in-plane symmetry axis, and N is perpendicular to the plane; [b]Values in parentheses refer to the perfect $P2_1/a$ lattice

cosines of the principal axes of a molecule in the chain and a molecule in the perfect $P2_1/a$ lattice. Although a number of various arrangements for two particular orientations were found in the relaxed structure, they did not favor the formation of the photodimer.

Thus, both the energetics and the structure of the *b*-stacking fault showed it as hardly responsible for the photodimerization of anthracene.

7.3 Planar Faults

7.3.1 Surface

A calculation of the energetics of a crystal surface was first undertaken by *Kitaigorodsky* and *Ahmed* [7.42] to explain the geometrical form of the anthracene crystal.

According to *Gibbs* [7.43], the equilibrium form of a crystal is determined by the condition

$$\sum_i f_i A_i = \text{minimum} \tag{7.25}$$

where A_i is the area of the ith face of the crystal and f_i is the surface free energy per unit area of the ith face. If the thermal contribution to the free energy is neglected, f_i becomes the surface potential energy.

Wulff [7.44] has shown that (7.25) is equivalent to the equation

$$f_1/d_1 = f_2/d_2 = \ldots = f_i/d_i = \text{const}, \tag{7.26}$$

where d_i is the central distance for the ith face (the distance between the center of the crystal and the ith face along the normal to the face).

The equilibrium form of a crystal may be derived from (7.26) by varying the central distance of each face until it is proportional to the specific surface energy f_i. Such an approach is usually known as the Wulff method. In two dimensions this method uses a polar diagram of the specific surface

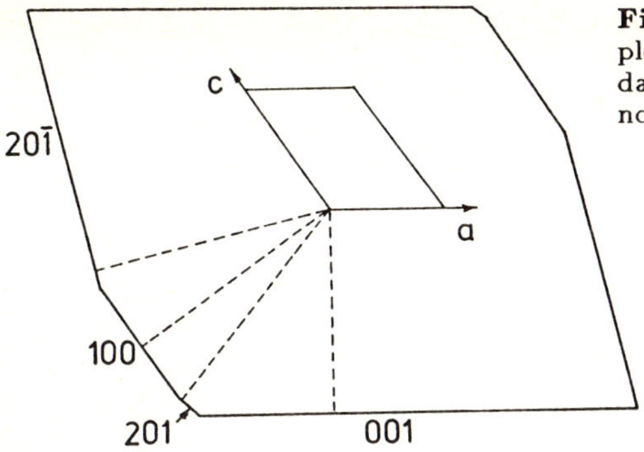

Fig. 7.8. The (010) projection of the Wulff plot for the anthracene crystal [7.42]. The dashed lines indicate the directions of the normals

energy of the various planes which belong to a particular zone (Fig. 7.8). The crystal form which corresponds to the lowest surface energy is obtained by selecting the planes that enclose the smallest area. These planes are orthogonal to the lines drawn from the origin and intersect these lines at distances proportional to the specific surface energy of a particular plane. In three-dimensions the specific surface energies of the planes are plotted in terms of spherical coordinates.

The specific surface energy f_i in (7.25,26) is defined to be half the energy per unit area required to cleave the crystal along the ith plane. If the structure disturbances appearing in the surface layers after the cleavage are neglected, as was done by *Kitaigorodsky* and *Ahmed* [7.42], the f_i's can be readily evaluated in a simple summation of the intermolecular interactions across the ith plane. This is schematically illustrated in Fig. 7.9 for the (100) plane of the anthracene crystal. Since the molecules in each

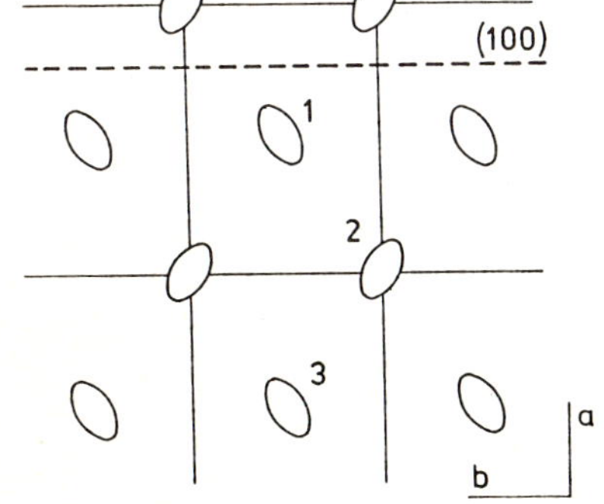

Fig. 7.9. A schematic representation of the (001) plane in anthracene. Projection of the cleavage plane (100) is indicated by the dashed line

layer parallel to the cleavage plane are translationally identical, it suffices to consider only one representative molecule in each layer (molecules 1, 2 and 3 in Fig. 7.9) and evaluate the interaction energies of these representative molecules with the molecules above the cleavage plane (including those in the third dimension).

The numerical values of f_i calculated by *Kitaigorodsky* and *Ahmed* [7.42] for several planes of simple indices of the anthracene crystal are given in Table 7.9. The resulting form that is obtained using the Wulff method is shown in Fig. 7.10. This form is in agreement with the observations of *Groth* [7.45] who described the anthracene crystals as plates containing the a and b axes, with the faces $\{001\}$, $\{110\}$, $\{20\bar{1}\}$ and $\{11\bar{1}\}$ being the most common.

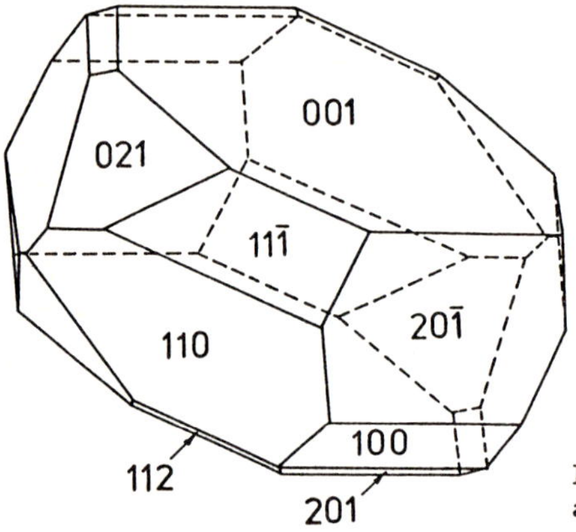

Fig. 7.10. Theoretical equilibrium form of the anthracene crystal [7.42]

The importance of structural distortions, which occur in the surface layers, has been examined by *Filippini* et al. [7.46] and, more thoroughly, by *Pawley* [7.9] for the crystal of naphthalene. The (001) surface was chosen by *Pawley* [7.9] for his calculations. It has been shown by *Firment* and *Somorjai* [7.47] to be the surface formed preferentially during vapor deposition.

The adopted surface model represented an array of monomolecular layers. The starting structure is identical to that of the (001) planes in the minimum-energy bulk crystal configuration. It was assumed that the glide plane of the naphthalene crystal (space group $P2_1/a$) was retained in each of the surface layers. Thus, the two-dimensional repeating unit of a layer contained only one symmetrically independent molecule. Since the dimensions of the repeating units were predetermined by the structure of the bulk of the crystal, only six parameters, i.e., the three translations and the three rotations of a representative molecule, were allowed to vary in each layer. The potential energy of the surface model was calculated as

$$\phi_{\mathrm{surf}} = \frac{1}{2} \sum_i \psi_i, \qquad (7.27)$$

Table 7.9. Calculated values of the specific surface potential energy for planes of simple indices in anthracene [erg/cm^2] [7.42]

Plane	f	Plane	f
100	87.98	201	90.14
010	117.50	$20\bar{1}$	91.94
001	75.70	021	111.70
110	90.34	111	98.87
011	103.20	$11\bar{1}$	105.50
101	90.80	112	94.20
$10\bar{1}$	119.25	$11\bar{2}$	122.00

Table 7.10. Translational and rotational molecular displacements as a function of the layer depth for the six- and eightlayer models of the (001) surface of napthalene [Å, deg] [7.9]

Layer	Δx	Δy	Δz	$\Delta\theta_x$	$\Delta\theta_y$	$\Delta\theta_z$
six-layer model						
6	0.0000	0.0000	0.0000	0.0000	0.0000	0.0000
5	-0.0128	-0.0003	0.0751	0.1127	0.7210	-0.0398
4	0.0237	0.0001	0.1614	0.1534	0.9191	-0.0624
3	0.1053	-0.0000	0.2900	0.2591	1.4712	-0.1260
2	0.2630	-0.0010	0.4699	0.4911	2.6512	-0.2716
1	0.5929	0.0014	0.6895	0.6888	4.8348	-0.4968
eight-layer model						
8	0.0000	0.0000	0.0000	0.0000	0.0000	0.0000
7	-0.0186	-0.0004	0.0627	0.1255	0.7499	-0.0504
6	0.0045	0.0001	0.1427	0.1402	0.8063	-0.0572
5	0.0394	0.0001	0.2368	0.1640	0.9294	-0.0694
4	0.0996	0.0000	0.3355	0.2142	1.2197	-0.1020
3	0.2157	-0.0004	0.4805	0.3473	1.8886	-0.1775
2	0.4184	-0.0008	0.6593	0.5648	3.0135	-0.3158
1	0.7659	0.0017	0.8742	0.6988	4.9084	-0.5082

where ψ_i is the interaction energy of a molecule in the ith layer and its surroundings and the summation is carried out over all layers in the model.

Two-, four-, six- and eight-layer models were examined by *Pawley* [7.9]. In each model the deepest layer was assumed fixed while the other layers were allowed to relax, so as to minimize the total packing energy.

Table 7.10 lists the translational and rotational molecular displacements as a function of the layer depth for the six- and eight-layer models [7.9]. All of the displacements were measured with respect to the bulk equilibrium positions of the corresponding molecules. The x and y coordinate axes are chosen parallel to the bulk crystallographic axes a and b, while the axis z is perpendicular to the surface and parallel to the c' crystallographic axis. The rotational displacements $\Delta\theta_x$, $\Delta\theta_y$ and $\Delta\theta_z$ are right-handed rotations about these axes and are applied in this order.

It is seen from Table 7.10 that new features in the surface structure are not observed on going from the six- to the eight-layer model. Both models predict rather small structural changes at the surface, so that the general pattern of the surface packing is essentially the same as in the bulk of the crystal. This is in agreement with the work by *Firment* and *Somorjai* [7.47] in which structural disturbances at the surface were not observed.

Table 7.11. Molecular packing energies [kcal/mol] of the surface layers in the naphthalene crystal [7.9]. (All of the values are given relative to the packing energy of the bulk)

	Layer 1	2	3	4	5	6	7	8	Total excess energy
Two-layer model	2.424	0.008[a]							2.432
Four-layer model	2.379	-0.109	-0.125	-0.041[a]					2.101
Six-layer model	2.408	-0.125	-0.156	-0.134	-0.119	-0.028[a]			1.846
Eight-layer model	2.410	-0.128	-0.166	-0.147	-0.130	-0.117	-0.109	-0.024[a]	1.589

[a] Fixed layer

The energetic characteristics of the various model surfaces are given in Table 7.11. The packing energies $\psi_i/2$ of the layers are presented relative to the packing energy of the bulk of the crystal. The last value in each row is the sum of the preceding ones and represents the excess energy of the model surface relative to the energy of the same number of layers in the bulk

$$\Delta\phi = \phi_{\text{surf}} - \frac{1}{2}n\psi_{\text{bulk}}, \tag{7.28}$$

where n is the number of layers in the model and ψ_{bulk} is defined similarly to ψ_i but with respect to a molecule in the bulk.

Scanning the excess surface energies given in Table 7.11 shows that the surface becomes progressively more stable as the number of layers allowed to relax is increased. It is also seen that most of the surface energy is concentrated in the outer layer. An interesting observation is that in all the models (except the crudest, two-layer one) the molecules just below the surface find potential minima which are more stable than that for a molecule in the bulk of the crystal.

Pawley [7.9] has also studied the dynamics of the (001) surface of the napthalene crystal. The chosen model consisted of a slab of twelve molecular layers, such that the two middle layers corresponded exactly to the bulk structure. The slab showed clear evidence of well defined surface modes, with typical frequencies lower than those of the bulk modes.

7.3.2 Stacking Faults

Another kind of planar fault, which has been treated using the atom-atom potential method, is a stacking fault produced by a uniform fractional slip of one half of a crystal with respect to the other half. Such a fault may serve as an idealized model for planar faults formed as a result of dissociation of edge dislocations into partials (Fig. 7.11). The idealization that the fault extends across the entire plane is introduced to avoid having to take into account the loss of periodicity about the dislocation core. The slip sys-

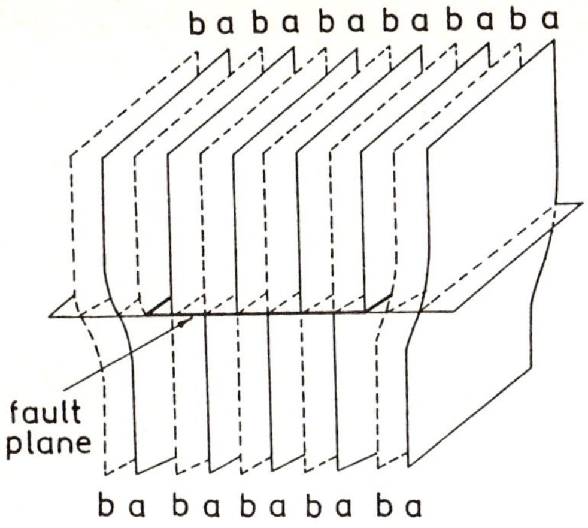

Fig. 7.11. Illustration of how a line defect, upon dissociation into two separated partial dislocations, gives rise to a planar stacking fault, across which there is mismatch between the upper and lower parts of the crystal. The symbols *a*, *b*, *a*, *b*, etc. denote the periodicity of the structure [7.48]

tems associated with planar faults of this kind are usually described by the notation $(hkl)\xi[uvw]$, which defines the slip plane and the Burgers vector.

A calculation of the detailed crystal structure about a planar stacking fault has been performed by *Ramdas* et al. [7.48] to explain the photoreactivity of 1,8-dichloro-9-methylanthracene (DMA). The behavior of DMA, which upon UV irradiation yields the trans dimers shown in Fig. 7.12, is

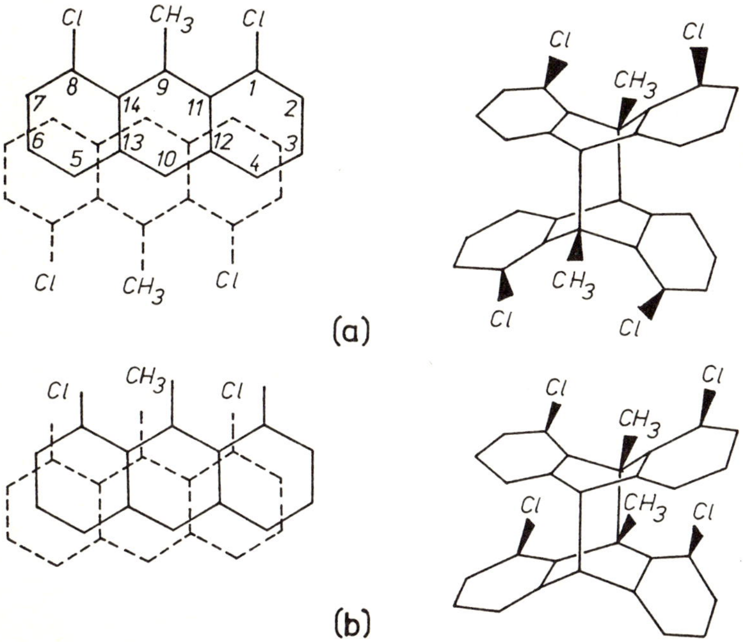

Fig. 7.12a,b. Schematic illustration of the arrangement of adjacent molecules of 1,8-dichloro-9-methylanthracene prior to the formation of the centro-symmetric (trans) photodimer (**a**) and the mirror-symmetric (cis) photodimer (**b**) [7.48]. The stereochemistry of the dimers is indicated on the right-hand side

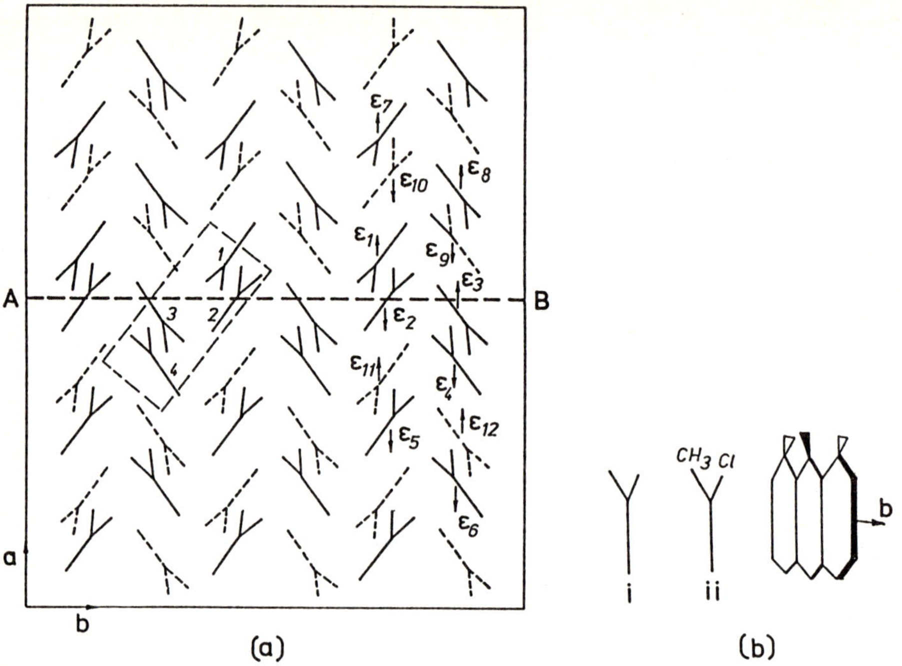

Fig. 7.13. (a) Representation of the (010) projection of the crystal structure of 1,8-dichloro-9-methylanthracene. The face of the planar fault is given by the dashed line *AB*. A displacement vector of 1/2[010] on the (100) plane brings the molecules 1 and 3 on one side of the fault into close contact with the molecules 2 and 4, respectively, on the other side. The pairs 1/2 and 3/4 are incipient dimers. The molecular translational relaxations, imposed in an attempt to minimize the energy (see text), are designated $\varepsilon_1, \varepsilon_2, \ldots, \varepsilon_{12}$. **(b)** Explanation of molecular symbols: (i) is an abbreviated version of (ii) which, in turn, represents the skeletal appearance of the molecule along the *b* axis [7.48]

inconsistent with topochemical preformation theory since the neighboring molecules in DMA are oriented unfavorably with respect to one another and are too far apart for the photodimerization to occur.

From the crystal structure outlined in Fig. 7.13, one can see that the application of a uniform shear defined by (100)(1/2)[010] brings the molecules on different sides of the slip plane into a favorable incipient trans orientation in which the molecular planes are essentially parallel.

The explanation of the photoreactivity of DMA in terms of the (100) (1/2) [010] slip system is supported by optical observations [7.49] which show that (100)[010] is a dominant slip system in the crystal and that the dimers are preferably formed along the [001] direction.

The calculations of the detailed structure of the fault were carried out in increasing order of complexity. At first, *Ramdas* et al. [7.48] examined the unrelaxed situation in which the entire half of the crystal was shifted along the (100) plane in the [010] direction. Such an unrelaxed fault was found to have too high an energy due to the shortened contacts between the neighboring molecules across the fault plane.

Ramdas et al. [7.48] then allowed the individual molecules to relax in either the [100] or the [$\bar{1}$00] direction. Twelve variable parameters, representing molecular displacements in the three layers on either side of the fault plane, were introduced (see ε_1, ε_2, ... and ε_{12} shown in Fig. 7.13). However, only shallow depressions could be imposed on the fault energy in this way. On the other hand, when the half-crystal that had already experienced a slip of (100)(1/2)[100] was shifted in the [$\bar{1}$00] direction by a translation xa, the fault energy was found to decrease sharply with increasing x, reaching a minimum at $x \approx 0.3$.

Although the fault structure corresponding to a 0.3 a shift in the [$\bar{1}$00] direction was energetically acceptable, it could not explain the photoreactivity of the crystal because the distances between active molecular sites ($C_9 \ldots C_{10'}$) proved to be about 1 Å larger than that required for the dimerization to take place (~4.5 Å). To make the system more adaptable to the fault, *Ramdas* et al. [7.48] explored the possibility of intramolecular relaxation, by allowing the molecules to fold across the molecular $C_9 - C_{10}$ axis. Such a folding markedly reduces the fault energy and allows the potentially dimerizable molecules to approach one another more closely. The most favorable configuration was found for a fold angle of 7°. In this configuration the shift in the [$\bar{1}$00] direction is reduced to $x = 0.2$ and the $C_9 \ldots C_{10'}$ distance is only 4.78 Å.

Thus, the final relaxed structure of the fault can be represented by two half-crystals that have been shifted relative to one another by (1/2)[010] + (2/10)[$\bar{1}$00] = (1/10)[$\bar{2}$50] along the (100) plane. In addition, the molecules at the fault are folded by 7° across the $C_9 - C_{10}$ axis.

Although the calculations by *Ramdas* et al. [7.48] are not flawless (e.g., they neglect rigid-molecule rotations and molecular movements in the [001] direction), they provide a plausible working model to explain the photoreactivity of the DMA crystal.

Calculations of the stacking-fault energies that include the relaxations of molecules in the vicinity of a slip plane have also been performed for anthracene itself [7.4,5,50]. The aim was to find slip systems which might bring neighboring molecules into orientations favorable for photodimerization. However, it has recently been shown [7.51–53] that most of the dislocations in the anthracene crystal do not function as sites of preferred photoreactivity, that instead the crystal owes its photoreactivity to volume defects [7.54–56].

7.4 Volume Defects

Volume defects are small inclusions of a metastable phase within a thermodynamically stable crystal. Defects of this kind have been found to exist in anthracene crystals subjected to mechanical stress perpendicular to the basal (001) plane [7.55]. Microelectron diffraction experiments have shown that the crystallites of the new phase (phase II) are in coherent contact with the parent monoclinic phase (phase I). The topotactic relationships between the parent and daughter phases are: $(010)_{II}\|(011)_I$, $(100)_{II}\|(101)_I$, and $(001)_{II}\|(001)_I$. The anthracene II crystallites proved to play an important role in photochemical dimerization, acting as nuclei for the photodimerization process [7.53,55,56].

Unfortunately, the electron diffraction data of the phase II crystallites were too meager to define a structure using standard crystallographic methods. Thus, it remained to be explained why the new phase did facilitate photodimerization.

An attempt to determine the detailed structure of anthracene II has been made by *Gramaccioli* et al. [7.54] based on potential-energy and lattice-dynamical calculations. Using the experimentally established topotactic relationships and assuming a perfect topotactic fit at the interface between the above planes, the unit-cell parameters of anthracene II were evaluated to be: $a = 8.56$, $b = 6.04$, $c = 11.22$ Å, $\alpha = 122.6$, $\beta = 101.2$ and $\gamma = 90.0°$. Clearly, these values can only be used as preliminary rough estimates of the actual unit-cell parameters in view of a possible mismatch between the lattices of the parent and daughter phases.

The search for the optimum crystal structure of anthracene II was carried out in a local minimization of the potential energy, starting from a selected set of unit-cell parameters, molecular orientations and translations. The crystal was initially assumed to possess the lowest possible symmetry ($P1$), with two symmetrically distinct molecules in the unit cell. The center of one molecule was kept fixed at the origin of the unit cell, while the center of the other was left free to move.

After the energy minimization, some starting points converged to the anthracene I structure or gave rise to dynamically unstable structures (i.e., with imaginary phonon frequencies). The other starting points converged to a structure which was only 0.5 kcal/mol less stable than anthracene I and possessed the following unit-cell parameters: $a = 8.34$, $b = 5.89$, $c = 11.28$ Å, $\alpha = 123.3$, $\beta = 96.7$ and $\gamma = 85.9°$. In this structure, identified as the structure of anthracene II, the center of the molecule which was allowed translational displacements shifted to (1/2, 1/2, 0) of the unit cell, thereby raising the crystal symmetry to $P\bar{1}$.

The reasonableness of the derived structure was further examined in thermodynamic calculations. These were performed using the quasiharmonic approximation. The equilibrium unit-cell parameters were determined

Table 7.12. Calculated unit-cell parameters and free energies of anthracene I and II at 95 and 290 K [Å, deg, kcal/mol] [7.54]

T[K]	a	b	c	α	β	γ	V_c	-F
Anthracene I								
95	8.228	5.994	11.117	90.0	124.4	90.0	452.3	24.98
290	8.445	6.014	11.314	90.0	124.4	90.0	474.1	29.28
Anthracene II								
95	8.340	5.888	11.281	123.3	96.7	85.9	459.4	24.51
290	8.350	5.995	11.479	123.3	96.7	85.9	476.7	28.78

by minimizing the free energy. The results are presented in Table 7.12. From this table one can see that anthracene I is ~0.5 kcal/mol more stable than anthracene II both at 95 and 290 K. This difference in free energy coincides with that found for the static potential energy, which suggests that the entropies of the two phases are approximately the same. It can, therefore, be concluded that temperature alone cannot bring about the bulk transformation from I to II.

A similar conclusion can be drawn with respect to the pressure. Since the volume of phase I is smaller than that of phase II, the thermodynamic potential of phase I should be lower than that of phase II and, hence, phase I should remain more stable at elevated pressures.

Thus, the thermodynamic calculations have shown that neither the temperature nor the pressure can stabilize phase II in the bulk, at least not at room temperature or below. Based on this result, *Gramaccioli* et al. [7.54] concluded that the actual stability of phase II is due to the interface interactions of the small islands of this phase with the surrounding matrix of anthracene I. This is a plausible explanation but not the only one. The fact is that the comparison of the thermodynamic stabilities of phases I and II was made for hydrostatic pressure conditions, not for the actual conditions of anisotropic stress. For this reason, it is not unlikely that the calculations would indeed be capable of predicting the I to II transition if the appropriate stress conditions were applied.

It is essential that there are molecular pairs in the suggested structure for anthracene II whose configurations satisfy the requirements of topochemical preformation theory. This throws some fresh light on the mechanism of the solid-state photodimerization of anthracene. Since phase II is even realized with small compressive forces, it is reasonable to suppose that even a normal handling of anthracene will promote the formation of small islands of phase II within the bulk of the parent crystal. Upon UV irradiation, these islands will convert to crystallites of dianthracene. This conversion, in turn, should create mechanical stress at the interface boundaries, thereby favoring a further transformation of I to II and a further development of the dimerization process.

References

Chapter 1

1.1 A.I. Kitaigorodsky: Organic Chemical Crystallography (Consultants Bureau, New York 1961)
1.2 B.K. Vainshtein: Modern Crystallography I, Springer Ser. Solid-State Sci., Vol. 15 (Springer, Berlin, Heidelberg 1981)
1.3 A.I. Kitaigorodsky: Molecular Crystals and Molecules (Academic, New York 1973)
1.4 A.I. Kitaigorodski: Molekülkristalle (Akademie Verlag, Berlin 1979)
1.5 A.I. Kitaigorodsky: Mixed Crystals, Springer Ser. Solid-State Sci., Vol. 33 (Springer, Berlin, Heidelberg 1984)
1.6. A.I. Kitaigorodsky: Izv. Akad. Nauk SSSR 15, 157 (1951)
1.7 A.I. Kitaigorodsky: Chem. Soc. Rev. 7, 133 (1978)
1.8 E. Clementi: Liquid Water Structure, Lecture Notes Chem., Vol. 2 (Springer, Berlin, Heidelberg 1976)
1.9 E. Clementi, G. Corongiu, B. Jonsson, S. Romano: J. Chem. Phys. 72, 260 (1980)
1.10 G. Corongiu, E. Clementi: J. Chem. Phys. 72, 3979 (1980)
1.11 S. Romano, E. Clementi: Int. J. Quantum Chem. 17, 1007 (1980)
1.12 G.C. Lie, E. Clementi: J. Chem. Phys. 64, 5308 (1976)
1.13 R.M. Metzger: Cohesion and Ionicity in Organic Semiconductors and Metals, in Crystal Cohesion and Conformational Energies, ed. by R.M. Metzger, Topics Current Phys., Vol. 26 (Springer, Berlin, Heidelberg 1982) Chap.4
1.14 E.L. Eliel, N.L. Allinger, S.L. Angyal, G.A. Morrison: Conformational Analysis (Wiley, New York 1965)
1.15 F.A. Momany: Conformational Analysis and Polypeptide Drug Design, in Crystal Cohesion and Conformational Energies, ed.by R.M. Metzger, Topics Current Phys., Vol. 26 (Springer, Berlin, Heidelberg 1982) Chap.3
1.16 V.G. Dashevsky: Konformatsionnyi analiz organicheskikh molekul (Khimia, Moscow 1982)
U. Burkert, N.L. Allinger: Molecular Mechanics (American Chemical Society, Washington, DC 1982)
1.17 A. Warshel, E. Huler: Chem. Phys. 6, 463 (1974)
1.18 A. Warshel, Z. Shakked: J. Am. Chem. Soc. 97, 5679 (1975)
1.19 M.D. Cohen, R. Haberkorn, E. Huler, Z. Ludmer, M.E. Michel-Beyerle, D. Rabinovich, R. Sharon, A. Warshel, V. Yakhot: Chem. Phys. 27, 211 (1978)
1.20 D.P. Craig, C.P. Mallett: Chem. Phys. 65, 129 (1982)

Chapter 2

2.1 H. Margenau, N.R. Kestner: Theory of Intermolecular Forces, 2nd ed. (Pergamon, Oxford 1971)
2.2 P. Hobza, R. Zahradnik: Weak Intermolecular Interactions in Chemistry and Biology (Elsevier, Amsterdam 1980)

2.3 T. Kihara: Intermolecular Forces (Wiley, Chichester 1978)
2.4 B. Pullman (ed.): Intermolecular Interactions: from Diatomics to Biopolymers (Wiley, New York 1978)
2.5 H. Ratajczak, W.J. Orville-Thomas (eds.): Molecular Interactions, Vol. 1 (Wiley, Chichester 1980)
2.6 W. Kolos, L. Wolniewicz: J. Chem. Phys. 43, 2429 (1965)
2.7 T.H. Dunning: J. Chem. Phys. 53, 2823 (1970)
2.8 M. Urban, P. Hobza: Theor. Chim. Acta 36, 215 (1975)
2.9a A. Johansson, P. Kollman, S. Rothenberg: Theor. Chim. Acta 29, 167 (1973)
b A.J. Sadlej: Theor. Chim. Acta 61, 1 (1982)
2.10 B. Jeziorski, M. Van Hemmert: Molec. Phys. 31, 713 (1976)
2.11 G.F.M. Diercksen, W.P. Kraemer, B.O. Roos: Theor. Chim. Acta 36, 249 (1975)
2.12 A. Meukier, B. Levy, G. Bertier: Theor. Chim. Acta 29, 49 (1973)
2.13 J.P. Daudey, J.P. Malrieu, O. Rojas: Int. J. Quantum Chem. 8, 17 (1974)
2.14 O. Matsuoka, E. Clementi, M. Yoshimine: J. Chem. Phys. 64, 1351 (1976)
2.15 G.C. Lie, E. Clementi, M. Yoshimine: J. Chem. Phys. 64, 2314 (1976)
2.16 K. Morokuma: J. Chem. Phys. 55, 1236 (1971)
2.17 K. Kitaura, K. Morokuma: Int. J. Quantum Chem. 10, 325 (1976)
2.18 P. Schuster: The Fine Structure of the Hydrogen Bond, in Ref. 2.2, Chap.4
2.19 L.A. Curtiss: Int. J., Quantum Chem., Quantum Chem. Symp. 11, 459 (1977)
2.20 S. Yamabe, K. Kitaura, K. Nishimoto: Theor. Chim. Acta 47, 111 (1978)
2.21 P.H. Smit, J.L. Derissen, F.B. Van Duijneveldt: Molec. Phys. 37, 501 (1979)
2.22 E. Clementi, F. Cavallone, R. Scordamaglia: J. Am. Chem. Soc. 99, 5531 (1977)
2.23 R. Scordamaglia, F. Cavallone, E. Clementi: J. Am. Chem. Soc. 99, 5545 (1977)
2.24 G.Bolis, E. Clementi: J. Am. Chem. Soc. 99, 5550 (1977)
2.25 B. Pullman, A. Pullman: Theor. Chim. Acta 50, 317 (1979)
2.26 A. Pullman, D. Perahia: Theor. Chim. Acta 48, 29 (1978)
2.27 M. Dreyfus, A. Pullman: Theor. Chim. Acta 19, 20 (1970)
2.28 B.D. Silverman: Slipped Versus Eclipsed Stacking of Tetrathiafulvalene (TTF) and Tetracyanoquinodimethane (TCNQ) Dimers, in Crystal Cohesion and Conformational Energies, ed. by R.M. Metzger, Topics Current Phys., Vol. 26 (Springer, Berlin, Heidelberg 1982) Chap.5
2.29 W.A. Lathau, K. Morokuma: J. Am. Chem. Soc. 97, 3615 (1975)
2.30 W.A. Lathau, G.R. Pack, K. Morokuma: J. Am. Chem. Soc. 97, 6624 (1975)
2.31 H. Umeyama, K. Morokuma: J. Am. Chem. Soc. 98, 7208 (1976)
2.32 R. Osman, H. Weinstein: Chem. Phys. Lett. 49, 69 (1977)
2.33 H. Weinstein, R. Osman, W.D. Edwards, J.P. Green: Int. J. Quantum Chem., Quantum Biol. Symp. 5, 449 (1978)
2.34 R. Osman, S. Topiol, H. Weinstein, J.E. Eilers: Chem. Phys. Lett. 73, 399 (1980)
2.35 W. Kolos, G. Ranghino, E. Clementi, O. Novaro: Int. J. Quantum Chem. 17, 429 (1980)
2.36 T. Aoyamo, O. Matsuoka, N. Nagakawa: Chem. Phys. Lett. 67, 508 (1979)
2.37a H. Umeyama, K. Morokuma: J. Am. Chem. Soc. 99, 1316 (1977)
b E. Kochanski: Theochem 16, 49 (1984)
2.38 W. Kolos: Theor. Chim. Acta 54, 187 (1980)
2.39 F.B. Van Duijneveldt: IBM Tech. Rpt. RJ945 (1971)
2.40 A. Dalgarno: Adv. Chem. Phys. 12, 143 (1967)

2.41 P.H. Smit, J.L. Derissen, F.B. Van Duijneveldt: J. Chem. Phys. 67, 274 (1977)
2.42 P.H. Smit, J.L. Derissen, F.B. Van Duijneveldt: J. Chem. Phys. 69, 4241 (1978)
2.43 International Tables for X-ray Crystallography, Vol. 1 (Kynoch, Birmingham 1974)
2.44 J.L. Derissen: J. Molec. Struct. 38, 177 (1977)
2.45 E. Clementi, W. Kolos, G.C. Lie, G. Ranghino: Int. J. Quantum Chem. 17, 377 (1980)
2.46 R. McWeeny, B.T. Sutcliffe: Methods of Molecular Quantum Mechanics (Academic, London 1969)
2.47 I.G. Kaplan: Simmetriya mnogoelektronnykh sistem (Nauka, Moscow 1969)
2.48 P.O. Lowdin: Phys. Rev. 97, 1474 (1955)
2.49 P.O. Lowdin: Adv. Quantum Chem. 5, 185 (1970)
2.50 W. Moffitt: Proc. Roy. Soc. London A218, 486 (1953)
2.51 I.G. Kaplan, O.B. Rodimova: Int. J. Quantum Chem. 7, 1203 (1973)
2.52 P.E.S. Wormer, A. Van der Avoird: J. Chem. Phys. 62, 3326 (1975)
2.53 P.E.S. Wormer, T. Van Berkel, A. Van der Avoird: Molec. Phys. 29, 1181 (1975)
2.54 P.E. Phillipson: Phys. Rev. 125, 1981 (1962)
2.55 F.A. Matseu: Adv. Quantum Chem. 1, 60 (1964)
2.56 C.M. Reeves: Commun. Assoc. Comput. Mach. 9, 276 (1966)
2.57 L.W. Flynn, G. Thodos: Am. Inst. Chem. Eng. J. 8, 362 (1962)
2.58 J.O. Hirschfelder, C.F. Curtiss, R.B. Bird: Molecular Theory of Gases and Liquids (Wiley, New York 1954)
2.59 T. Wasiutynski, A. Van der Avoird, R.M. Berns: J. Chem. Phys. 69, 5288 (1978)
2.60 F. Mulder, M. Van Hemmert, P.E.S. Wormer, A. Van der Avoird: Theor. Chim. Acta 46, 39 (1977)
2.61 P.E.S. Wormer, F. Mulder, A. Van der Avoird: Int. J. Quantum Chem. 11, 959 (1977)
2.62 P. Claverie: "Elaboration of Approximate Formulas for the Interactions between Large Molecules: Applications in Organic Chemistry", in Ref. 2.2, Chap.2
2.63 I.G. Kaplan, O.B. Rodimova: Usp. Fiz. Nauk 126, 403 (1978)
2.64 M.V. Basilevsky, M.M. Berenfeld: Int. J. Quantum Chem. 6, 23 (1972)
2.65a J.P. Daudey, P. Claverie, J.P. Malrieu: Int. J. Quantum Chem. 8, 1 (1974)
b I.C. Hayes, A.J. Stone: Mol. Phys. 53, 83 (1984)
c I.C. Hayes, G.J.B. Hurst, A.J. Stone: Mol. Phys. 53, 107 (1984)
2.66 D.M. Chipman, J.D. Bowman, J.O. Hirschfelder: J. Chem. Phys. 59, 2830 (1973)
2.67 B. Jeziorski, W. Kolos: Int. J. Quantum Chem. 12, Suppl. 1, 91 (1977)
2.68 A.T. Amos: Chem. Phys. Lett. 5, 587 (1970)
2.69 J.N. Murrell, G. Shaw: J. Chem. Phys. 46, 1768 (1967)
2.70 J.I. Musher, A.T. Amos: Phys. Rev. 164, 31 (1967)
2.71 W. Kutzelnigg: J. Chem. Phys. 73, 343 (1980)
2.72 P.R. Certain, J.O. Hirschfelder, W. Kolos, L. Wolniewicz: J. Chem. Phys. 49, 24 (1962)
2.73 I.K. Snook, T.H. Spurling: J. Chem. Soc., Faraday Trans. II 71, 852 (1975)
2.74 A.T. Amos, R.J. Crispin: Calculations of Intermolecular Interaction Energies, in Theoretical Chemistry, Advances and Perspectives, Vol.2, ed. by H. Eyring, D. Henderson (Academic, New York 1976) pp. 1-66
2.75 L. Jansen: Phys. Rev. 110, 661 (1958)

2.76 J.O. Hirschfelder (ed.): Intermolecular Forces, in Advances in Chemical Physics, Vol.12 (Wiley, New York 1967)
2.77 F. Mulder, C. Huiszoon: Molec. Phys. 34, 1215 (1977)
2.78 A. Van der Avoird, P.E.S. Wormer, F. Mulder, R.M. Berns: Ab Initio Studies of the Interactions in Van der Waals Molecules, in Van der Waals Systems, ed. by R. Zahradnik, Topics Current Chem., Vol.93 (Springer, Berlin, Heidelberg 1980) pp. 1-51
2.79 F. Mulder, G. Van Dijk, C. Huiszoon: Molec. Phys. 38, 577 (1979)
2.80 O. Chalvet, R. Daudel, S. Diner, J.P. Malrieu (eds.): Localization and Delocalization in Quantum Chemistry, Vol.1 (Reidel, Dordrecht 1975)
2.81 S.F. Boys: in Quantum Theory of Atoms, Molecules, and the Solid State, ed. by P.-O. Lowdin (Academic, New York 1966) p. 253
2.82 C. Edmiston, K. Ruedenberg: Rev. Mod. Phys. 35, 457 (1963)
2.83 R. Bonaccorsi, R. Cimiraglia, E. Serocco, J. Tomasi: Theor. Chim. Acta 33, 97 (1971)
2.84 P.H. Smit, J.L. Derissen, F.B. Van Duijneveldt: Molec. Phys. 37, 521 (1979)
2.85 G.G. Hall: Chem. Phys. Lett. 20, 501 (1973)
2.86 A.T. Amos, J.A. Yoffe: Chem. Phys. Lett. 39, 53 (1975)
2.87 A.T. Amos, J.A. Yoffe: Theor. Chim. Acta 42, 247 (1976)
2.88 A. Dalgarno, I.M. Morrison, R.M. Pengelly: Int. J. Quantum Chem. 1, 161 (1967)
2.89 A. Dalgarno: Adv. Phys. 11, 281 (1962)
2.90 A.T. Amos, R.J. Crispin: Molec. Phys. 31, 159 (1976)
2.91 R.J.W. LeFevre: Adv. Phys. Org. Chem. 3, 1 (1965)
2.92 L. Salem: Proc. Roy. Soc. London A264, 379 (1961)

Chapter 3

3.1a J.A. Pople, D.L. Beveridge: Approximate Molecular Orbital Theory (McGraw-Hill, New York 1970)
b H. Ratajczak, W.J. Orville-Thomas (eds.): Molecular Interactions, Vol.1 (Wiley, Chichester 1980)
3.2 R.L. Ellis: J. Chem. Phys. 64, 342 (1976)
3.3a D.P. Santry: in Physics and Chemistry of Ice, ed. by E. Whaley, S.J. Stones, L.W. Gold (Royal Society of Canada, Ottawa 1973)
b K.M. Middlemiss, D.P. Santry: Chem. Phys. Lett. 1, 128 (1973)
c R.W. Crowe, D.P. Santry: Chem. Phys. 2, 304 (1973)
d J.Bacon, D.P. Santry: J. Chem. Phys. 56, 2011 (1972)
3.4 V.A. Zubkov: Theor. Chim. Acta 66, 295 (1984)
3.5 V.M. Promyslov, I.A. Misurkin: Teor. Eksp. Khim. 20, 649 (1984)
3.6 O. Chalvet, R. Daudel, S. Diner, J.P. Malrieu (eds.): Localization and Delocalization in Quantum Chemistry, Vol. 1 (Reidel, Dordrecht 1975)
3.7 R.F.W. Bader, P.M. Beddall: J. Chem. Phys. 56, 3320 (1972)
3.8 R.F.W. Bader, P.M. Beddall, J. Peslak, Jr.: J. Chem. Phys. 58, 557 (1972)
3.9 R.F.W. Bader, P.M. Beddall: J. Am. Chem. Soc. 95, 305 (1973)
3.10 R.F.W. Bader, G.R. Runtz: Molec. Phys. 30, 117 (1975)
3.11 S. Srebrenik, R.F.W. Bader, T.T. Nguyen-Dang: J. Chem. Phys. 68, 3667 (1978)
3.12 R.F.W. Bader, S.G. Anderson, A.J. Duke: J. Am. Chem. Soc. 101, 1389 (1979)
3.13 E. Clementi, G. Corongiu, G. Ranghino: J. Chem. Phys. 74, 578 (1981)
3.14 F. Mulder, G. Van Dijk, C. Huiszoon: Molec. Phys. 38, 577 (1979)
3.15 A. Warshel, S. Lifson: J. Chem. Phys. 53, 582 (1970)

3.16 B.M. Powell, G. Dolling, H. Bonadeo: J. Chem. Phys. 69, 2428 (1978)
3.17 G.S. Pawley, S.J. Cyvin: J. Chem. Phys. 52, 4073 (1970)
3.18 G. Dolling, G.S. Pawley, B.M. Powell: Proc. Roy. Soc. London A333, 363 (1973)
3.19 A. Warshel: J. Chem. Phys. 54, 5324 (1971)
3.20a A.I. Kitaigorodsky, K.V. Mirskaya: Mater. Res. Bull. 7, 1271 (1972)
b A.I. Kitaigorodsky, K.V. Mirskaya: Kristallografia 6, 507 (1961)
c A.I. Kitaigorodsky: Molecular Crystals and Molecules (Academic, New York, 1973)
3.21 K.V. Mirskaya: Kristallografia 17, 67 (1972)
3.22 E. Giglio: Z. Kristallogr. 131, 385 (1970)
3.23 A. Di Nola, E. Giglio: Acta Crystallogr. A26, 144 (1970)
3.24 A.M. Liquori, E. Giglio, L. Mazzarella: Nuovo Cimento 55B, 476 (1968)
3.25 G. Capaccio, P. Giacomello, E. Giglio: Acta Crystallogr. A27, 229 (1971)
3.26 C. Dosi, E. Giglio, V. Pavel, C. Quagliata: Acta Crystallogr. A29, 644 (1973)
3.27 P. Giacomello, E. Giglio: Acta Crystallogr. A26, 324 (1970)
3.28 G. Taddei, E. Giglio: J. Chem. Phys. 53, 2768 (1970)
3.29 S. Lifson, A. Warshel: J. Chem. Phys. 49, 5116 (1968)
3.30 D.E. Williams: J. Chem. Phys. 45, 3770 (1966)
3.31 H. Bonadeo, E. D'Alessio: Chem. Phys. Lett. 19, 177 (1973)
3.32 H. Bonadeo, E. D'Alessio: The Refinement of Intermolecular Potentials Using Statical and Dynamical Properties of Molecular Crystals. Application to Chlorinated Benzenes,in Lattice Dynamics and Intermolecular Forces ed. by L.V. Corso (Academic, New York 1975) pp.136-148
3.33 Z. Gamba, H. Bonadeo: Chem. Phys. Lett. 69, 525 (1980)
3.34 E. Burgos, H. Bonadeo: Chem. Phys. Lett. 49, 475 (1977)
3.35 A.T. Hagler, S. Lifson: Acta Crystallogr. B30, 1336 (1974)
3.36 Leh-Yeh Hsu, D.E. Williams: Acta Crystallogr. A36, 277 (1980)
3.37 A.I. Kitaigorodsky: Dokl. Akad. Nauk SSSR 137, 116 (1961)
3.38 A.I. Kitaigorodsky: Tetrahedron 14, 230 (1961)
3.39 A.D. Crowell: J. Chem. Phys. 29, 446 (1958)
3.40 A.I. Kitaigorodsky, K.V. Mirskaya, A.B. Tovbis: Kristallografia 13, 225 (1968)
3.41 K.V. Mirskaya, I.E. Kozlova: Kristallografia 14, 412 (1969)
3.42 A.I. Kitaigorodsky: J. Chim. Phys. 1, 9 (1966)
3.43 V.F. Bereznitskaya, A.I. Kitaigorordsky, V.M. Kozhin, I.E. Kozlova, K.V, Mirskaya: Zh. Fiz. Khim. 46, 2492 (1972)
3.44 P. Weulersse: Compt. Rend. Acad. Sci. Paris B264, 327 (1967)
3.45 A.I. Kitaigorodsky, E.I. Mukhtarov: Kristallografia 14, 784 (1969)
3.46 A.I. Kitaigorodsky, K.V. Mirskaya, V.F. Bereznitskaya: Kristallografia 15, 405 (1970)
3.47 K.V. Mirskaya, I.E. Kozlova, V.F. Bereznitskaya: Phys. Status Solidi (b) 62, 291 (1974)
3.48 G.A. Mackenzie, G.S. Pawley, O.W. Dietrich: J. Phys. C 10, 3723 (1977)
3.49 D.E. Williams: J. Chem. Phys. 47, 4680 (1967)
3.50 D.E. Williams, T.L. Starr: Computers and Chem. 1, 173 (1977)
3.51 D.E. Williams: Acta Crystallogr. A30, 71 (1976)
3.52 G. Taddei, H. Bonadeo, M.P. Marzocchi, S. Califano: J. Chem. Phys. 58, 966 (1973)
3.53 V.V. Nauchitel', K.V. Mirskaya: Kristallografia 16, 1025 (1971) /English transl.: Sov. Phys. - Crystallogr. 16, 891 (1972)/

3.54 R.P. Rinaldi, G.S. Pawley: Nuovo Cimento B16, 55 (1973)
3.55 R.P. Rinaldi, G.S. Pawley: J. Phys. C 8. 599 (1975)
3.56 H.A.J. Govers: Acta Crystallogr. A35, 236 (1979)
3.57 A.I. Kitaigorodsky, K.V. Mirskaya, V.V. Nauchitel': Kristallografia 14, 900 (1969)
3.58 K.V. Mirskaya, V.V. Nauchitel': Sov. Phys. - Crystallogr. 17, 56 (1972)
3.59 H.A.J. Govers: Acta Crystallogr. A31, 380 (1975)
3.60 H.A.J. Govers: Calculation of Lattice Energies of Unary and Binary Molecular Crystals by the Atom-Atom Approximation; Thesis, University of Utrecht (1974)
3.61 T. Luty, A. Van der Avoird, A. Mierzejewsky: Chem. Phys. Lett. 61, 10 (1979)
3.62 P.A. Reynolds: J. Chem. Phys. 59, 2777 (1973)
3.63 S. Besnainou, D.L. Cummings: J. Molec. Struct. 34, 131 (1976)
3.64 Z. Gamba, H. Bonadeo: J. Chem. Phys. 75, 5059 (1981)
3.65a S. Califano, R. Righini, S.M. Walmsley: Chem. Phys. Lett. 64, 491 (1979)
b F. Mulder, C. Huiszoon: Molec. Phys. 34, 1215 (1977)
c D.E. Williams, S.R. Cox: Acta Crystallogr. B40, 404 (1984)
d S.L. Price, A.J. Stone: Mol. Phys. 51, 569 (1984)
3.66 A.I. Kitaigorodsky, V.G. Dashevsky: Tetrahedron 24, 5917 (1962)
3.67 P.A. Reynolds, J.K. Kjems, J.W. White: J. Chem. Phys. 60, 824 (1974)
3.68 J.B. Bates, W.R. Busing: J. Chem. Phys. 60, 2414 (1974)
3.69 K. Mirsky, M.D. Cohen: Chem. Phys. 28, 193 (1978)
3.70 K. Mirsky, M.D. Cohen: Acta Crystallogr. A34, 346 (1978)
3.71 G.L. Wheeler, S.D. Colson: J. Chem. Phys. 65, 1227 (1976)
3.72 W.R. Busing: Trans. Am. Crystallogr. Assoc. 6, 57 (1970)
3.73 Leh-Yeh Hsu, D.E. Williams: Inorg. Chem. 18, 79 (1979)
3.74 S. Lifson, A.T. Hagler, P. Dauber: J. Am. Chem. Soc. 101, 5111 (1979)
3.75 A.T. Hagler, E. Huler, S. Lifson: J. Am. Chem. Soc. 96, 5319 (1974)
3.76 T. Ooi, R.A. Scott, G. Vanderkooi, H.A. Scheraga: J. Chem. Phys. 46, 4410 (1967)
3.77 R. Schroeder, E.R. Lippincott: J. Am. Chem. Soc. 61, 921 (1957)
3.78a A.M. Liquori: Quart. Rev. Biophys. 2, 65 (1969)
b Z. Berkovitch-Yellin, L. Leizerowitz: J. Am. Chem. Soc. 104, 4052 (1982)
3.79 J.L. Derissen, P.H. Smit: Acta Crystallogr. A34, 842 (1978)
3.80 M. Simonetta: Conformation of Constituents in Molecular Crystals, in Electronic Structure of Polymers and Molecular Crystals, ed. by L. Andre, J. Ladik, J. Delhalle (Plenum, New York 1975) pp. 547-599
V. Burkert, N.L. Allinger: Molecular Mechanics (American Chemical Society, Washington, DC 1982)
E.L. Eliel, N.L. Allinger, S.L. Angyal, G.A. Morrison: Conformational Analysis (Wiley, New York 1965)
F.A. Momany: Conformational Analysis and Polypeptide Drug Design, in Crystal Cohesion and Conformational Energies, ed. by R. M. Metzger, Topics Current Phys., Vol.26 (Springer, Berlin, Heidelberg 1982) Chap.3
3.81 T.L. Hill: J. Chem. Phys. 14, 465 (1946)
3.82 T.L. Hill: J. Chem. Phys. 16, 938 (1948)
3.83 F.H. Westheimer, J.E. Mayer: J. Chem. Phys. 14, 733 (1946)
3.84 F.H. Westheimer, J. Chem. Phys. 15, 252 (1947)
3.85 A.I. Kitaigorodsky: Izv. Akad. Nauk SSSR, Ser. Fiz. 15, 157 (1951)

3.86 A.I. Kitaigorodsky: Dokl. Akad. Nauk SSSR 124, 1967 (1959)
3.87 N.L. Allinger, J.T. Sprague: J. Am. Chem. Soc. 95, 3893 (1973)
3.88 N.L. Allinger, M.T. Tribble, M.A. Miller, D.H. Werts: J. Am. Chem. Soc. 93, 1637 (1971)
3.89 N.L. Allinger, M.T. Tribble, M.A. Miller: Tetrahedron 28, 1173 (1972)
3.90 N.L. Allinger, J.T. Sprague: J. Am. Chem. Soc. 94, 5734 (1972)
3.91 A.I. Kitaigorodsky, V.G. Dashevsky: Tetrahedron 24, 5917 (1968)
3.92 A.I. Kitaigorodsky, V.G. Dashevsky: Teor. Eksp. Khim. 3, 35 (1967)
3.93 A. Warshel, M. Kaplus: J. Am. Chem. Soc. 94, 5612 (1972)
3.94 L. Salem: The Molecular Orbital Theory of Conjugated Systems (Benjamin, New York 1966)
3.95 R.A. Nemenoff, J. Snir, H.A. Scheraga: J. Phys. Chem. 82, 2513 (1978)
3.96 J. Snir, R.A. Nemenoff, H.A. Scheraga: J. Phys. Chem. 82, 2497 (1978)
3.97a R.A. Nemenoff, J. Snir, H.A. Scheraga: J. Phys. Chem. 82, 2504 (1978)
b R.A. Nemenoff, J. Snir, H.A. Scheraga: J. Phys. Chem. 82, 2521 (1978)
c J. Snir, R.A. Nemenoff, H.A. Scheraga: J. Phys. Chem. 82, 2527 (1978)
3.98 F.T. Marchese, P.H. Mehrotra, D.L. Beveridge: J. Phys. Chem. 86, 2592 (1982)
3.99 T. Wasiutynski, A. Van der Avoird, R.M. Berns: J. Chem. Phys. 69, 5288 (1978)
3.100 D. Hall, D.E. Williams: Acta Crystallogr. A31, 56 (1975)
3.101 P.H. Smit, J.L. Derissen, F.B. Van Duijneveldt: Molec. Phys. 37, 501 (1979)
3.102 C. Huiszoon, F. Mulder: Molec. Phys. 38, 1497 (1979)
C. Huiszoon, F. Mulder: Molec. Phys. 40, 249 (1980)
3.103 F.A. Momany, L.A. Carruthers, R.F. McGuire, H.A. Scheraga: J. Phys. Chem. 78, 1595 (1974)
3.104 G.E.W. Bauer, C. Huiszoon: Molec. Phys. 47, 565 (1982)
3.105a J. Van der Linden, F.B. Van Duijneveldt: unpublished results quoted in /3.104/
b G. Karlström, P. Linse, A. Wallquist, B. Jönsson: J. Am. Chem. Soc. 105, 3777 (1983)
c K. Kitaura, K. Morokuma: Int. J. Quantum Chem. 10, 325 (1976)
3.106a E. Clementi: Determination of Liquid Water Structure, Coordination Numbers for Ions and Solvation for Biological Molecules, Lecture Notes Chem., Vol.2 (Springer, Berlin, Heidelberg 1976)
b O. Matsuoka, E. Clementi, M. Yoshimine: J. Chem. Phys. 64, 1351 (1976)
3.107a H. Kistenmacher, H. Popkie, E. Clementi: J. Chem. Phys. 60, 4455 (1974)
b H. Kistenmacher, G.C. Lie, H. Popkie, E. Clementi: J. Chem. Phys. 61, 546 (1974)
3.108 G.C. Lie, E. Clementi: J. Chem. Phys. 62, 2195 (1975)
3.109 H. Popkie, E. Clementi: J. Chem. Phys. 57, 1077 (1972)
3.110 E. Clementi, G. Corongiu, B. Jonsson, S. Romano: J. Chem. Phys. 72, 260 (1980)
3.111 H. Kistenmacher, H. Popkie, E. Clementi: J. Chem. Phys. 58, 1689 and 5627 (1973)
3.112 G. Corongiu, E. Clementi: J. Chem. Phys. 69, 4885 (1978)
3.113 L. Carozzo, G. Corongiu, C. Petrongolo, E. Clementi: J. Chem. Phys. 68, 787 (1978)
3.114 M. Ragazzi, D. Ferro, E. Clementi: J. Chem. Phys. 70, 1040 (1979)

3.115a E. Clementi, F. Cavallone, R. Scordamaglia: J. Am. Chem. Soc. 99, 5531 (1977)
b R. Scordamaglia, F. Cavallone, E. Clementi: J. Am. Chem. Soc. 99, 5545 (1977)
c G. Bolis, E. Clementi: J. Am. Chem. Soc. 99, 5550 (1977)
3.116 G. Corongiu, E. Clementi: Gazz. Chim. Ital. 108, 687 (1978)
3.117 S. Swaminathan, D.L. Beveridge: J. Am. Chem. Soc. 99, 8392 (1977)
3.118 E. Clementi, H. Kistenmacher, W. Kolos, S. Romano: Theor. Chim. Acta 55, 257 (1980)
E. Clementi, G. Corongiu: Biopolymers 18, 2431 (1979)
3.119 F.F. Abraham: J. Chem. Phys. 61, 1221 (1974)
3.120 G. Corongiu, E. Clementi: J. Chem. Phys. 72, 3979 (1980)
3.121 S. Romano, E. Clementi: Int. J. Quantum Chem. 15, 849 (1978)
3.122a G. Ranghino, R. Scordamaglia, E. Clementi: Chem. Phys. Lett. 49, 218 (1977)
b K.S. Kim, E. Clementi: J. Am. Chem. Soc. 107, 227 (1985)
c E. Clementi, P. Habitz: J. Phys. Chem. 87, 2815 (1983)
d H.J. Böhm, R. Ahlrichs, P. Scharf, H. Schiffer: J. Chem. Phys. 81, 1389 (1984)
3.123 V.I. Poltev: A semiempirical Calculation of Intermolecular Interactions in the Double Helix of DNA; Thesis, Moscow (1968)
3.124 B.I. Sukhorukov, V.I. Poltev: Biofizika 9, 148 (1964)
3.125 S.S. Batsanov: Strukturnaya refraktometria (Moscow State University, Moscow 1959)
3.126 J. Hinze, H.H. Jaffe: J. Am. Chem. Soc. 84, 540 (1962)
3.127 G. Del Re: J. Chem. Soc. 1958, 4031 (1958)
3.128 H. Berthod, A. Pullman: J. Chim. Phys. 62, 942 (1965)
3.129 R.A. Scott, H.A. Scheraga: J. Chem. Phys. 42, 2209 (1965)
3.130 K.S. Pitzer: Adv. Chem. Phys. 2, 59 (1959)
3.131 A. Dalgarno: Adv. Phys. 11, 281 (1962)
3.132 R.F. McGuire, F.A. Momany, H.A. Scheraga: J. Phys. Chem. 76, 375 (1972)
3.133a J. Caillet, P. Claverie: Acta Crystallogr. A31, 448 (1975)
b J. Caillet, P. Claverie, B. Pullman: Acta Crystallogr. B32, 2740 (1976)
3.134 R.J.W. LeFevre: Adv. Phys. Org. Chem. 3, 1 (1965)
3.135 J. Caillet, P. Claverie: Biopolymers 13, 601 (1974)
3.136 A. Bondi: J. Phys. Chem. 68, 441 (1964)
3.137 J. Caillet, P. Claverie, B. Pullman: Acta Crystallogr. B34, 3266 (1978)
3.138 Y.S. Kim, R.G. Gordon: J. Chem. Phys. 61, 1 (1974)

Chapter 4

4.1 M. Born, K. Huang: Dynamical Theory of Crystal Lattices (Clarendon, Oxford 1954)
4.2 V. Chandrasekharan, S.H. Walmsley: Molec. Phys. 33, 573 (1977)
4.3 F.L. Hirshfeld: Acta Crystallogr. A24, 301 (1968)
4.4 I.M. Sobol': Mnogomernye kvadraturnye formuly i funktsii Haara (Nauka, Moscow 1969)
4.5 I.M. Sobol', R.B. Statnikov: " ΛΠ-poisk i zadachi optimal'nogo konstruirovaniya", In Problemy sluchainogo poiska, Vol. 1, ed. by L.A. Rastrigin (Zinatne, Riga 1972) pp. 117-135
4.6 A.J. Pertsin, A.I. Kitaigorodsky: Kristallografia 21, 587 (1976)

4.7 I.M. Gelfand, E.B. Vul, S.L. Ginzburg, Yu. G. Fedorov: Metod ovragov v zadachakh strukturnogo analiza (Nauka, Moscow 1966)
4.8 D.M. Himmelblau: Applied Nonlinear Programming (McGraw-Hill, New York 1972)
4.9 W.I. Zangwill: Nonlinear Programming: A Unified Approach (Prentice-Hall, Englewood Clifts, New York 1969)
4.10 O.L. Mangasarian: Nonlinear Programming (McGraw-Hill, New York 1969)
4.11 D.J. Wilde, C.S. Beightler: Foundations of Optimization (Prentice-Hall, Englewood Clifts, New York 1969)
4.12 G.W. Stewart: J. Assoc. Comput. Mach. 14, 72 (1967)
4.13 M.J.D. Powell: Comput. J. 7, 155 (1964)
4.14 H.H. Rosenbrock: Comput. J. 3, 175 (1960)
4.15 P.P. Ewald: Ann. Phys. (Leipzig) 64, 253 (1921)
4.16 F. Bertaut: J. Phys. Radium 13, 499 (1952)
4.17 D.E. Williams: Acta Crystallogr. A27, 452 (1971)
4.18 B.R.A. Nijboer, F.W. De Wette: Physica 23, 309 (1957)
4.19 D.E. Williams: Acta Crystallogr. A28, 629 (1972)
4.20 G. Von Leibfried: Gittertheorie der mechanischen und thermischen Eigenschaften der Kristalle, Handbuch der Physik, Band 7, Teil 2 (Springer, Berlin, Göttingen, Heidelberg 1955)
4.21 C.M. Grammaccioli, G. Filippini: Acta Crystallogr. A35, 727 (1979)
4.22 D.M. Templeton: J. Chem. Phys. 23, 1629 (1954)
4.23 F.A. Momany, L.M. Carruthers, H.A. Scheraga: J. Phys. Chem. 78, 1621 (1974)
4.24 J.L. Derissen, P.H. Smit, J. Voogd: J. Phys. Chem. 81, 1474 (1977)
4.25 A.I. Kitaigorodsky: Chem. Soc. Rev. 7, 133 (1978)
4.26 T.V. Timofeeva, N.Yu. Chernikova, P.M. Zorkii: Usp. Khim. 49, 966 (1980)
4.27 S. Ramdas, J.M. Thomas: Chem. Phys. Solids and Their Surfaces 7, 31 (1978)
4.28 A. Dworkin, P. Figuiere, M. Ghelfenstein, M. Szwarc: J. Chem. Thermodynamics 8, 835 (1976)
4.29 G.L. Wheeler, S.D. Colson: J. Chem. Phys. 65, 1227 (1976)
4.30 K. Mirsky, M.D. Cohen: Acta Crystallogr. A34, 346 (1978)
4.31 S. Ramdas, W. Jones, J.M. Thomas: Chem. Phys. Lett. 54, 490 (1978)
4.32 J. Caillet, P. Claverie: Acta Crystallogr. B36, 2642 (1980)
4.33 J. Bernstein, A.T. Hagler: J. Am. Chem. Soc. 100, 673 (1978)
4.34 A.T. Hagler, J. Bernstein: J. Am. Chem. Soc. 100, 6349 (1978)
4.35 D.E. Williams: Acta Crystallogr. A30, 71 (1974)
4.36 A.T. Hagler, S. Lifson: Acta Crystallogr. B30, 1336 (1974)
4.37 A.T. Hagler, E. Huler, S. Lifson: J. Am. Chem. Soc. 96, 5319 (1974)
4.38 E. Giglio: Nature (London) 222, 339 (1969)
4.39 I. Bar, J. Bernstein: J. Phys. Chem. 86, 3223 (1982)
4.40 I. Bar, J. Bernstein: J. Phys. Chem. 88, 243 (1984)
4.41 A.I. Kitaigorodsky, A.J. Pertsin, I.E. Kozlova: Kristallografia 20, 1035 (1975)
4.42 A.I. Kitaigorodsky: Organic Chemical Crystallography (Consultants Bureau, New York 1961)
4.43 A.B. Dzyabchenko: Zh. Strukt. Khim. 25 (3), 85 (1984)
4.44 A.B. Dzyabchenko: Zh. Strukt. Khim. 25 (4), 57 (1984)
4.45 A.T. Hagler, L. Leiserowitz: J. Am. Chem. Soc. 100, 5879 (1978)
4.46 J.L. Derissen, P.H. Smit: Acta Crystallogr. A33, 230 (1977)
4.47 Z. Berkovitch-Yellin, L. Leizerowitz: J. Am. Chem. Soc. 104, 4052 (1982)

4.48 M. Simonetta: Conformation of Constituents in Molecular Crystals, in Electronic Structure of Polymers and Molecular Crystals, ed. by J. Andre, J. Ladik, J. Delhalle (Plenum, New York 1975) pp. 547-599
4.49 A. Warshel, S. Lifson: J. Chem. Phys. 53, 582 (1970)
4.50 F.A. Momany, L.A. Carruthers, R.F. McGuire, H.A. Scheraga: J. Phys. Chem. 78, 1595 (1974)
4.51 W.R. Busing, G.M. Brown: Chem. Div. Annu. Progr. Rep. August 31, 1971, ORNL-5485, pp. 139-140
4.52 W.R. Busing: Chem. Div. Annu. Progr. Rep. February 29, 1980, ORNL-5665, pp. 133-135
4.53 W.R. Busing, W.E. Thiessen: Chem. Div. Annu. Progr. Rep. May 20, 1972, ORNL-4791, pp. 128-129
4.54 R.F. McGuire, G. Vanderkooi, F.A. Momany, R.T. Ingwall, G.M. Crippen, N. Lotan, R.W. Tuttle, K.L. Kashuba, H.A. Scheraga: Macromolecules 4, 112 (1971)
4.55 F.A. Momany, R.F. McGuire, A.W. Burgess, H.A. Scheraga: J. Phys. Chem. 79, 2361 (1975)
4.56 L.G. Dunfield, A.W. Burgess, H.A. Scheraga: J. Phys. Chem. 82, 2609 (1978)
4.57 G. Casalone, C. Mariani, A. Mugnoli, M. Simonetta: Molec. Phys. 15, 339 (1968)
4.58 L.S. Bartell: J. Chem. Phys. 32, 827 (1960)
4.59 H. Berthod, A. Pullman: J. Chim. Phys. 62, 942 (1965)
4.60 G. Casalone, C. Mariani, A. Mugnoli, M. Simonetta: Acta Crystallogr. B25, 1741 (1969)
4.61 G. Casalone, A. Gavezzotti, M. Simonetta: J. Chem. Soc., Perkin Trans. II 4, 342 (1973)
4.62 G. Casalone, C. Mariani, A. Mugnoli, M. Simonetta: Acta Crystallogr. 22, 228 (1967)
4.63 A. Warshel, E. Huler, D. Rabinovich, Z. Shakked: J. Molec. Struct. 23, 175 (1974)
4.64 A. Warshel: The Consistent Force Field and Its Quantum Mechanical Extension, in Semiempirical Methods of Electronic Structure Calculation, Part A, ed. by G.A. Segal (Plenum, New York 1977) pp. 133-172
4.65 J. Caillet, P. Claverie, B. Pullman: Acta Crystallogr. B32, 2740 (1976)
4.66 B. Pullman, H. Berthod, P. Courriere: Int. J. Quantum Chem., Quantum Biol. Symp. 1, 93 (1974)
4.67 W. Jones, M.D. Cohen: Mol. Cryst. Liq. Cryst. 41, 103 (1978)
4.68 R.M. Hockstrasser, A. Malliaris: Mol. Cryst. Liq. Cryst. 11, 331 (1970)
4.69 E. Halac, E. Burgos, H. Bonadeo, E. D'Alessio: Acta Crystallogr. A33, 845 (1977)
4.70 F.H. Herbstein: Acta Crystallogr. 18, 997 (1965)
4.71 A. Monfils: C.R. Acad. Sci. Paris 241, 561 (1955)
4.72 E. D'Alessio, H. Bonadeo: Chem. Phys. Lett. 22, 559 (1973)
4.73 C.M. Gramaccioli, G. Filippini, M. Simonetta, S. Ramdas, G.M. Parkinson, J.M. Thomas: J. Chem. Soc., Faraday Trans. II 76, 1336 (1980)
4.74 M.L. Cangeloni, V. Schettino: Mol. Cryst. Liq. Cryst. 31, 219 (1975)
4.75 D.E. Williams: Acta Crystallogr. A25, 464 (1969)
4.76 V.M. Coiro, P. Giacomello, E. Giglio: Acta Crystallogr. B27, 2112 (1971)
4.77 V.M. Coiro, E. Giglio, A. Lucano, R. Puliti: Acta Crystallogr. B29, 1404 (1973)
4.78 E. Gavuzzo, F. Mazza, E. Giglio: Acta Crystallogr. B30, 1351 (1974)

4.79 S. Candeloro De Sanctis, V.M. Coiro, E. Giglio, S. Pagliuca, N. V. Pavel, C. Quaglita: Acta Crystallogr. B34, 1928 (1978)
4.80 S. Candeloro De Sanctis, E. Giglio, F. Petri, C. Quagliata: Acta Crystallogr. B35, 226 (1979)
4.81 S. Candeloro De Sanctis, E. Giglio: Acta Crystallogr. B35, 2650 (1979)
4.82 V.M. Coiro, A. D'Andrea, E. Giglio: Acta Crystallogr. B36, 848 (1980)
4.83 A. D'Andrea, W. Fedeli, E. Giglio, F. Mazza, N.V. Pavel: Acta Crystallogr. B37, 368 (1981)
4.84 J.M. Adams, S. Ramdas: Acta Crystallogr. B35, 679 (1979)
4.85 A.V. Dzyabchenko, V.E. Zavodnik, V.K. Belsky: Acta Crystallogr. B35, 2250 (1979)
4.86 N.S. Magomedova, A.V. Dzyabchenko, V.E. Zavodnik, V.K. Belsky: Cryst. Struct. Commun. 9, 713 (1980)
4.87 A.I. Kitaigorodsky, R.M. Myasnikova, I.E. Kozlova, A.J. Pertsin: Kristallografia 21, 511 (1976)
4.88 M.J.S. Dewar, A.J. Harget: Proc. Roy. Soc. London A315, 443 (1970)
4.89 J.M. Thomas: Pure Appl. Chem. 51, 1065 (1979)
4.90 K.S. Kim, E. Clementi: J. Am. Chem. Soc. 107, 227 (1985)
4.91 G.S. Murthy, K. Venkatesan: Acta Crystallogr. C40, 1581 (1984)
4.92 H. Nakanishi, W. Jones, J.M. Thomas, F.R.S., M. Hasegawa, W.L. Rees: Proc. Roy. Soc. London A369, 307 (1980)
4.93 B.K. Vainshtein, V.M. Fridkin, V.L. Indenbom: Modern Crystallography II, Springer Series in Solid-State Sciences, Vol. 21 (Springer, Berlin, Heidelberg, New York 1980)
4.94 P. Zugenmayer, A. Kuppel, E. Husemann: In Cellulose Chemistry and Technology, ACS Symposium Series, No. 48, ed. by J.C. Arthur (Am. Chem. Soc. 1977) pp. 115-132
4.95 T.L. Bluhm. G. Rappenecker, P. Zugenmayer: Carbohydr. Res. 60, 241 (1978)
4.96 T.L. Bluhm, P. Zugenmayer: Polymer 20, 23 (1979)
4.97 C. Woodcock, A. Sarko: Macromolecules 13, 1183 (1980)
4.98 A. Stipanovic, A. Sarko: Macromolecules 9, 851 (1976)
4.99 A. Sarko, R. Muggli: Macromolecules 7, 486 (1974)
4.100 H. Sugeta, T. Miyazawa: Biopolymers 5, 673 (1967)
4.101 S. Arnott, A. Wonacott: Polymer 7, 157 (1966)
4.102 S. Arnott, W.E. Scott: J. Chem. Soc., Perkin Trans. II 3, 324 (1972)
4.103 A.J. Pertsin, O.K. Nugmanov, V.F. Sopin, G.N. Marchenko, A.I. Kitaigorodsky: Vysokomol. Soed. A23, 2147 (1981)
4.104 P. Zugenmayer, A. Sarko: The Variable Virtual Bond Method, in Fiber Diffraction Methods, ACS Symposium Series, No 141, ed. by A.D. French, K.C.M. Gardner (Am. Chem. Soc. 1980) pp. 226-237
4.105 J.D. Dunitz: X-ray Anaysis and the Structure of Organic Molecules (Cornell University, Ithaca, London 1979)
4.106 Yi-Chang Fu, R.F. McGuire, H.A. Scheraga: Macromolecules 7, 468 (1974)
4.107 L. D'Ilario, E. Giglio: Acta Crystallogr. B30, 372 (1974)
4.108 P. De Sanctis, E. Giglio, A.M. Liquori, A. Ripamonti: Nuovo Cimento, Ser. X 26, 616 (1962)
4.109 M. Cesari, L. D'Ilario, E. Giglio, G. Perego: Acta Crystallogr. B31, 49 (1975)
4.110 G. Conte, L. D'Ilario, N.V. Pavel, E. Giglio: J. Polymer Sci.: Polymer Phys. Edition 17, 753 (1979)
4.111 G. Conte, L. D'Ilario, N.V. Pavel, E. Giglio: J. Polymer Sci.: Polymer Phys. Edition 14, 1553 (1976)
4.112 E. Giglio, A.M. Liquori, L. Mazzarella: Lett. Nuovo Cimento, Ser. I 1, 135 (1969)

4.113 E. Cernia, G. Conte, L. D'Ilario, N.V. Pavel, E. Giglio: J. Polymer Sci.: Polymer Chem. Edition 13, 125 (1975)
4.114 A.V. Sidorovich, I.S. Milevskaya, Yu.G. Baklagina: Vysokomol. Soed. A26, 1390 (1984)
4.115 F.J. Kolpak, J. Blackwell: Macromolecules 9, 273 (1976)
4.116 A.J. Pertsin, O.K. Nugmanov, G.N. Marchenko, A.I. Kitaigorodsky: Polymer 25, 107 (1984)
4.117 V.G. Dashevsky: Konformatsii organicheskikh molekul (Khimia, Moscow 1974)
4.118 F.J. Kolpak, M. Weih, J. Blackwell: Polymer 19, 123 (1978)
4.119 W.C. Hamilton: Acta Crystallogr. 18, 502 (1965)
4.120 M.D. Newton, G.A. Jeffrey, S. Takagi: J. Am. Chem. Soc. 101, 1997 (1979)
4.121 G.A. Jeffrey, H. Maluszynska: Int. J. Quantum Chem., Quantum Biol. Symp. 8, 231 (1981)
4.122 R.E. Hunter, N.E. Dweltz: J. Appl. Polymer Sci. 23, 249 (1979)

Chapter 5

5.1 M. Born, T. Von Karman: Phys. Z. 13, 297 (1912)
5.2 A. Warshel, S. Lifson: J. Chem. Phys. 53, 582 (1970)
5.3 E. Huler, A. Warshel: Acta Crystallogr. B30, 1822 (1974)
5.4 E. Huler, A. Warshel: Chem. Phys. 8, 239 (1975)
5.5 G. Taddei, H. Bonadeo, M.P. Marzocchi, S. Califano: J. Chem. Phys. 58, 966 (1973)
5.6 S. Califano: in Vibrational Spectroscopy - Modern Trends, ed. by A.J. Barnes, W.J. Orville-Thomas (Elsevier, Amsterdam, Oxford, New York 1977) Chap. 18, p. 285
5.7 S. Califano, V. Schettino, N. Neto: Lattice Dynamics of Molecular Crystals, Lecture Notes in Chemistry, Vol. 26 (Springer, Berlin, Heidelberg, New York 1981)
5.8 G.S. Pawley: Phys. Status Solidi 49, 475 (1972)
5.9 A.A. Maradudin, S.H. Vosko: Rev. Mod. Phys. 40, 1 (1968)
5.10 G. Venkataraman, V.C. Sahni: Rev. Mod. Phys. 42, 409 (1970)
5.11 G.S. Pawley: Acta Crystallogr. A30, 585 (1974)
5.12 G.S. Pawley, G.A. Mackenzie, E.L. Bokhenkov, E.F. Sheka, B. Dorner, J. Kalus, U. Schmelzer, I. Natkaniec: Molec. Phys. 39, 251 (1980)
5.13 G.A. Mackenzie, G.S. Pawley, O.W. Dietrich: J. Phys. C 10, 3723 (1977)
5.14 W. Cochran, G.S. Pawley: Proc. Roy. Soc. A280, 1 (1964)
5.15 R.F. Dyer, G.G. Low: in Inelastic Scattering of Neutrons in Solids and Liquids (IAEA, Vienna 1961)
5.16 P.A. Reynolds, J.K. Kjems, J.W. White: J. Chem. Phys. 56, 2928 (1972)
5.17 E. Burgos, H. Bonadeo, E. D'Alessio: J. Chem. Phys. 63, 38 (1975)
5.18 A.I. Kitaigorodsky: Molecular Crystals and Molecules (Academic, New York 1973)
5.19 G.S. Pawley: in Crystallographic Computing, ed. by F.R. Ahmed (Munksgaard, Copenhagen 1970) p. 243
5.20 D.W.J. Cruickshank: Acta Crystallogr. 9, 754 (1956)
5.21 V. Schomaker, K.N. Trueblood: Acta Crystallogr. B24, 63 (1968)
5.22 G.S. Pawley: Acta Crystallogr. 20, 631 (1966)
5.23 A.I. Kitaigorodsky, B.D. Koreshkov, A.G. Kulkin: Fiz. Tverd. Tela 7, 643 (1965)
5.24 A.I. Kitaigorodsky: J. Chim. Phys. 1, 9 (1966)

5.25 D.A. Oliver: Thesis, London Univ., London (1967)
5.26 D.A. Oliver, S.H. Walmsley: Molec. Phys. 17, 617 (1969)
5.27 G.S. Pawley: Phys. Status Solidi 20, 347 (1967)
5.28 P. Weulersse: Compt. Rend. Acad. Sci. Paris B264, 327 (1967)
5.29 A.I, Kitaigorodsky, E.I. Mukhtarov: Kristallografia 14, 789 (1969)
5.30 E.I. Mukhtarov: Determination of Physical Characteristics of Molecular Crystals by Atom-Atom Potentials; Thesis, Baku (1969)
5.31 I. Harada, T. Shimanouchi: J. Chem. Phys. 46, 2708 (1967)
5.32 G.W. Chantry, M.A. Gebbie, B. Lassier, G. Wyllie: Nature (London) 214, 163 (1967)
5.33 B. Wyncke, A. Hadni: C.R. Acad. Sci. Ser. B275, 825 (1972)
5.34 A. Fruhling, Ann. Phys. (Paris) 6, 40 (1951)
5.35 M. Ito, T. Shigeoka: Spectrochim. Acta 22, 1029 (1966)
5.36 A.K. Gee, G.W. Robinson: J. Chem. Phys. 46, 4847 (1967)
5.37 H. Bonadeo, M.P. Marzocchi, E. Castellucci, S. Califano: J. Chem. Phys. 57, 4299 (1972)
5.38 W.D. Ellenson, M.Nicol: J. Chem. Phys. 61, 1380 (1974)
5.39 M.P. Marzocchi, H. Bonadeo, G. Taddei: J. Chem. Phys. 53, 867 (1970)
5.40 J.J. Rush: J. Chem. Phys. 47, 3936 (1967)
5.41 K.W. Logan, S. Trevino, H.J. Prask, J.D. Gilat: J. Chem. Phys. 53, 3417 (1970)
5.42 E.L. Bokhenkov, V.G. Fedorov, E.F. Sheka, I. Natkaniec, M. Sudnik-Hrynkiewicz: Nuovo Cimento 44B, 324 (1978)
5.43 B.M. Powell, G. Dolling, H. Bonadeo: J. Chem. Phys. 69, 2428 (1978)
5.44 S. Califano, R. Righini, S.H. Walmsley: Chem. Phys. Lett. 64, 491 (1979)
5.45 H. Bonadeo, G. Taddei: J. Chem. Phys. 58, 979 (1973)
5.46 M. Suzuki, T. Yokoyama, M. Ito: Spectrochim. Acta 24A, 1091 (1968)
5.47 K. Tsuji, H. Yamada: J. Phys. Chem. 76, 260 (1972)
5.48 A. Bree, R.A. Kydd: Spectrochim. Acta 26A, 1791 (1970)
5.49 A. Hadni: In Excitons, Magnons and Phonons in Molecular Crystals ed. by A.B. Zahlan (Cambridge University, London 1968) p.31
5.50 M. Hineno, H. Yoshinaga: Spectrochim. Acta 31A, 617 (1975)
5.51 E.L. Bokhenkov, E.F. Sheka, B. Dorner, I. Natkaniec: Solid State Commun. 23, 89 (1977)
5.52 E.L. Bokhenkov, E.F. Sheka, I. Natkaniec, B. Dorner, W. Drexel: Proc. Conf.on Lattice Dynamics, ed. by M. Balkanski (Paris 1978) p.471
5.53 I. Natkaniec, E.L. Bokhenkov, B. Dorner, J. Kalus, G.A. Mackenzie, G.S. Pawley, U. Schmelzer, E.F. Sheka: J. Phys. C 13, 4265 (1980)
5.54 E.L. Bokhenkov, T. Natkaniec, E.F. Sheka: Zh. Eksp. Teor. Fiz. 70, 1027 (1976)
5.55 E.L. Bokhenkov, I. Natkaniec, E.F. Sheka: Phys. Status Solidi (b) 75, 105 (1976)
5.56 E.L. Bokhenkov, E.M. Rodina, E.F. Sheka, I. Natkaniec: Phys. Status Solidi (b) 85, 331 (1978)
5.57 A.V. Belushkin, E.L. Bokhenkov, A.I. Kolesnikov, I. Natkaniec, R. Righini, E.F. Sheka: Fiz. Tverd. Tela 23, 2607 (1981)
5.58 E.F. Sheka, E.L. Bokhenkov, B. Dorner, J. Kalus, G.A. Mackenzie, I. Natkaniec, G.S. Pawley, U. Schmeltzer: J. Phys. C17, 5893 (1984)
5.59 J. Kalus: Spez. Ber. Kernforschungsanlage Juelich, Juel-Spez-250, 188 (1984)
5.60 R. Righini: Chem. Phys. 84, 97 (1984)

5.61 G.K. Afanas'eva: Kristallografia 13, 1024 (1968)
5.62 G.S. Pawley, S.J. Cyvin: J. Chem. Phys. 52, 4073 (1970)
5.63 T.A. Krivenko, E.F. Sheka: Fiz. Tverd. Tela 26, 164 (1984)
5.64 S.J. Cyvin, B.N. Cyvin: J. Phys. Chem. 73, 1430 (1969)
5.65 G. Filippini, C.M. Gramaccioli, M. Simonetta, G.B. Suffritti: J. Chem. Phys. 59, 5088 (1973)
5.66 R. Righini, S. Califano, S.H. Walmsley: Chem. Phys. 50, 113 (1980)
5.67 R.G. Della Valle, G.S. Pawley: Acta Crystallogr. A40, 297 (1984)
5.68 H.N.V. Temperly, J.S. Rowlinson, G.S. Rushbrooke (eds.): Physics of Simple Liquids (North-Holland, Amsterdam 1968)
5.69 C. Kittel: Introduction to Solid State Physics (Wiley, New York 1967)
5.70 B. Dorner, E.L. Bokhenkov, S.L. Chaplot, J. Kalus, I. Natkaniec, G.S. Pawley, U. Schmelzer, E.F. Sheka: J. Phys. C: Solid State Phys. 15, 2353 (1982)
5.71 D.J. Evans, D.B. Scully: Spectrochim. Acta 20, 891 (1964)
5.72 S.L. Chaplot, G.S. Pawley, E.L. Bokhenkov, E.F. Sheka, B. Dorner, J. Kalus, V.K. Jindal, I. Natkaniec: Chem. Phys. 57, 407 (1981)
5.73 S.L. Chaplot, G.S. Pawley, B. Dorner, V.K. Jindal, J. Kalus, I. Natkaniec: Phys. Status Solidi (b) 110, 445 (1982)
5.74 T.A. Krivenko, V.A. Dement'ev, E.L. Bokhenkov, A.I. Kolesnikov, E.F. Sheka: Mol. Cryst. Liq. Cryst. 104, 207 (1984)
5.75 G.K. Afanas'eva, K.S. Aleksandrov, A.I. Kitaigorodsky: Phys. Status Solidi 24, K61 (1967)
5.76 G. Dolling, B.M. Powell: Proc. Roy. Soc. London A319, 209 (1970)
5.77 G. Dolling, G.S. Pawley, B.M. Powell: Proc. Roy. Soc. London A333, 363 (1973)
5.78 P.A. Reynolds, J.K. Kjems, J.W. White: J. Chem. Phys. 60, 824 (1974)
5.79 H. Bonadeo, E. D"Alessio: Chem. Phys. Lett. 19, 117 (1973)
5.80 X. Gerbeaux, A. Hadni: J. Chem. Phys. 49, 955 (1968)
5.81 M. Ito, T. Shigeoka: J. Chem. Phys. 44, 1001 (1966)
5.82 P.A. Reynolds: J. Chem. Phys. 59, 2777 (1973)
5.83 T. Luty: Mol. Cryst. Liq. Cryst. 17, 327 (1972)
5.84 E. Burgos, H. Bonadeo, E. D'Alessio: J. Chem. Phys. 65, 2460 (1976)
5.85 I. Natkaniec, A.V. Belushkin, T. Wasiutynsky: Phys. Status Solidi (b) 105, 413 (1981)
5.86 A.V. Belushkin, I. Natkaniec, J. Wasicki, T. Zaleski: Phys. Status Solidi (b) 123, K115 (1984)
5.87 N.M. Plakida, A.V. Belushkin, I. Natkaniec, T. Wasiutynsky: Phys. Status Solidi (b) 118, 129 (1983)
5.88 M.M. Szostak, A.V. Belushkin, I. Natkaniec: Phys. Status Solidi (b) 127, K1 (1985)
5.89 I.A. Remizov, V.G. Podoprigora, A.N. Botvich, T.A. Kharitonova, V.F. Shabanov: Chem. Phys. 92, 163 (1985)
5.90 G. Filippini, C.M. Gramaccioli: Chem. Phys. Lett. 104, 50 (1984)
5.91 S.L. Chaplot, A. Mierzejewski, G.S. Pawley, J. Lefevre, T. Luty: J. Phys. C16, 625 (1983)
5.92 G. Filippini, C.M. Gramaccioli, M. Simonetta: J. Chem. Phys. 73, 1376 (1980)
5.93 Z. Gamba, H. Bonadeo: J. Chem. Phys. 76, 6215 (1982)

Chapter 6

6.1 P.A. Reynolds: Molec. Phys. 30, 1165 (1975)

6.2 M. Born, K. Huang: Dynamical Theory of Crystal Lattices (Clarendon, Oxford 1954)
6.3 A.I. Kitaigorodsky, E.I. Mukhtarov: Fiz. Tverd. Tela 13, 1654 (1971)
6.4 H. Bonadeo, E. D'Alessio, E. Halac, E. Burgos: J. Chem. Phys. 68, 4714 (1978)
6.5 G. Filippini, C.M. Gramaccioli, M. Simonetta, G.B. Suffritti: Acta Crystallogr. A32, 259 (1976)
6.6 A.J. Pertsin: J. Phys. Chem. Solids 46, 1019 (1985)
6.7 J.A. Barker: Lattice Theories of the Liquid State (Pergamon, Oxford, London, New York, Paris 1963)
6.8 J.E. Lennard-Jones, A.F. Devonshire: Proc. Roy. Soc. London A163, 53 (1937)
6.9 C. Kittel: Introduction to Solid State Physics (Wiley, New York 1967) p. 285
6.10 J.G. Kirkwood: J. Chem. Phys. 18, 380 (1950)
6.11 A.J. Pertsin: Evaluation of the Thermodynamic Properties of Organic Crystals; Thesis, Moscow (1976)
6.12 H.J. Mandell, J.P. McTague, A. Rahman: J. Chem. Phys. 66, 3070 (1977)
6.13 A.C. Holt, W.G. Hoover, S.G. Gray, D.R. Shortle: Physica (Utr.) 49, 61 (1970)
6.14 T.G. Gibbons, M.L. Klein: J. Chem. Phys. 60, 112 (1974)
6.15 N.N. Godnev: Vychisleniye termodynamicheskikh funktsii po molekulyarnym dannym (Mir, Moscow 1956)
6.16 D.R. Stull, E.F. Westrum, Jr., G.C. Sinke: The Chemical Thermodynamics of Organic Compounds (Wiley, New York, London, Syndey, Toronto 1969)
6.17 H. Kahn: Nucleonics 6, 27 (1950)
6.18 A.J. Pertsin, A.I. Kitaigorodsky: Molec. Phys. 32, 1781 (1976)
6.19 A.P. Ryzhenkov, V.M. Kozhin: Kristallografia 12, 1079 (1967)
6.20 J.P. McCullough, M.L. Finke, J.F. Messerly, S.S. Todd, T.C. Kincheloe, G. Waddington: J. Phys. Chem. 61, 1105 (1957)
6.21 M. Suzuki, T. Yokoyama, M. Ito: Spectrochim. Acta A24, 1091 (1968)
6.22 G.S. Pawley, S.J. Cyvin: J. Chem. Phys. 52, 4073 (1970)
6.23 S.S. Mitra, H.J. Bernstein: Canad. J. Chem. 37, 553 (1959)
6.24 E.P. Krainov: Optika i Spectroskopia 16, 763 (1964)
6.25 G.K. Afanas'eva: Kristallografia 13, 1024 (1968)
6.26 A.I. Kitaigorodsky: Kristallografia 9, 171 (1964)
6.27 D.W.J. Cruickshank: Acta Crystallogr. 10, 504 (1957)
6.28 V.I. Ponomariov, O.S. Filipenko, L.O. Atovmyan: Sov. Phys. Crystallogr. 21, 215 (1976)
6.29 V.M. Kozhin, A.I. Kitaigorodsky: Zh. Fiz. Khim. 29, 2075 (1955)
6.30 G.E. Bacon, N.A. Curry, S.A. Wilson: Proc. Roy. Soc. London A279, 98 (1964)
6.31 A.I. Ryzhenkov, V.M. Kozhin, R.M. Myasnikova: Kristallografia 13, 1028 (1968)
6.32 Yu.P. Ivanov, M.Yu. Antipin, A.J. Pertsin, Yu.T. Struchkov: Mol. Cryst. Liq. Cryst. 71, 181 (1981)
6.33 A.J. Pertsin, Yu.P. Ivanov, A.I. Kitaigorodsky: Kristallografia 26, 115 (1981)
6.34 Yu.P. Ivanov, M.Yu. Antipin, A.J. Pertsin: Kristallografia 27, 627 (1982)
6.35 C.E. Nordman, D.L. Schmitkons: Acta Crystallogr. 18, 764 (1965)
6.36 D. Fox, M.M. Labes, A. Weissberger (eds.): Physics and Chemistry of the Organic Solid State, Vol. 1 (Wiley, New York, London 1963) Chap. 1, pp. 56-72
6.37 W.-K. Wong, E.F. Westrum: J. Chem. Thermodynamics 3, 105 (1971)

6.38 H.L. Finke, J.F. Messerly, S.M. Lee, A.G. Osborn, D.R. Douslin: J. Chem. Thermodynamics 9, 937 (1977)
6.39 P. Goursot, H.L. Girdhar, E.F. Westrum, Jr.: J. Phys. Chem. 74, 2538 (1970)
6.40 S.S. Chang, E.F. Westrum, Jr.: J. Phys. Chem. 64, 1547 (1960)
6.41 L.M. Sverdlov, M.A. Kovner, E.P. Krainov: Kolebatel'nye spektry mnogoatomnykh molekul (Nauka, Moscow 1970)
6.42 M.A. Kovner: Dokl. Akad. Nauk SSSR 91, 499 (1953)
6.43 A. Bree, R.A. Kydd, T.N. Misra, V.V.B. Vilkos: Spectrochim. Acta A27, 2315 (1971)
6.44 K. Witt, R. Mecke: Z. Naturforsch. A22, 1247 (1967)
6.45 A. Bree, R.A. Kydd, T.N. Misra: Spectrochim. Acta A25, 1815 (1969)
6.46 R.T. Bailey: Spectrochim. Acta A27, 1447 (1971)
6.47 A.J. Pertsin, Yu.P. Ivanov, A.I. Kitaigorodsky: Acta Crystallogr. A37, 908 (1981)
6.48 Yu.P. Ivanov: The Cell Model in Calculations of the Thermodynamic Properties of Organic Crystals; Thesis, Moscow (1984)
6.49 G.L. Wheeler, S.D. Colson: J. Chem. Phys. 65, 1227 (1976)
6.50 P.D. Andre, R. Fourme, M. Rewaund: Acta Crystallogr. B27, 2371 (1971)
6.51 G.M. Brown, O.A.W. Strydom: Acta Crystallogr. B30, 801 (1974)
6.52 H.J. Milledge, L.M. Pant: Acta Crystallogr. 13, 285 (1960)
6.53 D.G. Gafner, F.M. Herbstein: Acta Crystallogr. 13, 702 (1960)
6.54 C. Dean, M. Pollak, B.M. Craven, G.A. Jeffrey: Acta Crystallogr. 11, 710 (1958)
6.55 E.M. Moiseeva: Physico-Chemical Analysis of Multicomponent Mixtures of Polyvinyltrimethylsilane with Methylenechloride, Chlorobenzene and Isobutanol; Thesis, Gorkii (1978)
6.56 D.L. Hildenbrand, W.R. Kramer, D.R. Stull: J. Phys. Chem. 62, 958 (1958)
6.57 A. Dworkin, P. Figuiere, M. Ghelfenstein, M. Szwarc: J. Chem. Thermodynamics 8, 835 (1976)
6.58 J.R. Scherer, J.C. Evans: Spectrochim. Acta 19, 1739 (1963)
6.59 D.H. Whiffen: J. Chem. Soc. 6, 1350 (1956)
6.60 J.B. Bates, D.M. Thomas, A. Brandy, G. Leppincott: Spectrochim. Acta A27, 637 (1971)
6.61 J.R. Scherer, J.C. Evans, W.W. Muelder, J.V. Overend: Spectrochim. Acta 18, 57 (1962)
6.62 F.A. Mason, W.E. Rice: J. Chem. Phys. 22, 843 (1954)
6.63 G. Filippini, C.M. Gramaccioli, M. Simonetta, G.B. Suffritti: Chem. Phys. Lett. 39, 14 (1976)
6.64 J.F. Messerly, H.L. Finke: J. Chem. Thermodynamics 2, 867 (1970)
6.65 J.M.S. Green, D.J. Harrison: J. Chem. Thermodynamics 8, 529 (1976)
6.66 N. Boden, P.P. Davis: Molec. Phys. 25, 81 (1973)
6.67 S. Chang, E.F. Westrum: J. Phys. Chem. 64, 1547 (1960)
6.68 T. Eleverbredd, S.J. Cyvin: Molecular Structure and Vibrations (Elsevier, Amsterdam 1972)
6.69 L.N. Becka, D.W.J. Cruickshank: Proc. Roy. Soc. A273, 435 (1963)
6.70 K.V. Mirskaya: Kristallografia 8, 225 (1963)
6.71 A.J. Pertsin, Yu.P. Ivanov, A.I. Kitaigorodsky: J. Phys. Chem. 85, 1546 (1981)
6.72 M. Anderson. L. Basio, J. Bruneaux-Paulle, R. Fourme: J. Chem. Phys. 74, 68 (1977)
6.73 C. La Lau, R.G. Snyder: Spectrochim. Acta A27, 2073 (1971)
6.74 D.W. Scott, G.B. Gurthrie, J.F. Messerly, S.S. Todd, W.T. Berg, I.A. Hossenlopp, J.P. McCullough: J. Phys. Chem. 66, 911 (1962)
6.75 S.H. Hustings, D.E. Nicholson: J. Phys. Chem. 61, 730 (1957)

6.76 M. Frankosky, J.G. Aston: J. Phys. Chem. 69, 3126 (1965)
6.77 S.Seki, H. Chihara: Collect. Pap. Fac. Sci., Osaka Imp. Univ., Ser. C 11, 1 (1950)
6.78 L.O. Brockway, J.M. Robertson: J. Chem. Soc. 24, 1324 (1939)
6.79 I. Woodward: Acta Crystallogr. 11, 441 (1958)
6.80 O. Schnepp, D.M. McClure: J. Chem. Phys. 26, 83 (1957)
6.81 M. Born, T. Von Karman: Phys. Z. 13, 297 (1912)
6.82 A.J. Pertsin, A.I. Kitaigorodsky: in preparation
6.83 P.W. Bridgman: J. Chem. Phys. 9, 794 (1941)
6.84 P.W. Bridgman: Proc. Am. Acad. Arts Sci. 77, 129 (1949)
6.85 D. Hall, T.H. Starr, D.E. Williams, M.K. Wood: Acta Crystallogr. A36, 494 (1980)
6.86 E. Neusy, S. Nose, M. Klein: Mol. Phys. 52, 269 (1984)
6.87 M. Parrinello, A. Rahman: Phys. Rev. Lett. 45, 1196 (1980)
6.88 S. Nose, M.L. Klein: Mol. Phys. 50, 1055 (1983)
6.89 D.A. Badalyan: Kristallografia 14, 48 (1969)
6.90 D.A. Badalyan, A.G. Khachaturyan, A.I. Kitaigorodsky: Kristallografia 14, 404 (1969)
6.91 S.M. Breiting, A.D. Jones, R.M. Boyd: J. Chem. Phys. 54, 3959 (1971)
6.92 C.W.F.T. Pistorius: Mol. Cryst. Liq. Crist. 5, 353 (1969)
6.93 Yu.S. Genshaft, L.B. Larionov: Zh. Fiz. Khim. 43, 625 (1969)
6.94 K.V. Mirskaya: Kristallografia 8, 225 (1963)

Chapter 7

7.1 J.M. Thomas: Pure Appl. Chem. 51, 1065 (1979)
7.2 H. Nakanishi, W. Jones, J.M. Thomas, F.R.S., M. Hasegawa, W.L. Rees: Proc. Roy. Soc. London A369, 307 (1980)
7.3 F.L. Hirshfeld, G.M.J. Schmidt: J. Polymer Sci. A2, 2181 (1964)
7.4 J. Sworakowski: Local Ionized Levels in Molecular Crystals, Scientific Papers of the Institute of Organic and Physical Chemistry of Wroclaw Technical University, No. 6 (Wroclaw 1974)
7.5 E.A. Silinsh: Organic Molecular Crystals. Their Electronic States, Springer Series in Solid-State Sciences, Vol. 16 (Springer, Berlin, Heidelberg, New York 1980)
7.6 J.P. Desvergne, J.M. Thomas: J. Chem. Soc., Perkin Trans. II 1975, 584 (1975)
7.7 S. Ramdas, W. Jones, J.M. Thomas: Chem. Phys. Lett. 57, 468 (1978)
7.8 R.H. Boyd: J. Polymer Sci.: Polymer Phys. Edition 13, 2345 (1975)
7.9 G.S. Pawley, S.L. Chalpot: Phys. Status Solidi (b) 99, 517 (1980)
7.10 D.P. Craig, B.R. Markey, A.O. Griewank: Chem. Phys. Lett. 62, 223 (1979)
7.11 T.R. Koehler: J. Chem. Phys. 72, 3389 (1980)
7.12 A.I. Kitaigorodsky: Organic Chemical Crystallography (Consultants Bureau, New York 1961)
7.13 A.I. Kitaigorodsky: Molecular Crystals and Molecules (Academic, New York 1973)
7.14 R.M. Myasnikova, I.E. Kozlova, A.J. Pertsin, L.G. Radchenko: Kristallografia 21, 964 (1976)
7.15 V.D. Samarskaya, R.M. Myasnikova, A.I. Kitaigorodsky: Kristallografia 13, 616 (1968)
7.16 G.W. Frank, R.M. Myasnikova, A.I. Kitaigorodsky: Kristallografia 16, 329 (1971)
7.17 D.P. Craig, B.R. Markey: Mol. Cryst. Liq. Cryst. 58, 77 (1980)

7.18 B.R. Markey: in Computer Simulation in the Physics and Chemistry of Solids, Proc. Daresbury Study Weekend, Daresbury, UK, 9-11 May, 1980, comp. by C.R. Catlow, W.C. Mackrodt, V.R. Saunders, pp.104-106
7.19 H. Dorner, R. Hundhausen, D. Schmid: Chem. Phys. Lett. 53, 101 (1978)
7.20 B. McCool, B.R. Markey, R. Bramsley: Mol. Phys. 51, 935 (1984)
7.21 J. Ferguson, A.W.-H. Mau: Molec. Phys. 27, 377 (1974)
7.22 M. Ehrenberg: Acta Crystallogr. 20, 177 (1966)
7.23 G.W. Cauters, J. Van Egmond, T.J. Schaafsma, J.H. Van der Waals: Molec. Phys. 24, 1203 (1972)
7.24 W.G. Van Dorp, M. Soma, J.A. Koofer, J.H. Van der Waals: Molec. Phys. 28, 1551 (1974)
7.25 J.A. Koofer, J.H. Van der Waals: Molec. Phys. 37, 977 (1979)
7.26 S. Voelker, R.M. Macfarlane, J.H. Van der Waals: Chem. Phys. Lett. 53, 8 (1978)
7.27 S.A. Remyga, R.M. Myasnikova, A.I. Kitaigorodsky: Kristallografia 12, 900 (1967)
7.28 A.V. Nikitin, A.J. Pertsin: Proc. 6th Symp. on Intermolecular Interactions and Molecular Conformation, Vilnius, USSR, June 22-23, 1982, p. 176
7.29 R.K. Boyd, C.A. Fyfe, D.A. Wright: J. Phys. Chem. Solids 35, 1355 (1974)
7.30 C.A. Fyfe, D. Harold-Smith: Canad. J. Chem. 54, 769 (1976)
7.31 A. Gavezzotti, M. Simonetta: Acta Crystallogr. A31, 645 (1975)
7.32 A. Gavezzotti, M. Simonetta: Acta Crystallogr. A32, 997 (1976)
7.33 W.E. Sanford, R.K. Boyd: Canad. J. Chem. 54, 2773 (1976)
7.34 W.E. Sanford, R.K. Boyd, J.A. Ripmeester: Canad. J. Chem. 56, 714 (1978)
7.35 J.A. Ripmeester, R.K. Boyd: Canad. J. Chem. 57, 128 (1979)
7.36 J.A. Ripmeester, R.K. Boyd: J. Chem. Phys. 71, 5167 (1979)
7.37 R.K. Boyd, J. Comper, G. Ferguson: Canad. J. Chem. 57, 3056 (1979)
7.38 W.E. Sandford, G.J. Kupferschmidt, C.A. Fyfe, R.K. Boyd, J.A. Ripmeester: Canad. J. Chem. 58, 906 (1980)
7.39 R.K. Boyd, C.A. Fyfe: J. Phys. Chem. Solids 39, 693 (1978)
7.40 D.P. Craig, J.F. Ogilvie, P.A. Reynolds: J. Chem. Soc., Faraday Trans. II 72, 1603 (1976)
7.41 B.R. Markey, S.H. Walmsley: Chem. Phys. Lett. 74, 354 (1980)
7.42 A.I. Kitaigorodsky, N.A. Ahmed: Acta Crystallogr. A28, 207 (1972)
7.43 J.W. Gibbs: Collected Works (Longman Green, New York 1928)
7.44 G. Wulff: Z. Kristallogr. 34, 449 (1901)
7.45 P. Groth: Chemische Kristallographie, Teil 5 (Engelman, Leipzig 1919)
7.46 G. Filippini, C.M. Gramaccioli, M. Simonetta: J. Chem. Phys. 71, 89 (1979)
7.47 L.E. Firment, G.A. Somorjai: J. Chem. Phys. 63, 1037 (1975)
7.48 S. Ramdas, J.M. Thomas, M.J. Goringe: J. Chem. Soc., Faraday Trans. II 73, 551 (1977)
7.49 J.P. Desvergne, J.M. Thomas, J.O. Williams, H. Bouas-Laurent: J. Chem. Soc. Perkin Trans. II 1974, 363 (1974)
7.50 K. Mirsky, M.D. Cohen: J. Chem. Soc., Faraday Trans. II 72, 2155 (1976)
7.51 G.M. Parkinson, W. Rees, M.J. Goringe, W. Jones, S. Ramdas, J.M. Thomas, J.O. Williams: Inst. Phys. Conf. Ser. No. 41, Chap. 3, p. 172 (1978)
7.52 G.M. Parkinson, M.J. Goringe, W. Jones, J.M. Thomas, J.O. Williams: in Developments in Electron Microscopy and Analysis, ed. by J.A. Venables (Academic, London 1975) p. 325

7.53 J.M. Thomas, S.E. Morsi, J.P. Desvergne: Adv. Phys. Org. Chem. 15, 63 (1977)
7.54 C.M. Gramaccioli, G. Filippini, M. Simonetta, S. Ramdas, G.M. Parkinson, J.M. Thomas: J. Chem. Soc., Faraday Trans. II 76, 1336 (1980)
7.55 G.M. Parkinson, M.J. Goringe, S. Ramdas, J.O. Williams, J.M. Thomas: J. Chem. Soc. Chem. Commun. 1978, 134 (1978)
7.56 S. Ramdas, G.M. Parkinson, J.M. Thomas, C.M. Gramaccioli, G. Filippini, M. Simonetta, M.J. Goringe: Nature 284, 153 (1980)

Subject Index